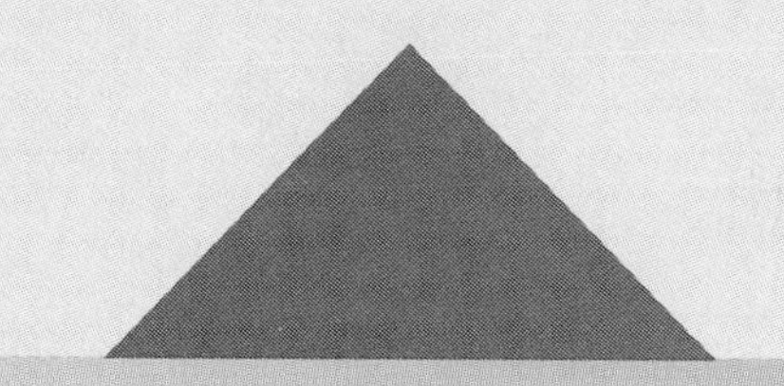

National Center for Construction Education and Research

Plumbing Level One

Upper Saddle River, New Jersey
Columbus, Ohio

This information is general in nature and intended for training purposes only. Actual performance of activities described in this manual requires compliance with all applicable operating, service, maintenance, and safety procedures under the direction of qualified personnel. References in this manual to patented or proprietary devices do not constitute a recommendation of their use.

Copyright © 2000 by the National Center for Construction Education and Research, Gainesville, FL 32614-1104, and published by Prentice-Hall, Inc., Upper Saddle River, New Jersey 07458. All rights reserved. Printed in the United States of America. This publication is protected by Copyright and permission should be obtained from the NCCER prior to any prohibited reproduction, storage in a retrieval system, or transmission in any form or by any means, electronic, mechanical, photocopying, recording, or likewise. For information regarding permission(s), write to: NCCER, Curriculum Revision and Development Department, P.O. Box 141104, Gainesville, FL 32614-1104.

Prentice Hall

nccer

10 9 8 7 6 5 4 3 2 1
ISBN 0-13-030926-5

Preface

This volume was developed by the National Center for Construction Education and Research (NCCER) in response to the training needs of the construction and maintenance industries. It is one of many in the NCCER's standardized craft training program. The program, covering more than 30 craft areas and including all major construction skills, was developed over a period of years by industry and education specialists. Sixteen of the largest construction and maintenance firms in the U.S. committed financial and human resources to the teams that wrote the curricula and planned the nationally-accredited training process. These materials are industry-proven and consist of competency-based textbooks and instructor's guides.

The NCCER is a nonprofit educational entity affiliated with the University of Florida and supported by the following industry and craft associations:

PARTNERING ASSOCIATIONS

- American Fire Sprinkler Association
- American Society for Training & Development
- American Welding Society
- Associated Builders and Contractors, Inc.
- Association for Career and Technical Education
- Associated General Contractors of America
- The Business Roundtable
- Carolinas AGC, Inc.
- Carolinas Electrical Contractors Association
- Construction Industry Institute
- Design-Build Institute of America
- Merit Contractors Association of Canada
- Metal Building Manufacturers Association
- National Association of Minority Contractors
- National Association of Women in Construction
- National Insulation Association
- National Ready Mixed Concrete Association
- National Utility Contractors Association
- National Vocational Technical Honor Society
- North American Crane Bureau
- Painting & Decorating Contractors of America
- Portland Cement Association
- SkillsUSA–VICA
- Steel Erectors Association of America
- Texas Gulf Coast Chapter ABC
- U.S. Army Corps of Engineers
- University of Florida
- Women Construction Owners & Executives, USA

Some of the features of the NCCER's standardized craft training program include:

- A proven record of success over many years of use by industry companies.
- National standardization providing portability of learned job skills and educational credits that will be of tremendous value to trainees.
- Recognition: upon successful completion of training with an accredited sponsor, trainees receive an industry-recognized certificate and transcript from the NCCER.
- Compliance with Apprenticeship, Training, Employer, and Labor Services (ATELS) (formerly BAT) requirements for related classroom training (CFR 29:29).
- Well-illustrated, up-to-date, and practical information.

COMMEMORATE 2000 WITH COLOR!

To kick off the new millennium, the NCCER and Prentice Hall are publishing select textbooks in color. *Plumbing Level One* is one of the first titles to incorporate a brand-new design and layout. In addition to full-color photos and illustrations, also look for the following new pedagogical features designed to augment the technical material and make the book even more user friendly:

Did You Know? Explains fun facts and interesting tidbits about the plumbing trade from historical to modern times.

On the Level provides helpful hints for those entering the field by presenting tricks of the trade from plumbers in a variety of disciplines.

Profiles in Success introduce trainees to plumbers who share their experience and advice after years in the field.

We're excited to offer you these new features and hope they help to foster a more rewarding learning experience. As always, your feedback is welcome! Visit the NCCER at www.NCCER.org or e-mail us at info@NCCER.org.

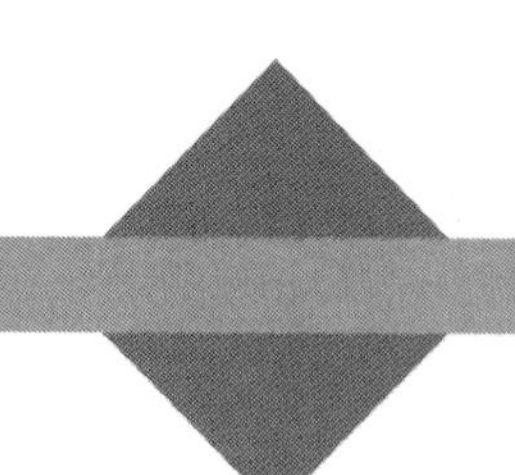

Acknowledgments

This program was revised as a result of the farsightedness and leadership of the following sponsors:

City of Phoenix, AZ
PMA/CEFGA Apprentice Program
TDIndustries
Yeagers, Inc.

This manual would not exist were it not for the dedication and unselfish energy of those volunteers who served on the Authoring Team. A sincere thanks is extended to:

Ed Cooper
Cary Mandeville
Roger Rotundo
Dan Warnick

We would also like to thank the following reviewers for contributing their time and expertise to this endeavor:

George Benoit
Ron Braun
James Lee
Charles Owenby
Wilford Seilhamer

A final note: This book is the result of a collaborative effort involving the production, editorial, and development staff at Prentice-Hall, Inc., and the National Center for Construction Education and Research. Thanks to all of the dedicated people involved in the many stages of this project.

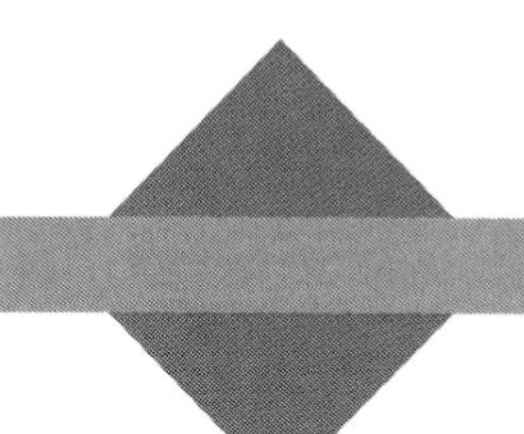

Contents

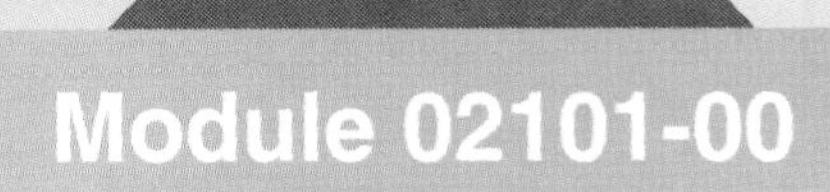

Introduction to the Plumbing Trade

Course Map

This course map shows all of the modules in the first level of the Plumbing curriculum. The suggested training order begins at the bottom and proceeds up. Skill levels increase as you advance on the course map. The local Training Program Sponsor may adjust the training order.

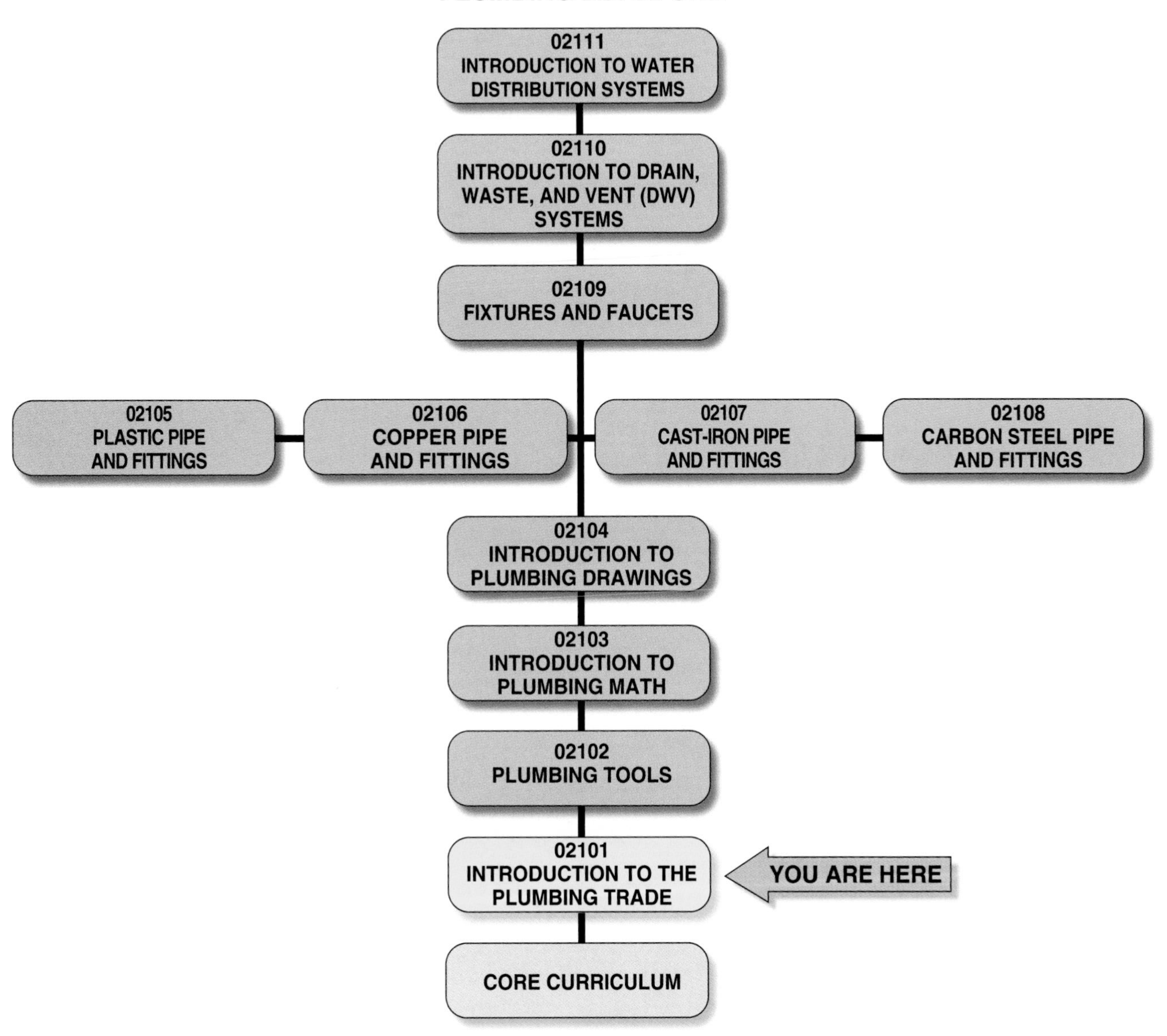

101CMAP.EPS

Copyright © 2000 National Center for Construction Education and Research, Gainesville, FL 32614-1104. All rights reserved. No part of this work may be reproduced in any form or by any means, including photocopying, without written permission of the publisher.

MODULE 02101 CONTENTS

Figures

MODULE 02101

Introduction to the Plumbing Trade

Objectives

When you finish this module, you will be able to:

1. Describe the history of the plumbing trade.
2. Identify the stages of progress within the plumbing trade.
3. Identify the responsibilities of a person working in the construction industry.
4. State the personal characteristics of a professional.
5. Explain the importance of safety in the construction industry.

Prerequisites

Before you begin this module, it is recommended that you successfully complete the Core Curriculum.

Materials Required for Trainees

1. This module
2. Appropriate personal protective equipment
3. Sharpened pencil and paper

1.0.0 ◆ INTRODUCTION

Opportunity is driven by knowledge and ability, which in turn are driven by education and training. This program of the National Center for Construction Education and Research (NCCER) was designed and developed by the construction industry for the construction industry. It is the only nationally accredited, competency-based construction training program in the United States. A *competency-based* program requires the trainee to demonstrate the ability to perform specific job-related tasks safely in order to receive credit. This approach is unlike other apprenticeship programs that merely require a trainee to put in a number of hours in the classroom and on the job.

The primary goal of NCCER is to standardize construction craft training throughout the country so that employers and employees will benefit from it, no matter where they are located. As a trainee in an NCCER program, you will become part of a national registry. You will receive a certificate for each level of training you complete. If you apply for a job with any participating contractor in the country, a transcript of your training will be available. If your training is incomplete when you make a job transfer, you can pick up where you left off, because every participating contractor is using the same training program. Many technical schools and colleges also use this program.

2.0.0 ◆ HISTORY OF PLUMBING

Plumbing has profoundly influenced the development of modern society. The earliest plumbing systems allowed individuals, villages, and towns to have fresh water for everyday use. By improving sanitation, plumbing reduced the spread of disease caused by waterborne wastes in villages and towns, contributing to increased life expectancy worldwide. In some places, plumbing also allowed citizens to fight fires in the days when most buildings were made of wood.

As early as 2900 B.C.E., humans were using plumbing (see *Figure 1*). We know this because in southwest Asia, in the area known as Mesopotamia, archaeologists have found ancient evidence of earthenware pipes and masonry sewers, water closets (toilets), and drainage systems.

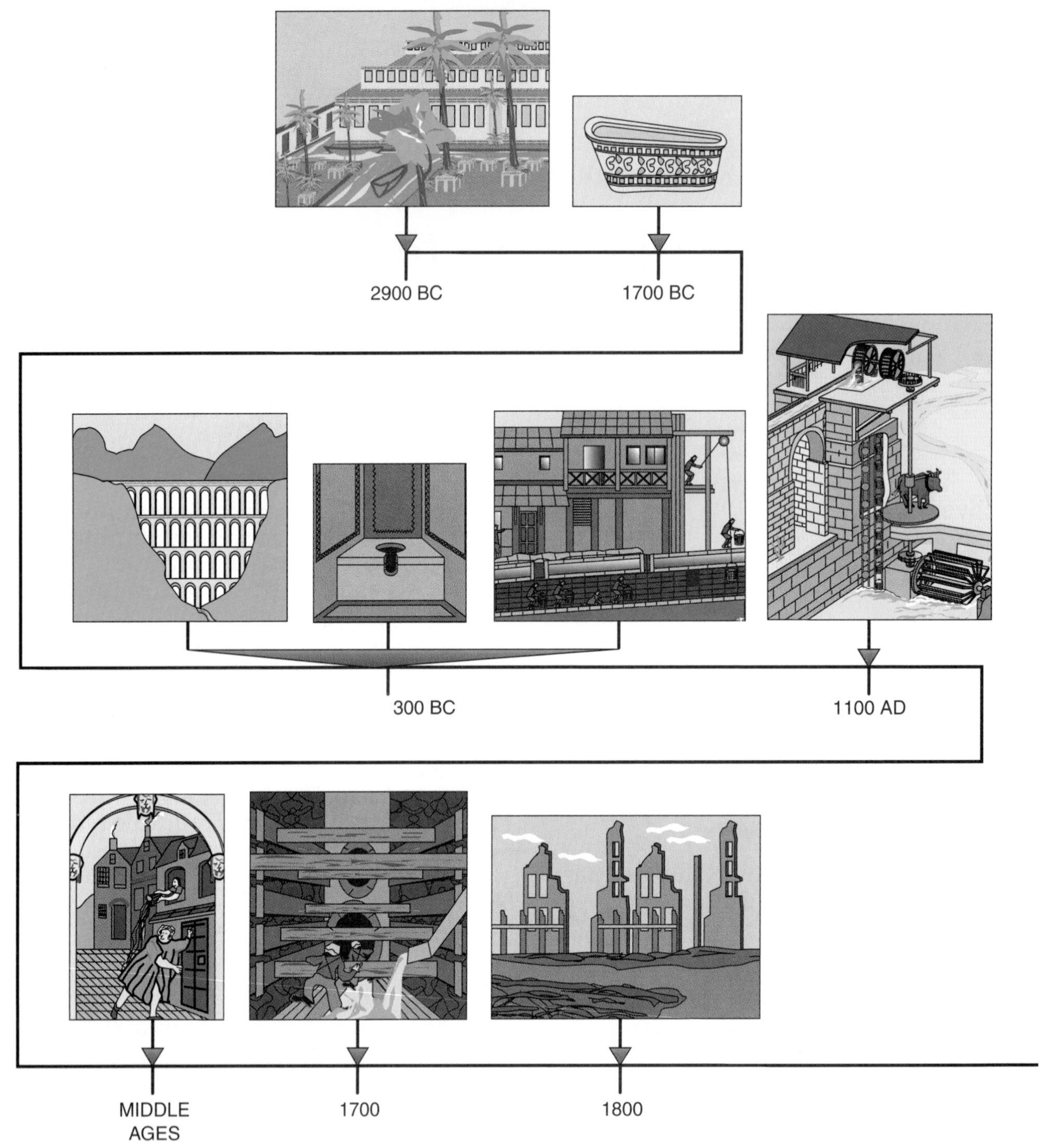

Figure 1 ◆ The evolution of plumbing.

In 312 B.C.E., the Romans began bringing water into the city of Rome through **aqueducts.** The first aqueducts were open trenches that operated by gravity (water runs downhill). The beautiful arched stone aqueducts that you see in photos were not common. Most aqueducts were simply dug into the ground and lined with stone.

By 100 C.E., this constant supply of water allowed the Romans to build and maintain public bathhouses and fountains throughout the city. It also provided a way to remove wastes. Unfortunately, there was no water treatment system, so sewage was discharged back into the river downstream from the city.

In addition, the use of lead pipe in Roman and Greek cities during this time may have caused widespread illness. We now know that lead has side effects on the human brain, nervous system, and vital organs such as the liver. Some historians blame the use of lead pipes for the eventual decline of the Roman Empire. The Latin word for lead—**plumbum**—is the basis for our word **plumber.** A person working with lead pipes was called a **plumbarius.**

During the fifth century C.E., the Goths invaded Rome. The population and culture suffered, and much of the sophistication of the Roman and Greek golden age was forgotten. The period after the fall of Rome is generally referred to as the Dark Ages (500 C.E. to 1400 C.E.). Many of the technological and cultural advances made by the Greeks and Romans were lost. For example, little was done to maintain the aqueducts, and eventually the systems began to deteriorate.

Disease was a bigger killer than wars and conquests during this period. Some historians estimate that one-third of the population of Europe died because of unsanitary living conditions that led to the quick spread of diseases such as bubonic plague and smallpox. Because the basic theory of freshwater supply and waste removal had been forgotten, people lived in sewage-strewn streets and practiced unhealthful food storage and preparation.

The light began to dawn, however, by the mid-1300s, when London, England, laid its first water supply pipe. As early as 1625, the first apprenticeship laws for plumbing were in effect in England. Other countries began to develop their own water supply systems, and the era of sanitation began. Toilets were developed in the 1770s.

Disasters often sparked new developments in plumbing systems. In 1666, after the Great Fire of London nearly destroyed the city, citizens organized the first firefighters and brought about an expansion of the water supply system. In the United States, deadly cholera epidemics in Philadelphia (1793) and New York City (1832) spurred construction of improved water supply systems.

Despite these advances in supplying cities with clean, fresh water, little was done to treat wastewater. The first advancement in water purification was a slow sand filter installed in Richmond, Virginia, in 1832. Scientists showed cities in England and the United States how to treat water supplies with **chlorine** to destroy deadly bacteria.

By the mid-1800s, though, plumbing systems had not advanced or expanded enough to meet the needs of growing populations clustered in cities. In most American cities at that time, people were still dumping sewage and garbage into the street for pigs to scavenge. The industrial revolution, with its power-driven machinery, was spurring the rapid growth of factories, populations, and houses in both American and European cities. By 1855, Chicago had a population of 75,000, but its only water sources were individual wells and the river. The increase in the size of industrial cities like Chicago and the growth of densely-populated housing districts near factories contributed to pollution of wells and increased waste disposal problems. These same conditions were found in industrial areas in Europe.

To appreciate how far plumbing has advanced, compare those conditions with what you can observe about plumbing systems today. In the United States, the greatest progress in the development of plumbing has been made since 1910. Advancements in piping materials such as copper, cast iron, and **CPVC;** technological advancements

DID YOU KNOW?

Diseases like cholera, typhoid fever, typhus, and dysentery are caused by bacteria. These bacteria attack the stomach, intestines, and other organs and may cause death if victims are not treated with antibiotics. The bacteria can pass from an infected person to a healthy person by various means. They can get into the water supply from outdoor toilets or septic tanks, for example. This is a danger in areas where people get their water from shallow wells, rivers, or streams. Eating shellfish that lived in water contaminated with feces can also infect a person. Often, in times of war or other circumstances when people must live in crowded unsanitary conditions, lice, fleas, mites, and ticks that carry bacteria transfer diseases among humans.

Proper sanitation methods can prevent these means of contamination and infection. Control of sewage through drain, waste, and vent (DWV) systems discourages rats and other rodents, which carry fleas, from living close to homes and businesses. Raw sewage must be treated before its by-products can be released back into the ecosystem.

Sanitary water supply and distribution systems make sure that the water that goes to homes and businesses is safe to drink. Water used by restaurants to prepare food must be safe or people will get sick and possibly die.

in fixtures such as efficient faucets, toilets, showers, and tubs; and sophisticated treatment processes have led the way to improved sanitation. Plumbing science has changed the way households and entire populations live. Sanitation, along with medical science, continues to be largely responsible for the health of the human race.

3.0.0 ◆ MODERN PLUMBING

Plumbers today design, install, repair, and maintain water supply lines, DWV systems, drainage systems, and gas systems. They may specialize in installation, repair, and maintenance of high- and low-pressure pipe systems used in manufacturing; in the generation of electricity; and in heating and cooling buildings. Some related trades involve specializing as a sprinkler fitter to install fire sprinkler systems in buildings; as a pipefitter to install piping for steam, hot water heating, cooling, lubrication, and other systems; or as a steamfitter to install pipe systems to move liquids or gases under high pressure. A plumber may also specialize in residential design, construction, remodeling, renovation, or maintenance. Plumbers can own and operate their own contracting business. Related fields include plumbing inspection and plumbing instruction.

Whatever the specialty, plumbers work with the latest technology, tools, and equipment. Installing pipe systems requires plumbers to work with precise measuring and testing tools. The old stereotype of a plumber unclogging a sink is not the norm anymore. Working in new construction means very clean working conditions. Only plumbers who work in repair and maintenance are likely to encounter situations where mud, dirt, or waste are involved.

Plumbers must understand blueprints and construction drawings and basic math. They must be familiar with a variety of local laws, state and local **building codes,** and standards that apply to the plumbing trade. These requirements are designed to protect the safety and health of the public.

All states require plumbers to be licensed. Many states require completion of an apprenticeship program and/or on-the-job training before testing for a license. When you apply for a license (see *Figure 2*), you will take a test on your knowledge of the trade and of local plumbing codes.

Advances in technology affect plumbing as much as any other trade. Plumbers will always have to learn about new and better methods for doing their work, the qualities and advantages of newly developed materials, and changing codes and standards.

Plumbing can be divided into three broad phases: **ground rough-in, stack-out,** and **trim-out.** During the ground rough-in phase of construction, plumbers cut holes in walls, ceilings, or floors to attach or hang pipes for connection to fixtures. To join pipe lengths, they may use welding tools, soldering equipment, or even special chemicals for plastic pipes. Pipes are commonly made from plastic, copper, steel, and cast iron. Plumbers may operate power threading machines, propane torches, and other power tools during the installation of the systems.

In the stack-out phase, the various types of pipe are installed in the structural walls. This involves running the pipe through the holes cut during the rough-in phase and securing it to the structural members using supports and hangers. All of the DWV piping and the water supply piping is put in place during this phase.

In the trim-out phase of construction, plumbers install fixtures and appliances. They may install sprinkler heads, sinks, showers, and toilets. Plumbers install appliances such as dishwashers, purification systems, and water heaters. Increasingly, plumbers are expected to install the automatic controls that are used to regulate pressurized pipe systems. As technology develops, so will the skills that a plumber will be required to master.

Service and maintenance of plumbing systems are a large part of a plumber's work. Once systems are installed, they must be monitored, maintained, and repaired periodically. Way beyond fixing a clogged toilet, today's plumbers are required to check lubrication levels in pumps, test gauges, and meters; repair faulty fixtures and components; verify operating systems; and regulate flow and usage rates. Many plumbers develop specialties in servicing specific types of systems, such as air conditioning, heating, and distribution. This allows them to become experts in troubleshooting and repairing malfunctioning equipment. If you specialize, you will always have an active and growing customer base.

4.0.0 ◆ CAREER OPPORTUNITIES IN THE CONSTRUCTION INDUSTRY

The construction industry employs more people and contributes more to the nation's economy than any other industry. Our society will always need new homes, roads, airports, hospitals, schools, factories, and office buildings. This means that there will always be a source of well-paying jobs and career opportunities for construction

Board of Plumbers Examination and Registration of Alabama

Certificate of Competency

No. 2813 Date January 5, 1991

This is to certify that

John A. Smith of Mobile, Alabama

Has Been Duly Registered As A

Master Plumber

Under the Provisions of Alabama Law, Act No. 529 (Regular Session 1949)

In witness whereof, are herewith affixed the Seal of the Board of Plumbers Examination and Registration of Alabama and the signature of the Secretary-Treasurer of the Board.

SAMPLE

101F02A.EPS

City of Montgomery, Alabama

MASTER PLUMBER'S CERTIFICATE

No. 1136

This is to certify that

Maria Sanchez

has passed a satisfactory examination with the

BOARD OF PLUMBING EXAMINERS

as required by ordinance of the City of Montgomery, Alabama

Given under our hand and seal this 16th March, 19 91

BOARD OF PLUMBING EXAMINERS

Robert C. Stevens chairman

Ben E. Morrison

Mollie D. Chen

SAMPLE

101F02B.EPS

VALID THRU DECEMBER 31 2002

STATE OF ALABAMA CERTIFIED MASTER PLUMBER/GAS FITTER Certificate of Competency MPMG

3112

YOUSSEF I. MUHAMMED
2345 MAIN ST.
COASTAL CITY, AL 36601

MEDICAL GAS

SAMPLE

101F02C.EPS

Figure 2 ◆ Examples of plumbing licenses.

trade professionals. As shown in *Figure 3*, the opportunities are not limited to work on construction projects. A skilled, knowledgeable plumber can work in a number of areas. As a construction worker, a plumber can progress from apprentice through many levels:

- Journey plumber
- Master plumber
- Superintendent plumber
- Plumbing supervisor
- Safety manager
- Project manager/administrator
- Plumbing estimator
- Architect
- General contractor
- Construction manager
- Contractor/owner
- Plumbing inspector
- Plumbing instructor

Journey plumber—After successfully completing an apprenticeship, a trainee becomes a journey plumber. The term comes from the 15th century word *journeyman,* which referred to a worker who journeyed away from the master to work alone or for someone else. A person can remain a journey plumber or advance in the trade. Journey plumbers may have additional duties such as supervising or estimating. With larger companies and on larger jobs, journey plumbers often become specialists.

Master plumber—A master craftsperson is one who has achieved and continuously demonstrates the highest skill levels in the trade. The master is a mentor and teacher of others. Master plumbers often start their own businesses and become contractors/owners.

Superintendent plumber—This person is a front-line leader who directs the work of a crew of craftworkers and laborers.

Plumbing supervisor—Large construction projects require supervisors who oversee the work of crews made up of superintendents, apprentices, and journey plumbers. They are responsible for assigning, directing, and inspecting the work of construction crew members.

Safety manager—This person is responsible for safety and health-related issues on a project, including development of the safety plan and procedures, safety training for workers, and regulatory compliance.

Project manager/administrator—A manager/administrator deals with controlling the scope and

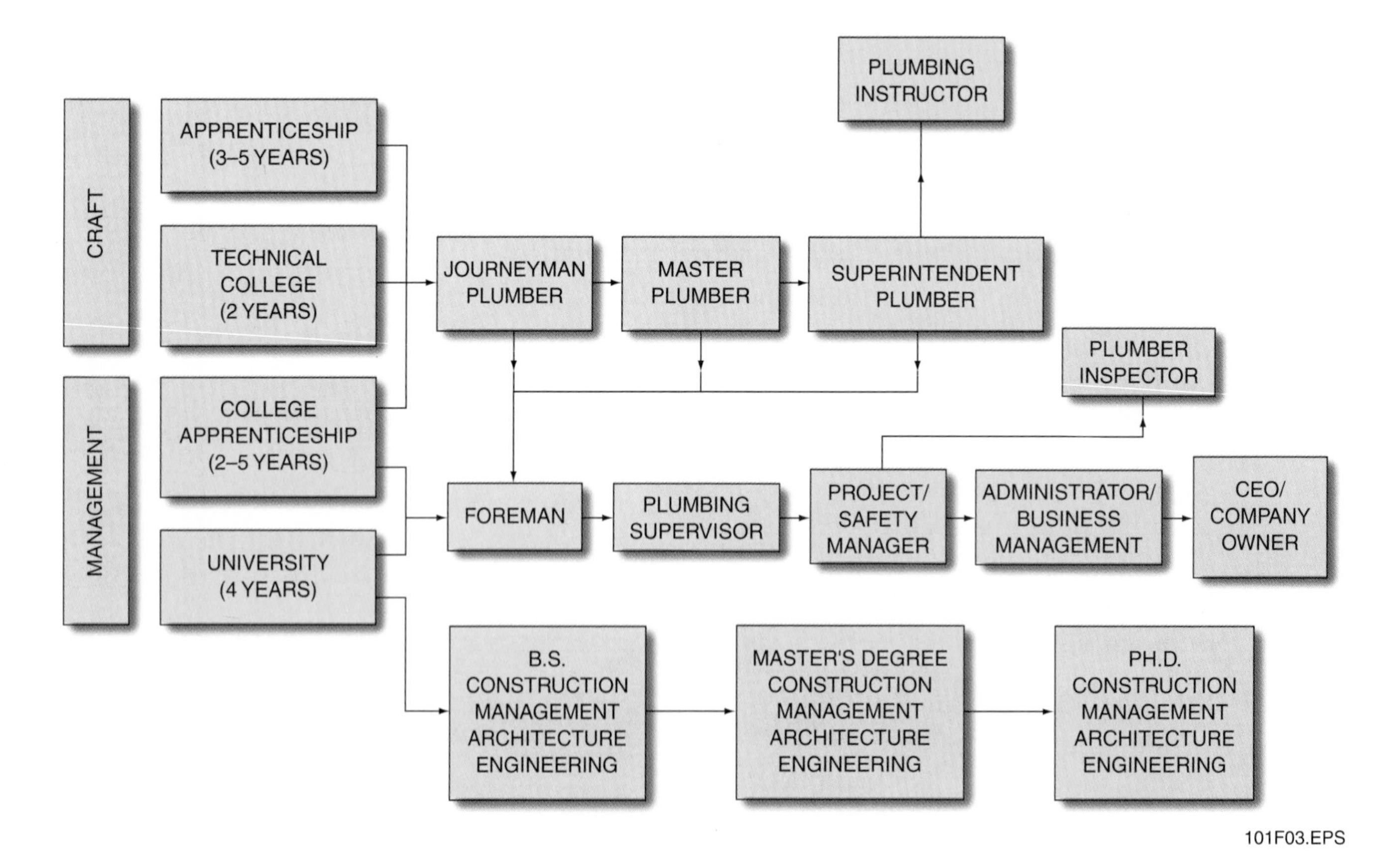

Figure 3 ◆ Opportunities in the construction industry.

direction of a business. This person is responsible for worker output and must determine the best methods to use on a job and how to apply workers' skills to accomplish the job. This person is also responsible for a contractor's support operations, such as accounting (payroll, taxes, and employee benefits), finance, and secretarial work. Larger contracting firms may have several managers/ administrators.

Plumbing estimator—Estimators work for contractors and building supply companies. They make careful estimates of the materials and labor required for a job. On the basis of these estimates, the contractor submits bids for jobs. Estimating requires a complete understanding of construction methods as well as the materials and supplies required. Only experienced plumbers who have good math skills and the patience to prepare detailed, accurate estimates are employed to do this work. This is a highly responsible position because errors in estimates can result in financial losses to the contractor. Depending on the size and type of the business, the job of estimating may be done by the owner, manager, or administrator, or by an estimating specialist. Today's estimators need solid computer skills because advances in computer software have revolutionized the field of estimating.

Architect—An architect is a person who is licensed to design buildings and oversee their construction. A person normally needs a degree in architecture to qualify as an architect.

General contractor—A general contractor is an individual or company that manages an entire construction project. The general contractor plans and schedules the project, buys the materials, and usually contracts with plumbing, carpentry, electrical, and other trade contractors to perform the work. The general contractor may work with architects, engineers, and clients in planning and implementing a project. The general (prime) contractor is also responsible for safety on site.

Construction manager—The role of the construction manager is different from that of the general contractor. The construction manager is usually hired by the building owner to represent the owner's interests on the project. The construction manager works with the general contractor and architect to make sure that the building meets the owner's requirements.

Contractor/owner—Construction contractors/ owners are persons who have established a contracting business. Generally, they hire apprentices, journey plumbers, and master plumbers to work for them. Depending on the size of the business, contractors may work with the crew or they may manage the business full-time. Very small contractors may have only one or two people do everything, including managing the business, preparing estimates, obtaining supplies, and working on the job. This group includes specialty subcontractors, who perform specialized tasks such as installing sprinkler systems, high-pressure pipe systems, and municipal water treatment facilities.

Plumbing inspector—An inspector is an employee of a company, local government, or state who makes sure that plumbing installations meet code and quality requirements. An inspector must review and approve work at various stages of construction.

Plumbing instructor—Training apprentices is one way to give back to the industry. It allows a person to pass on knowledge and skill to the next generation of craftspeople. Dedicated craftworkers find that instructing is a natural development in their career. Often master plumbers become instructors.

The important thing to learn is that, regardless of the path you choose, a career is a lifelong learning process. To be an effective plumber, you need to keep up-to-date with new tools, materials, and methods. If you choose to work your way into management or to start your own construction business someday, you need to learn management and administrative skills on top of keeping your plumbing skills honed. Every successful manager and business owner started the same way you are starting, and they all have one thing in common: a desire and willingness to keep on learning. The learning process begins with apprentice training.

4.1.0 Formal Construction Training

Over the past 20 years, the rate of formal training in the construction industry has been declining. Until the establishment of the NCCER, the only opportunity for formal construction training was through the federal Department of Labor's Apprenticeship, Training, Employer, and Labor Services (ATELS). ATELS programs rely on mandatory classroom instruction and on-the-job training (OJT). The classroom instruction required is 144 hours per year and the OJT requirement is 2,000 hours per year. In a typical ATELS program, you would spend 8,000 hours in OJT and 576 hours in related classroom training before you received the journey plumber certificate.

Today, education and training throughout the country are undergoing significant change. Educators and researchers have been learning and applying new techniques to adjust to the way today's students learn and apply their education. One result has been the establishment of NCCER.

NCCER is an independent, private educational foundation, established and funded by the construction industry to solve the training problem in the industry today. The basic idea of NCCER is to provide *industry-driven* training and education programs. NCCER departs from traditional classroom learning and has adopted a pure competency-based training system. Competency-based training means that instead of needing to complete a certain number of hours of classroom training and OJT, you simply have to prove that you know what is required and demonstrate that you can perform specific skills. All completion information for every trainee is then sent to NCCER and kept in the National Registry. The National Registry can then confirm training and skills for workers as they move from state to state, company to company, or even within a company (see *Appendix A*).

The dramatic shortage of skills in the construction workforce, combined with the shortage of new workers coming into the industry, is forcing the industry to design new kinds of training. Whether you enroll in an ATELS program, an NCCER program, or both, make sure that you work for an employer who supports a nationally-standardized training program that includes credentials to confirm your skill development.

4.1.1 Apprenticeship Program

Apprentice training goes back thousands of years; its basic principles have not changed. First, it is a way for individuals entering the craft to learn from those who have mastered it. Second, it focuses on learning by doing: real skills versus theory. Although some theory is presented in the classroom, it is always presented in a way that helps the trainee understand the purpose behind the skill that is to be learned.

4.1.2 Youth Apprenticeship Program

A Youth Apprenticeship Program is available that allows students to begin their apprentice training while they are still in high school (see *Figure 4*). A student entering the plumbing program in 11th grade may complete as much as one year of the NCCER four-year program by high school graduation. In addition, the program, in cooperation with local craft employers, allows students to work in the trade and earn money while they are still in school. Upon graduation, the student can enter the industry at a higher level and with more pay than someone just starting the apprenticeship program.

This training program is similar to the one used by NCCER learning centers, contractors, and colleges

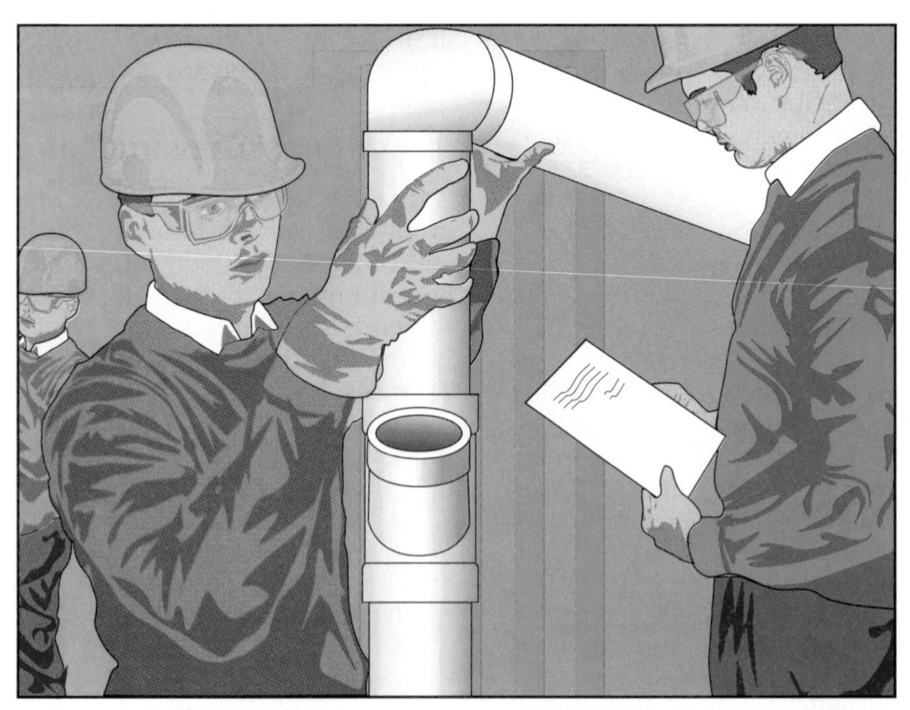

101F04A.EPS

Figure 4 (1 of 2) ◆ Youth Apprenticeship Program.

101F104B.EPS

101F04C.EPS

Figure 4 (2 of 2) ◆ Youth Apprenticeship Program.

across the country. With their official transcripts, students can enter the second year of the program wherever it is offered. They may also have the option of applying the credits at a two-year or four-year college that offers degree or certificate programs in the construction trades.

4.1.3 Apprenticeship Standards

All apprenticeship standards call for certain work-related or on-the-job training. This OJT is broken down into specific tasks in which the apprentice receives hands-on training during the apprenticeship. In addition, a specified number of hours are required in each task. The total number of hours for the plumbing apprenticeship program is traditionally 8,000, which amounts to about four years of training. In a competency-based program, it may be possible to shorten this time by testing out of specific tasks through a series of performance exams.

In a traditional program, you may complete the required OJT at a rate of 2,000 hours per year. If you are laid off or get sick, it may take longer.

You must log all work time and turn it in to the Apprenticeship Committee (discussed later), so that accurate time control can be maintained. Another important aspect of keeping work records up-to-date is that after each 1,000 hours of related work, the standards call for a pay increase.

Because of layoffs, as well as the type of work needed in the field, the classroom instruction and OJT will not always run at the same time. Apprentices with special job experience or coursework may obtain credit toward their classroom requirements. This reduces the total time required in the classroom while maintaining the total 8,000-hour OJT requirement. These special cases will depend on the type of program and its regulations and standards.

Informal OJT provided by employers is usually less thorough than that provided through a formal apprenticeship program. The degree of training and supervision in an informal program often depends on the size of the firm. A small contractor who specializes in home building may provide training in only one area, such as rough plumbing. A large general contractor may be able to provide training in several areas.

For people entering an apprenticeship program, a high school or technical school education is desirable, with courses in plumbing, shop, mechanical drawing, and general mathematics. Manual dexterity, good physical condition, and a good sense of balance are important. The ability to solve arithmetic problems quickly and accurately and to work closely with others is essential.

If you want to become an apprentice, you must submit to the Apprenticeship Committee certain information, which may include the following (actual requirements may vary by state):

- Aptitude test (General Aptitude Test Battery [GATB] or GATB Form Test) results (normally administered by the local Employment Security Commission)
- Proof of educational background (candidate should have school[s] send transcripts to the committee)
- Letters of reference from past employers and friends
- Results of a physical examination
- Proof of age
- If the candidate is a veteran, a copy of Form DD214
- A record of technical training received that relates to the construction industry and/or a record of any pre-apprenticeship training
- High school diploma or General Equivalency Diploma (GED)

Once you are in an apprenticeship program, you must:

- Wear proper safety equipment on the job
- Buy and maintain tools of the trade as needed and required by the contractor
- Submit a monthly OJT report to the committee
- Report to the committee if there is a change in your employment status
- Attend classroom instruction and follow all classroom regulations, such as those for attendance

4.2.0 Responsibilities of the Employee

To be successful as a professional plumber, you must have the skills to use current trade materials, tools, and equipment to produce a finished product of high quality in a minimum amount of time. You must be able to change methods to meet each situation. To be a successful plumber, you must continuously train to remain knowledgeable about the technical advancements in trade materials and equipment and to gain the skills to use them. You must never take chances with your own safety or the safety of others. *Appendix B* at the end of this module lists characteristics and values that most people associate with ethical behavior.

4.2.1 Professionalism

Professionalism is a broad term that describes the overall behavior and attitude expected in the workplace. Professionalism is too often absent from the construction site and the various trades. Ideally, professionalism starts at the top. Management should set the example and should support professionalism among all workers, but it is even more important that individual workers recognize their own responsibility for professionalism.

Professionalism includes honesty, productivity, safety, civility, cooperation, teamwork, clear and concise communication, arriving at the job on time and prepared for work, eagerness to do the job right, and sensitivity to your impact on co-workers. It can be demonstrated in a variety of ways every minute you are at the workplace. Most important is that you do not tolerate the unprofessional behavior of co-workers. This is not to say that you shun the unprofessional worker; instead, you work to demonstrate the benefits of professional behavior.

Professionalism is a benefit to both the employer and the employee. *It is a personal responsibility.* Our industry is what each individual chooses to make of it. Choose professionalism and the industry image will follow.

4.2.2 Honesty

Honesty and personal integrity are important traits of the successful professional. Professionals take pride in performing a job well and in being punctual and dependable. Each job is completed in a professional way, never by cutting corners or reducing materials. A valued professional maintains work attitudes and ethics that protect property such as tools and materials belonging to employers, customers, and other trades from damage or theft at the shop or job site.

Honesty and success go hand in hand. It is not simply a choice between good and bad, but a choice between success and failure. Dishonesty will *always* catch up with you. Whether you are stealing materials, tools, or equipment from the job site or simply lying about your work, it will not take long for your employer to find out. Of course, you can find another employer, but sooner or later this option will run out on you.

If you plan to be successful and enjoy continuous employment and consistent earnings, and if you want to be sought after by employers as opposed to always looking for work, then start out with the basic understanding of honesty in the workplace. You will reap the benefits.

Honesty means more than just not taking things that do not belong to you. It means giving a fair day's work for a fair day's pay; it means carrying out your side of a bargain; it means that your words convey true meanings. Your thoughts as well as your actions should be honest. Employers place a high value on an employee who is strictly honest.

4.2.3 Loyalty

As an employee, you expect your employers to look out for your interests, provide you with steady employment, and promote you to better jobs as openings occur. Employers feel that they, too, have a right to expect you to be loyal to them—to keep their interest in mind, to speak well of them to others, to keep any minor troubles strictly within the plant or office, and to keep absolutely confidential all matters that pertain to the business. Both employers and employees should keep in mind that loyalty is not something to be demanded; rather, it is something to be earned.

4.2.4 Willingness to Learn

Every office and plant has its own way of doing things, and your employer will expect you to be willing to learn these ways. Also, you must adapt to change and be willing to learn new methods and procedures as quickly as possible. Sometimes the installation of a new machine or the purchase of new tools makes it necessary for even experienced employees to learn new methods and operations (see *Figure 5*). You may resent the retraining, but you can be sure that employers will expect you to put forth the necessary effort. Keeping methods

101F05.EPS

Figure 5 ◆ Learning new tools.

and operations up-to-date is necessary to compete with other companies and show a profit. And it is this profit that allows the employer to stay in business and provide jobs.

4.2.5 Willingness to Take Responsibility

Most employers expect their employees to see what needs to be done, then go ahead and do it. If they ask you to do a specific task, they expect you to do it without being asked again. This is your responsibility. You should be alert to boxes that need to be moved out of the way, materials that should be stacked, or tools that need to be put away. And you should assume the responsibility for doing these tasks. If your employer asks you to do a certain job, take the responsibility for doing it, and doing it well, without being asked again.

4.2.6 Willingness to Cooperate

To cooperate means to work together. In our modern business world, cooperation is the key to getting things done. Learn to work as a member of a team with your employer, supervisor, and fellow workers in a common effort to get the work done efficiently, safely, and on time.

4.2.7 Rules and Regulations

People can work together well only if there is some understanding about what work is to be done, when it will be done, and who will do it. All employees must recognize that rules and regulations are necessary in any work situation. They may vary from employer to employer, so it is important to make sure you understand them and follow them each time you start a new job.

4.2.8 Tardiness and Absenteeism

Tardiness means being late for work. Absenteeism means being off the job for one reason or another. A lot of tardiness and absence is an indication of poor work habits, unprofessional conduct, and a lack of commitment.

We are all creatures of habit. What we do once, we tend to do again unless the results are too unpleasant. Habits developed in our childhood may be comfortable, but they may not serve us in the working world. Bad habits can be changed if we make an effort to do so.

Your work life is governed by the clock. You are required to be at work and ready to start work at a definite time (see *Figure 6*). So is everyone else.

101F06.EPS

Figure 6 ◆ Arriving at work on time.

Your failure to get to work on time results in confusion, lost time, and resentment from co-workers who do come in on time. In addition, it may lead to penalties, including being fired. Your obligation is to be at work on time. You agree to the terms of work when you accept the job. Maybe it will help you to see things more clearly if you look at the situation from the boss's point of view. Supervisors cannot keep track of workers who come in any time they please. Supervisors are not being fair to your fellow workers if they ignore your tardiness. Your failure to be on time keeps the rest of the workers from getting started on the job.

Better planning of your morning routine will keep you from being delayed and prevent a late arrival. In fact, arriving a little early shows your interest and enthusiasm for your work, which employers appreciate. The habit of being late can stand in the way of promotion.

Sometimes it is necessary to take time off from work. No one should be expected to work when ill or when there is serious trouble at home. However, it is possible to get into the habit of letting unimportant and unnecessary matters keep you from the job. This results in lost production and hardship on those who try to carry on the work with less help. Again, a principle is involved. Your employer has a right to expect you to be on the job unless there is some very good reason for staying away. You should not let a trivial reason keep you home. For instance, you should not stay up so late that you are too tired to go to work the next day. If you are ill, you should use the time at home to do all you can to recover quickly. After all, this is what you would expect of a person you had hired to work for you and on whom you depended to do a certain job.

If you do have to stay home for a very good reason, phone the office early in the morning so that the boss will know that you are not coming and can make arrangements for someone else to replace you while you are out. Failing to phone is the worst possible way to handle the matter. If you don't phone, your boss and fellow workers have no way of knowing whether you have merely been held up and will be in later, or whether they need to assign your work to someone else. You must let the boss know if you cannot come to work.

The most frequent causes of absenteeism are legitimate reasons such as illness, death in the family, and accidents, or reasons that you can control, such as personal business and dissatisfaction with the job. You have no control over accidents, but you can usually plan to take care of personal business after working hours. Frequent absences, for whatever reason, will reflect unfavorably on you when promotions are being considered.

Employers sometimes resort to docking pay, demotion, and even firing in an effort to control tardiness and absenteeism. No employer likes to impose restrictions of this kind. However, in fairness to the workers who do show up on time every day, an employer is sometimes forced to discipline workers who will not follow the rules.

4.3.0 What You Should Expect from Your Employer

After the Apprenticeship Committee has selected an applicant, the employer agrees to give the apprentice a job under conditions that will result in normal advancement. In return, the employer requires the apprentice to make satisfactory progress in OJT and related classroom instruction. The employer agrees not to employ the apprentice in a way that may violate the apprenticeship standards. The employer also pays a share of the cost of operating the apprenticeship program.

4.4.0 What You Should Expect from a Training Program

First and foremost, it is important to select an employer that has a training program. The program should be comprehensive, standardized, and based on competency, not on the amount of time you spend in a classroom.

When employers take the time and initiative to provide quality training, it is a sign that they are willing to invest in their workforce and improve the abilities of workers. It is important that the training program is national and participants receive transcripts and completion credentials. Employees in the trades may work for several different contractors, so it is important that the training program help the worker move from company to company, city to city, or state to state, and that the worker not have to start at the beginning after each move. Before you enroll in a training program, ask how many employers in the area use the same program. Make sure you will have access to transcripts and certificates to document your status and level of completion.

Training should be rewarded. The training program should include a well-defined compensation ladder. Successful completion and mastery of skill sets should be accompanied by wage increases.

Finally, the curriculum should be complete and up-to-date. Any training program has to be committed to maintenance of its curriculum, development of new delivery mechanisms (such as CD-ROM and the Internet), and constant vigilance for new techniques, materials, tools, and equipment in the workplace.

4.5.0 What You Should Expect from the Apprenticeship Committee

The Apprenticeship Committee is the local administrative body to which the apprentice is assigned. It is responsible for the appropriate training of apprentices. Every apprenticeship program, whether state or federal, is covered by standards that have been approved by appropriate government agencies. The committee is responsible for enforcing them.

- The committee is responsible for enforcing standards and for making sure that proper training is conducted so that a craftsperson graduating from the program is fully qualified in those areas of training designated by the standards.

- The committee screens and selects individuals for apprenticeship and refers them to participating forms for training.
- The committee places apprentices under written agreement for participation in the program.
- The committee establishes minimum standards for related classroom instruction and OJT and monitors the apprentice to see that these standards are followed during the training period.
- The committee hears all complaints of violations of apprenticeship agreements, whether by employer or apprentice, and takes action within the guidelines of the standards.
- The committee notifies the registration agencies of all enrollments, completions, and terminations of apprentices.

5.0.0 ◆ HUMAN RELATIONS

Most people underestimate the importance of working well with others. There is a tendency to pass off human relations as nothing more than common sense. What exactly is involved in human relations? One answer is that part of human relations is being friendly, pleasant, courteous, cooperative, adaptable, and sociable. As important as those characteristics are for personal success, they are not the whole story.

5.1.0 Making Human Relations Work

Human relations is more than just getting people to like you. It is also knowing how to handle difficult situations.

Human relations is knowing how to work with supervisors who may be demanding and sometimes unfair. It is understanding the personality traits of others as well as your own. Human relations is building sound working relationships with people you may not like. Human relations is knowing how to restore working relationships that have broken down. It is learning how to handle frustrations without hurting others. In short, human relations involves working with all kinds of people, even those who are hard for you to get along with.

5.2.0 Human Relations and Productivity

Effective human relations is directly related to productivity, and productivity is the key to business success. Every employee is expected to produce at a certain level. Employers quickly lose interest in an employee who is unable to produce. There are work schedules to be met and jobs that must be completed.

All employees, both new and experienced, are measured by the amount of quality work they can safely turn out. Employers expect all employees to do their share of the workload.

However, doing your share (or more than your share) is not enough in itself. To be productive, you must do your share without antagonizing your fellow workers. You cannot be successful without learning how to work with your peers.

Do everything you can to build strong, professional working relationships with fellow employees, supervisors, and clients.

5.3.0 Attitude

A positive attitude is essential to a successful career for several reasons. First, being positive means being energetic, highly motivated, attentive, and alert. A positive attitude is essential to safety on the job. Second, a positive employee contributes to the productivity of others. Both negative and positive attitudes are transmitted to others on the job. It is very difficult to maintain a high level of productivity if you are working next to a person with a negative attitude. Third, people favor a person who is positive. Being positive makes the job more interesting and exciting. Fourth, employers pay attention to your attitude. Supervisors can see your attitude in your approach to the job, your reactions to instructions, and the way you handle problems.

A positive attitude is far more than a smile. As a matter of fact, some people transmit a positive attitude even though they seldom smile. They do this by the way they treat others, the way they deal with their responsibilities, and the approach they take when they are faced with problems.

Here are a few suggestions to help you maintain a positive attitude:

- Remember that your attitude follows you wherever you go. If you try to be more positive in your social and personal life, it will automatically help you on the job. And a positive attitude on the job will carry over to your social and personal life as well.
- Your fellow workers do not welcome negative comments. The solution: talk about positive things and be complimentary.
- Look for the good things in people on the job, especially your supervisor. Nobody is perfect, but almost everyone has a few worthwhile qualities. If you dwell on people's good fea-

tures rather than their failings, you will find it easier to work with them.

- Look for the good things where you work. What makes it a good place to work? Is it the hours, the physical environment, the people, the actual work being done? Or is it the atmosphere? You do not have to like everything. No job is perfect, but if you concentrate on the good things, the negative factors will seem less important and bothersome.
- Look for the good things in the company. Just as there are no perfect jobs, there are no perfect companies. But almost all organizations have good features. Is the company progressive? What about promotional opportunities? Are there chances for self-improvement? What about the wage and benefit package? Is there a good training program? You cannot expect to have everything you would like, but there should be enough to keep you positive. In fact, if you decide to stick with a company for a long time, it is wise to look at the good features and think about them. If you think positively, you will act the same way.

6.0.0 ◆ EMPLOYER AND EMPLOYEE SAFETY OBLIGATIONS

An obligation is like a promise or a contract. In exchange for the benefits of your employment and your own well-being, you agree to work safely. In other words, you are *obligated* to work safely. You are also obligated to make sure anyone you supervise or work with is working safely. Your employer is also obligated to maintain a safe workplace for all employees. Safety is everyone's responsibility (see *Figure 7*).

Some employers have safety committees. If you work for such an employer, you are obligated to that committee to maintain a safe working environment. This means you must:

- Follow the safety committee's rules for proper working procedures and practices
- Report any unsafe equipment and conditions directly to the committee or your supervisor

Here is a basic rule to follow every working day:

If you see something that is not safe, report it! Do not ignore it. It will not fix itself.

Suppose you see a faulty electrical hookup. You know enough to stay away from it to remain safe, but then you continue with your work and forget about the hazard. The danger does not harm you, but later a co-worker accidentally touches the live wire.

Figure 7 ◆ Safety is everyone's responsibility.

In the long run, even if you do not think an unsafe condition affects you—it does. Report what is not safe. Do not think your employer will be angry because your productivity suffers while the condition is corrected. On the contrary, your employer will be more likely to criticize you for not reporting a problem.

Your employer knows that the small amount of time lost in making conditions safe is nothing compared with shutting down the whole job because of a major disaster. If that happens, you are out of work anyway. So do not ignore unsafe conditions. In fact, Occupational Safety and Health Administration (OSHA) regulations require you to report hazardous conditions to your supervisor.

This requirement applies to every part of the construction industry. Whether you work for a large contractor or a small subcontractor, you are obligated to report unsafe conditions. The easiest way to do this is to tell your supervisor. If that person ignores the unsafe condition, report it to the next highest supervisor. If it is the owner who is being unsafe, let him or her know your concerns.

If nothing is done about it, report it to OSHA. If you are worried about your job being on the line, think about it in terms of your life, or someone else's, being on the line.

Congress passed the Occupational Safety and Health Act in 1970. This act also created OSHA, which is part of the U.S. Department of Labor. The job of OSHA is to set occupational safety and health standards for all places of employment, enforce these standards, make sure that employers provide and maintain a safe workplace for all employees, and provide research and educational programs to support safe working practices.

OSHA requires each employer to provide a safe and hazard-free working environment. It also requires employees to comply with OSHA rules and regulations that relate to their conduct on the job.

According to OSHA standards, you are entitled to on-the-job safety training. If you are a new employee, your employer must:

- Show you how to do your job safely
- Provide you with required personal protective equipment
- Warn you about specific hazards
- Supervise you for safety while you work

This act of Congress is enforced by federal and state safety inspectors who have the legal authority to make employers pay fines for safety violations. The law allows states to have their own safety regulations and agencies to enforce them, but they must first be approved by the U.S. Secretary of Labor. In states that do not develop such regulations and agencies, federal OSHA standards must be obeyed.

These standards are listed in *OSHA Safety and Health Standards for the Construction Industry* (29 CFR, Part 1926), sometimes called *OSHA Standards 1926* (see *Figure 8*). Other safety standards that apply to the plumbing trade are published in *OSHA Safety and Health Standards for General Industry*.

The most important general requirements that OSHA places on employers in the construction industry are as follows:

- The employer must perform frequent and regular job site inspections of equipment.
- The employer must instruct all employees to recognize and avoid unsafe conditions, and to know the regulations that pertain to the job so they may control or eliminate any hazards.
- No one may use any tools, equipment, machines, or materials that do not comply with *OSHA Standards 1926*.

OCCUPATIONAL SAFETY and HEALTH STANDARDS for the CONSTRUCTION INDUSTRY

(29 CFR PART 1926)

Promulgated by the
OCCUPATIONAL SAFETY AND HEALTH ADMINISTRATION
UNITED STATES DEPARTMENT OF LABOR

101F08.EPS

Figure 8 ◆ OSHA standards govern workplace safety.

- The employer must be sure that only qualified individuals operate tools, equipment, and machines.
- The employer must have a written hazard communications (HazCom) program in place to give employees information about safety hazards in the workplace and how to control them.
- The employer must have material safety data sheets (MSDSs) for each hazardous material used in the workplace. The MSDS contains all manufacturer data, and lists possible hazards, first-aid and treatment requirements, and contact information for each product.
- A **competent person** is required to supervise, control, test, or inspect certain functions, including but not limited to excavations, scaffolding, and asbestos work.

Summary

Plumbing is very important for sanitation and public health. There are many job and career opportunities for skilled plumbers. An apprenticeship program that combines competency-based, hands-on training with classroom instruction is the most effective way for a person to learn and advance in the plumbing craft. One part of that task is to develop the job skills. It is just as important to learn good work habits, convey a positive, cooperative attitude to those around you, and practice good safety habits every day.

Review Questions

1. Modern plumbers design, install, repair, and maintain water supply lines; drain, waste, and vent systems; drainage systems; and _____.
   a. aqueducts
   b. petroleum pipelines
   c. lead piping
   d. gas systems

2. The step that immediately follows an apprenticeship as a plumber is _____.
   a. master plumber
   b. journey plumber
   c. general contractor
   d. estimator

3. A competency-based training program requires the trainee to _____.
   a. receive at least four years of classroom training
   b. receive on-the-job training for at least two years
   c. demonstrate the ability to perform specific job-related tasks
   d. put in a certain number of OJT hours

4. A(n) _____ is hired by the building owner to oversee the construction project and make sure it meets the owner's requirements.
   a. construction manager
   b. general contractor
   c. estimator
   d. architect

5. The main purpose of the Youth Apprenticeship Program is to _____.
   a. make sure all young people know how to use basic carpentry tools
   b. provide job opportunities for people who quit high school
   c. allow students to start in an apprenticeship program while they are still in high school
   d. make sure that people under 18 have proper supervision on the job

6. The total of on-the-job training needed for a plumbing apprentice to advance to the next step after apprenticeship is _____ hours.
   a. 2,000
   b. 4,000
   c. 6,000
   d. 8,000

7. An honest employee believes _____.
   a. it is okay to borrow tools from the job site as long as you return them before anyone notices
   b. you are doing your company a favor by using lower-grade materials than those listed in the specifications
   c. it is okay to take materials or tools from your employer if you feel the company owes you for past efforts
   d. arriving on time and doing a full day's work is an obligation if you accept a full day's pay

8. If one of your co-workers complains about your company, you should _____.
   a. contribute your own complaints to the conversation
   b. find some good things to say about the company
   c. suggest that the person look for another job
   d. agree with the person to avoid conflict

9. If you see an unsafe condition on the job, you should _____.
   a. ignore it because it is not your job
   b. report it to OSHA
   c. call ATELS
   d. report it to a supervisor

10. The purpose of OSHA is to _____.
    a. catch people breaking safety regulations
    b. make rules and regulations governing all aspects of construction projects
    c. make sure that the employer provides and maintains a safe workplace
    d. assign a safety inspector to every project

PROFILE IN SUCCESS

George Benoit

Plumbing
Senior Field Coordinator, Southern California PHCC
Sacramento, CA

George was born and raised in Winnipeg, Canada. He left school early to work with his dad in the plumbing trade. When he was 17, his father died, and George moved to the United States. He settled down and started a family in Illinois, then moved to California. He worked for more than 15 years with a plumbing company before starting his own company. After 31 years in the trade he is retired and now works in the state-approved apprenticeship training program for California.

How did you become interested in the construction industry?
I started working with my dad in plumbing and HVAC. I worked for a union company that didn't have any training opportunities. I asked to be trained anyway. I was the only apprentice out of about 150 people working in the shop. At that time, the Canadian government required one week's training each year. I ended up going to school 8 hours a day for 5 straight days. It got me started.

By the time I got to California it was tough. I tried to get into the apprenticeship program through the union, but they told me I had to be a union member. I would go to the union hall and was told they couldn't get me a job. I had enough experience to qualify for my journeyman's license. So I decided to go ahead and test for my license. I got it.

I took a job with a company that built apartment houses. After a time, I started working in the part of the firm that built commercial buildings and high-end custom homes. We did a lot of work for the movie industry, working on remodels for movie stars and directors. I worked on Cary Grant's house for about a year. After I went out on my own, I hooked up with a contractor doing government work. We remodeled Ronald Reagan's St. Cloud home to provide quarters for the Secret Service agents assigned to the former president. I worked on some Hughes Aircraft jobs as well.

What do you think it takes to be a success in your trade?
Being good at your trade is not enough. You have to be a good businessperson to make it these days. You have to understand the financial component of working a business. That means you have to know operating costs, cash flow, and how to price jobs correctly. Too many people don't know how to run a company, and they don't last long.

The trades are an honest living. There is room and opportunity enough to make a decent income to support your family. You have to have skill, drive, and aptitude. We interview and pretest our apprentices to help them make the decision to pursue a career. Our focus on training is the best start someone can have. We work with contractors and the NCCER's craft training materials to ensure that apprentices get the skills and experience they need to succeed.

What are some of the things you do in your role of supervisor?
I find participating contractors and recruit apprentices. At the start we had about 180 apprentices. Now we have between 400 to 500 apprentices and operate 13 schools. From the time we find enough contractors, it takes about four to five months to get a new school site up and running. We are also working on an instructor credential program.

As the senior field coordinator, I work with company owners and young people to develop and start training programs. It's a great challenge.

What do you like most about your job?
As a plumber, I liked being able to see a job through—to take pride in a job well done. The best part of my job was working with people and meeting the challenge of different work. Customer satisfaction is a great motivator. I really like working with people.

After 31 years in the trade, I now focus on helping young people get a good start on a career. The trades are—and always have been—an honorable way to make a decent living. The best part of working with apprentices is that they challenge me. They come up with questions that keep me learning all the time. They show such enthusiasm and promise; it is rewarding for that reason.

What would you say to someone entering the trades today?

I tell people that not everyone is going to be a computer technician. Not everyone is going to go through college. A lot of people are very good at hands-on work. If you want the opportunity to make a decent living and support your family, don't overlook the trades—there are many wonderful people working in them.

GLOSSARY

Trade Terms Introduced in This Module

Aqueduct: Man-made channel for carrying water.

Building codes: Requirements published by state and local governments to establish minimum safety standards for various types of construction.

Chlorine: A heavy, greenish-yellow gas used as a disinfectant in water treatment.

Competent person: A person who is capable of identifying existing and predictable hazards in the surroundings or working conditions that are unsanitary, hazardous, or dangerous, and who has authorization to act promptly to eliminate the hazards.

CPVC: Chlorinated polyvinylchloride (CPVC) pipe, used on hot- and cold-water distribution lines. It is designated for water service up to and including 180°F. It can be cut with a handsaw or tube cutter.

Ground rough-in: The phase of construction when the plumber cuts holes in the structure for the pipes that will be installed during stack-out.

Plumbarius: Latin term for someone who works with lead. An ancient term for a plumber.

Plumber: One who installs or repairs plumbing systems and plumbing fixtures.

Plumbing: The installation, maintenance, and making of alterations to piping or tubing used to carry and control the flow of waste, gases, and water to and from fixtures and appliances.

Plumbum: Latin word for lead. Origin of the English term *plumber.*

Stack-out: The phase of construction when pipes are installed in the structural walls.

Trim-out: The phase of construction when plumbers install fixtures and appliances.

APPENDIX A

Samples of NCCER Apprentice Training Recognition

December 29, 2000

Joseph Smith
c/o Training Sponsor, Inc.
1234 South Main Street
Anytown, US 12345-6789

Dear Joseph,

On behalf of the National Center for Construction Education and Research, I congratulate you for successfully completing the NCCER's standardized craft training program.

As the NCCER's most recent graduate, you are a valuable member of today's skilled construction and maintenance workforce. The skills that you have acquired through the NCCER craft training programs will enable you to perform quality work on construction and maintenance projects, promote the image of these industries and enhance your long-term career opportunities.

We encourage you to continue your education as you advance in your construction career. Please do not hesitate to contact us for information regarding our Management Education and Safety Programs or if we can be of any assistance to you.

Enclosed please find your diploma, transcript and wallet card. Once again, congratulations on your accomplishments and best wishes for a successful career in the construction and maintenance industries.

Sincerely,

Daniel J. Bennet
President, NCCER

National Center for Construction Education and Research

NCCER National Registry

Joseph Smith

Core Curricula 12/29/00
Plumbing Level One 12/29/00

SAMPLE

Sponsor

Daniel J. Bennet
President, NCCER

P.O. Box 141104
Gainesville, FL
32614-1104
352.334.0911
Fax 352.334.0932

Affiliated with the University of Florida

www.nccer.org

101A01.EPS

Figure A ◆ NCCER Registry Card and letter.

101A02.EPS

Figure B ◆ NCCER diploma.

NATIONAL CENTER FOR CONSTRUCTION EDUCATION AND RESEARCH

Affiliated with the University of Florida
Post Office Box 141104 Gainesville, Florida 32614-1104
352.334.0911 Fax 352.334.0932 www.nccer.org

12/12/00
Page: 1

Joseph Smith
1234 South Main Street
Anytown, US 12345-6789

SSN: 123123123

Course/Description	Instructor/Sponsor	Date Completed
00101 Basic Safety	Jones/Training Sponsor, Inc.	3/17/00
00102 Construction Math	Jones/Training Sponsor, Inc.	4/7/00
00103 Introduction to Hand Tools	Brown/Training Sponsor, Inc.	4/21/00
00104 Introduction to Power Tools	Brown/Training Sponsor, Inc.	5/5/00
00105 Introduction to Blueprints	Reyes/Training Sponsor, Inc.	5/26/00
00106 Basic Rigging	Jansen/Training Sponsor, Inc.	6/16/00
02101 Introduction to the Plumbing Trade	Maczka/Training Sponsor, Inc.	7/28/00
02102 Plumbing Tools	Harris/Training Sponsor, Inc.	8/11/00
02103 Introduction to Plumbing Math	Harris/Training Sponsor, Inc.	9/1/00
02104 Introduction to Plumbing Drawings	Harris/Training Sponsor, Inc.	9/15/00
02105 Plastic Pipe and Fittings	Chang/Training Sponsor, Inc.	9/29/00
02106 Copper Pipe and Fittings	Chang/Training Sponsor, Inc.	9/29/00
02107 Cast-Iron Pipe and Fittings	Spinelli/Training Sponsor, Inc.	10/13/00
02108 Carbon Steel Pipe and Fittings	Spinelli/Training Sponsor, Inc.	10/13/00
02109 Fixtures and Faucets	Gavin/Training Sponsor, Inc.	10/27/00
02110 Introduction to DWV Systems	Gavin/Training Sponsor, Inc.	11/17/00
02111 Introduction to Water Distribution	Vasquez/Training Sponsor, Inc.	12/8/00

SAMPLE

President

101A03.EPS

Figure C ◆ NCCER transcript.

APPENDIX B

Ethical Principles for Members of the Construction Trades

Honesty: Be honest and truthful in all dealings. Conduct business according to the highest professional standards. Faithfully fulfill all contracts and commitments. Do not deliberately mislead or deceive others.

Integrity: Demonstrate personal integrity and the courage of your convictions by doing what is right even where there is great pressure to do otherwise. Do not sacrifice your principles because it seems easier.

Loyalty: Be worthy of trust. Demonstrate fidelity and loyalty to companies, employers, co-workers, and trade institutions and organizations.

Fairness: Be fair and just in all dealings. Do not take undue advantage of another's mistakes or difficulties. Fair people are open-minded and committed to justice, equal treatment of individuals, and tolerance for and acceptance of diversity.

Respect for others: Be courteous and treat all people with equal respect and dignity, regardless of age, religion, sex, race, or national origin.

Obedience: Abide by laws, rules, and regulations relating to all personal and business activities.

Commitment to excellence: Pursue excellence in performing your duties, be well-informed and prepared, and constantly try to increase your proficiency by gaining new skills and knowledge.

Leadership: By your own conduct, seek to be a positive role model for others.

Answers to Review Questions

Answer	Section Reference
1. d	3.0.0
2. b	4.0.0
3. c	4.1.0
4. a	4.0.0
5. c	4.1.2
6. d	4.1.0
7. d	4.2.2
8. b	5.3.0
9. d	6.0.0
10. c	6.0.0

NCCER CRAFT TRAINING USER UPDATES

The NCCER makes every effort to keep these textbooks up-to-date and free of technical errors. We appreciate your help in this process. If you have an idea for improving this textbook, or if you find an error, a typographical mistake, or an inaccuracy in the NCCER's Craft Training textbooks, please write us, using this form or a photocopy. Be sure to include the exact module number, page number, a detailed description, and the correction, if applicable. Your input will be brought to the attention of the Technical Review Committee. Thank you for your assistance.

Instructors – If you found that additional materials were necessary in order to teach this module effectively, please let us know so that we may include them in the Equipment/Materials list in the Instructor's Guide.

Write: Curriculum Revision and Development Department
National Center for Construction Education and Research
P.O. Box 141104, Gainesville, FL 32614-1104

Fax: 352-334-0932

E-mail: curriculum@nccer.org

Craft ______________________ Module Name ______________________

Copyright Date ____________ Module Number ____________ Page Number(s) ____________

Description

____

____

____

____

(Optional) Correction

____

____

____

(Optional) Your Name and Address

____

____

____

Module 02102-00

Plumbing Tools

Course Map

This course map shows all of the modules in the first level of the Plumbing curriculum. The suggested training order begins at the bottom and proceeds up. Skill levels increase as you advance on the course map. The local Training Program Sponsor may adjust the training order.

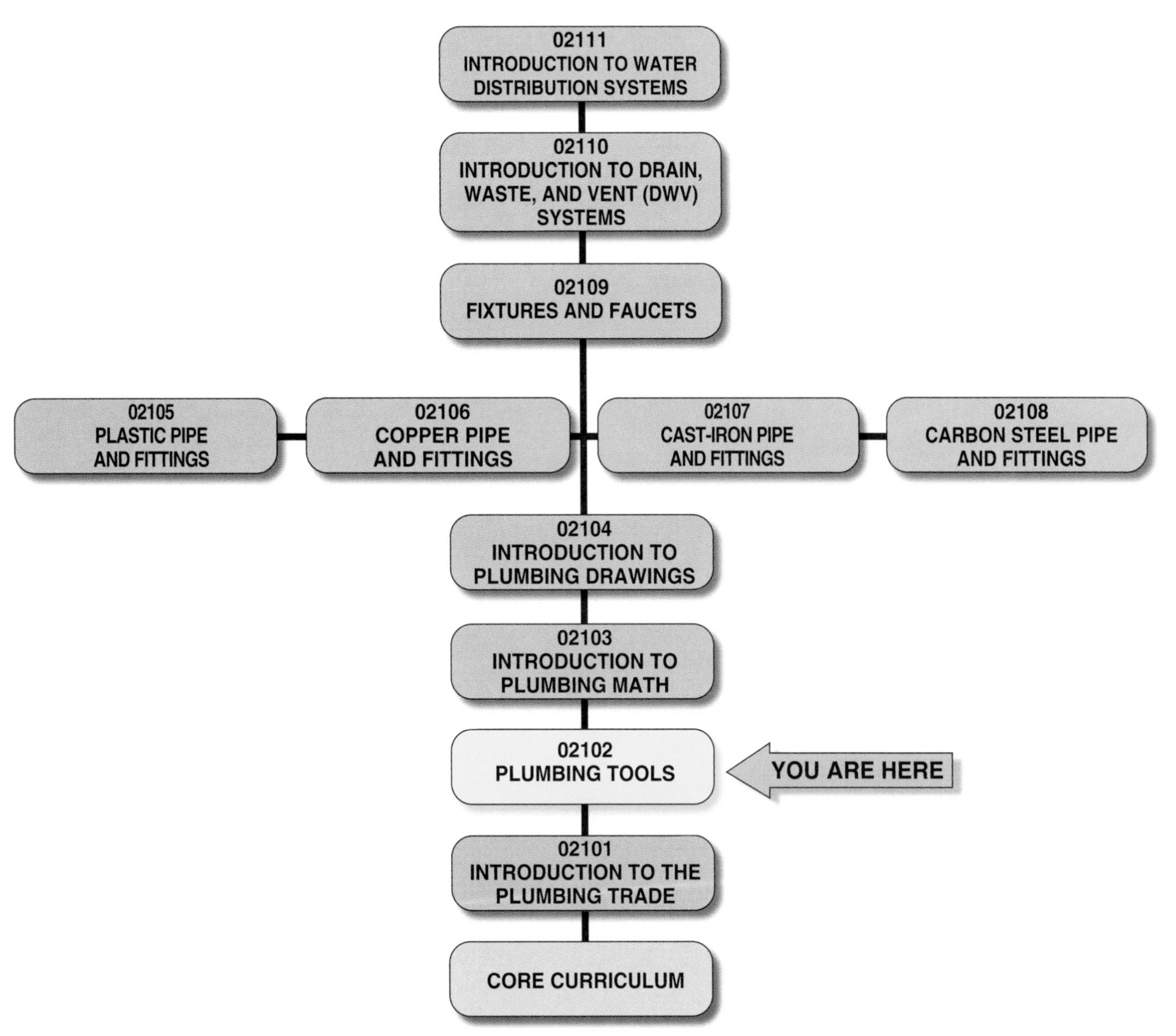

Copyright © 2000 National Center for Construction Education and Research, Gainesville, FL 32614-1104. All rights reserved. No part of this work may be reproduced in any form or by any means, including photocopying, without written permission of the publisher.

MODULE 02102 CONTENTS

Figures

Table

MODULE 02102

Plumbing Tools

Objectives

When you finish this task module, you will be able to:

1. Identify the basic hand and power tools used in the plumbing trade.
2. Demonstrate the proper maintenance procedures to be used for hand and power tools.
3. Explain safety as it applies to plumbing tools.

Prerequisites

Before you begin this module, it is recommended that you successfully complete task modules: Core Curriculum; Plumbing Level One, Module 02101.

Materials Required for Trainees

1. This module
2. Appropriate personal protective equipment
3. Sharpened pencil and paper

1.0.0 ◆ INTRODUCTION

Plumbing, like any other skilled trade, has its own tools. This module introduces you to the hand and power tools that are specific to plumbing, as well as to many other tools that are used by all construction trades. Most important, this module teaches tool safety and maintenance.

2.0.0 ◆ CARE AND USE OF TOOLS

Your tools are your livelihood. As a beginning plumber, you are starting to build a toolbox that is filled with the instruments of your trade. Give your tools the respect they deserve.

Tools cost money. Quality tools cost more money than bargain-bin tools. Buy the best quality you can afford and treat your tools well. Treat the tools your company supplies as well as you treat your personal tools. Just as you would never abuse your own drill motor, don't abuse the one the company lends you for your daily use.

3.0.0 ◆ SAFETY

Safe work habits are important to all workers. Plumbers must learn to use their tools safely. Jobsite safety protects everyone on the job from injury and possibly death. Developing safe work habits is really developing a safe attitude. You may see signs posted on job sites that urge workers to "Think Safety." These signs are posted for a reason: safety benefits everyone.

To list every possible safety precaution in one place would create a book so large that no one would bother to read it. Let's just start with some simple points to remember about safety:

- Know what tools are available and what work each one is designed to perform.
- Select the proper tool for the job.
- Use each tool as it was designed to be used.
- Make certain the tool is in good working condition. This includes:
  - Checking to see that cutting edges are sharp
  - Cleaning and lubricating the tools
  - Checking to see that handles are secure

Plumber Apprentice Toolbox

"A craftsperson is only as good as his or her tools." This adage is as true for plumbing as it is for all crafts. Tools allow you to perform your job. All your knowledge and skill mean little if you can't get the job done because you don't have the right tools.

Selecting and maintaining your tools is your responsibility. You must know how to choose high-quality, durable tools, and how to care for them. Preventive maintenance prolongs the life of any tool. This saves you money over time. Cheap tools are easily broken and have to be replaced. It is often a better investment to get the best tools you can afford now, rather than hope to get by with a poor-quality tool.

Plumbers follow the same safety precautions with tools as other craftworkers do. Always use the appropriate personal protective equipment for the task you perform. Never skip safety in favor of speeding up your work. An accident will cost you more time than the few minutes it takes to prepare to do a job the right way.

As with all the tasks you are performing as an apprentice, plumbing tasks may require supervision and direction from more experienced workers. Use that as an opportunity to learn more about your craft. Working with and watching accomplished plumbers will strengthen your own skills. Learning is a daily activity.

An apprentice plumber should have the following basic tools in his or her toolbox (see *Figure 1*):

Figure 1 ◆ Basic tools for a plumber's toolbox.

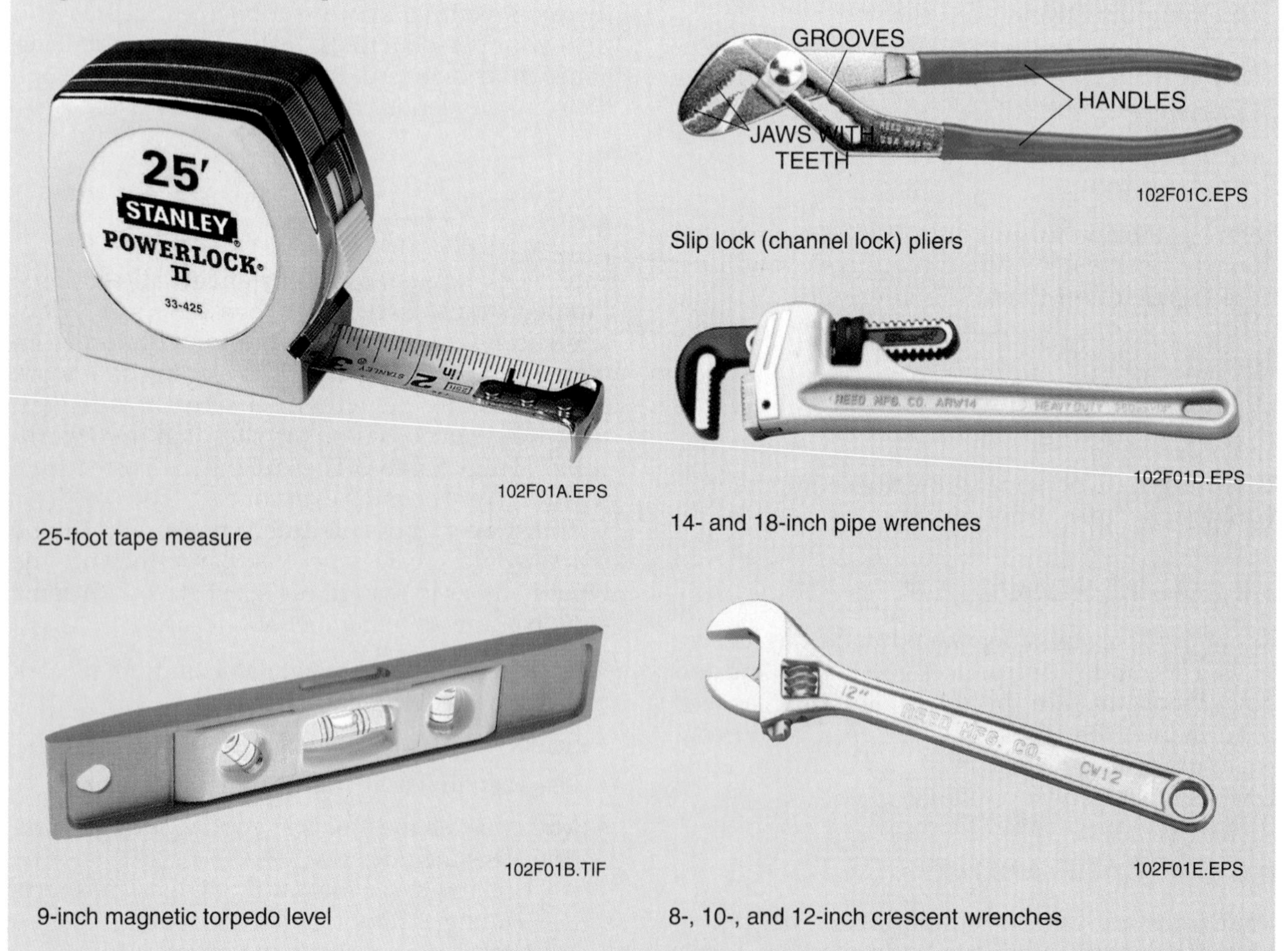

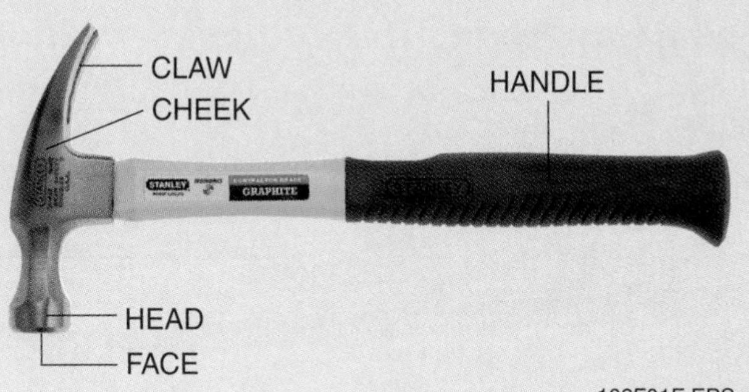

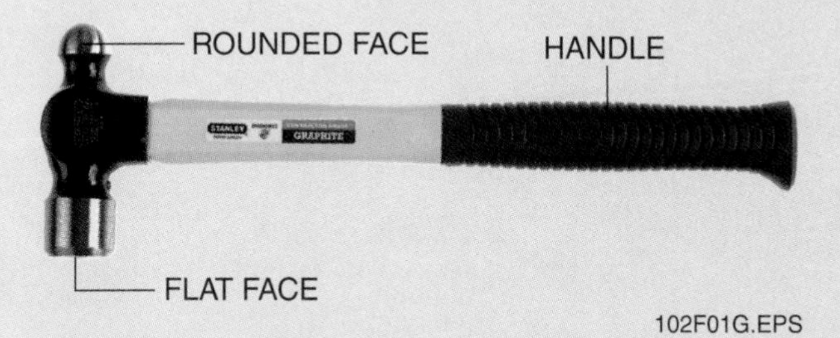

Hammers (claw and 24-ounce ball peen)

Torch kit (with striker) and tank

Plumber Apprentice Toolbox (continued)

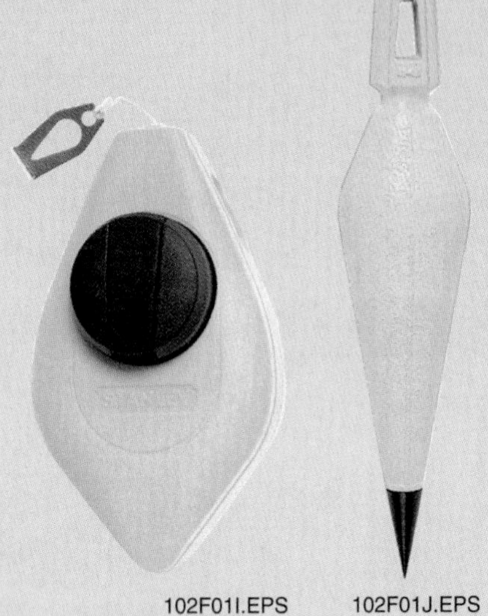

102F01I.EPS 102F01J.EPS

100-foot chalk box 8-ounce plumb bob

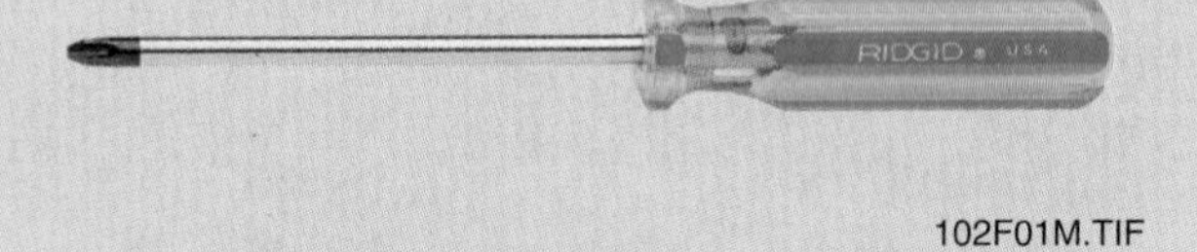

102F01M.TIF

#2 Phillips head screwdriver

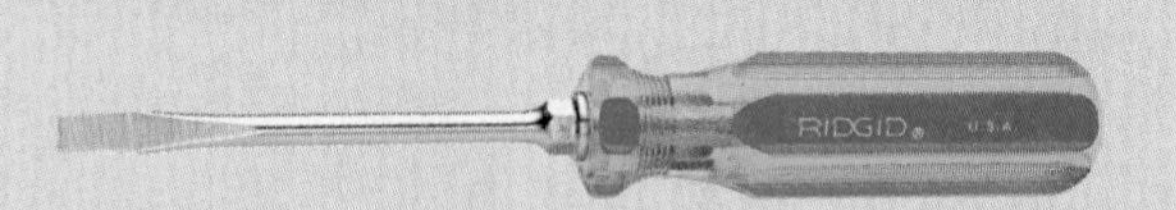

102F01N.TIF

6- and 8-inch slot screwdrivers

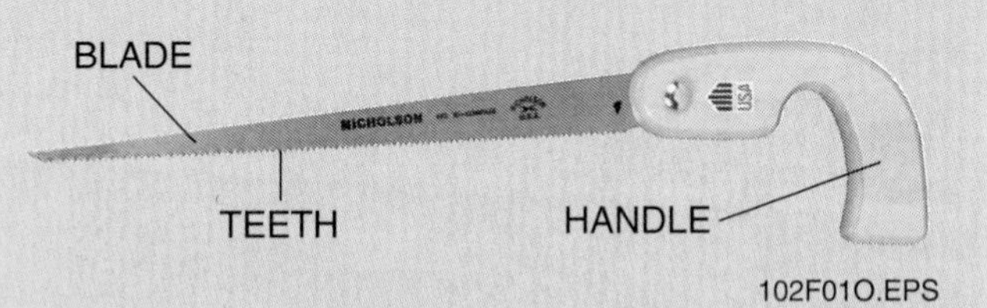

102F01O.EPS

Keyhole saw

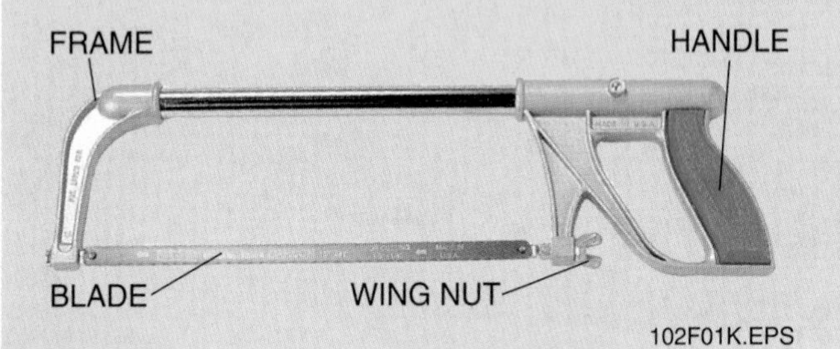

102F01K.EPS

Hacksaw

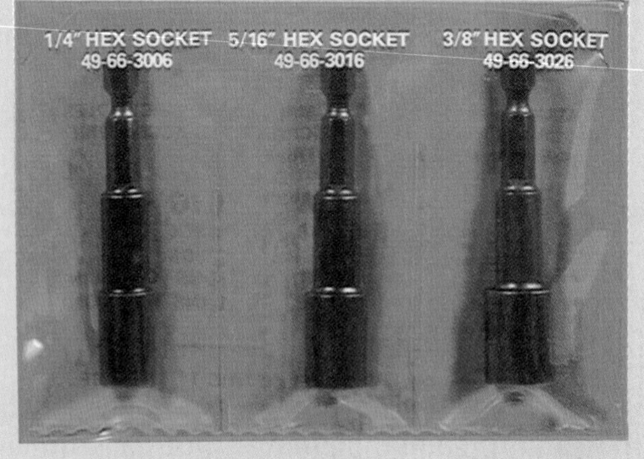

102F01L.TIF

$\frac{3}{16}$-inch to $1\frac{1}{8}$-inch tube cutter
$\frac{5}{8}$-inch to $2\frac{1}{8}$-inch tube cutter

102F01P.EPS

$\frac{1}{4}$-inch and $\frac{5}{16}$-inch magnetic drivers

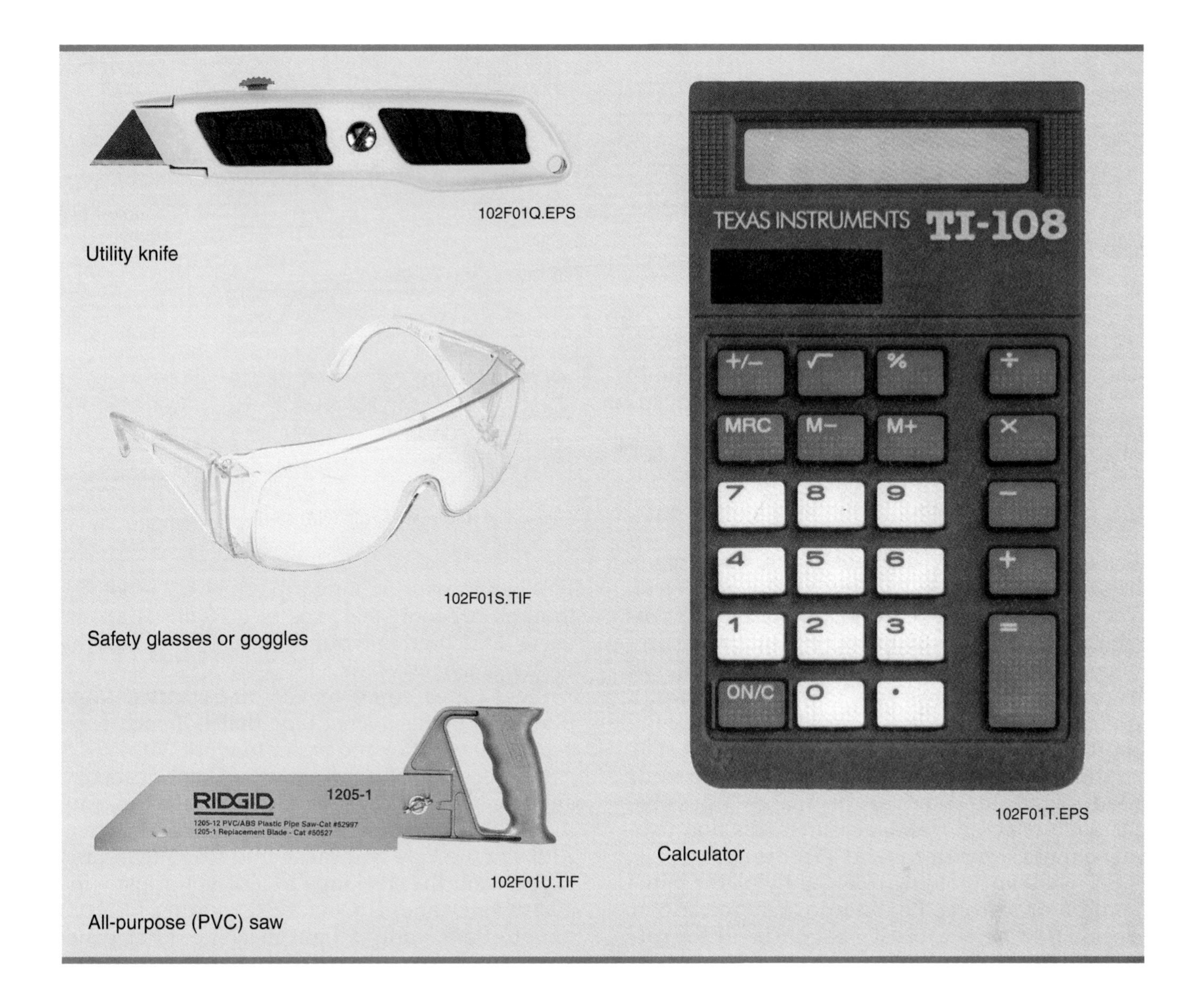

Utility knife

Safety glasses or goggles

All-purpose (PVC) saw

Calculator

- Check electrical tools periodically for damage to power cords. Also test three-prong electrical plugs with a meter to make sure that the safety ground is properly connected. Three-prong electrical plugs are effective only if properly grounded. Any repairs should be made under supervision.
- Wear proper clothing. Avoid loose shirt sleeves, jewelry, long hair, and other dangling items that might get caught in the moving parts of power tools.
- Wear safety goggles. **Soldering** and welding also require the use of eye protection equipment.
- Do not use electrical tools in damp or wet locations, even if they are equipped with a safety ground. Never use electrical tools near flammable liquids, either—one spark is all it takes to ignite these flammables.
- Keep the work area clean. Short lengths of pipe and other debris scattered around can cause you or someone else to fall.
- Concentrate on your work. Avoid distracting fellow workers. Giving your undivided attention to what you're doing reduces the chances of careless mistakes.

Safety can be expressed in two words: *common sense.* As you work, use your common sense. Your life may depend on it.

4.0.0 ◆ MEASURING AND LAYOUT TOOLS

Measuring tools give you information about the length, width, height, and **diameter** of objects.

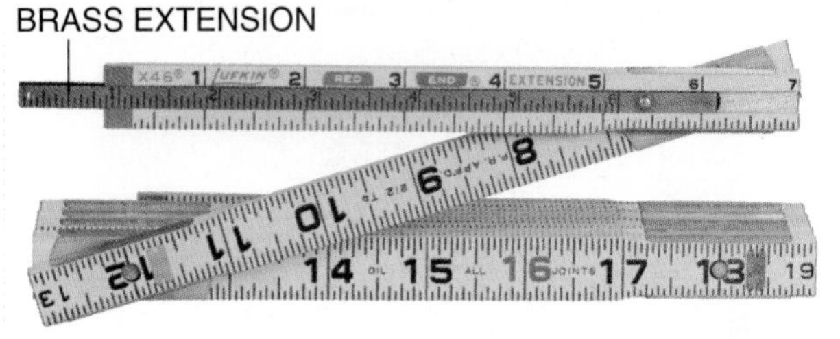

102F02.EPS

Figure 2 ◆ Folding rule.

They also tell you if a surface is level or plumb. Layout tools are used to make accurate lines, circles, or curves. This section discusses the common measuring and layout tools you are likely to use.

4.1.0 Folding Rule

Folding rules, shown in *Figure 2,* are usually made of wood and come in lengths of 6 feet and 8 feet. They are often equipped with a 6-inch brass extension that can be used to make inside measurements between two walls or to measure the depth of an opening. The brass extension allows a plumber to measure in tight places or to measure short distances beyond the end of the folding rule.

You can extend the folding rule above your head to measure heights. This makes it possible for one person to make measurements that usually require two persons and a ladder.

To maintain a folding rule, lightly oil the joints with oil or silicone. Be careful when opening or closing the rule, because it is easy to break the rule at the joints. Dropping the folding rule may loosen the joints enough to give inaccurate measurements. Small inaccuracies in measuring can add up to a large error over the span of several feet.

4.2.0 Plumber's Rule

The plumber's rule (see *Figure 3*) is a variation on the wooden folding rule. It has measurements on one side and a 45-degree scale on the other. The scale is used to measure **offsets** when a plumber must install a pipe or vent around an obstacle.

4.3.0 Steel Tape Measure

Plumbers carry steel tape measures (see *Figure 4*) that retract into a case at the push of a button. Retractable steel tape measures are available in lengths up to 30 feet. Steel tapes are popular because they are easy to carry. They can clip onto a belt or fit into special leather holders.

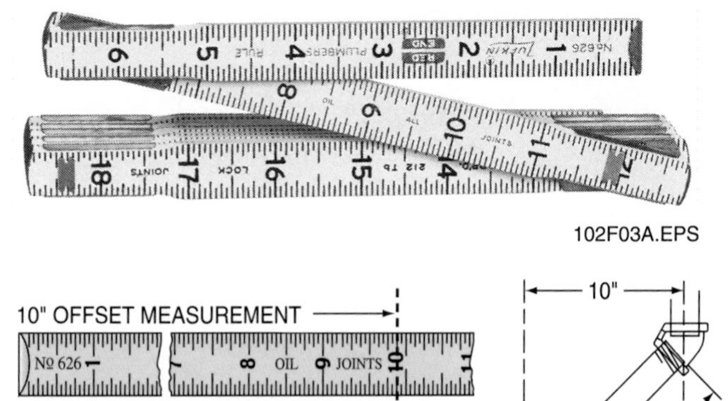

Figure 3 ◆ Plumber's rule with 45-degree scale.

When measuring long pipe runs, a 25-foot, 50-foot, or 100-foot steel tape (see *Figure 5*) can be used. This tape is wound back into its case by means of a hand crank.

Steel tapes come in various widths. Most plumbers prefer a steel tape that is 1 inch wide because it is stiffer and easier to work with.

All steel tape measures must be kept clean, dry, and free from kinks. Dirt and sand will wear away the markings. Water will cause the tape to rust. Kinks in the tape will cause it to stop rewinding. If you want the steel tape to last, you must wipe away water and dirt before rewinding it. From time to time, apply a light machine oil or penetrating oil to keep the tape lubricated and operating smoothly.

4.4.0 Square

Squares are used for marking, testing, and measuring. The type of square to be used depends on the job and the plumber's preference. Plumbers most often use the speed square, the combination square, and the framing square.

4.4.1 Speed Square

The speed square, also known as a rafter angle or magic square, is used for marking and measuring. It has a 6-inch to 12-inch blade (see *Figure 6*) and is made of cast aluminum. This square is a combination of a protractor, try miter, and framing square. It is marked with degree gradations for fast, easy layout. The square is small, making it easy to store and carry on the job site.

102F04A.EPS

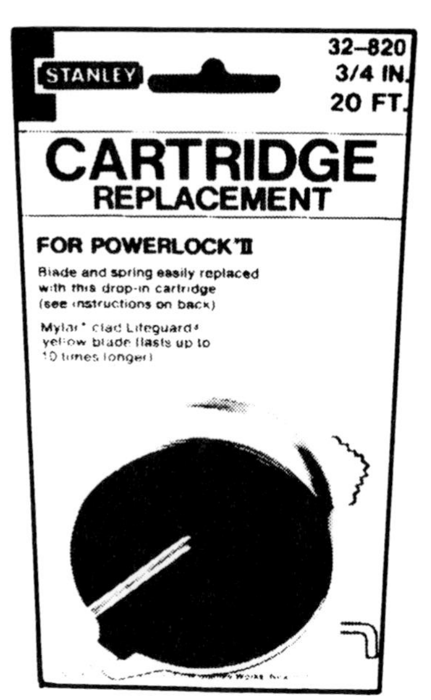

102F04B.TIF

Figure 4 ◆ Retractable steel tape measure and replacement tape.

102F05.EPS

Figure 5 ◆ Steel tape measure.

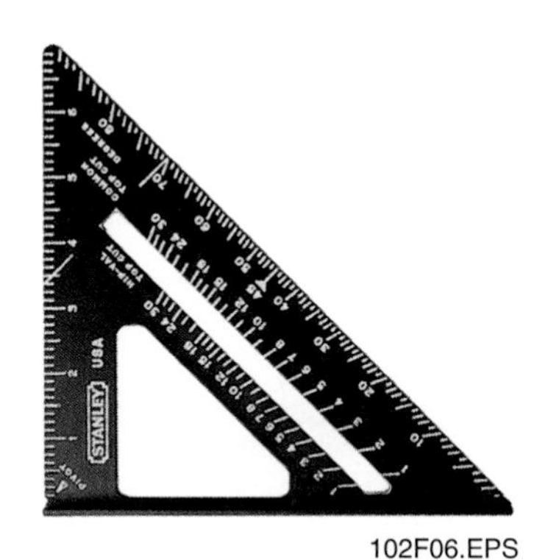

102F06.EPS

Figure 6 ◆ Speed square.

4.4.2 Combination Square

The combination square (see *Figure 7*) has a 12-inch blade that moves through a head. The head contains a 45- and 90-degree angle measure. Some squares also contain a small spirit level and a carbide **scribe** for marking metal.

4.4.3 Framing Square

The framing square (see *Figure 8*) is used for marking, testing squareness, and measuring. It has a 24-inch blade and a 16-inch tongue.

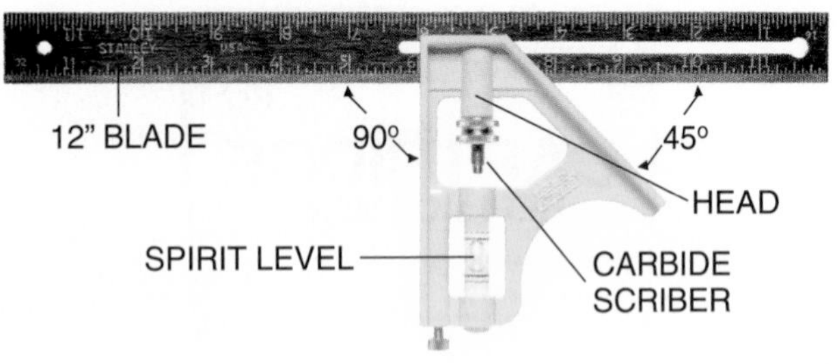

Figure 7 ◆ Combination square.

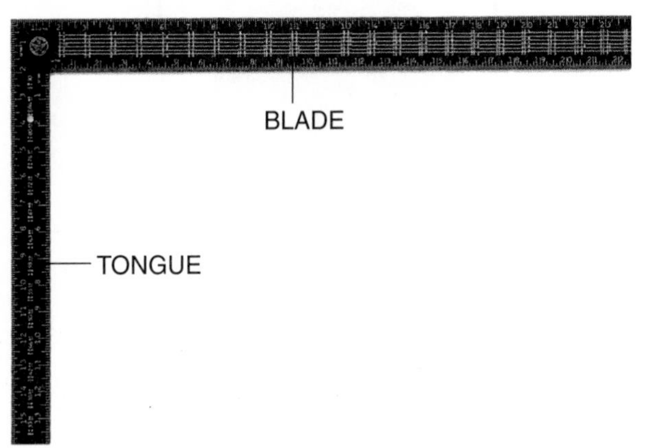

Figure 8 ◆ Framing square.

4.4.4 Care of Squares

Be careful not to drop or strike a square hard enough to change the angle between the blade and the head or tongue. Keep squares dry to prevent them from rusting. A periodic coat of light machine oil or penetrating oil will make them last longer.

Review Questions

Section 4.0.0

1. To measure heights, a plumber will often use a _____.
   a. square
   b. folding rule
   c. chalk box
   d. tape measure

2. A plumber's rule has ____ markings to measure offsets.
   a. 60-degree
   b. 90-degree
   c. 45-degree
   d. 30-degree

3. A _____ has a 45- and 90-degree angle measure and often a small spirit level and a scribe for measuring and marking.
   a. combination square
   b. builder's level
   c. framing square
   d. speed square

4. Hand-wound steel tape comes in lengths up to _____.
   a. 25 feet
   b. 50 feet
   c. 75 feet
   d. 100 feet

5. A _____ square includes a try miter, protractor, and framing square in one tool.
   a. framing
   b. carpenter's
   c. plumber's
   d. speed

5.0.0 ◆ LEVELING TOOLS

The term *level* describes the straightness of horizontal members. The term *plumb* describes the straightness of vertical members. Level and plumb meet at 90-degree angles. Many tools have been designed to measure level and plumb conditions. They range from simple spirit levels to very sophisticated electronic and laser instruments. This section discusses the most common leveling tools a plumber uses.

5.1.0 Spirit Level

The most common tool used to check both level and plumb is the spirit level, shown in *Figure 9*. Spirit levels are made of wood, aluminum, or a lightweight alloy such as magnesium. Plumbers generally buy aluminum or magnesium spirit levels because of their resistance to moisture.

DID YOU KNOW?
The word *spirit* refers to alcohol—the fluid used in the level.

Figure 9 ◆ Spirit level.

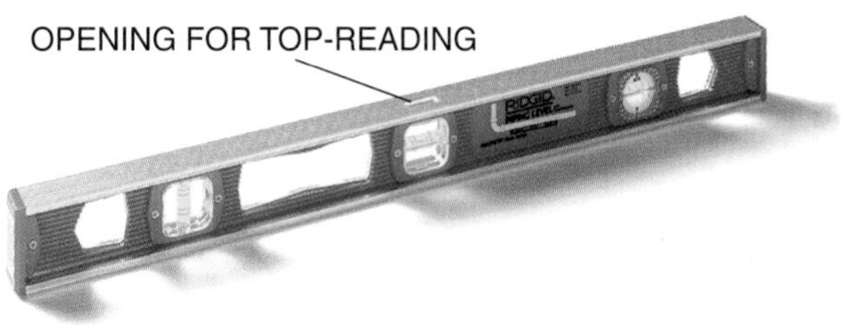

Figure 10 ◆ Plumber's level.

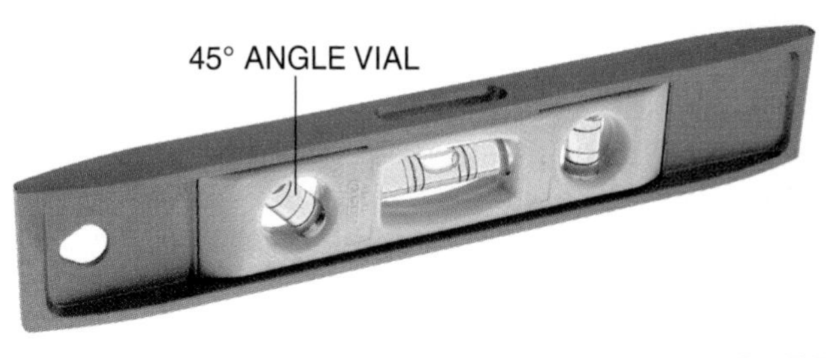

Figure 11 ◆ Torpedo level.

Spirit levels come in a variety of sizes. The 24-inch spirit level is considered standard, but these levels also come in 28-inch, 48-inch, 72-inch, and 78-inch sizes.

Spirit levels generally contain three vials. Inside each vial is a specific amount of alcohol and an air bubble. The outside of each vial is marked with lines. When the air bubble centers between the lines, the level is either level (on a horizontal member) or plumb (on a vertical member).

Vials are replaceable. Some, as on the plumber's level, are also adjustable. The center vial is used to verify level. The two end vials are used to verify plumb.

Spirit levels are accurate only up to their lengths. For accuracy over greater lengths, you must use a longer level.

Another way to increase the length of accuracy of a spirit level is with a **straightedge.** A straightedge is a length of wood or metal that does not bow or twist. Place the straightedge across the span to be checked and put the spirit level in the center of the straightedge.

5.2.0 Plumber's Level

The engineer's level, or plumber's level (see *Figure 10*), is a special level that can measure slope (angle) as well as plumb and level. One of the vials is adjustable to different angles. When you adjust the vial and hold the level at the desired angle, the bubble in the vial is centered.

5.3.0 Torpedo Level

The torpedo level (see *Figure 11*) is a small spirit level that is useful in confined areas or for leveling short runs of pipe. Most torpedo levels have a vial that indicates 45-degree angles. Plumbers often use a torpedo level with a magnetic base. This is handy because the magnetic base will stick to iron pipe.

5.4.0 Line Level

Plumbers use a line level (see *Figure 12*) by hooking it on a tightly stretched nylon string or line

Figure 12 ◆ Line level and nylon string.

Top-Reading

You will take most readings by looking directly at the spirit level as it lies on the material being leveled. But in some cases, you will need to place the spirit level on a pipe below you and you can only look down at the level. In this situation, you need a spirit level that allows you to top-read. This type of spirit level has an opening in the top that allows you to see the vial and bubble from above.

between two points. This level can transfer vertical dimensions (such as height or elevation) over long distances without using a leveling instrument called a transit (see next section). This is how plumbers make sure that pipe laid between two distant points is level along its entire distance.

5.5.0 Leveling Instruments

If distances are too great to be measured accurately by the levels already mentioned, or if more accuracy is required, the plumber must use a builder's (surveyor's) level or a transit and a stadia rod (see *Figure 13*).

Both tools work on the principle that a line of sight is perfectly straight. The telescope is mounted on a tripod and rotated in a complete circle without changing its horizontal position. The plumber sights through the telescope to a stadia rod.

The main difference between a builder's level and a transit is that the telescope of the transit moves up and down, but the telescope of the builder's level is fixed. This means that a builder's level can work only in the horizontal plane. The transit works in either the horizontal or the vertical plane.

You can use either the builder's level or the transit to measure differences in elevation between two distant points. Once the level or transit is set up on a tripod, the legs of the tripod must not be moved, because this will affect the accuracy of the measurement. A stadia rod, with graduated measurements in feet and tenths of a foot, is located at the point of interest and sighted through the telescope to determine vertical dimensions.

Another tool plumbers use is a cold beam laser (see *Figure 14*). This leveling device projects a thin beam of low-wattage light. Plumbers installing gravity flow pipelines can use this instrument for alignment by laying pipe along the beam of light.

The light emitted by a cold beam laser will not cause instant injury to humans, but it is possible that long-term exposure could cause injury to the eyes.

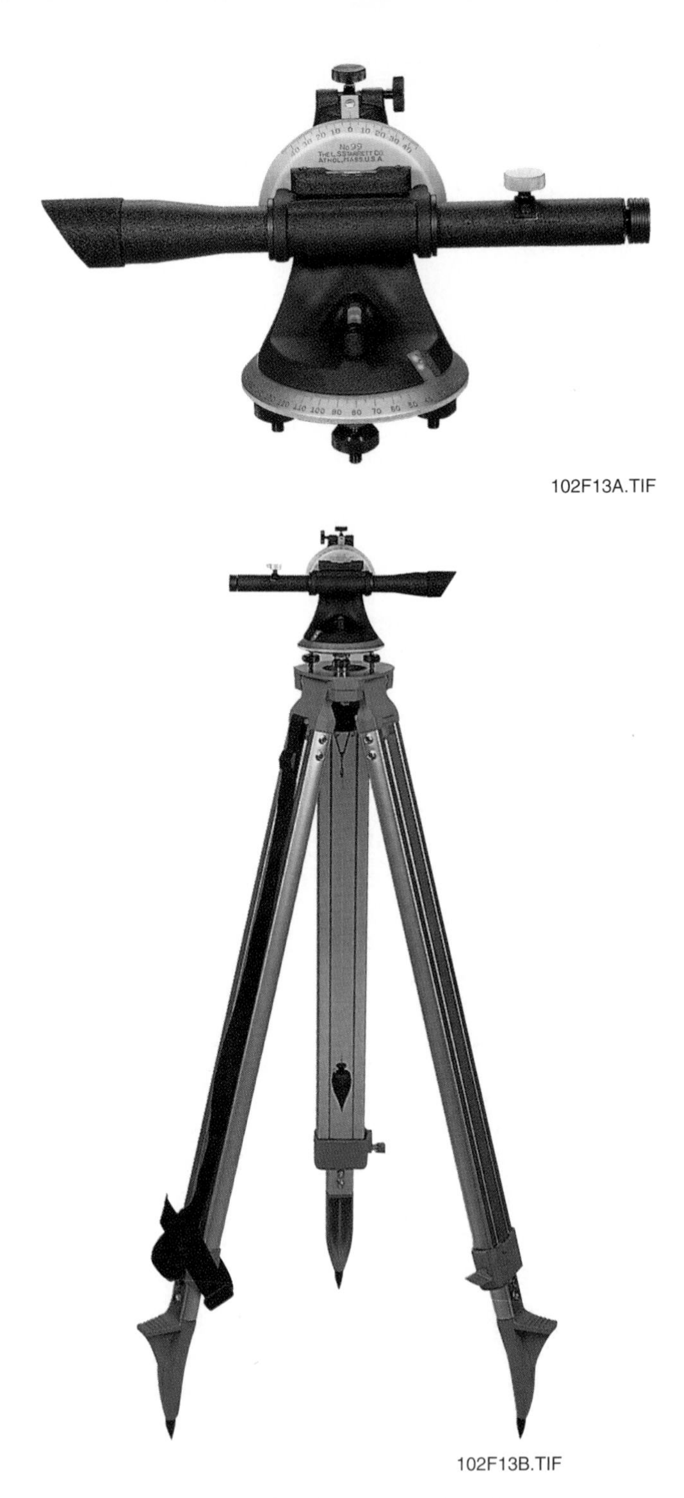
102F13A.TIF

102F13B.TIF

Figure 13 ◆ Builder's level, transit, and stadia rod.

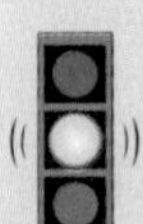

CAUTION

Never point a laser level at another person. This equipment should be used by qualified and trained employees only.

Levels and transits are precision instruments, and they are very expensive. Take care not to drop them or bump them against other materials, because loose or broken vials make these tools useless. You also need to keep levels and transits clean and dry so they will retain their true measure.

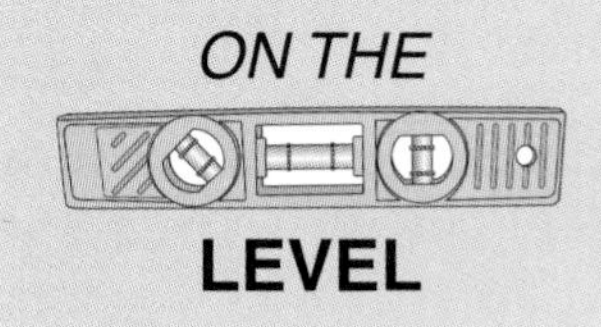

Calibrating a Builder's Level

A leveling vial is used on the builder's level. It is attached directly to the telescope. To make sure the telescope measures grade accurately, you must level it using the leveling vial. The vial is a spirit level design, but manufactured to be much more accurate. Four leveling screws are used to level the vial.

The plumber rotates the telescope until it lies over two opposing leveling screws. The plumber then adjusts the two screws in opposite directions until the bubble is centered in the leveling vial. Rotating the telescope 90 degrees, the plumber next adjusts the remaining two screws. This process should be repeated at least once to make sure that the level is accurately calibrated. Once the level is properly set up, the plumber is ready to measure grade.

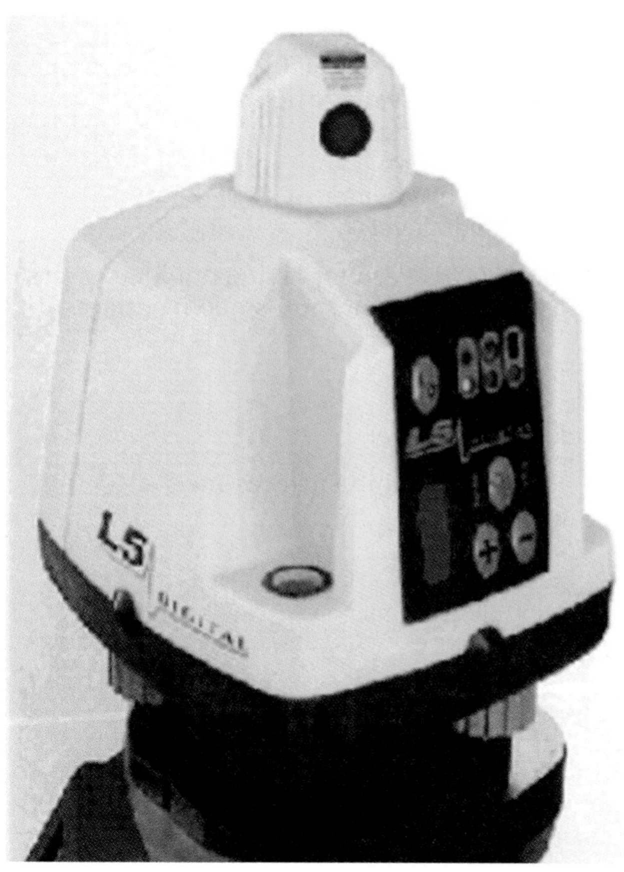

102F14.TIF

Figure 14 ◆ Cold beam laser.

102F15.EPS

Figure 15 ◆ Plumb bob.

5.6.0 Plumb Bob

A plumb bob is a precision measuring tool used to transfer locations from one floor to another floor below (see *Figure 15*). A plumber who must find the center point on a vertical run of pipe hangs the plumb bob from a point in the top plate of the framed wall. When the plumb bob hangs straight and free to the sole plate below, the point directly below the tip is exactly vertical with the point above. This means the pipe will pass through the framing members at precisely the same spot from floor to floor.

Be careful not to drop the plumb bob on its point. A bent or rounded point causes inaccurate readings. If the string is not properly attached at the center point of the plumb bob, you will get inaccurate readings as well because the plumb bob will hang at a slight angle to the string (see *Figure 16*).

5.7.0 Chalk Box

The chalk box (see *Figure 17*) is used for laying out long, straight lines of chalk on smooth or semi-smooth surfaces.

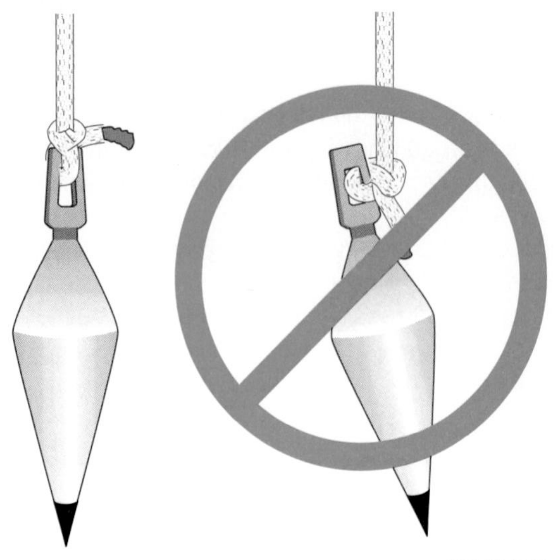

Figure 16 ◆ String attached to a plumb bob.

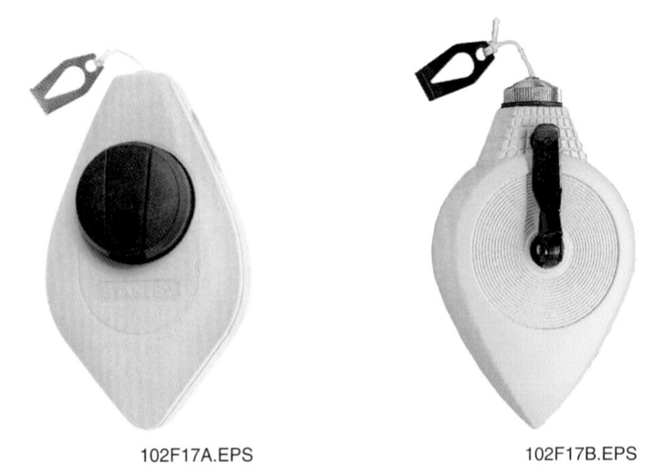

Figure 17 ◆ Chalk boxes.

You pull the chalk line taut between two points. Then you carefully stretch the line away from the surface and let it snap back to produce a straight line of chalk. Maintenance includes adding powdered chalk and replacing the string line when it is worn.

Because of the way most chalk boxes are designed, you can use them instead of a plumb bob. When you hang the chalk box overhead from the metal clip and allow the box to hang free and straight, the pointed end at the base of the chalk box indicates the exact point below on the vertical plane. You then transfer the mark to the floor to match the mark on the overhead framing member.

Review Questions

Section 5.0.0

1. A _____ is used for transferring locations from one floor to another floor below.
   a. speed square
   b. plumber's square
   c. plumb bob
   d. chalk box

2. A _____ has a movable telescope to allow measurements in both the horizontal and vertical plane.
   a. carpenter's level
   b. builder's level
   c. plumb bob
   d. transit

3. A leveling device called a _____ projects a thin beam of low-wattage light.
   a. cold beam laser
   b. straightedge
   c. warm beam laser
   d. torpedo level

4. The _____ is used for laying out long, straight lines on smooth or semi-smooth surfaces.
   a. chalk box
   b. plumb bob
   c. transit
   d. framing level

5. Level and plumb surfaces meet at _______ angles.
   a. 30-degree
   b. 45-degree
   c. 60-degree
   d. 90-degree

6.0.0 ◆ TOOTH-EDGED CUTTING TOOLS

Sometimes changes must be made to a commercial or residential building structure so that plumbing materials and fixtures can be installed. You use tooth-edged cutting tools to make these changes.

6.1.0 Hacksaw

The hacksaw (see *Figure 18*) is a multipurpose tool used to cut metals. Plumbers, however, should not use a hacksaw to cut anything that needs a square cut. A crooked cut can cause problems with threading and other operations that require accurate measurements. If a fixture must be installed

ON THE LEVEL

Tolerances

A tolerance is the allowable variation from a given dimension or quantity. Tolerances are expressed as "+/–" (plus or minus). If a pipe can be installed in a location "+/– ⅛ inch," you can vary the measurement by no more than one-eighth of an inch and still be within acceptable specifications.

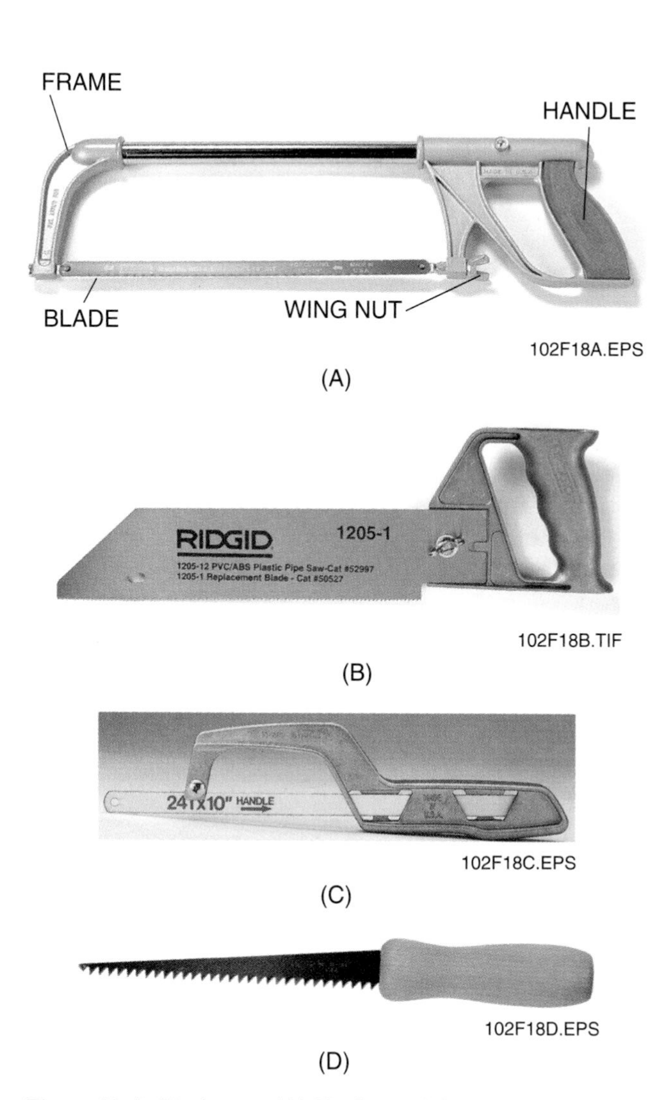

Figure 18 ◆ Hacksaws. (A) Hacksaw. (B) PVC saw. (C) Mini-hacksaw. (D) Sheetrock saw.

KERF

BLADE

TEETH ARE SET

102F19.EPS

Figure 19 ◆ Hacksaw blade kerf cut.

to meet close **tolerances** (little room for extra space), a crooked cut can make it nearly impossible to fit the fixture in the prepared space.

Hacksaw blades come in different styles for different uses. Blades differ in length, flexibility, the set (or bend) of their teeth, and their coarseness (number of teeth per inch), depending on the material they are made to cut through.

The blade used most often for cutting pipe is a flexible-back blade. Its teeth are set, which means each tooth is bent to alternating sides so the blade cuts a **kerf** that is slightly wider than the blade is thick, as shown in *Figure 19.* This prevents the blade from binding, and it reduces the friction that results from cutting. *Figure 20* illustrates the coarseness of a blade. Keep the blades sharp to make a good cut.

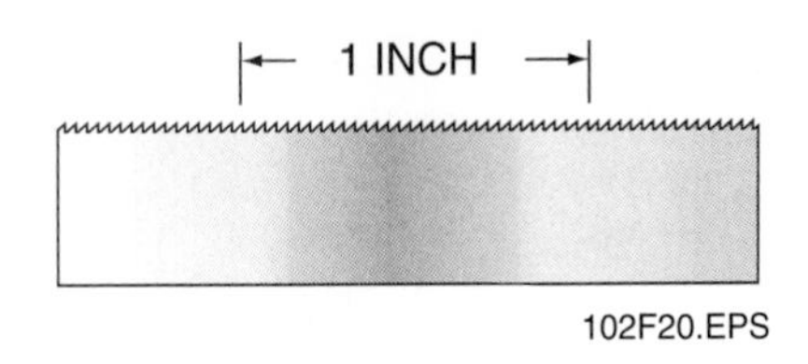

Figure 20 ◆ Blade coarseness.

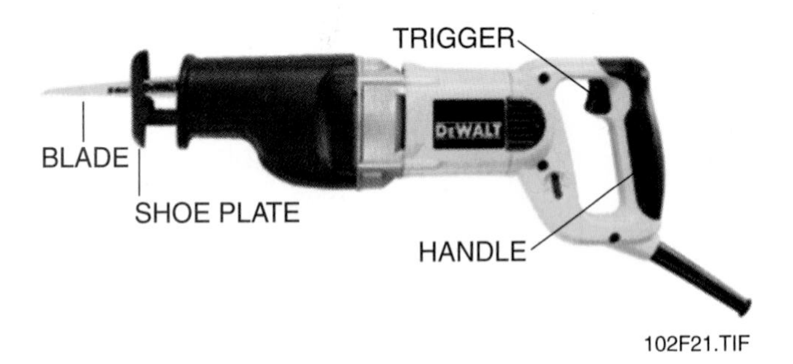

Figure 21 ◆ Reciprocating saw.

The more teeth per inch, the finer the cut. Generally, a blade with 32 teeth per inch is used for cutting pipe.

Always use hacksaw blades with the angle of the teeth pointing away from you. Always replace a hacksaw blade when it becomes dull. Worn or dull blades can cause accidents.

6.2.0 Reciprocating Saw

The reciprocating saw (see *Figure 21*) is an electric saw with various blades. A shaft pushes the blade forward and back. On the backstroke, it also lifts the blade to clear the cut and keep the teeth from being worn. Single-speed, two-speed, and variable-speed models are available. Strokes per minute range from 1,700 (the slow speed on two-speed models) to 2,400 (the highest speed on most variable-speed models).

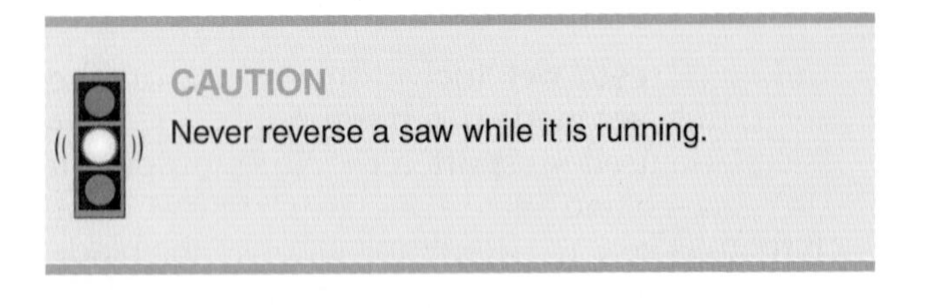

CAUTION

Never reverse a saw while it is running.

The reciprocating saw is used for brute power and is designed to cut hard-to-cut materials. It is used in practically every trade. Plumbers use it to cut cast iron, metal, plastic, and wood. It can saw through walls or ceilings and create openings for windows and plumbing lines. It is a basic necessity in any demolition work.

At the front of every reciprocating saw is a shoe plate. Never use a reciprocating saw without the shoe plate because, for a good cut, the shoe plate must rest against the work surface. Offset blade adapters can be used to move the blade to the top of the shoe plate for cutting flush or even with the material.

Always be sure to use the types of blades specified by the manufacturer for the kind of work you are doing with the reciprocating saw.

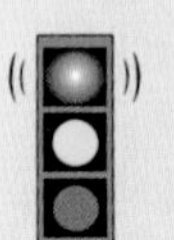

WARNING!

Always unplug any saw before you change the blades.

The reciprocating saw, like all electrical tools, must be grounded with an **electrical ground.** Inspect the body and the cord for damage before you operate the saw. Do not operate any electric tools when you are in contact with water or near flammable gases.

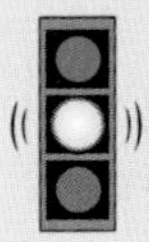

CAUTION

Do not remove the safe ground pin on the plug of the reciprocal saw.

Some manufacturers make a cordless reciprocating saw. This saw works the same as a regular reciprocating saw but uses battery power instead of electrical current. This means you do not have a cord dangling while you are working. In tight spaces, this could be an advantage. You must use all the same precautions when operating a cordless tool as you would with an electric tool. Be sure not to stand in water when using any power tool.

6.3.0 Portable Band Saw

The portable band saw, often called a portaband saw (see *Figure 22*), is used to cut both **ferrous** and **nonferrous** pipe, **bar stock, tube stock,** plastics, and materials with irregular shapes. This saw has a continuous blade, much like a hacksaw blade, that revolves around guides located at either end of the saw.

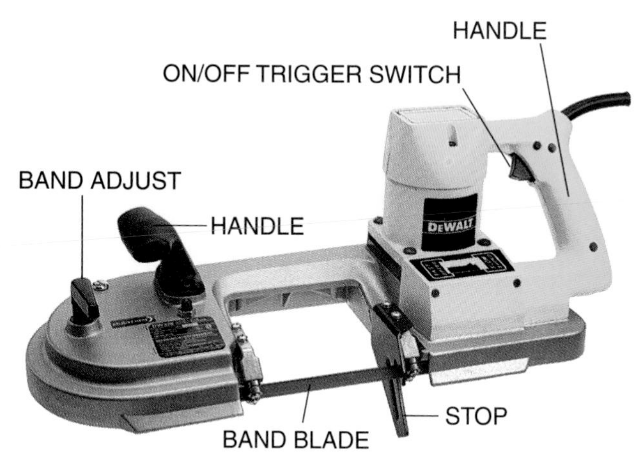

Figure 22 ◆ Portable band saw.

For best results, lubricate the blade to reduce friction and prolong the life of the saw.

6.4.0 Abrasive Saws

Plumbers also may use abrasive saws to cut pipe and other materials. These saws use a special wheel that can cut either metal or masonry. The most common types of abrasive saws are the chop saw and the demolition saw. The main difference between the two is that the demolition saw is not mounted on a base.

6.4.1 Demolition Saw

Demolition saws (see *Figure 23*) are either electric or gasoline powered. They can be used to cut through most materials on a construction site. This saw is a dangerous, unforgiving tool that requires the full attention and concentration of the operator. Be sure to use the proper personal protective equipment for your eyes, ears, and hands. See additional precautions in the *Saw Safety* section later in this module.

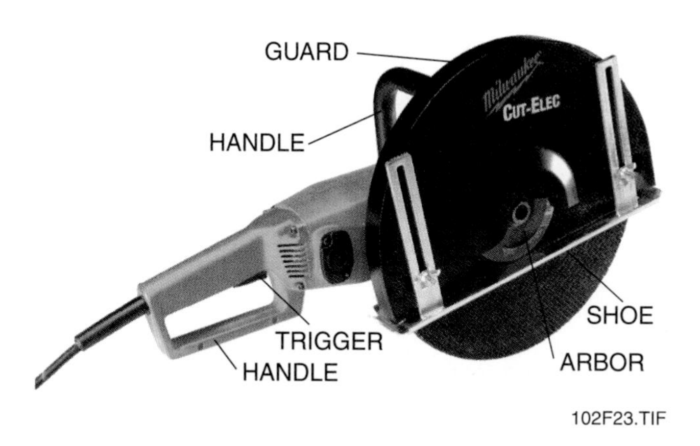

Figure 23 ◆ Demolition saw.

> **CAUTION**
>
> Because a demolition saw is handheld, there is a tendency to use it in positions that can easily compromise your balance and footing. Be very careful not to use the saw in any awkward positions.

6.4.2 Chop Saw

The chop saw (see *Figure 24*) uses a wheel to cut pipe, angle iron, channel, tubing, conduit, or other light-gauge materials. The most common wheel uses an abrasive metal blade, but blades made of other materials such as PVC or wood can also be used. The wheel and motor are located on a spring-loaded arm. A vise on the base of the chop saw holds the material securely and pivots to allow miter cuts.

Chop saws are sized according to the diameter of the largest abrasive wheel they accept. Two common sizes are 12 inch and 14 inch. Each abrasive wheel has a maximum safe speed. Never exceed that speed. Typically, the maximum safe speed is 5,000 revolutions per minute (rpm) for 12-inch wheels and 4,350 rpm for 14-inch wheels.

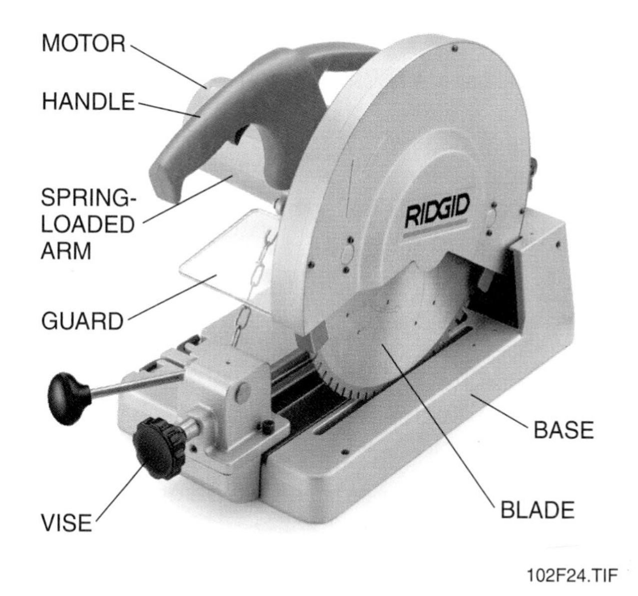

Figure 24 ◆ Chop saw.

6.4.3 Saw Safety

To avoid injuring yourself when you are using an abrasive saw, follow these guidelines:

- Always wear personal protective gear to protect your eyes and ears.
- Use abrasive wheels rated at higher rpms than the tool can produce. This way, you will never exceed the maximum safe speed for the abrasive wheel you are using. The blades are matched to a safety rating based on their rpms.
- Inspect the wheel before you use the saw. If you see damage, throw the wheel away.
- Before you start the tool, be sure the wheel is secured on the arbor (see *Figure 23*). Point the wheel of a demolition saw away from yourself and others before you start the tool.
- Always use two hands when operating a demolition saw.
- Always secure the materials you are cutting in a vise. The jaws of the vise hold the pipe, tube, or other material firmly and prevent it from turning while you cut. (Vises are discussed in a later section.)
- Before using a demolition saw, be sure the guard is in place and the adjustable shoe is secured (see *Figure 23*).
- Keep the tool clean.
- Inspect the saw each time you use it. Never operate a damaged saw (see *Figure 25*). Ask your supervisor if you have a question about the condition of a saw.

Review Questions

Section 6.0.0

1. Do not use a hacksaw when you need a _____ cut.
   a. fast
   b. square
   c. crooked
   d. flexible

2. A _____ has a continuous blade that revolves around guides located at either end of the saw.
   a. chop saw
   b. hacksaw
   c. reciprocating saw
   d. portaband saw

3. The more teeth per inch on a blade, _____ .
   a. the coarser the cut
   b. the finer the cut
   c. the more dangerous the saw is
   d. the longer the blade lasts

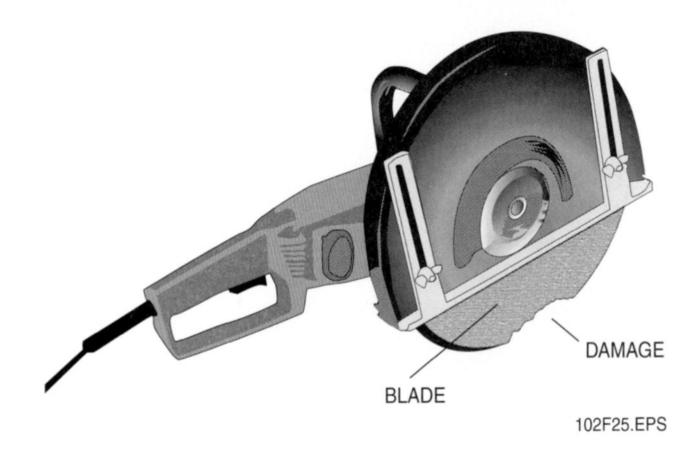

Figure 25 ◆ Damaged saw.

4. A _____ mounted on the base of a chop saw holds materials securely and pivots to allow miter cuts.
   a. vise
   b. grip
   c. spring-loaded arm
   d. guard

5. A _____ is a necessity for any demolition work because of its brute power.
   a. hacksaw
   b. reciprocating saw
   c. portable band saw
   d. chop saw

7.0.0 ◆ SMOOTH-EDGED CUTTING TOOLS

Smooth-edged cutting tools have a smooth (as opposed to toothed) cutting edge. Chisels and pipe cutters are the most common plumbing tools in this category.

7.1.0 Chisels

Plumbers use two kinds of chisels: wood chisels and cold chisels. Both chisels are made from heat-treated steel to increase the hardness of their cutting blades. By hitting the opposite end, or handle, of a chisel with a hammer, you force the chisel into and through the material you are working on.

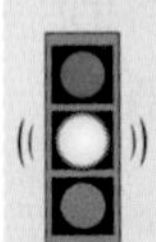

CAUTION

Always wear appropriate gear to protect your eyes when you are using chisels.

7.1.1 Wood Chisel

The wood chisel (see *Figure 26*) is used to make openings or notches in wooden structural material so that pipes can be installed. For the wood chisel to cut well, the blade needs to be beveled at a precise 25-degree angle (see *Figure 27*). Also, the cutting edge must be honed on an oilstone to produce a keen edge.

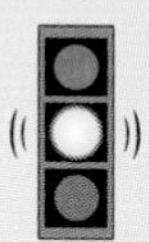

CAUTION

Rusty or dull cutting tools not only cut poorly but can cause serious accidents if the tool slips.

7.1.2 Cold Chisel

The cold chisel (see *Figure 28*) is used for cutting cold metal, trimming concrete, and enlarging holes. The blade of a cold chisel is beveled on both sides to a 60-degree angle, as shown in *Figure 29.*

Mushrooming, which results from hammering, is a common problem for cold chisels. Mushrooming can cause serious injuries to hands or eyes if the metal fragments break off. When the handle end of a tool mushrooms, grind off the mushroom at a slight bevel, as shown in *Figure 30.*

7.2.0 Pipe Cutters

Pipe cutters come in a number of sizes, shapes, and designs. Common sizes range from ⅛ inch to 4 inches. This section discusses three manual pipe cutters: the tube cutter, the pipe cutter, and the soil pipe cutter.

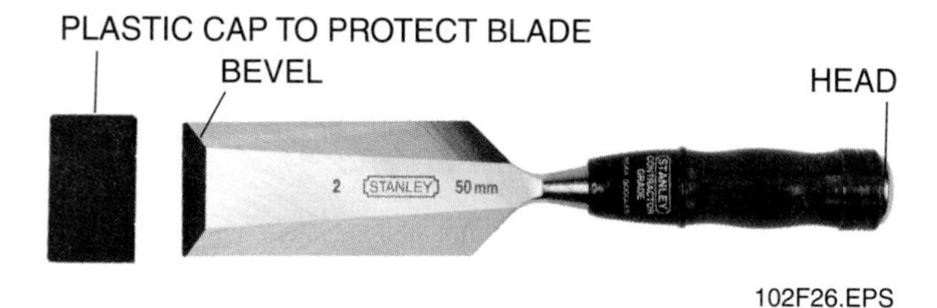

Figure 26 ◆ Wood chisel.

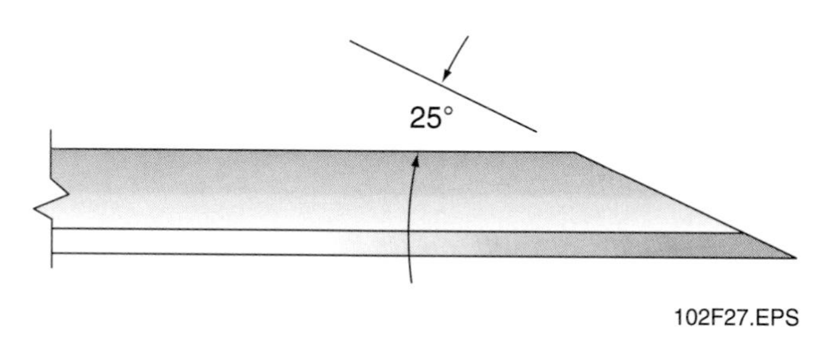

Figure 27 ◆ 25-degree angle cutting edge.

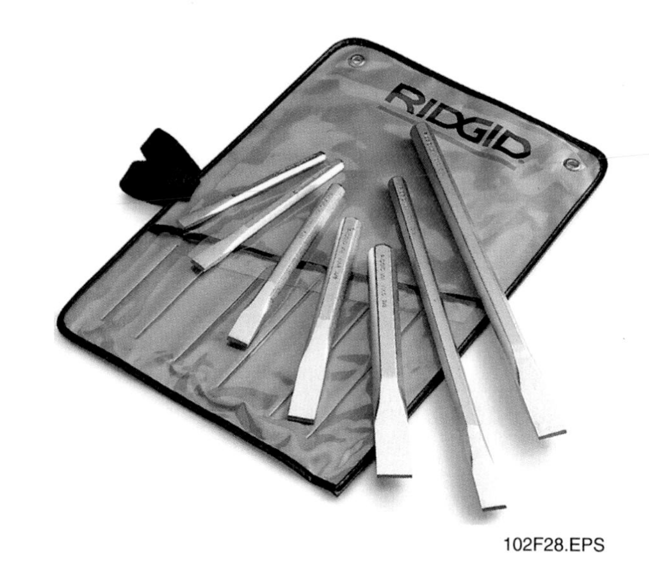

Figure 28 ◆ Cold chisel.

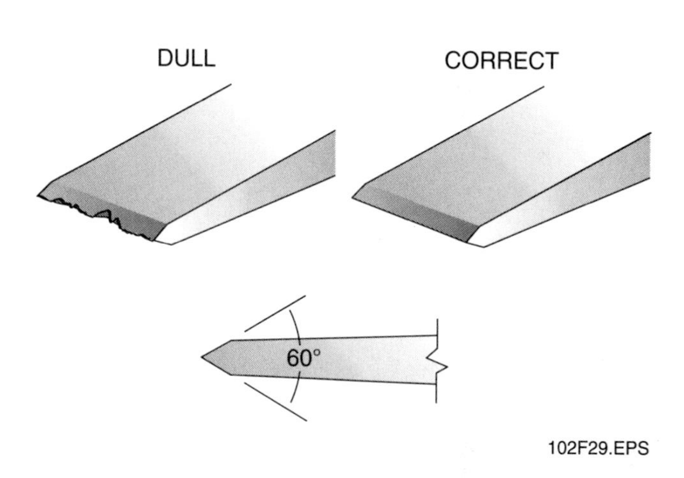

Figure 29 ◆ 60-degree angle cutting edge.

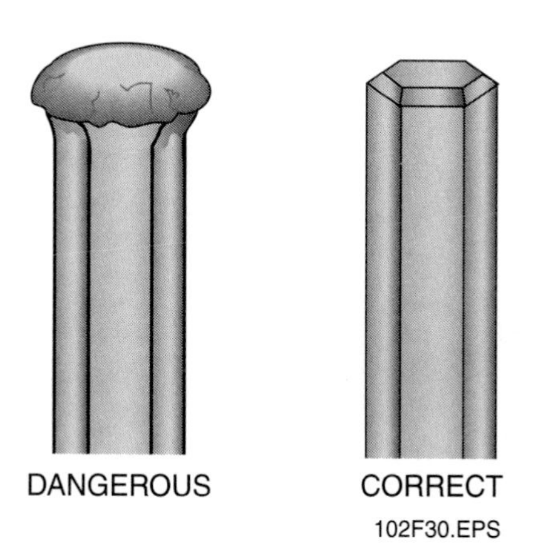

Figure 30 ◆ Repairing a mushroomed chisel.

7.2.1 Tube Cutter

The tube cutter (see *Figure 31*) is used for cutting small, soft materials such as copper, brass, and aluminum tubing or thin-wall conduit that is ⅛ inch to 4 inches in diameter. The tube cutter is made up of four movable parts: a cutter wheel, an adjusting screw, and at least two guiding wheels. A specialized *midget cutter* is used for cutting copper tubing, sizes ⅛ inch to ⅞ inch, when working in tight places. Quick adjust tube cutters have a sliding adjustment instead of an adjusting screw.

A plumber often uses a waxy crayon (keel crayon) to mark a cutting line on the surface of the tube. The plumber will set the cutting wheel to cut on this line.

7.2.2 Pipe Cutter

The pipe cutter (see *Figure 32*) is basically the same tool as the tube cutter. However, it is heavier, it often has more than one cutting wheel, and it is used to cut larger sizes of pipe.

7.2.3 Soil Pipe or Snap Cutter

Soil pipe or snap cutters are used to cut cast iron or clay (see *Figure 33*). The soil pipe cutter has a length of chain that wraps around the pipe. The user tightens the chain by ratcheting a handle, which pushes cutters into the pipe and creates a clean break. The snap cutter, which is more popular, cuts with a scissors action. Two long handles provide leverage to the cutting chain wrapped around the pipe. The chain is set with cutting rollers at each link. The plumber tightens the chain and turns it around the pipe to be cut. Because the chain tightens and cuts evenly, the pipe breaks cleanly with a snap.

The plumber often marks a line on the surface of the pipe with soapstone to guide the cut. Soapstone is a soft, talc-based rock.

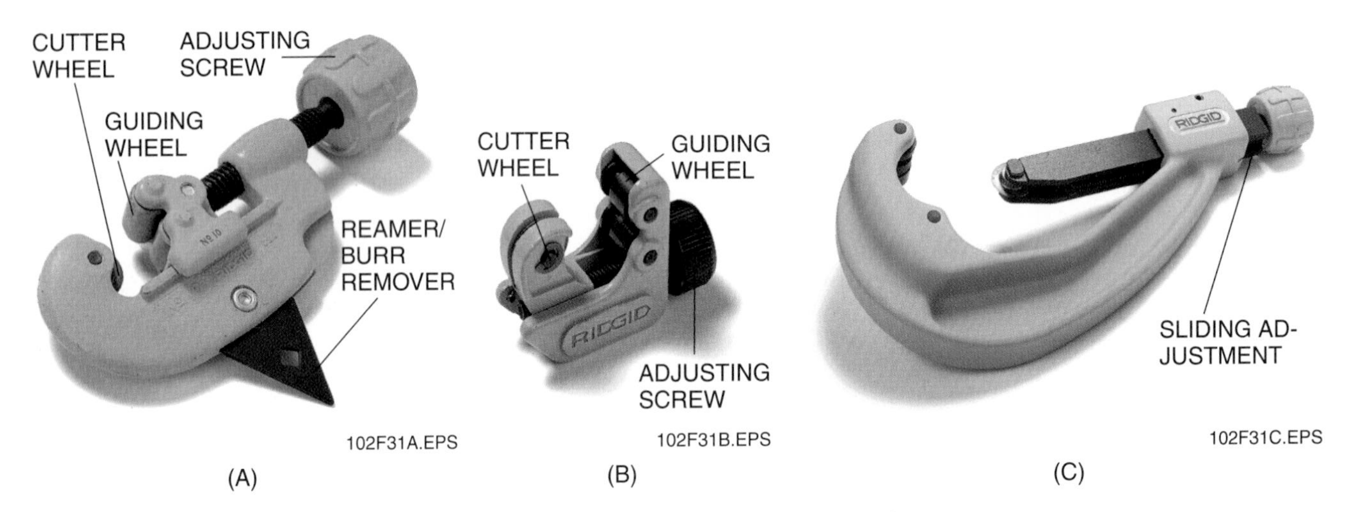

Figure 31 ◆ Tube cutters. (A) Tube cutter. (B) Midgit tube cutter. (C) Quick-adjust tube cutter.

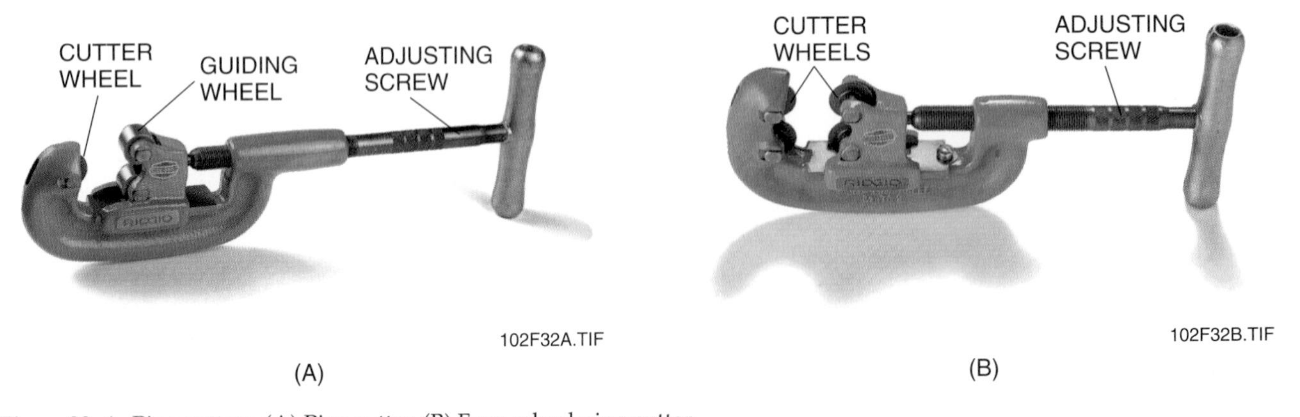

Figure 32 ◆ Pipe cutters. (A) Pipe cutter. (B) Four-wheel pipe cutter.

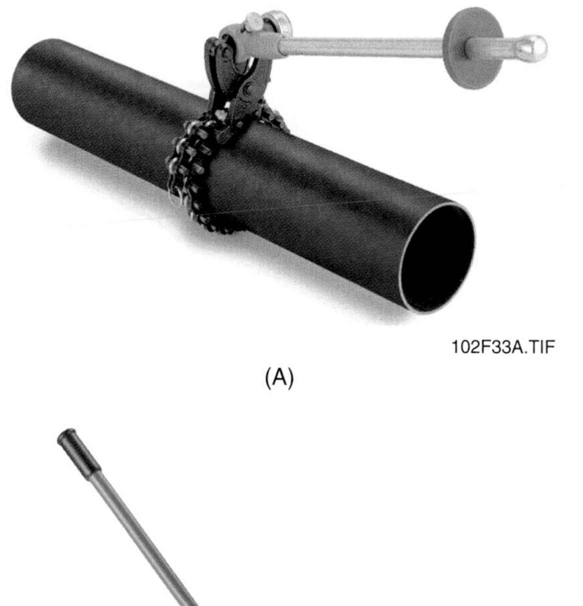

Figure 33 ◆ Soil pipe cutters. (A) Soil pipe rachet cutter. (B) Soil pipe snap cutter.

7.2.4 Care of Cutters

If the tube or pipe you are cutting looks as if it has been mashed after you cut it, you probably need to replace the cutting wheel or wheels. You also need to apply lubricating oil periodically to all movable parts on the pipe cutter so that it will operate smoothly.

Review Questions

Section 7.0.0

1. For the _____ chisel to cut well, the blade needs to be beveled at a precise 25-degree angle.
a. wood
b. mushroomed
c. cold
d. snap

2. Chisel blades are made from ____ to increase their hardness.
a. cast iron
b. heat-treated steel
c. wrought iron
d. copper steel

3. _____ is a common result of hammering a cold chisel.
a. Smashing
b. Mushrooming
c. Spreading
d. Splaying

4. As each link of chain in a _____ cutter gets tight and pushes the cutting rollers in, the pipe breaks cleanly.
a. tube
b. ratchet
c. soil pipe
d. midget

5. A _____ is used for cutting small-diameter, soft materials such as copper, brass, and aluminum tubing or thin-wall conduit.
a. pipe cutter
b. snap cutter
c. reamer
d. tube cutter

8.0.0 ◆ DRILLING AND BORING TOOLS

Electric drilling and boring tools allow the plumber to make openings in the building structure for installation of pipe. These tools have changed the plumbing trade. They have greatly shortened the time it takes to install pipe.

8.1.0 Portable Electric Drill

The portable electric drill, sometimes called a *pistol drill* (see *Figure 34*), is used to drill holes in structural materials so that pipe can be run and fixtures can be installed. If a plumber needs to drill between joists or in tight corners where a pistol drill will not fit, a right angle portable electric drill (see *Figure 35*) can come in handy. The offset design of this drill allows the plumber to drill holes where space is limited. Models that have extended handles give the plumber a better grip to guide the drill when boring through the material. Milwaukee's Hole Hawg® is a type of offset drill that is legendary for its torque. The rpm of the Hole Hawg® is geared down to allow the use of larger, self-feeding bits. This drill is used in tight places.

Two other kinds of drills are the hammer drill and the rotary hammer drill (see *Figure 36*). The rotary hammer drill combines the force of a hammer with the boring capability of a drill. The hammering action penetrates dense material like concrete or masonry, and the rotating bit removes the chips. A variety of drill bits are available for different types of jobs. The rotary hammer drill is used for larger jobs; the regular hammer drill is used for drilling holes to set small anchors that secure piping to a structure.

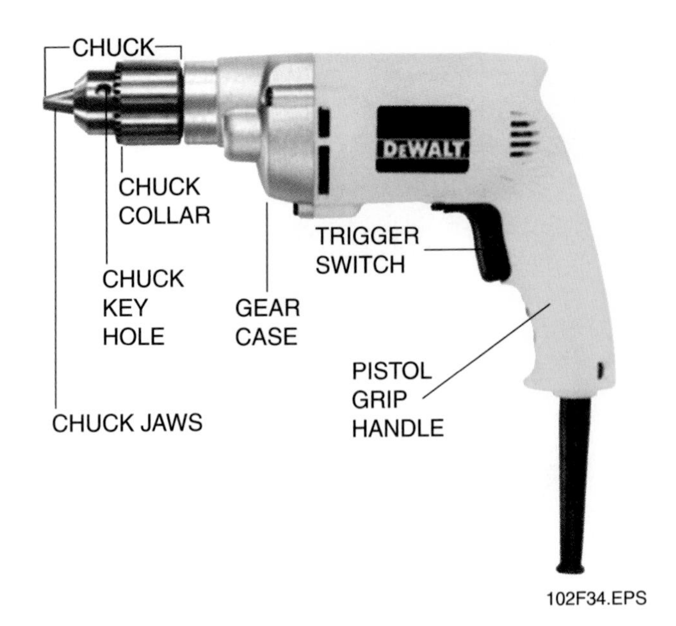

Figure 34 ◆ Portable electric drill (pistol drill).

8.1.1 Uses of Pistol Drills

Pistol drills are perhaps the most widely used and most versatile of all power tools. Their primary function is drilling holes in steel, wood, concrete, plastic, and virtually any other material. The great number of accessories available for pistol drills also turns them into grinders, sanders, polishers, sheet-metal cutters, and, with the addition of wire or flapper wheels, metal finishers. No other power tool can do so much.

Many styles of pistol drills are available, made by a variety of manufacturers. Some are good, industrial-quality tools, while others are better suited to the homeowner who occasionally needs to drill a hole. Of course, price often reflects quality. An industrial model rated for continuous use can be quite an investment, either for individual plumbers or for the company that employs them.

Even though there are many different styles of pistol drills, they have much in common. Each has a specified capacity. Common capacities are ¼ inch, ⅜ inch, and ½ inch. This capacity refers to the maximum opening of the chuck and the size of the drill bit that will fit into the chuck.

ON THE

LEVEL

Torque

Torque is a measure of the force exerted in a rotating motion. Torque wrenches measure this force or resistance to turning. You need torque wrenches when you are installing fasteners that must be tightened in sequence without distorting the work piece. You will use a torque wrench only when a torque setting is specified for a particular bolt.

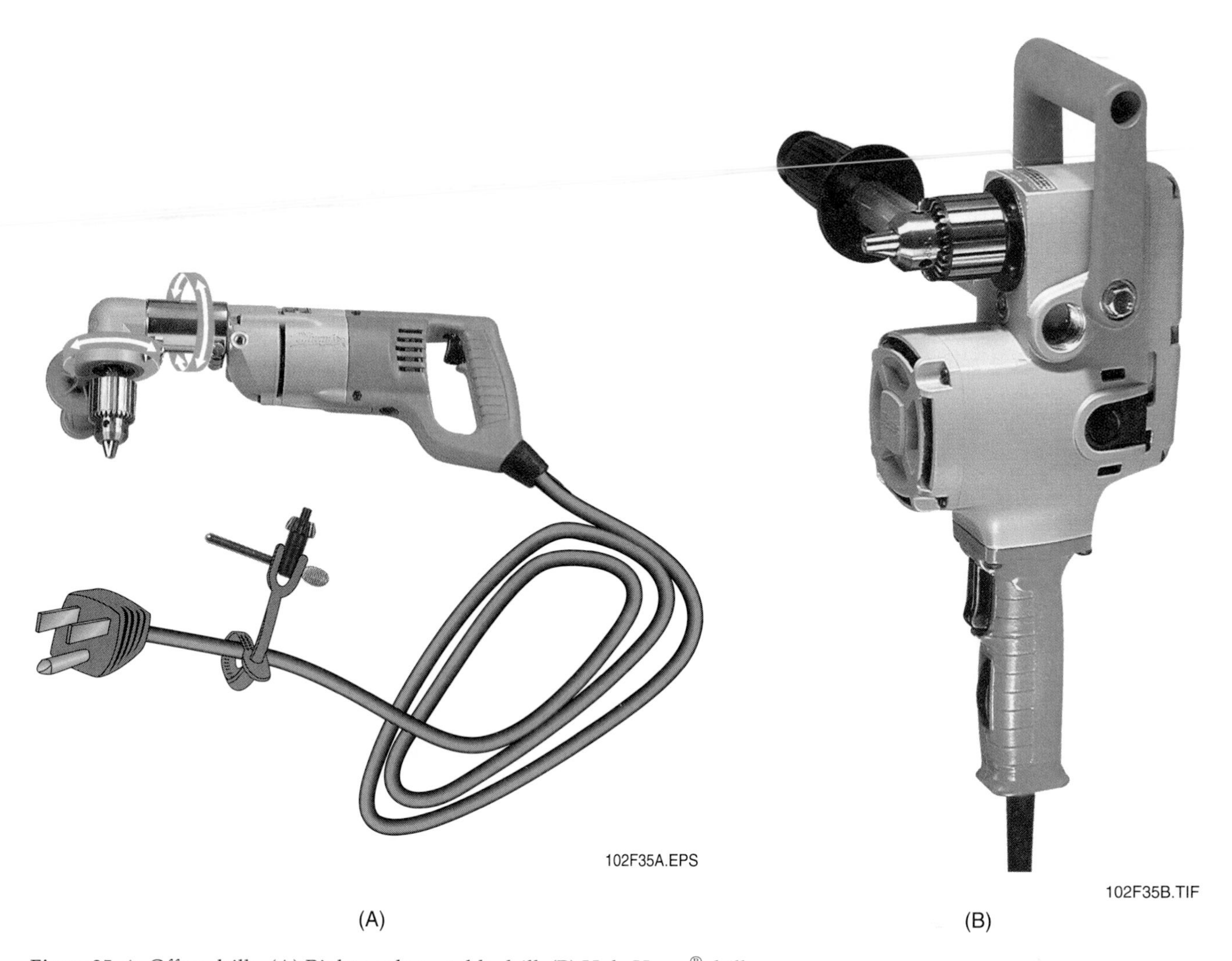

Figure 35 ◆ Offset drills. (A) Right angle portable drill. (B) Hole Hawg® drill.

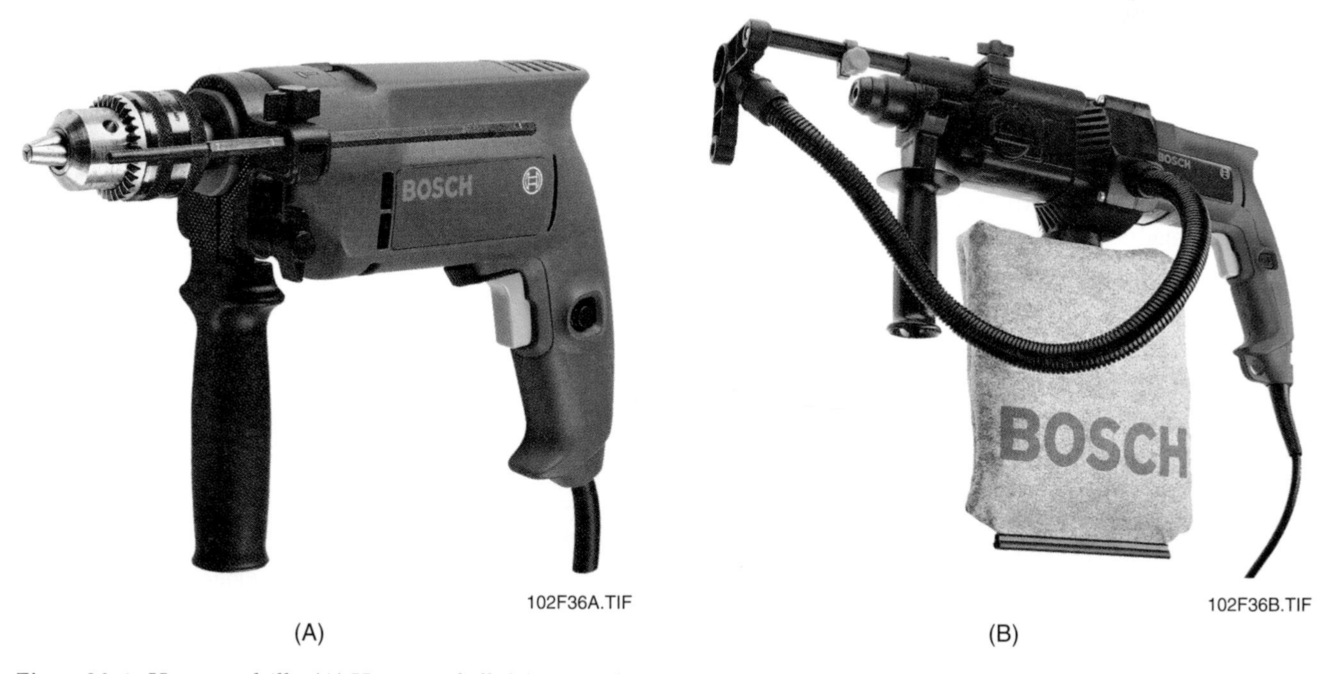

Figure 36 ◆ Hammer drills. (A) Hammer drill. (B) Rotary hammer drill.

8.1.2 Drill Chuck

The chuck is the mechanism at the end of the drill motor (see *Figure 37*). Chucks usually contain three jaws that open and close simultaneously, either to grip or to release a drill bit or other accessory. The chuck is worked with a chuck key. The chuck key is really a gear that meshes with another gear at the base of the drill chuck. The key is attached to the cord with a piece of rubber.

The size of the pistol drill is stated in terms of the diameter of the chuck opening when the jaws are fully open. For instance, if the chuck opening measures ¼ inch in diameter when the jaws are fully open, the pistol drill is said to be a ¼-inch drill. This ¼-inch drill uses drill bits with shanks up to ¼ inch in diameter.

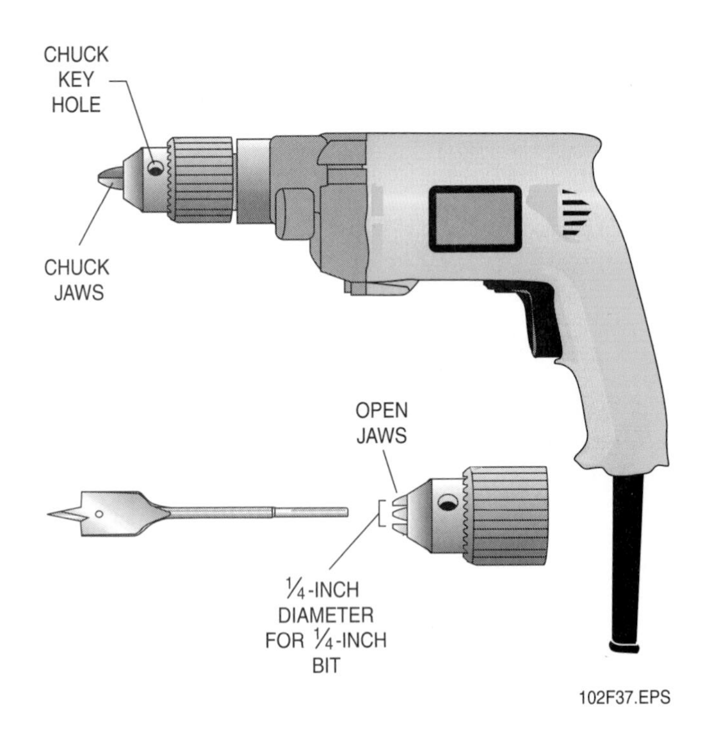

Figure 37 ◆ Drill chuck.

8.1.3 Drilling Capacity

Each drill motor also has a maximum no-load rating in rpm. No-load means that the pistol drill is not doing any work and the motor is running freely at its maximum speed. No-load ratings vary depending on the capacity of the drill.

A ¼-inch drill, for example, may have a no-load rating of 2,000 to 3,500 rpm. A ⅜-inch drill may be rated from 650 to 1,200 rpm. A ½-inch drill may be rated from 350 to 850 rpm. These ratings vary by manufacturer, but not much, and one fact remains constant: as drill capacities increase, speeds decrease.

Simply stated, bigger drills turn more slowly. But as a drill's capacity increases, so does its torque. Torque is a measure of a tool's ability to do work. More torque is required, for example, to drill a ½-inch hole through an I-beam than to drill a ⅛-inch hole through a metal stud. Therefore, you should choose a ½-inch drill motor for the I-beam and a ¼-inch drill motor for the metal stud. Another reason that larger-capacity drill motors turn more slowly is that the larger drill bits cannot stand up to high speeds without overheating and losing their **temper** (strength).

8.1.4 Drill Rotation

Another standard feature of all drill motors is the rotation of their motors. The basic pistol drill turns clockwise at a single speed. This speed is achieved almost instantly as the trigger is pressed.

Other drills have an infinite range of speeds, from slightly above zero to their full rated speed. Manufacturers realize that not all drilling applications are standard. Some materials are drilled best at high speeds, others at lower speeds. So manufacturers began to introduce variable-speed drill motors. With these tools, the position of the trigger determines the speed of the drill.

ON THE

LEVEL

Tempering

Certain metals and special glass are tempered (heated and cooled) to increase their hardness or strength. Alternately heating and cooling a metal such as steel prestresses the metal and greatly increases its strength. Drill bits are often tempered to make them stronger than the materials they will be drilling. But overheating a tempered metal causes it to lose this special hardness. Glass that is prestressed by heating and then rapidly cooled is two to four times stronger than ordinary glass.

Another variation on the basic drill motor was the introduction of the reversing motor. These drill motors can turn either clockwise or counterclockwise. This is useful if a plumber needs to back out of a drilled hole where the drill bit may have become pinched. Usually, you choose the direction of rotation by toggling a lever located on the body of the drill. Most reversing drill motors also have variable speed capabilities. Most manufacturers designate these tools as VSR, for variable-speed reversing.

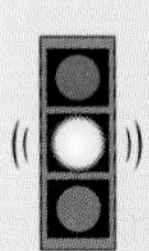

CAUTION

Never reverse drill motors while they are running.

8.2.0 Cordless Drill

Cordless power tools are becoming more popular because of their versatility. They are especially useful for working in awkward areas or in areas where a power source is difficult to find.

Cordless drills (see *Figure 38*) contain a rechargeable battery pack that runs the motor. The pack is detachable and can be plugged into a battery charger whenever the drill is not in use. Some chargers can recharge the battery pack in an hour; others require more time. Workers who use cordless drills a lot usually carry an extra battery pack with them.

All the features found on pistol drills are also found on the cordless models. Many are VSR. Some have adjustable clutches that shift the force (torque) to allow the drill to double as a power screwdriver.

8.3.0 Bits

Drill bits are the cutting tools used in electric drills. They are mounted in the drill chuck. Different bits have different uses.

Figure 38 ◆ Cordless drill.

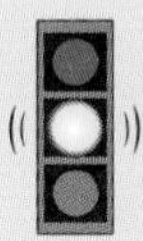

CAUTION

Bits should be sharpened only by qualified personnel. Don't do it yourself—you could hurt yourself and possibly damage the bit.

8.3.1 Twist Bit

The twist bit (see *Figure 39A*) is an all-purpose bit used for drilling almost any material. It is used most often for drilling small holes for installing fixtures. Twist bits are commonly sized at ½ inch.

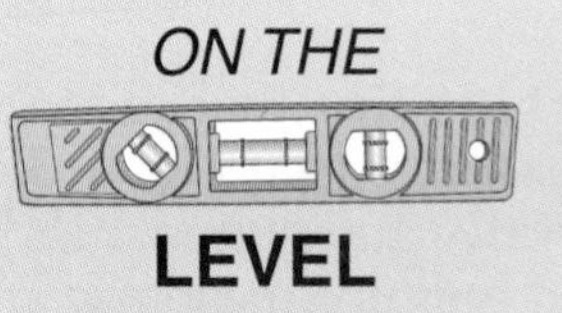

Battery-Operated Tools

Battery-operated tools are becoming common on construction sites. These tools eliminate the need for extension cords and electrical outlets. Most tools come with a replacement battery and a charger. The nickel-cadmium (Ni-Cad) batteries are plugged into the charger when they run low. This way, the worker can be sure that he or she always has a fully charged battery ready to use.

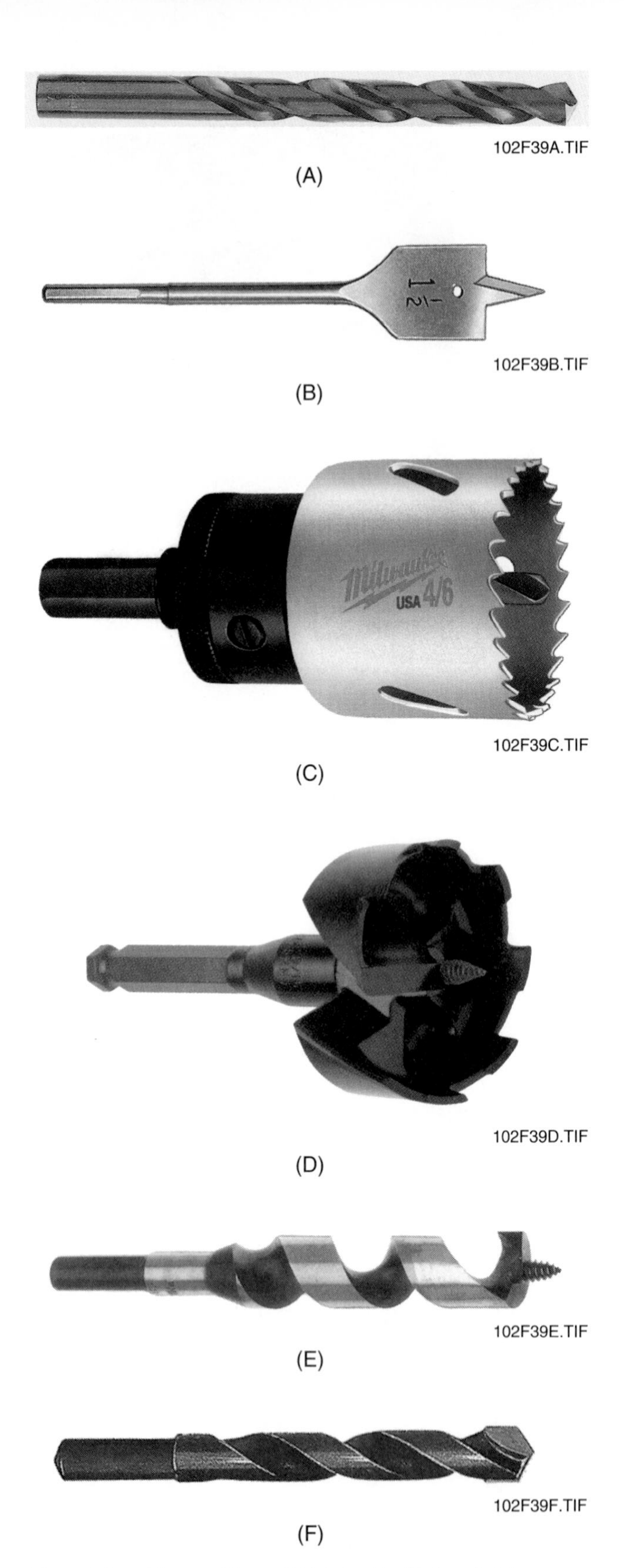

Figure 39 ◆ Types of bits. (A) Twist bit. (B) Spade bit. (C) Hole saw bit. (D) Self-feeding bit. (E) Auger bit. (F) Masonry bit.

Two types of twist bits are available. Those made from high-speed steel are of higher quality and can be used to drill metal as well as wood. The cheaper, high-carbon steel bits are not suitable for drilling hard metals such as iron and steel.

8.3.2 Spade Bit

Spade bits (see *Figure 39B*) are relatively inexpensive. They cut fast and are long enough to drill through fairly thick materials. They produce holes that vary from ¼ inch to 2 inches in diameter.

8.3.3 Hole Saw Bit

To drill holes from 1 inch to 3½ inches in diameter, a plumber uses a hole saw bit. This bit is set with teeth like a saw (see *Figure 39C*). As the drill turns the bit, the teeth cut a hole.

8.3.4 Self-Feeding Bit

Self-feeding bits, or multispur bits (see *Figure 39D*), will bore holes in wood fast. They have diameters ranging from ½ inch to 4⅝ inches. These bits are good for drilling through floor joists and wall studs and for drilling in hard-to-reach areas since you will not need to move the drill to bore the hole.

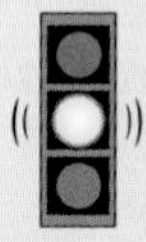

CAUTION

Self-feeding bits should be used only at right angles to the surface being drilled, otherwise they might slip.

8.3.5 Auger Bit

These bits are designed to cut only wood (see *Figure 39E*). They come in diameters of ¾ inch to 1½ inches. Do not use them to cut metal—that will damage the bit.

8.3.6 Masonry Bit

Some drill bits have carbide tips that allow for easy drilling through masonry (see *Figure 39F*). Do not use these specially hardened bits on wood—they are designed only for masonry.

8.3.7 Care of Bits

If you want your drill bits to work well, you must protect them from moisture and keep them sharp. A thin coat of all-purpose oil can help prevent rust. If you notice that a bit isn't sharp, have a qualified person sharpen it.

8.4.0 Burring Reamer

After you cut a pipe, you will find a **burr** on the inside of the pipe. A reamer (see *Figure 40*) is used to remove the burr. This process is called *reaming.* If the burr (see *Figure 41*) is left in the pipe, it will collect deposits and slow the flow of liquid within the pipe. Because reamers are tapered, one reamer can de-burr many sizes of pipes. A pencil reamer (see *Figure 42*) is used with copper tubing. It removes burrs from copper tubing and is approximately the size of a pencil.

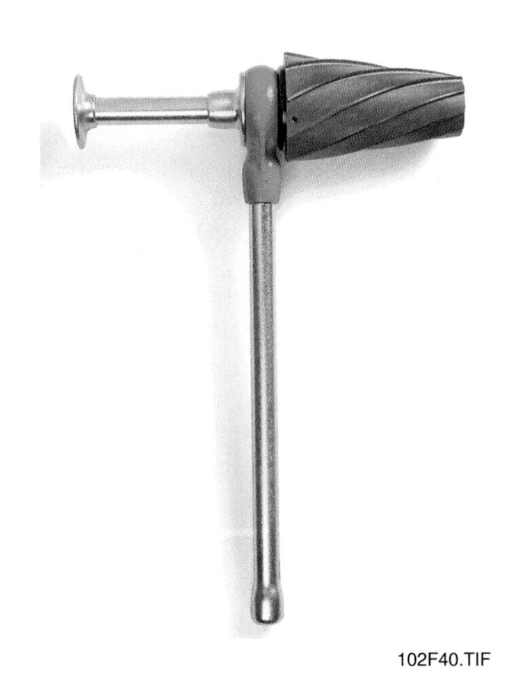

Figure 40 ◆ Burring reamer.

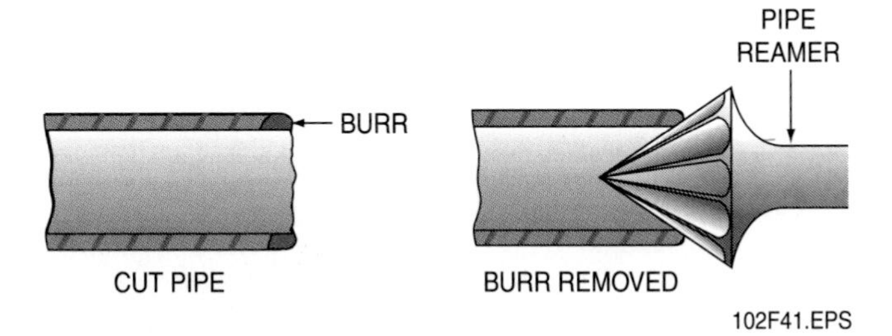

Figure 41 ◆ Burr.

Figure 42 ◆ Pencil reamer.

Review Questions

Sections 8.0.0–8.4.0

1. The size of the drill is stated in terms of the diameter of the _____ opening when the jaws are fully open.
   a. bit
   b. chuck
   c. jaws
   d. pistol
2. _____ drills are useful because some materials are drilled best at high speeds; others at lower speeds.
   a. Alternating-speed
   b. Electric-speed
   c. Variable-speed
   d. Reversible-speed
3. The _____ bit combines a drill bit with a cylindrical saw blade.
   a. hole saw
   b. twist
   c. variable
   d. pistol
4. _____ bits will bore holes in wood quickly and are good for drilling through floor joists and wall studs and for drilling in hard-to-reach areas.
   a. Masonry
   b. Hole saw
   c. Twist
   d. Self-feeding
5. A _____ is the cutting tool that is mounted in the drill chuck.
   a. reamer
   b. blade
   c. bit
   d. twist

8.5.0 Dies

Plumbers join some kinds of pipe by screwing one pipe end into another pipe end. To do so, they must prepare each end to join the pipes together. Galvanized pipe and black iron (cast iron) pipe need to be threaded like a screw on the outside. Plumbers use dies (see *Figure 43*) to cut the

102F43A.TIF

(A)

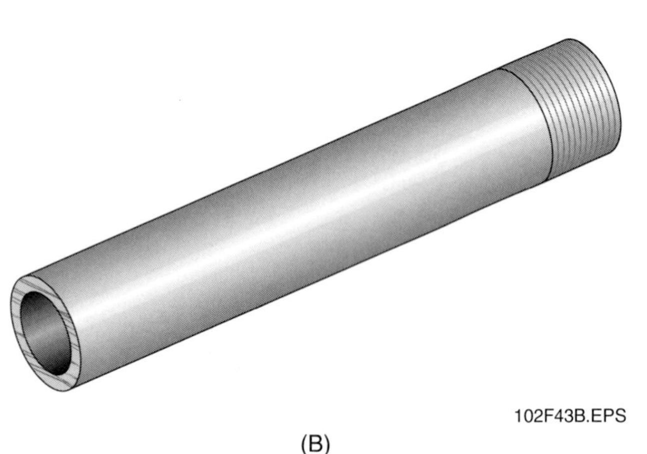

102F43B.EPS

(B)

Figure 43 ◆ (A) Die set. (B) Pipe with threaded end.

threads. Dies must be sharp so that the threads they cut will be correctly tapered (shaped).

Applying cutting oil to the die will reduce the friction and heat caused by cutting. If there is too much friction or heat, the die will push the metal that is being threaded, instead of making a clean cut. Heat and friction can also break the teeth off the die.

Dies are made to fit all standard pipe sizes and piping materials. Dies cut threads in a variety of sizes, depending on the size of thread needed to connect the pipe lengths. Typically, dies cut threads in ½-, ¾-, 1-, and 1¼-inch sizes. Be sure that the dies are compatible with the type of material being worked. Never use steel pipe dies to thread stainless steel.

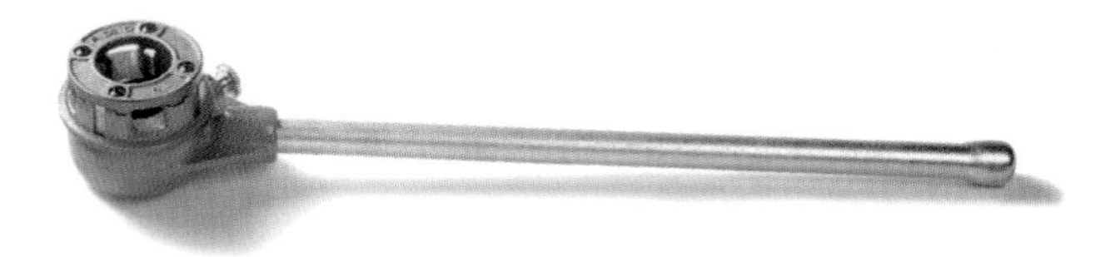

102F44.EPS

Figure 44 ◆ Die and ratchet stock.

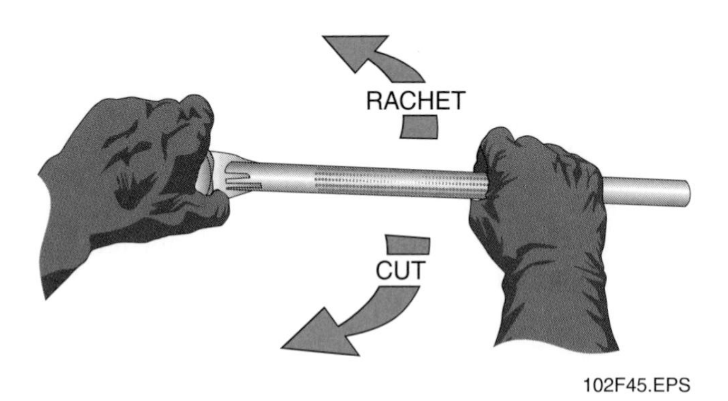

102F45.EPS

Figure 45 ◆ Operating a die and ratchet stock.

8.5.1 Die and Ratchet Stock

The die is locked or clamped into a ratchet stock (handle) (see *Figure 44*). This ratchet stock helps the plumber turn the die by increasing leverage on it. Because the stock includes a ratchet mechanism, it is easy to operate by applying pressure in a downward direction to cut the thread (see *Figure 45*) and by ratcheting the handle up to prepare for another downward stroke.

9.0.0 ◆ ELECTRIC PIPE THREADING MACHINE

The electric pipe threading machine (see *Figure 46*) rotates, threads, cuts, and reams pipe. (The machine can also be used to thread bolts, but it is cheaper just to buy the bolts.) The pipe is inserted into the chuck of the machine. After the chuck is tightened, this tool performs one or more of those operations. For trouble-free operation, be sure to follow the preventive maintenance schedule provided by the manufacturer.

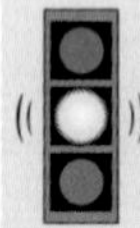

CAUTION

When you are working with a power threading machine, do not wear loose clothing, jewelry, or gloves that could easily get caught in the rotating machine parts. Also, be sure to tie back long hair to prevent it from getting caught in the machine.

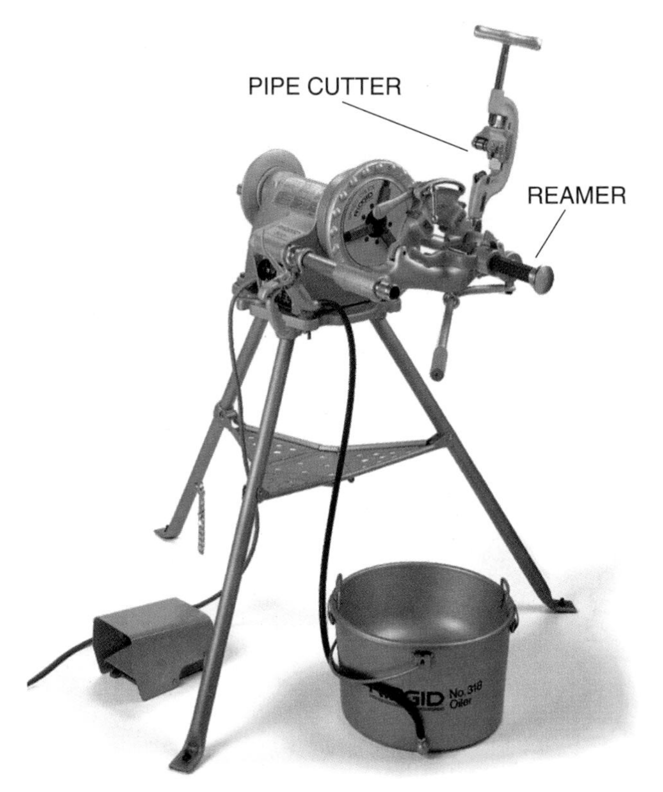

Figure 46 ◆ Electric pipe threading machine.

Figure 47 ◆ Propane gas torch.

10.0.0 ◆ SOLDERING TOOLS

Soldering is a method of using heat to form joints. Plumbers generally use soldering to join rigid copper pipe and fittings. The filler material (**solder**) is distributed evenly between the close-fitting surfaces of the pipes when the pipes reach the correct temperature.

Plumbers use a small, portable propane gas torch for soldering (see *Figure 47*). The torch heats the metal and the **flux** to join the pipe.

> **WARNING!**
> The droplets of solder are molten metal. They will cause a severe burn if they come in contact with skin. When you are using a welding torch, always wear welding goggles and protective clothing. Weld only under proper supervision.

Review Questions

Sections 8.5.0–10.0.0

1. _____ are used to cut threads into galvanized and black iron (cast-iron) pipes.
   a. Augers
   b. Bits
   c. Dies
   d. Tube cutters

2. A _____ stock provides the plumber with increased leverage to turn the die.
   a. die
   b. ratchet
   c. pipe
   d. reamer

3. An electric pipe threading machine cuts, threads, and _____ the pipe.
   a. joins
   b. drills
   c. ratchets
   d. reams

4. Plumbers use a propane gas torch when they are soldering to _____.
   a. thread the pipe
   b. oxidize the pipe
   c. heat the pipe
   d. remove any burrs

Brazing

In plumbing, a process called **brazing** is used to join pipe and fittings in saltwater pipelines, oil pipelines, refrigeration systems, vacuum lines, chemical-handling systems, air lines, and low-pressure steam lines. Brazing requires a much higher temperature than that used for soldering copper pipe.

To get a sufficiently hot flame, the plumber must use an oxyacetylene torch unit (see *Figure 48*) capable of reaching 1,400° F (760° C) or higher. This job requires a torch with the correct tip, regulator valves, and oxygen and acetylene tanks. Tanks come in two sizes. Plumbers commonly use a B-tank, which is smaller and has two compartments to hold the pressurized acetylene and oxygen.

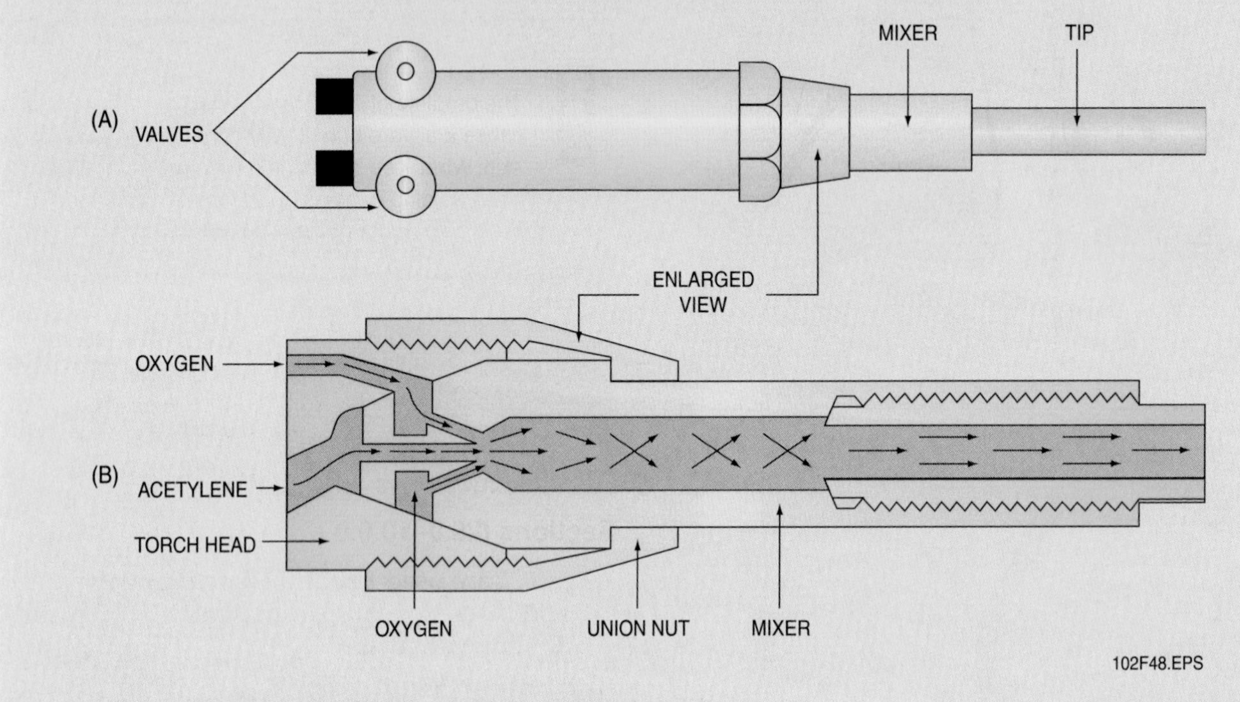

Figure 48 ◆ Oxyacetylene torch unit.

5. If droplets of solder come in contact with skin, the solder will cause _____.
   a. a severe burn
   b. a deep cut
   c. electric shock
   d. a stinging sensation

11.0.0 ◆ EXTENSION CORDS

In order for a power tool to function properly, it must receive enough electrical energy. As electricity travels through the length of an extension cord, it tends to lose voltage. This loss is called **voltage drop.** When voltage drops, the tool has to draw more **amperage,** causing the motor to run hotter. Under prolonged use, the motor can overheat and be ruined.

DID YOU KNOW?

Amperage is the strength of an electrical current that is expressed in amperes (amps). An ampere is a unit of electrical current. The higher the amps, the more current is flowing through a wire. All electrical wire is rated by the number of amps it is designed to carry. For example, 20-amp, 40-amp, and 80-amp wires are available. Always be sure that you are using the right wire for the job. Too many amps through a wire will cause the wire to heat up, which could cause a fire.

Table 1 Amperage Ratings

Amps Rating (on nameplate)	0–2.0	2.10–3.4	3.5–5.00	5.10–7.0	7.10–12	12.1–16	
Ext. Cord Length	**Wire Size**						
25'	18	18	18	18	16	14	
50'	18	18	18	16	14	12	
75'	18	18	16	14	12	10	
100'	18	16	14	12	10	8	
150'	16	14	12	12	8	8	
200'	16	14	12	10	8	6	**Not normally available as flexible extension cords**
300'	14	12	10	8	6	4	
400'	12	10	8	5	4	4	
500'	12	10	8	5	4	2	
600'	10	8	6	4	2	2	
800'	10	8	6	4	2	1	
1,000'	8	6	4	2	1	0	

Table 1 shows the wire gauge required for extension cords at given lengths for given amperage draws. The smaller the gauge number, the heavier the wire. Be sure that the extension cords you use—especially the long ones—are heavy enough to minimize voltage drop.

12.0.0 ◆ TOOLS FOR ASSEMBLY AND HOLDING

The plumber must have a variety of wrenches for turning pipe, fittings, and fasteners. Some wrenches are used for turning finish fasteners, where the slightest mark will ruin the appearance. Others are used to grip with holding power on materials that will be hidden in the structure.

The tools used for holding include vises and pliers. These tools are used to hold plumbing materials while the plumber performs operations on the material.

12.1.0 Wrenches

All wrenches are used for assembly and disassembly. Many wrenches also have special uses. Be sure to keep the jaws of wrenches clean by using a wire brush.

12.1.1 Pipe Wrench

The pipe wrench (see *Figure 49*) is used to grip and turn round pipe and tubing. Wrenches come in various lengths such as 6, 12, 14, 18, 24, 36, and 48 inches. Length determines the size of pipe that can be worked with the wrench. The straight pipe wrench is the most common. When working in tight quarters, plumbers can use a 45-degree or a 90-degree offset wrench. The angles of these wrenches make it easier to reach when there is not much room to work. The jaws of pipe wrenches always leave jaw marks on the pipe.

The force (see *Figure 50*) should always be directed toward the open side of the jaws. This gives you the best grip and leverage.

Another variation on the straight pipe wrench is the compound leverage pipe wrench, shown in *Figure 51.* The design increases the leverage on a pipe. This wrench is usually used on pipe joints that have frozen or locked together.

A chain wrench is a type of pipe wrench that has a length of chain permanently attached to a handle at one end. The other end of the chain can be secured at various positions on the handle (see *Figure 52*). The chain can be looped around the pipe to grip and secure it.

12.1.2 Pipe Tongs

Pipe tongs, shown in *Figure 53,* are the wrenches typically used on large pipe. Smaller sizes of chain wrenches are also used. The chain must be oiled often to prevent it from becoming stiff or rusty.

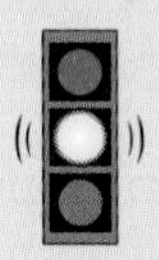

CAUTION

Never use cheaters (pipes used to extend wrenches). Always use the right-sized tool.

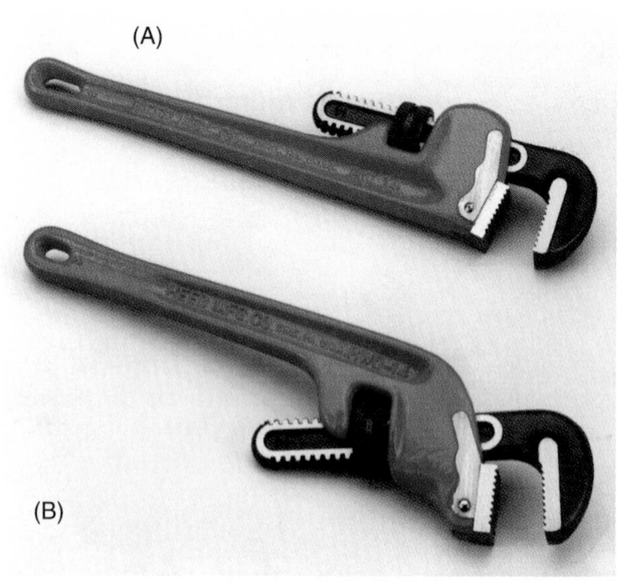

102F49AB.EPS

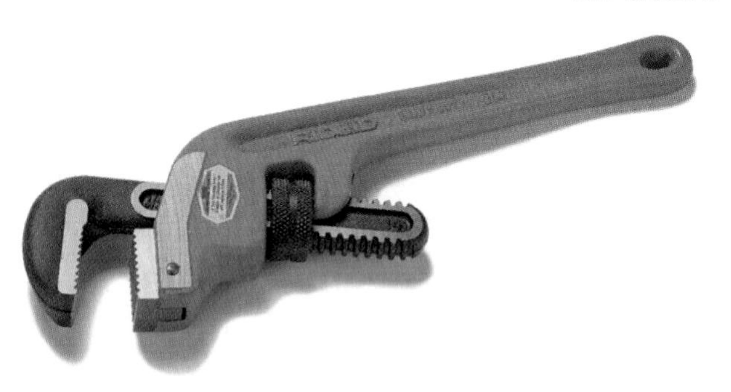

102F49C.TIF

(C)

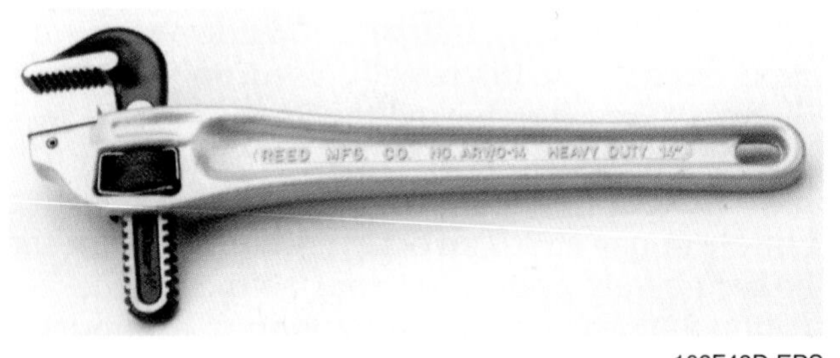

102F49D.EPS

(D)

Figure 49 ◆ Pipe wrenches. (A) Heavy-duty pipe wrench. (B) 45° offset wrench. (C) Heavy-duty end wrench. (D) 90° offset pipe wrench.

12.1.3 Strap Wrench

The strap wrench (see *Figure 54*) is used to hold chrome-plated or other types of finished pipe. The strap wrench does not leave jaw marks or scratches on the pipe.

Some straps need rosin to keep the strap from slipping. Other strap wrenches have a vinyl strap that doesn't require rosin.

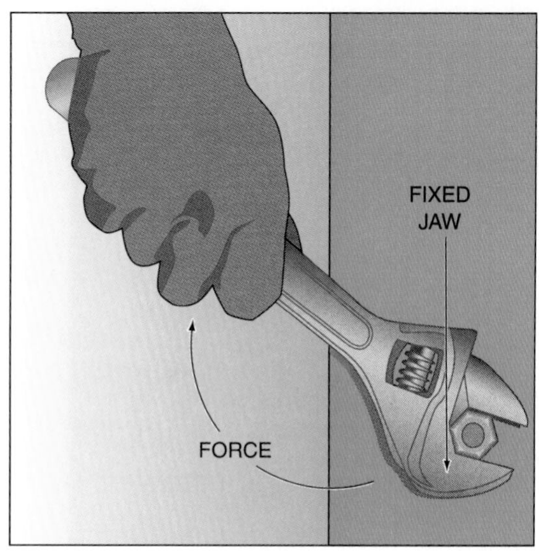

102F50.EPS

Figure 50 ◆ Applying force with a wrench.

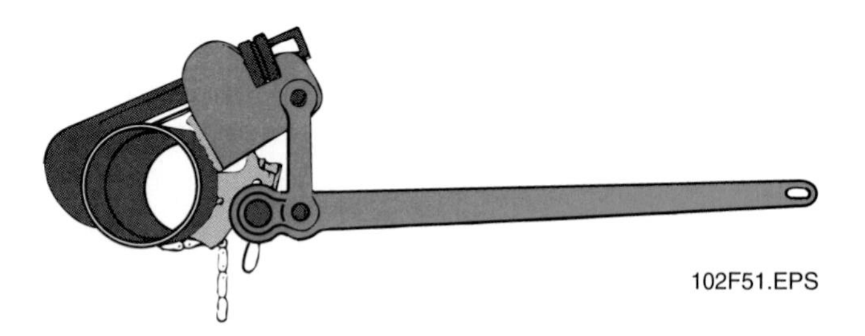

102F51.EPS

Figure 51 ◆ Compound leverage pipe wrench.

102F52.TIF

Figure 52 ◆ Chain wrench.

12.1.4 Spud Wrench

The spud wrench (see *Figure 55*) is similar to a pipe wrench but has no teeth in its jaws. Because it is difficult to adjust the spud wrench, plumbers often use other wrenches, such as the adjustable wrench and the open-end wrench.

Spud wrenches loosen and tighten fittings on drain traps, sink strainers, toilet connections, and large, odd-shaped nuts. The wrench features narrow jaws to fit into tight places.

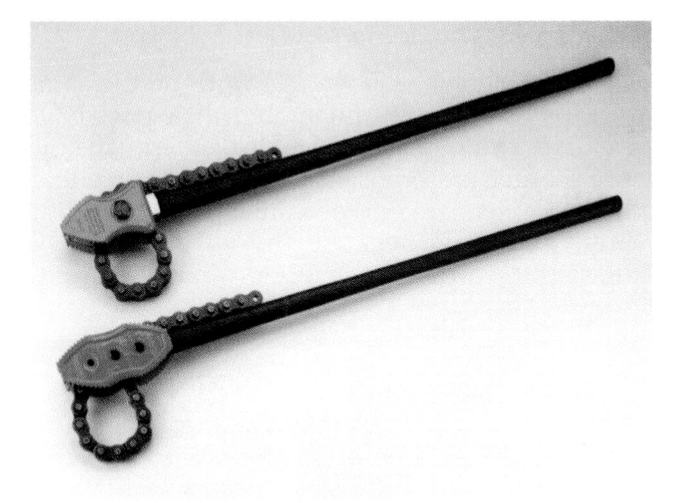

102F53.EPS

Figure 53 ◆ Pipe tongs.

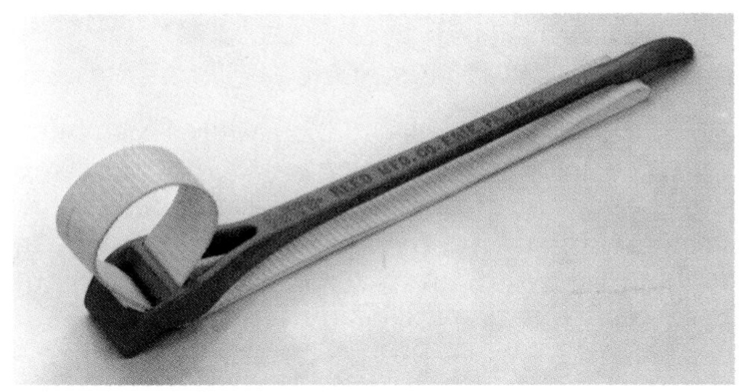

102F54.EPS

Figure 54 ◆ Strap wrench.

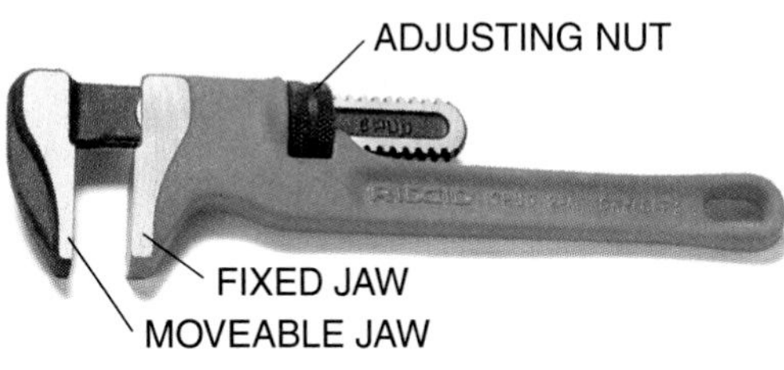

102F55.EPS

Figure 55 ◆ Spud wrench.

12.1.5 Open-End Wrench

The open-end wrench (see *Figure 56*) usually has two different-sized openings. The size of each opening is stamped on the handle. This tool is usually used for assembling fittings and fasteners that are less than 1 inch across.

12.1.6 Adjustable (Crescent) Wrench

The adjustable wrench (see *Figure 57*) is like the open-end wrench except that it has an adjustable jaw. Because it is so easy to adjust, this wrench is very popular. Various sizes of adjustable wrenches are available, from 6 inches to 18 inches.

Adjustable wrenches are smooth jawed for turning nuts, bolts, small pipe fittings, and chrome-plated pipe fittings. Using an adjustable wrench may save time by avoiding the need to switch to a box wrench or open-end wrench when you are working with different sizes of nuts and bolts.

102F56.EPS

Figure 56 ◆ Open-end wrench.

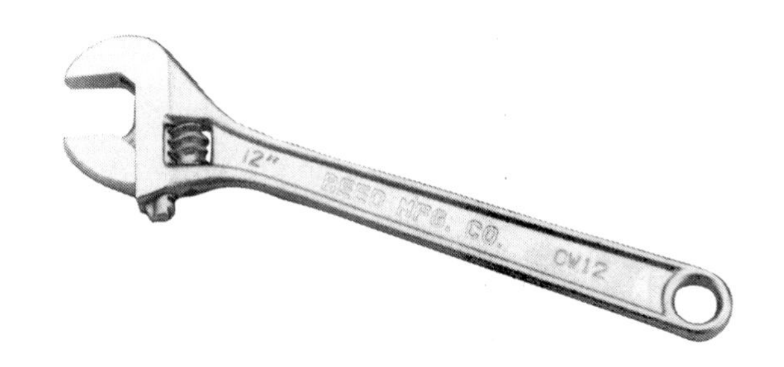

102F57.EPS

Figure 57 ◆ Adjustable wrench.

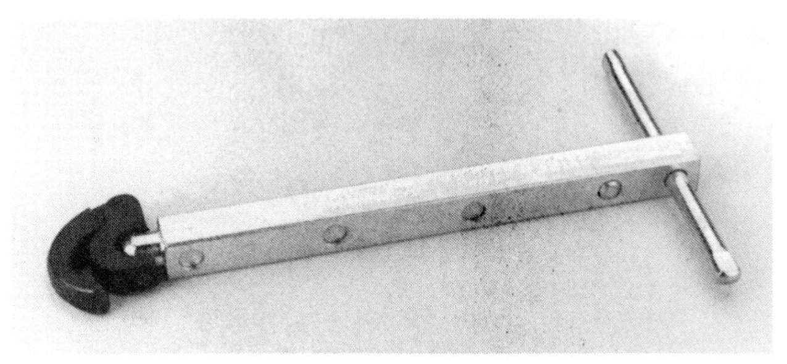

102F58.EPS

Figure 58 ◆ Basin wrench.

12.1.7 Basin Wrench

The basin wrench (see *Figure 58*) was developed to work in small or hard-to-get-at areas such as the recesses where sink and lavatory faucets are secured. This specialty wrench, with its offset jaws, permits the plumber to turn a nut or a fastener that ordinarily could not be reached with regular wrenches.

12.2.0 Pliers

Pliers are used for bending, gripping, and holding. Pliers cannot be used on exposed pipe or fasteners because they will leave jaw marks.

12.2.1 Slip Joint Pliers

Slip joint pliers (see *Figure 59*) are very popular. They are used for bending wire and for gripping and holding during assembly operations. The adjustable jaws make it easy to change the distance between the two jaws. There are two jaw settings: one for small materials and one for larger materials.

12.2.2 Lineman's Pliers

Lineman's pliers (or side cutters) (see *Figure 60*) have wider jaws than slip joint pliers. Lineman's pliers are mainly used for cutting heavy or large-gauge wire and for holding work. The shape of the jaws reduces the chance that wires will slip, and the hook bend in the handles gives you a better grip.

102F59.TIF

Figure 59 ◆ Slip joint pliers.

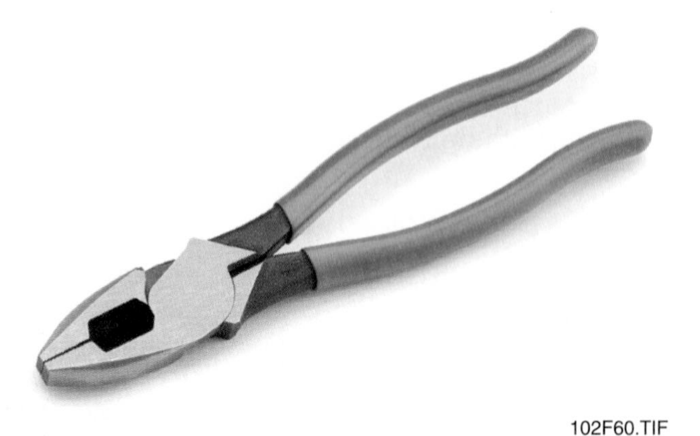

102F60.TIF

Figure 60 ◆ Lineman's pliers.

12.2.3 Slip Lock Pliers

Slip lock pliers (see *Figure 61*) work on the same principle as slip joint pliers, but their jaws have five or more adjustments. Slip lock pliers can hold much larger materials than channel lock pliers. Slip lock pliers are not recommended for holding small materials because the jaws cannot be adjusted flat against each other.

DID YOU KNOW?

Many common tools have different names, depending on where you work in the country. For example, slip lock pliers are called pump pliers or channel lock pliers in some regions of the United States.

12.2.4 Locking (Vise-Grip®) Pliers

Locking pliers (see *Figure 62*) operate as both pliers and a vise. Their jaws can be adjusted by the screw attached to one of the handles. Because of their fine jaw adjustment and tremendous clamping power, locking pliers are an excellent tool for removing bolts that have had the head rounded off.

12.2.5 Care of Pliers

Pliers need very little preventive maintenance. A little oil on all movable joints and some oil on the locking pliers' adjustment screw will prevent rust and keep them working smoothly.

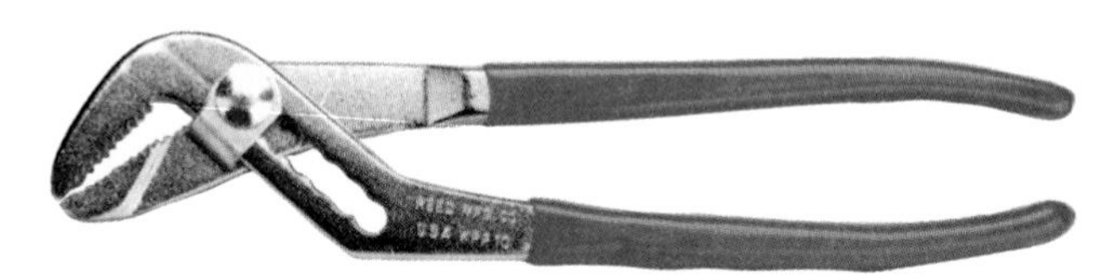

102F61.EPS

Figure 61 ◆ Slip lock pliers.

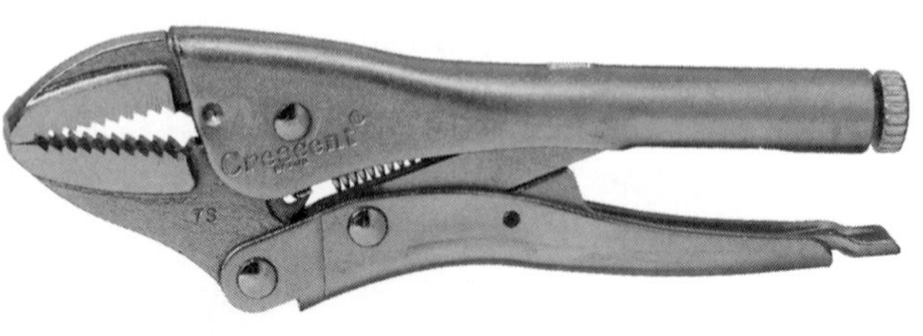

102F62.EPS

Figure 62 ◆ Locking pliers.

13.0.0 ◆ HAMMERS

The plumber, like all other skilled workers, uses a hammer. There are many types of hammers including claw hammers, ball-peen hammers, and sledgehammers.

13.1.0 Claw Hammer

The most common type of hammer is the carpenter's curved claw hammer (see *Figure 63*). This hammer is used for pulling or driving nails or striking a chisel. To draw a nail out of material, slip the claw of the hammer under the nail head and pull until the handle is nearly straight up (vertical) and the nail is partly drawn out of the wood. Then, pull the nail straight up from the wood (see *Figure 64*).

Hammers come in many sizes. The size of a hammer is determined by the weight of the head. Usually a 14- or 16-ounce hammer is considered medium weight.

Hammers also come with different types of handles. Wooden-handle hammers are the least expensive and the most easily broken. Fiberglass-reinforced plastic handles produce nearly the same "feel" as a wood handle and are unbreakable in normal use. Steel handles are very strong, but they transfer more shock to the hand than either wooden or fiberglass-reinforced plastic handles.

13.2.0 Ball-Peen Hammer

The ball-peen hammer (see *Figure 65*) has a flat face for striking and a rounded section that is used to align brackets and drive out bolts. This hammer is also used with chisels. The ball peen is classified by weight. Its head weighs from 6 ounces to 2½ pounds.

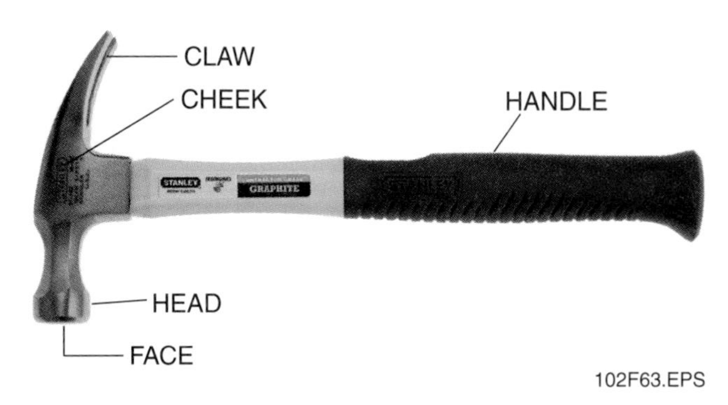

Figure 63 ◆ Claw hammer.

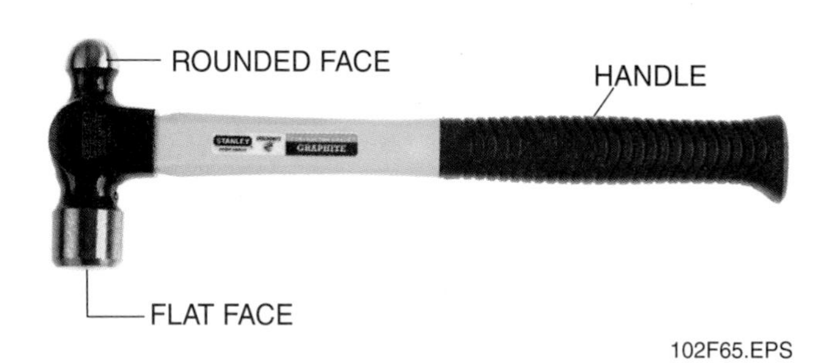

Figure 65 ◆ Ball-peen hammer.

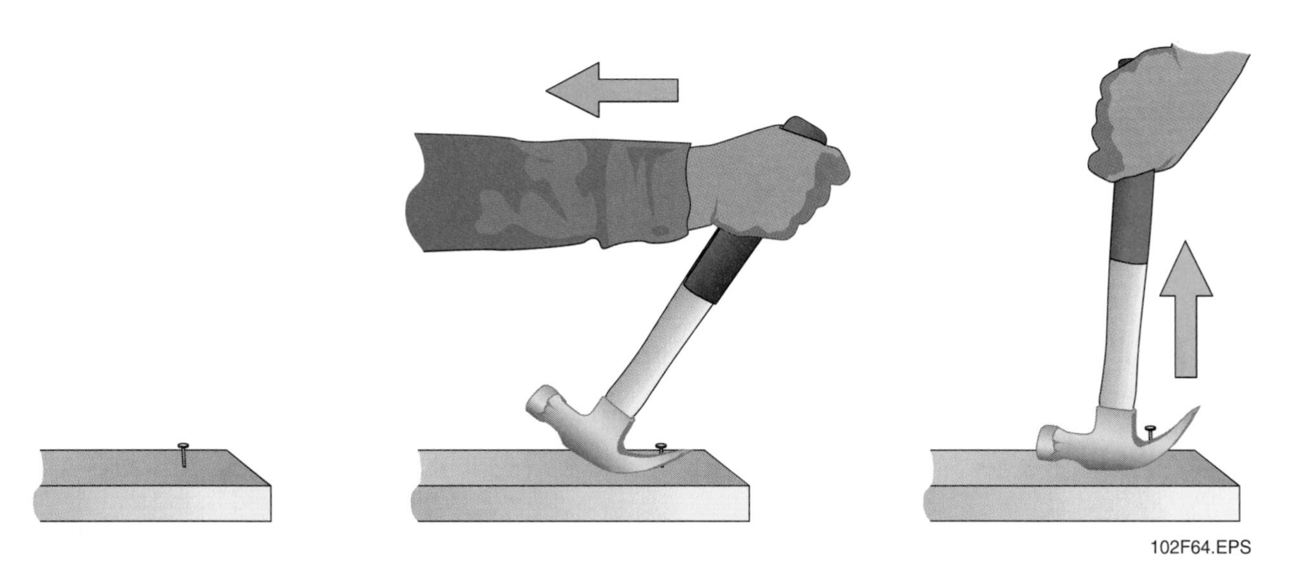

Figure 64 ◆ Using a claw hammer to pull a nail.

13.3.0 Sledgehammer

A very heavy hammer is called a sledgehammer (see *Figure 66*). The sledgehammer is a heavy-duty tool used for driving posts or other large stakes. It can also be used for breaking up cast iron or concrete. The head of the sledgehammer is made of high-carbon steel and weighs from 2 to 20 pounds. The shape of the head depends on the job the sledgehammer is designed to do. Sledgehammers can be either long- or short-handled, depending on the job they are designed for.

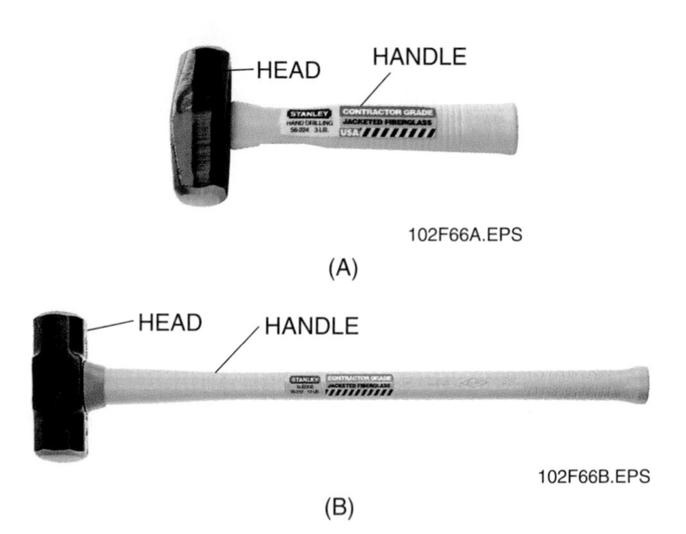

Figure 66 ◆ Sledgehammers. (A) Double-face short-handled sledgehammer. (B) Double-face long-handled sledgehammer.

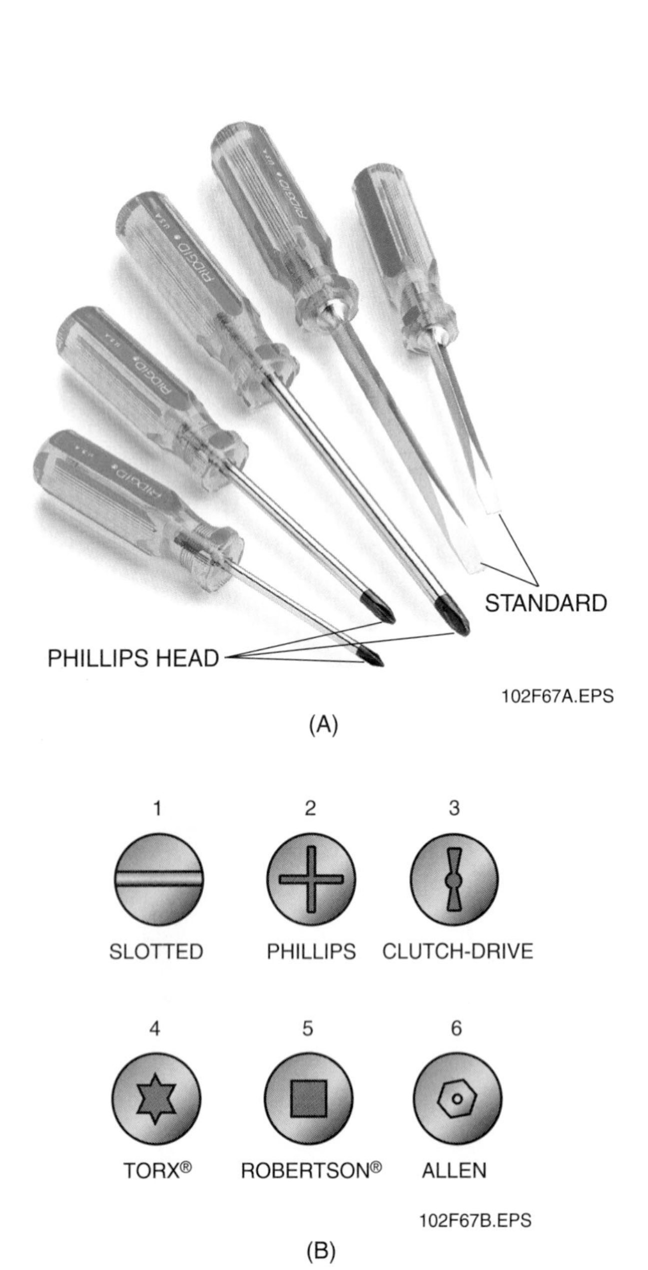

Figure 67 ◆ (A) Common screwdrivers. (B) Screw head shapes.

14.0.0 ◆ SCREWDRIVERS

Screwdrivers are used by most skilled trades to install and remove a wide variety of screws. The most common types are the standard screwdriver and the Phillips screwdriver (see *Figure 67*). The screwdriver's handle is most often made of wood or plastic. The size of a screwdriver is determined by the length of its blade and the width of its tip. A correct fit between the screwdriver tip and the screw head will prevent the marring of the screw head.

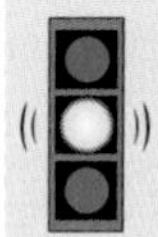

CAUTION

Do not use the screwdriver as a lever, chisel, or punch. Always replace worn screwdriver tips to protect the screws.

14.1.0 Nut Driver

A nut driver is similar to a screwdriver (see *Figure 68*). Instead of a blade, the nut driver has a socket end for tightening hex-head screws and bolts. It comes in all bolt sizes.

14.2.0 Hollow-Shank Screwdriver

Some screwdrivers come with interchangeable bits (see *Figure 69*). You simply remove one bit from the end of the shank and replace it with another bit. The bits are stored in the handle of the screwdriver. The bits typically have a flat blade on one end and a Phillips head on the other. Specialty bits are also included.

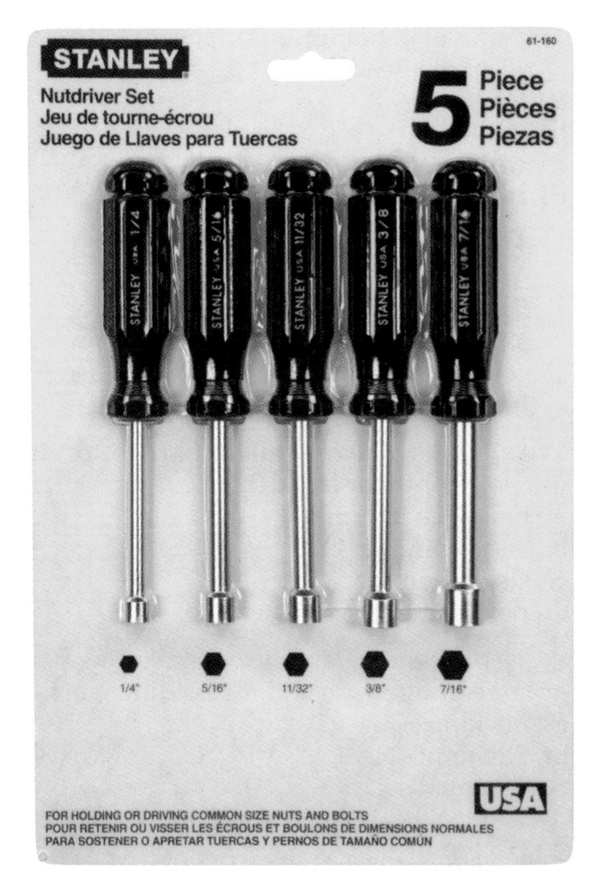

102F68.TIF

Figure 68 ◆ Nut driver.

102F69.EPS

Figure 69 ◆ Hollow-shank screwdriver.

15.0.0 ◆ VISES

Vises are holding tools. Plumbing processes such as cutting, threading, and reaming would be difficult to perform without the vise. The vise permits one plumber to do work that otherwise would require two people—one to hold the material to be cut, threaded, or reamed, and one to perform the process.

15.1.0 Standard Yoke Vise

The standard yoke vise (see *Figure 70*) is probably the most common vise used by plumbers. Its jaws hold pipe firmly and prevent it from turning. This vise can handle pipe ⅛ inch to 3½ inches in diameter, depending on the size of the vise. Because the teeth usually leave jaw marks, only pipes that will not be exposed are secured in this vise.

15.2.0 Chain Pipe Vise

The chain pipe vise (see *Figure 71*) is used in the same way as the standard yoke vise, but the chain vise can hold much larger pieces of pipe. The chain must be kept oiled or it will become stiff. A stiff chain will make the chain vise operate poorly.

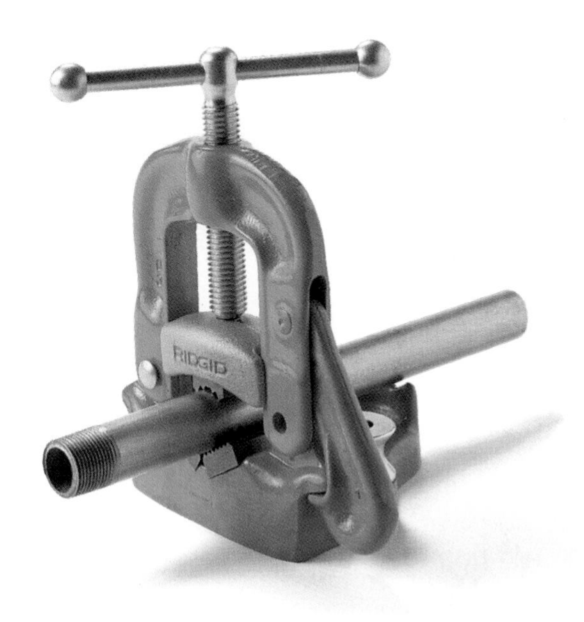

102F70.TIF

Figure 70 ◆ Yoke vise.

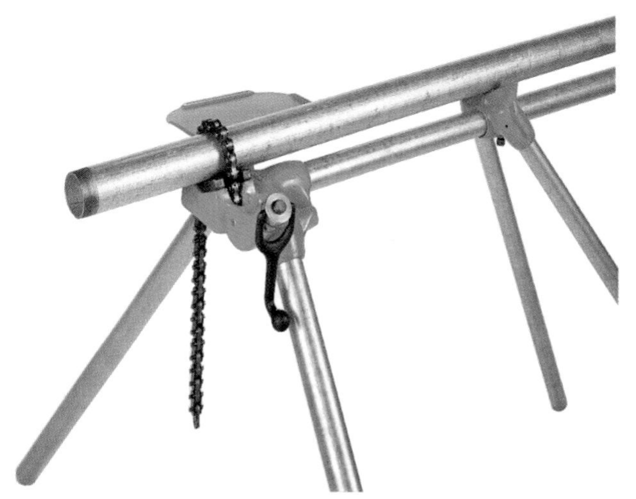

102F71.TIF

Figure 71 ◆ Chain pipe vise.

15.3.0 Bench Vise

The bench vise (see *Figure 72*) has two sets of jaws: one to hold flat work and another to hold pipe. You will not find this vise very often on the job site, but you may find it mounted in some plumbing trucks.

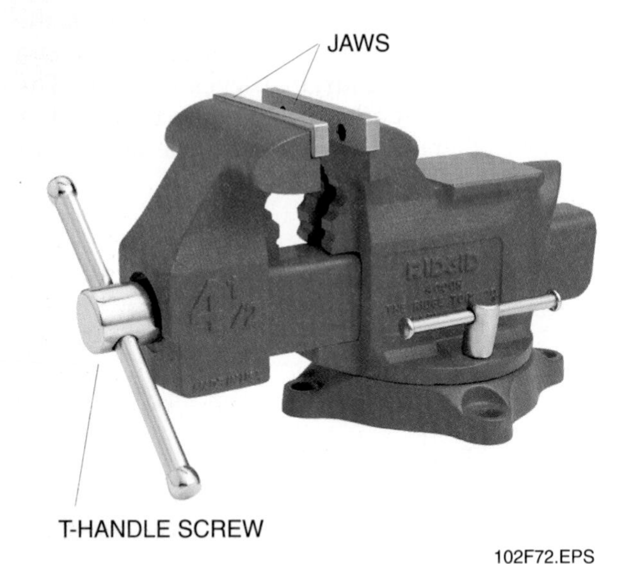

Figure 72 ◆ Bench vise.

Review Questions

Sections 11.0.0–15.3.0

1. The _____ wrench is used to hold chrome-plated or other types of finished pipe.
   a. spud
   b. pipe
   c. strap
   d. open-end

2. When using a wrench, always direct the force _____ the open side of the jaws.
   a. toward
   b. away from
   c. above
   d. below

3. _____ wrenches are smooth jawed for turning nuts, bolts, small pipe fittings, and chrome-plated pipe fittings.
   a. Basin
   b. Strap
   c. Pipe
   d. Adjustable

4. The size of a hammer is determined by the _____ of the head.
   a. length
   b. width
   c. weight
   d. shape

5. Make sure the extension cords you use are _____.
   a. light enough to increase voltage drop
   b. long enough to increase voltage drop
   c. color-matched to the tool you are using
   d. heavy enough to minimize voltage drop

6. The _____ is a heavy-duty tool used for driving posts or other large stakes.
   a. ball peen hammer
   b. claw hammer
   c. sledgehammer
   d. post hammer

7. _____ pliers are mainly used for cutting heavy or large-gauge wire and for holding work.
   a. Locking
   b. Slip joint
   c. Lineman's
   d. Slip lock

8. A _____ permits one plumber to do work that otherwise would require two people, such as cutting, threading, and reaming a pipe.
   a. pipe wrench
   b. strap wrench
   c. vise
   d. tube cutter

9. The chain pipe vise is used to hold _____ pieces of pipe.
   a. small
   b. large
   c. several
   d. two

10. You must put _____ on vise chains to keep them from becoming stiff and unable to wrap around a pipe.
    a. water
    b. oil
    c. solder
    d. rosin

Summary

Plumbing is a sophisticated and complex craft. It is full of challenges for the dedicated worker. Being the best you can be involves learning how to perform the different tasks assigned. Plumbers do not just fix toilets and sinks. Plumbers use a variety of tools to design, install, and maintain new pipe systems, renovate existing systems, and repair broken systems.

As you begin to learn your craft, you will become familiar with the tools used in the plumbing trade. The hand tools you will use every day include many varieties of measuring, marking, and cutting tools. Each tool has specific use and care requirements as well as safety rules for its use. Hand tools can cause just as serious an injury as power tools if you are careless. Some tools are designed to speed up the work by combining several tools in one. A combination square is a good example: you can use it to measure both 45- and 90-degree angles, check for square and level, and also mark your measurements.

Plumbers need to lay pipe that is straight and even. To do this, you may use simple tools like a plumb bob and a spirit level. You may find yourself using sophisticated electronic and laser equipment for some employers. As technology advances, tool requirements will change. You will be expected to understand how to use and care for all the different types of tools on the job site.

You will spend a lot of your time measuring, marking, cutting, and installing different types of pipe and tubing. You will use special cutting tools designed to cut specific materials like cast iron, various types of plastic, and copper. Each type of tool has a job to do. You must learn how to choose the right tool for the job.

Plumbing is closely tied to other trades. Your work will affect how the carpenters and electricians complete their jobs. You will work with Heating, Ventilating, and Air Conditioning (HVAC) technicians to install systems in new buildings. To be successful, you must have some of the same skills as each of these other trades. After a few years, you will find that you know a little bit about each of these crafts.

PROFILE IN SUCCESS

Jerome Sabol

Vice President
Plumb Works
Atlanta, GA

Jerome grew up in upstate New York. In the summers during high school, he traveled to Atlanta to work with his brother in his small plumbing business. In high school, Jerome also attended vocational education classes in carpentry, wiring, heating, and plumbing. He was exposed to many crafts, but he enjoyed plumbing the most. Following high school, he attended a technical college. After graduation, Jerome joined his brother.

How did you become interested in the construction industry?
My older brother had a plumbing company in Atlanta, so I worked for him during summers in high school. It seemed a logical choice to go into the business with him after I graduated from college. I had tried my hand at some of the other trades, but found I really liked plumbing the best.

We see a lot of plumbers who get into the family business. The family connection is a great support to someone starting out in the trades. You have not only the tradition but also the encouragement from the family. You've got someone pulling for you and teaching you each day. For apprentices who are forging new ground, a mentor can serve the same purpose. It helps to have someone who takes a special interest in your success.

What do you think it takes to be a success in your trade?
It's simple: hard work and interest. You have to show up every day. You see this in any apprentice or vocational education situation. There are always people who stand out because of their interest in the craft. It is obvious who is there to work and who is there just to get by. Instructors know this, too. Their time and effort will naturally flow to the people who want to make plumbing a career.

What are some of the things you do in your role of vice president?
As the estimator, I price out the job and then take over the reins once the project gets under way. I manage the work of groups of plumbers. This involves scheduling, assigning jobs, monitoring progress, and ensuring quality work. I don't do much plumbing any more. I am a contact point between the crews and the customers. This is great because it keeps me in touch with the people we work for. But it is also a lot of responsibility. At first, I was just working on different projects. Now, I am in charge of about 15 other people and their work. It is up to me to make sure that everything is done right.

What do you like most about your job?
The satisfaction of doing a good job. The best reward is a happy customer. I like working with my hands and seeing the results of my efforts. It is a real ego boost when a customer pays you a compliment for fine work. Of course, I like the feeling of accomplishment—knowing that the job is done and done right.

What would you say to someone entering the trades today?
Today there is a real shortage of skilled labor. My advice would be to go to school. Learn everything you can about the craft. There is a big advantage to learning the basics in the technological environment of an apprenticeship program. There is so much more to plumbing than meets the eye. We have guys show up and say, "I'm a plumber." But they know only *how* to do something, not *why* it should be done. If you learn a task simply because you have done it over and over again, you may not understand why it should be done that way. You must understand the *why* if you are to use the skill in all the possible situations that come up. Real understanding means you can apply

what you know to different jobs. You can learn the hands-on part of plumbing—the how-to—pretty fast in the field. It is critical that you also know why you are doing it.

Apprenticeship programs introduce you to many aspects of a trade. You might be working for a residential company. In a program you will be exposed to commercial and industrial skills. This diversity helps you find your area of special interest and expertise. There are so many different paths you can take in plumbing. Get all the exposure you can while you are learning.

I would also say, "Think about the future." To some people, it makes no difference if they are flipping burgers or learning a trade that will last them a lifetime. So long as the money is good for the moment, they don't care. That is a real waste.

GLOSSARY

Trade Terms Introduced in This Module

Amperage: A measure of electrical current.

Bar stock: Uncut bar material on hand for use on a project.

Brazing: Joining two pieces of metal by soldering them together with a nonferrous metal such as brass.

Burr: Uneven or jagged edge left on metal by certain cutting tools.

Diameter: The distance across the center of a circle.

Electrical ground: A conductive connection that provides a path for electrical current to pass from an electrical component into the earth.

Ferrous: Containing iron.

Flux: A substance that facilitates the fusion (joining) of metals and helps prevent surface oxidation (rusting, tarnishing) during welding, brazing, and soldering.

Joist: A piece of lumber used horizontally as a support for a ceiling or a floor.

Kerf: The cut or groove made by a saw blade, determined by the way the teeth are set on the blade.

Nonferrous: Not containing iron and therefore not magnetic.

Offset: A change in alignment; in plumbing, a ⅛-bend offset routes a drain line around an obstacle.

Scribe: A sharply pointed and hardened steel tool used for marking a surface to be cut by etching a line or a point into the surface.

Solder: An alloy (tin plus antimony, copper, and silver) with a low melting point used to join metals or seal joints.

Soldering: Method of joining metals or sealing joints using solder and heat.

Straightedge: A length of wood or metal that does not bow or twist along its length.

Temper: The strength of a metal. Tempering is a process of heating and rapidly cooling metal to increase its strength.

Tolerance: Allowable variation in a given measurement or quantity.

Torque: Twisting or turning force applied in a rotating motion. Measurements are given in either inch-pounds or foot-pounds.

Tube stock: Uncut tubing material on hand for use on a project.

Voltage drop: The tendency for electricity traveling through the length of an extension cord to lose voltage.

ANSWER KEY

Answers to Review Questions

Section 4.0.0

1. b
2. c
3. a
4. d
5. d

Section 5.0.0

1. c
2. d
3. a
4. a
5. d

Section 6.0.0

1. b
2. d
3. b
4. a
5. b

Section 7.0.0

1. a
2. b
3. b
4. c
5. d

Sections 8.0.0–8.4.0

1. b
2. c
3. a
4. d
5. c

Sections 8.5.0–10.0.0

1. c
2. b
3. d
4. c
5. a

Sections 11.0.0–15.3.0

1. c
2. a
3. d
4. c
5. d
6. c
7. c
8. c
9. b
10. b

NCCER CRAFT TRAINING USER UPDATES

The NCCER makes every effort to keep these textbooks up-to-date and free of technical errors. We appreciate your help in this process. If you have an idea for improving this textbook, or if you find an error, a typographical mistake, or an inaccuracy in the NCCER's Craft Training textbooks, please write us, using this form or a photocopy. Be sure to include the exact module number, page number, a detailed description, and the correction, if applicable. Your input will be brought to the attention of the Technical Review Committee. Thank you for your assistance.

Instructors – If you found that additional materials were necessary in order to teach this module effectively, please let us know so that we may include them in the Equipment/Materials list in the Instructor's Guide.

Write: Curriculum Revision and Development Department
National Center for Construction Education and Research
P.O. Box 141104, Gainesville, FL 32614-1104

Fax: 352-334-0932

E-mail: curriculum@nccer.org

Craft ______ Module Name ______

Copyright Date ______ Module Number ______ Page Number(s) ______

Description ______

______

______

______

(Optional) Correction ______

______

______

(Optional) Your Name and Address ______

______

______

Module 02103-00

Introduction to Plumbing Math

Course Map

This course map shows all of the modules in the first level of the Plumbing Curriculum. The suggested training order begins at the bottom and proceeds up. Skill levels increase as you advance on the course map. The local Training Program Sponsor may adjust the training order.

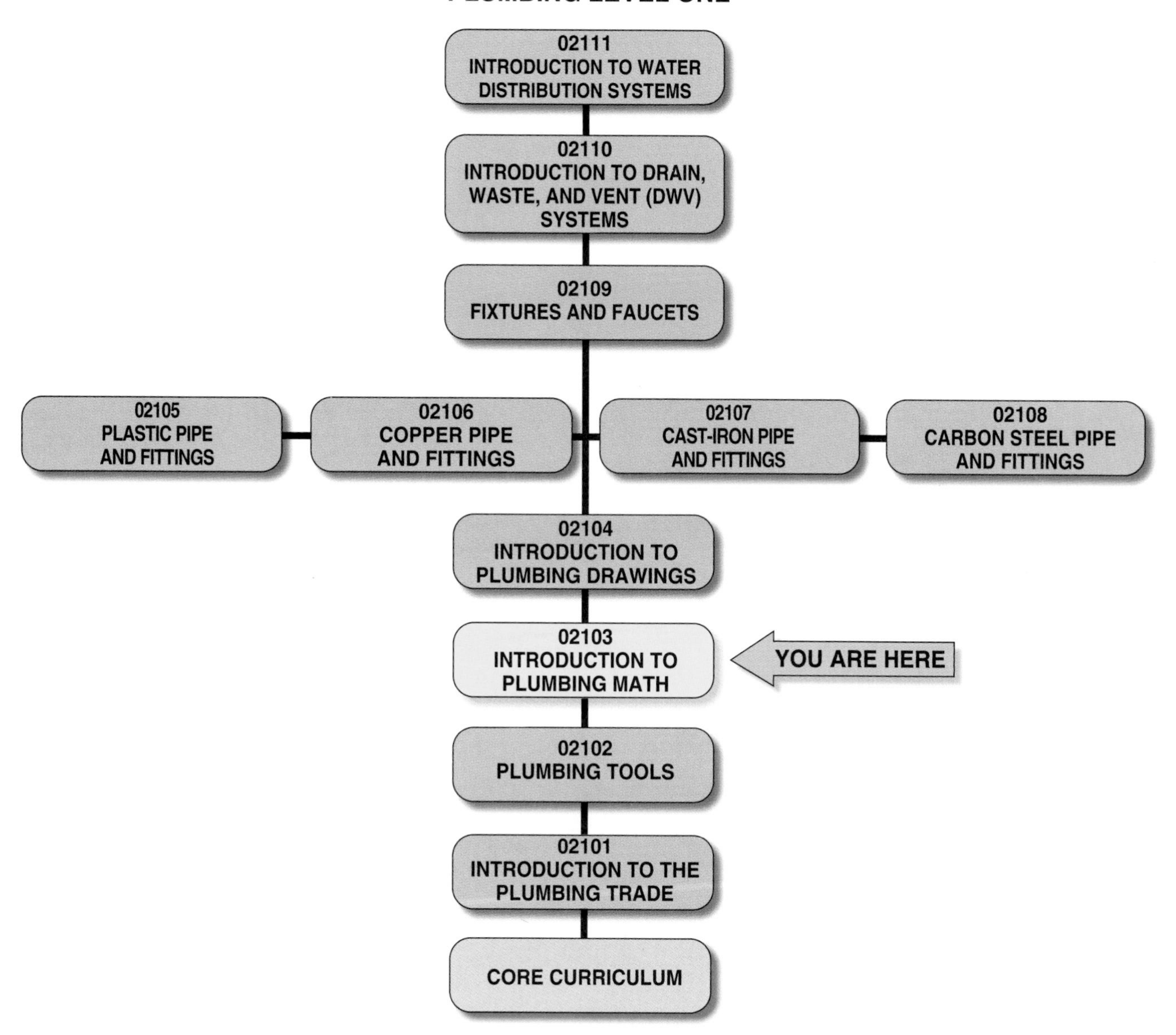

103CMAP.EPS

Copyright © 2000 National Center for Construction Education and Research, Gainesville, FL 32614-1104. All rights reserved. No part of this work may be reproduced in any form or by any means, including photocopying, without written permission of the publisher.

MODULE 02103 CONTENTS

Figures

Table

MODULE 02103

Introduction to Plumbing Math

Objectives

When you finish this module, you will be able to:

1. Identify the parts of a fitting and use common pipe measuring techniques.
2. Use fitting dimension tables and a framing square to determine fitting allowances and pipe makeup.
3. Calculate end-to-end measurements by figuring fitting allowances and pipe makeup.
4. Use a framing square to find the center of fittings.
5. Figure 45-degree offsets and travel using the Pythagorean theorem.
6. Figure 45-degree offsets and travel using the constant method.
7. Figure 45-degree offsets and travel using a framing square and a wooden rule or tape measure.

Prerequisites

Before you begin this module, it is recommended that you successfully complete task modules: Core Curriculum; Plumbing Level One, Modules 02101 and 02102.

Materials Required for Trainees

1. This module
2. Appropriate personal protective equipment
3. Sharpened pencil and paper

1.0.0 ◆ INTRODUCTION

Why use math? Math is used in all phases of construction. Plumbers use math to read plans, calculate pipe length, lay out fixtures, and much more. Good math skills will help you advance in the plumbing trade.

Mathematics is the best tool you'll ever own. It's free, comes with a lifetime guarantee, and actually gets better the more you use it. It is also your daily companion. This module introduces some of the basic mathematics used by plumbers in the field, beginning with how to measure pipe and ending with how to calculate simple 45-degree **offsets.**

2.0.0 ◆ BASIC MATH REVIEW

You learned basic math in the Core Curriculum. The following sections highlight what you covered there.

2.1.0 Review of Whole Numbers

Whole numbers are complete units without any fractions or decimals. Examples of whole numbers are:

15 32 60 144 2,436

2.1.1 Place Value

Each digit in a whole number represents a place value. Each digit has a value that depends on its place, or location, in the whole number. In the whole number 5,679, for example, the place value of the 5 is five thousand, and the place value of the 6 is six hundred.

Numbers larger than zero are called positive (+) numbers (such as 1, 2, 3). Numbers less than zero are negative (−) numbers (such as −1, −2, −3).

Zero is neither positive nor negative. Numbers without a minus sign in front of them are positive.

2.1.2 Addition

To *add* means to combine the value of two or more numbers. The total when you add two or more numbers is called the *sum*. The sign for addition is the plus sign (+). For example, add these numbers:

$$56 + 32 =$$

The answer is 88.

2.1.3 Subtraction

Subtraction means finding the difference between two numbers, or taking away one number from another. The subtraction sign (−) is also called the *minus* sign. The result (answer) of a subtraction problem is called the *difference*. For example, subtract these numbers:

$$56 - 32 =$$

The answer is 24.

2.1.4 Multiplication

Multiplication is the quick way to add the same number together many times. The symbol for multiplication is the × sign. For example, you must deliver 5 boxes each to 8 job sites. How many boxes do you deliver in all?

$$5 \times 8 =$$

$$(5 + 5 + 5 + 5 + 5 + 5 + 5 + 5)$$

The answer is 40.

2.1.5 Division

Division is the opposite of multiplication. The symbol for division is the ÷ sign. Instead of adding a number several times ($5 + 5 + 5 = 5 \times 3 = 15$), when you are dividing you subtract it several times. But just as with using multiplication instead of addition, you can solve a problem faster by using division instead of subtracting the same number over and over. For example, you have 40 boxes to deliver to 8 different job sites. How many boxes go to each site?

$$40 \div 8 =$$

$$(40 - 8 - 8 - 8 - 8 - 8)$$

The answer is 5.

2.2.0 Review of Fractions

A fraction divides whole units into parts. Common fractions are written as two numbers, separated by a slash or by a horizontal line, like this:

$$1/2 \text{ or } \frac{1}{2}$$

The top number is the *numerator*. The bottom number is the *denominator*.

The slash or horizontal line means the same thing as the ÷ sign. So think of a fraction as a division problem. The fraction ½ means 1 divided by 2, or one divided into two equal parts. Read this fraction as "one-half."

2.2.1 Equivalent Fractions

Equivalent fractions have the same value or are equal. For example, ½, 2/4, 4/8, and 8/16 are equivalent fractions. If you cut off a piece of wood 8/16-inch long and another trainee cuts off a piece ½-inch long, the two pieces would be the same length.

2.2.2 Lowest Terms

When you are working with fractions, it is often best to reduce them to lowest terms. To reduce a fraction, ask yourself, "What is the largest number that I can divide evenly into both the numerator and the denominator?" If you're unsure about the largest number that divides evenly into both the numerator and the denominator, try using the number 2 or 3 to start, and then continue until the fraction is at its lowest form. If there is no number (other than 1) that will divide evenly into both numbers, the fraction is already in its lowest terms. For example, the following fractions reduce to their lowest forms:

2/4 = ½
4/16 = ¼
8/32 = ¼
3/8 = 3/8 (already in lowest terms)
7/16 = 7/16 (already in lowest terms)

2.2.3 Common Denominator

A common denominator means the bottom number, or denominator, in a group of fractions is the same: Some examples are ¼, 2/4, ¾. It is important that numbers have the same denominators to compare fractions. For example, which of these fractions is larger?

$$\frac{3}{4} \text{ or } \frac{5}{8}$$

To compare, you need to find a common denominator for the fractions. The common denominator is a number that both denominators can go into evenly. Here's one way to find a common denominator.

Step 1 Multiply the two denominators ($4 \times 8 = 32$). The result is a common denominator for $\frac{3}{4}$ and $\frac{5}{8}$.

You found a common denominator so that you can compare the fractions more easily. Now convert the two fractions so that they will have the same denominator by multiplying the numerator and denominator by the same number.

Step 2 Convert $\frac{3}{4}$ and $\frac{5}{8}$ to fractions having the common denominator of 32.

$$\frac{3 \times 8}{4 \times 8} = \frac{24}{32}$$

$$\frac{5 \times 4}{8 \times 4} = \frac{20}{32}$$

Now it's easy to compare the two fractions to see which is larger: $\frac{3}{4}$ is larger than $\frac{5}{8}$. In construction many of our denominators will be 2, 4, 8, or 16. In the earlier example it's easier to use 8 as the common denominator and multiply $\frac{3}{4}$ by a number that will convert the denominator to an 8. That number is 2.

$$\frac{3 \times 2}{4 \times 2} = \frac{6}{8}$$

Now you can see that $\frac{3}{4}$, which is equivalent to $\frac{6}{8}$, is more than $\frac{5}{8}$.

2.2.4 Improper Fractions

An improper fraction has a numerator larger than the denominator. For example, $\frac{11}{8}$ is an improper fraction.

In this case, you need to reduce the improper fraction $\frac{11}{8}$ to its lowest terms. Improper fractions can be converted to mixed numbers. For example, $\frac{11}{8} = 1\frac{3}{8}$. The answer is a combination of a whole number and a fraction, or a mixed number.

Sometimes, you will need to use improper fractions. To change a whole number into an improper fraction, simply place the number over 1. Remember, fractions express division. When you divide any number by 1, the result is the same number. Here is an example:

$$4 = 4/1$$

To change a mixed number (for example, $2\frac{1}{8}$) to an improper fraction, follow these steps:

Step 1 Multiply the whole number by the denominator.

Step 2 Add the result to the numerator and put it over the denominator.

$$2\frac{1}{8} = (2 \times 8) + 1 = \frac{17}{8}$$

2.2.5 Addition

How many total inches will you have if you add $\frac{3}{4}$ inch plus $\frac{5}{8}$ inch? To answer this question, you will have to add two fractions using the following steps.

Step 1 Find the common denominator of the fractions you wish to add (32).

Step 2 Convert the fractions to equivalent fractions with the same denominator.

Step 3 Add the numerators of the fractions. Place this sum over the denominator.

$$\frac{24}{32} + \frac{20}{32} = \frac{44}{32}$$

Step 4 Reduce the fraction to its lowest terms. For this example, it is $\frac{11}{8}$.

2.2.6 Subtraction

Subtracting fractions is very much like adding fractions. You must find a common denominator before you subtract. For example, subtract the following fractions.

$$\frac{7}{8} - \frac{1}{4}$$

Step 1 Find the common denominator. Multiply the entire fraction ($\frac{1}{4}$) by 2 to get a new fraction with a denominator of 8.

$$\frac{1 \times 2}{4 \times 2} = \frac{2}{8}$$

Step 2 Rewrite the equation and subtract the numerators. The difference is $\frac{5}{8}$.

$$\frac{7}{8} - \frac{2}{8} = \frac{5}{8}$$

2.2.7 Multiplication and Division

Multiplying and dividing fractions is different from adding and subtracting them. You do not have to find a common denominator when you multiply or divide fractions.

Using $\frac{3}{4} \times \frac{5}{6}$ as an example, follow these steps:

Step 1 Multiply the numerators together to get a new numerator. Multiply the denominators together to get a new denominator.

$$\frac{3 \times 5}{4 \times 6} = \frac{15}{24}$$

Step 2 Reduce if possible ($\frac{15}{24}$ reduces to $\frac{5}{8}$).

Dividing fractions is very much like multiplying fractions, with one difference. You must invert, or flip, the fraction you are dividing by. Using $\frac{3}{8} \div \frac{3}{4}$ as an example, follow these steps:

Step 1 Invert the fraction you are dividing by ($\frac{3}{4}$).

$$\frac{3}{4} \text{ becomes } \frac{4}{3}$$

Step 2 Change the division sign (÷) to a multiplication sign (×).

$$\frac{3 \div 3}{8 \div 4} = \frac{3 \times 4}{8 \times 3}$$

Step 3 Multiply the fraction.

$$\frac{3 \times 4}{8 \times 3} = \frac{12}{24}$$

Step 4 Reduce to lowest terms. In this example, $\frac{12}{24}$ equals $\frac{1}{2}$.

If you are working with a whole number or a mixed number (for example, $2\frac{1}{8}$), you must convert it to a fraction before you invert it. In the following example, $2\frac{1}{8}$ is divided by 4.

$$2\tfrac{1}{8} \div 4 = ?$$

Step 1 Convert the 2 to an improper fraction and then multiply it by the lowest common denominator.

$$2 = \frac{2}{1} \text{ then } \frac{2 \times 8}{1 \times 8} = \frac{16}{8}$$

Step 2 Add that to the fraction portion of the number.

$$\frac{16}{8} + \frac{1}{8} = \frac{17}{8}$$

Step 3 Now you have the following equation:

$$\frac{17}{8} \div 4 = ?$$

Step 4 Convert 4 to an improper fraction.

$$4 = \frac{4}{1}$$

Step 5 Invert $\frac{4}{1}$ and multiply by $\frac{17}{8}$.

$$\frac{17 \times 1}{8 \times 4} = \frac{17}{32}$$

Step 6 Determine if the fraction can be further reduced to lower terms. In this case, $\frac{17}{32}$ is in its lowest form.

2.3.0 Review of Decimals

Decimals represent values less than one whole unit. They are fractions expressed in a different form. You are already familiar with decimals in the form of money.

25¢ = 0.25 or $\frac{25}{100}$
10¢ = 0.10 or $\frac{10}{100}$
50¢ = 0.50 or $\frac{50}{100}$

On the job, you may need to use decimals to read instruments or calculate flow rates.

The following chart compares whole number place values with decimal place values:

Whole Numbers		Decimals	
1	ones		
10	tens	0.1	tenths
100	hundreds	0.01	hundredths
1000	thousands	0.001	thousandths

2.3.1 Addition and Subtraction

There is one major rule to remember when adding and subtracting decimals:

Keep your decimal points lined up!

Suppose you want to add 4.76 and 0.834. Line up the problem like this:

```
  4.760   You can add a 0 to help keep the numbers
+ 0.834   lined up.
  5.594
```

The same is true for subtraction. Line up the decimal points.

```
  5.6       5.600   Notice that two zeros were added
- 2.724   - 2.724   to the end of the first number to
            2.876   make it easier to subtract.
```

2.3.2 Multiplication

To multiply decimals, set up the problem as you would with whole numbers as follows:

```
  4.5
×   7
```

Step 1 Multiply.

```
   4.5
 ×   7
   315
```

Step 2 Once you have the answer, count the number of digits to the right of the decimal point in both numbers being multiplied. (In this example, there is only one decimal point [4.5] and only one number to the right of it.)

Step 3 In the answer, count over the same number of digits (from right to left) and place the decimal point there.

```
  4.5   (count one total digit to the right of the
×   7   decimal point in the two numbers)
 31.5   (count in one digit from right to left in
        the answer, and place the decimal
        point there)
```

2.3.3 Division

There are three types of division problems involving decimals:

- Those that have a decimal point in the number being divided (the dividend)

  22)44.5

- Those that have a decimal point in the number you are dividing by (the divisor)

  0.22)4,450

- Those that have decimal points in both numbers (the dividend and the divisor)

  0.22)44.5

For the first type of problem, let's use 44.5 ÷ 22 as our example.

Step 1 Place a decimal point directly above the decimal point in the dividend.

22)44.5

Step 2 Divide as usual.

```
     2.0
 22)44.5
    44
    00.5r
```

How many 22-inch pieces of pipe will you have? The answer: two, with a little (.5 inch) left over. This leftover portion is called the **remainder** and is expressed as *r*.

For the second type of problem, let's use 4,450 ÷ .22 as our example.

Step 1 Move the decimal point in the divisor to the right until you have a whole number.

.22)4450 (The decimal point in the divisor will be moved two places to the right.)

Step 2 Move the decimal point in the dividend the same number of places to the right. You may have to add zeros first. Then divide as usual.

```
     20227.2
  22)4450.00.
     44
     0050
     0044
     00060
     00044
     000160
     000154
     0000060
     0000044
     0000016r
```

(After you have added the zeros and moved the decimal in the dividend two places to the right, the number becomes 445,000.)

For the third type of problem, let's use 44.5 ÷ .22 as our example.

Step 1 Move the decimal point in the divisor to the right until you have a whole number.

.22)44.5 (After you have moved the decimal point in the divisor to the right, .22 becomes 22.)

Step 2 Move the decimal point in the dividend the same number of places to the right. Then divide as usual.

22)4450 (Moving the decimal point in the dividend two places to the right so 44.5 becomes 4,450.)

2.3.4 Rounding Decimals

Often, calculations with decimals produce very precise answers like 25.29411764. But in most practical measurements, you probably only need an answer to the nearest tenth (0.1). For this exercise, you will round 25.29411764 to the nearest tenth:

Step 1 Underline the place to which you are rounding.

25.2̲9411764

Step 2 Look at the digit one place to its right.

25.2̲**9**411764

Step 3 If the digit to the right is 5 or more, you will round up by adding 1 to the underlined digit. If the digit is less than 5, leave the underlined digit the same. In this example, the digit to the right is 9, which is more than 5, so you round up by adding 1 to the underlined digit.

25.3**9**411764

Step 4 Drop all other digits to the right.

25.3

2.4.0 Review of Conversion Processes

In some situations, you need to convert the numbers you want to work with so that all your numbers are in the same form. For example, you may have some numbers that appear as decimals, some that appear as percents, and some that appear as fractions. Decimals, percents, and fractions are all just different ways of expressing the same thing. The decimal 0.25, the percent 25%, and the fraction ¼ all mean the same thing. To work successfully with the different forms of numbers like these, you will need to know how to convert them from one form into another.

2.4.1 Decimals to Percents and Percents to Decimals

What are percents? Think of a whole number divided into 100 parts. You can express any part of the whole as a percent. Percents are an easy way to express parts of a whole. Decimals and fractions also express parts of a whole. Let's look at the relationship among percents, decimals, and fractions.

Sometimes you may need to express decimals as percents or percents as decimals. Suppose you are preparing a gallon of cleaning solution. The mixture should contain 10 to 15 percent of the cleaning agent. The rest should be water. You have 0.12 gallon of cleaning agent. Will you have enough to prepare a gallon of the solution? To answer the question, you must convert a decimal (0.12) to a percent. You will change 0.12 to a percent for this exercise.

Step 1 Multiply the decimal by 100. (Move the decimal point two places to the right.)

$$0.12 \times 100 = 12$$

Step 2 Add a % sign.

12%

You have enough cleaning agent to make the solution. Recall that the mixture should be from 10 to 15 percent cleaning agent. You have 12 percent.

You may also need to convert percents to decimals. Let's say that another mixture should contain 22 percent of a certain chemical by weight. You're making 1 pound of the mixture. You weigh the ingredients on a digital scale. How much of the chemical should you add? To answer this, you must convert a percent (22%) to a decimal. You will change 22 percent to a decimal in the following exercise:

Step 1 Drop the % sign.

22

Step 2 Divide the number by 100. (Move the decimal point two places to the left.)

$$22 \div 100 = 0.22$$

The answer to the problem is that you would add 0.22 pounds of the chemical to make a 22 percent mixture.

2.4.2 Fractions to Decimals

You will often need to change a fraction to a decimal. For example, you need ¾ of a pound of material. How do you convert ¾ to its decimal equivalent?

Step 1 Divide the numerator of the fraction by the denominator.

4)3.0

In this example, you need to put the decimal point and the zero after the number 3, because you need a number large enough to divide by 4.

Step 2 Put the decimal point directly above its location within the division symbol.

```
  .?
4)3.0
```

Step 3 Once the decimal point is in its proper place above the line, you can divide as you normally would. The decimal point *holds* everything in place.

```
  .75
4)3.00
  2.8
  0.20
  0.20
  0.00
```

Step 4 Read the answer. The fraction ¾ converted to a decimal is 0.75, so ¾ of a pound is the same as 0.75 pound.

2.4.3 Decimals to Fractions

Let's say you have 0.25 of a pound of nails. What fraction of a pound is that? Follow these steps to find out:

Step 1 Say the decimal in words.
0.25 is expressed as "twenty-five hundredths"

Step 2 Write the decimal as a fraction.
0.25 is written as a fraction as $^{25}/_{100}$

Step 3 Reduce it to its lowest terms.

$$\frac{25}{100} = \frac{25 \div 25}{100 \div 25} = \frac{1}{4}$$

Step 4 You can see that 0.25 converted to a fraction is ¼. If you have 0.25 of a pound of nails, you have ¼ of a pound.

When calculating offsets, plumbers need to be able to convert a decimal to the nearest $^{1}/_{16}$ of an inch. You'll learn more about calculating offsets later in this module.

2.4.4 Inches to Decimal Equivalents in Feet

What happens if you need to convert inches to their decimal equivalents in feet? For example: 3 inches equals what decimal equivalent in feet?

Here's a hint: First, express the inches as a fraction that has 12 as the denominator. You use 12 because there are 12 inches in a foot. Then reduce the fraction, and convert it to a decimal.

In this example, the fraction $^{3}/_{12}$ reduces to ¼.

You convert the fraction ¼ to a decimal by dividing the 4 into 1.00:

```
     .25
  4)1.00
    0.8
    0.20
    0.20
```

Thus, 3 inches converts to 0.25 feet.

Review Questions

Section 2.0.0

1. 32 rolls of insulation
+75 rolls of insulation

a. 108
b. 107
c. 97
d. 100

2. 26 neoprene gaskets
−17 neoprene gaskets

a. 19
b. 9
c. 8
d. 10

3. 9 fixtures
×6 apartments

a. 45
b. 63
c. 15
d. 54

4. Include the remainder (if there is one) for the following division problem.

$$16 \div 5 =$$

a. 11
b. 9
c. 3 *r*1
d. 2 *r*5

5. Find the equivalent of the following measurement: ⅜ inch = ___/32 inch.

a. 12
b. 14
c. 16
d. 18

6. The lowest form of the fraction $^{4}/_{16}$ is _____.

a. 1
b. ¼
c. ⅛
d. ½

7. Find a common denominator for the following pair of fractions:

$$\frac{2}{6} \quad \frac{3}{8}$$

The answer is _____.

a. 6
b. 24
c. 12
d. 16

8. Find the lowest common denominator for the following pair of fractions:

$$\frac{4}{12} \quad \frac{8}{12}$$

The answer is _____.

a. 6
b. 4
c. 3
d. 8

9. $\frac{5}{8} + \frac{1}{16} =$

a. $^{11}/_{16}$
b. $^{14}/_{16}$
c. $^{10}/_{24}$
d. $^{24}/_{24}$

10. Find the answer to the following subtraction problem and reduce to its lowest terms.

$$\frac{12}{16} - \frac{4}{8} =$$

a. $^{4}/_{16}$
b. $^{1}/_{16}$
c. $^{1}/_{8}$
d. $^{1}/_{4}$

11. Find the answer to the following multiplication problem and reduce to its lowest terms.

$$\frac{3}{4} \times \frac{5}{8} =$$

a. $^{15}/_{32}$
b. $^{5}/_{6}$
c. $^{21}/_{12}$
d. $^{6}/_{7}$

12. Find the answer to the following division problem and reduce it to its lowest terms.

$$\frac{6}{8} \div 3 =$$

a. $^{3}/_{24}$
b. $^{1}/_{4}$
c. $^{1}/_{8}$
d. $^{9}/_{8}$

13. 2.8
4.2
+6.0

a. 13.0
b. 12.8
c. 113
d. 130

14. Yesterday, a supply yard contained 6.7 tons of sand. Since then, 2.3 tons were removed. The supply yard now contains _____ tons of sand.
a. 9
b. 44
c. 0.9
d. 4.4

15. If cast-iron pipe weighs 4.75 pounds per foot, 128 feet weighs _____ pounds.
a. 6,080
b. 608
c. 60.8
d. 90

16. 55.35 ÷ 18 = ____
a. 307.5
b. 0.307
c. 3.075
d. 30.75

17. 90 ÷ 0.45 = _____
a. 20.02
b. 2.20
c. 200
d. 22

18. 28.9 ÷ 3.4 = _____
a. 85
b. 8.5
c. 850
d. 0.85

19. If tubing costs $4.20 per foot and you pay a total of $120.95, you have purchased _____ feet of tubing. (Round your answer to the nearest tenth.)
a. 28.6
b. 2.8
c. 287.9
d. 28.8

20. 0.42 = _____
a. 4.2 percent
b. 42 percent
c. 420 percent
d. 0.42 percent

21. $^{3}/_{16}$ = _____
a. 0.1875
b. 3.187
c. 0.316
d. 1.875

22. 0.12 = _____
a. $^{12}/_{10}$
b. $^{6}/_{5}$
c. $^{0.12}/_{100}$
d. $^{3}/_{25}$

23. Convert the following mixed decimal to its equivalent mixed number (whole number with fraction expressed in lowest terms):

2.8 = _____

a. $2^{4}/_{5}$
b. $2^{1}/_{4}$
c. $2^{8}/_{10}$
d. $2^{8}/_{25}$

24. Convert the following to its decimal equivalent in feet:

6.85 inches = _____ feet

a. 0.75
b. 0.60
c. 0.57
d. 0.50

25. Convert the following to its decimal equivalent in feet:

3 inches = _____ feet

a. 0.50
b. 0.05
c. 0.52
d. 0.25

3.0.0 ◆ PIPE MEASURING METHODS

A plumber must be able to measure pipe accurately and quickly. Measuring is basic to the plumbing trade and lies at the heart of many other trade skills as well. In the fabrication and installation of piping systems, plumbers often use the center-to-center (C-C) measurements between two fittings.

The length of pipe and fittings is measured along **centerlines.** The extension of the centerline of the pipe and the centerline of the fitting meet inside the fitting to create a **center point.**

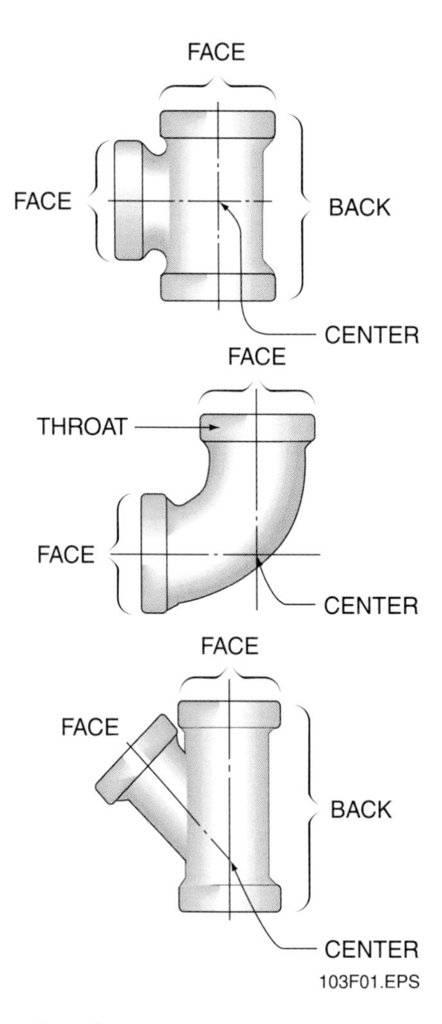

Figure 1 ◆ Basic fitting parts.

3.1.0 Fitting Terminology

The basic parts of a fitting are shown in *Figure 1.* The terms **face, center,** and **back** are used to describe these parts. These elements are important when you measure pipe length, because they define the beginning and ending points of the measurement. The **throat** of the fitting, also shown in *Figure 1,* is not used as often in measuring pipe.

Figure 2 shows the various methods that are used to measure pipe lengths.

Figure 3 shows the same information in schematic form. **Schematic drawings** are simple, single-line representations of pipe and fittings. It is more convenient to draw this way than to draw the whole pipe or fitting. Compare *Figures* 2 and 3 and note the differences.

When you are sketching pipe assemblies, be sure to place the **dimension** and **extension lines** accurately to prevent any errors in communication.

The dimensions of the pipe assemblies shown in *Figures* 2 and 3 also can be expressed verbally, as *Table 1* shows. Even though speaking this way may seem strange at first, this is part of the language of plumbing. To be a plumber, you must be able to understand, speak, and write this language.

Read each statement out loud and relate each to one of the sketches in *Figure 3.*

3.2.0 Pipe Makeup

Pipe makeup is the distance a pipe goes into the fitting. For threaded pipe this pipe makeup is also called *thread-in.* This is the distance the pipe screws into the fitting (see T in *Figure 4*).

3.2.1 Fitting Allowance

Fitting allowance or take-off is the distance from the end of the pipe that goes into the fitting to the center of the fitting. In *Figure 4,* fitting allowance or take-off is calculated by subtracting T, the thread-in measurement, from C in the center-to-face (C-F) measurement. For example, the fitting take-off for a threaded steel 1-inch 90-degree

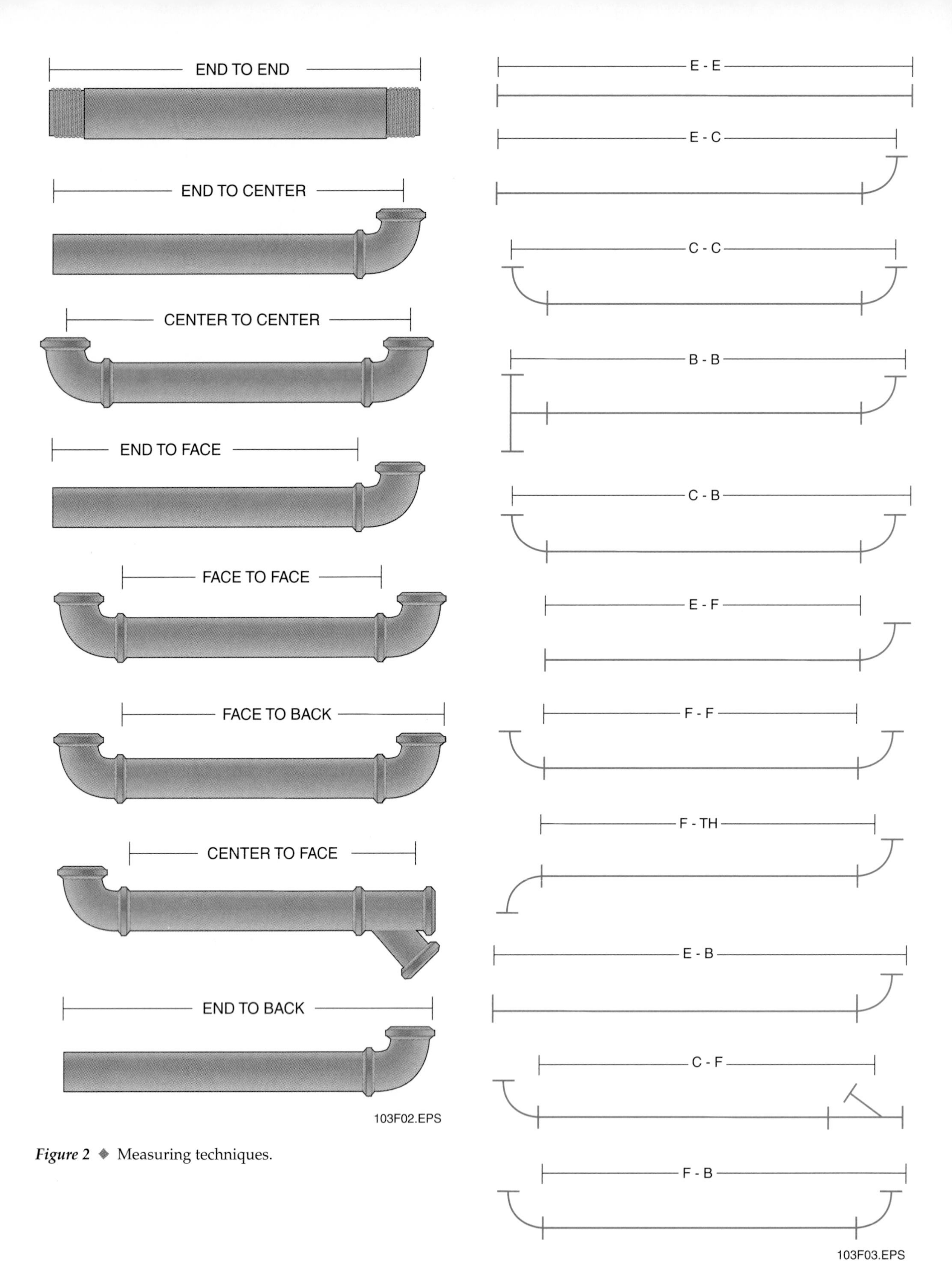

Figure 2 ◆ Measuring techniques.

Figure 3 ◆ Schematic view.

Table 1 Verbal Expression of Pipe Measurements

Pipe Size	Statement	Abbreviation
½"	8 inches end to end	½" 8" E–E
¾"	10 inches end to center of ell	¾" 10" E–C ell
1"	8 inches end to face of ell	1" 8" E–F ell
3"	17 inches end to back of ell	3" 17" E–B ell
1"	11 inches center of ell to center of tee	1" 11" C ell–C tee
2"	8½ inches face of ell to throat of ell	2" 8½" F ell–TH ell
4"	8 inches face of ell to face of wye	4" 8" F ell–F wye
¾"	42 inches center of wye to face of ell	¾" 42" C wye–F ell
6"	14 inches center of tee to back of ell	6" 14" C tee–B ell
1½"	69" face of ell to back of ell	1½" 69" F ell–B ell

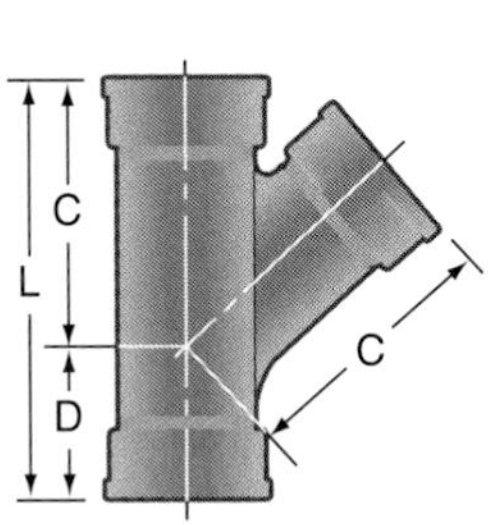

THREADED WYE-45-DEGREE ANGLE

L IS FACE TO FACE
C IS CENTER TO FACE
D IS CENTER TO FACE
T IS THREAD-IN OR MAKE-UP

NOMINAL PIPE SIZE	L	C	D	T
1½"	4¾"	3 5/16"	1 7/16"	½"
2"	5⅞"	4 1/16"	1 13/16"	½"
2" × 1½"	5⅞"	4 ¼"	1⅝"	½"
2½"× 2"	6¼"	4 ⅝"	1⅝"	¾" × ½"

103F04A.EPS

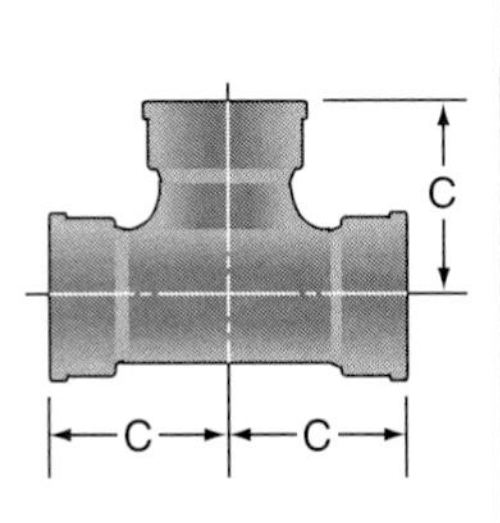

THREADED TEES

C IS CENTER-TO-FACE MEASURE
T IS THREAD-IN OR MAKEUP

NOMINAL PIPE SIZE	C	T
⅜"	1"	⅜"
½"	1⅛"	½"
¾"	1⅜"	½"
1"	1½"	½"
1¼"	1¾"	½"
1½"	1⅞"	½"
2"	2¼"	½"
2" × 1½"	2¼"	½"
2½"	2¾"	¾"

103F04B.EPS

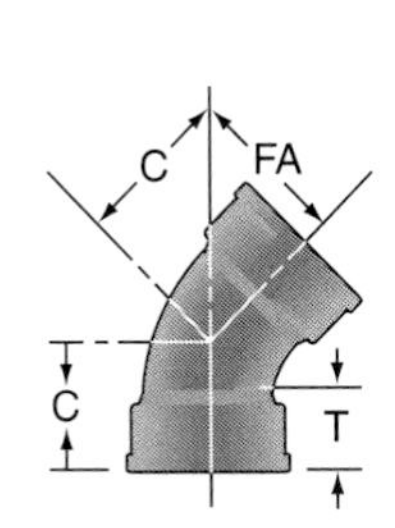

THREADED ELBOWS

FA IS FITTING ANGLE
C IS CENTER-TO-FACE MEASURE
T IS THREAD-IN OR MAKEUP MEASURE

NOMINAL PIPE SIZE	C					T
	FITTING ANGLES					
	90°	60°	45°	22½°	11¼°	
⅜"	1"		¾"			⅜"
½"	1⅛"		¾"			½"
¾"	1⅜"		1"			½"
1"	1½"		1⅛"			½"
1¼"	1¾"	1¼"	1¼"	1⅛"	1"	½"
1½"	1⅞"	1¾"	1½"	1¼"	1⅛"	½"
2"	2¼"	2¼"	1⅝"	1⅜"	1¼"	½"
2½"	2¾"	2½"	1¾"	1½"	1⅜"	¾"

103F04C.EPS

Figure 4 ◆ Fitting dimensions for threaded pipe.

elbow fitting would be 1 inch because C = 1½" and T = ½", so 1½ − ½ = 1.

3.2.2 Threaded Pipe

Pipe manufacturers publish tables of dimensions for their fittings. *Figure 4* shows fitting dimensions for threaded wyes (45-degree angle), threaded tees, and threaded elbows. A capital letter on the drawing of the fitting is keyed to a column on the table. To use the tables, first find the pipe size and then find the dimensions in the columns to the right.

The letter "T" in the last column stands for **thread makeup,** also known as thread-in or thread engagement. For threaded fittings, this is the distance that the pipe screws into the fitting.

3.2.3 PVC Pipe

When joining PVC (polyvinyl chloride) pipe, you must also determine fitting allowances. *Figure 5* shows fitting dimensions for wyes, tees, and elbows.

3.2.4 Cast-Iron (Soil) Pipe

Cast-iron (soil) pipe is not threaded. Two types of cast-iron are cast-iron pipe with hub and spigot fittings and cast iron pipe with no-hub fittings. *Figure 6* shows fitting dimensions for both types of cast-iron (soil) pipe.

3.2.5 Copper DWV

Copper DWV (drain, waste, and vent) pipe is not threaded. *Figure 7* shows fitting dimensions for copper DWV pipe.

3.3.0 Field Measuring Techniques

Even though you can determine pipe makeup by referring to tables published by fitting manufacturers, the most convenient way is simply to measure the fitting. To determine pipe makeup, measure the depth of the socket or thread. To find the approximate center of a fitting when you are working in the field, use a **framing square,** as shown in *Figure 8*. Notice how the inner edge of the **tongue** and **blade** of the framing square have been placed along the centerlines of the fitting.

3.4.0 Calculating Pipe Length

Calculating the length of pipe to cut is determined by what measuring technique you used. You will use fitting dimensions as part of your calculations in most pipe measuring techniques.

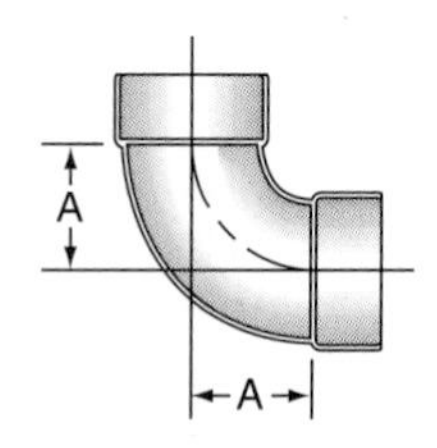

PART NO. 300 BEND (SANITARY 90° Ell) ALL HUB			
SIZE	A	B	C
1½	1¾		
2	2 5/16		
3	3 1/16		
4	3 7/8		
6	5		
8	6		

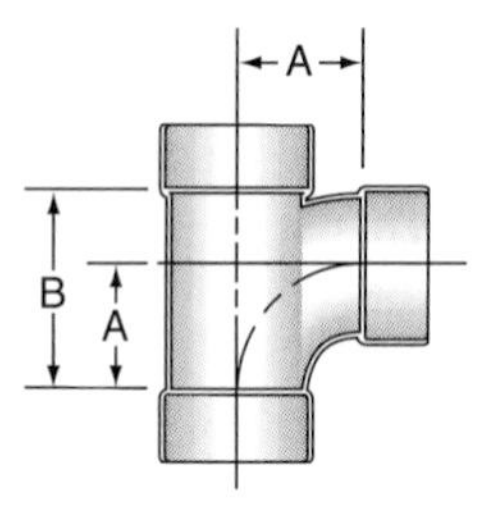

PART NO. 400 SANITARY TEE ALL HUB			
SIZE	A	B	C
1½	1¾	2¾	
2	2 5/16	3 11/16	
3	3 1/16	4 7/8	
4	3 7/8	6 1/8	
6	5	8½	
8	6	10½	

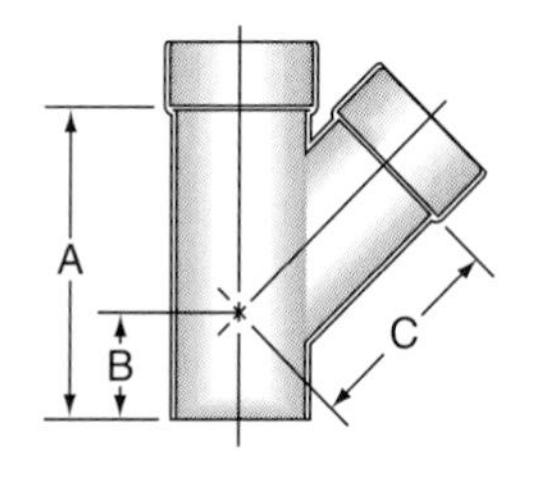

PART NO. 602 WYE, STREET (45° WYE) SPIGOT × HUB × HUB			
SIZE	A	B	C
1½	4¾	1 7/8	2 7/8
2	5 7/8	2¼	3 5/8
3	8 1/8	3 1/8	5
4	10	3 5/8	6 3/8

AVAILABLE IN PVC ONLY

103F05.EPS

Figure 5 ◆ Fitting dimensions for PVC pipe.

3.4.1 Center-to-Center Measurement

Say you want to know how much pipe to cut to fill a space between two fittings. The center-to-center measurement between the two fittings is 50 inches. You are using a 1¼-inch threaded steel pipe, which, as you can see by referring to *Figure 4*, has a ½-inch thread-in (T) and a 1¾-inch center-to-face measurement (C). First, determine the fitting allowance (or take-off) by subtracting T from C. So your equation reads

$$C - T = \text{fitting allowance}$$

$$1\tfrac{3}{4} - \tfrac{1}{2} = 1\tfrac{1}{4}$$

Now, because you have two fittings, you must add the allowances together to get the total fitting allowance. Subtract that total from the 50-inch center-to-center measurement. Your equation reads

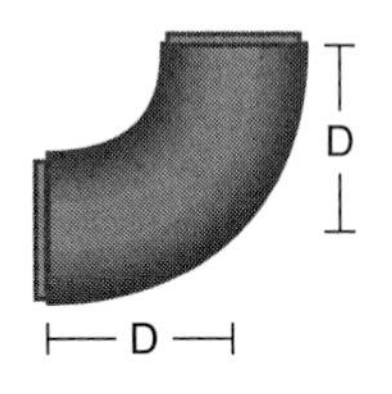

QUARTER BEND		
SIZE (IN.)	D	WEIGHT (LB.)
1 1/2	4 1/4	1.9
2	4 1/2	2.4
3	5	3.9
4	5 1/2	6.0
5	6 1/2	10.0
6	7	12.0
8	8 1/2	25.0
4 × 3	5 1/2	6.0

103F06A.EPS

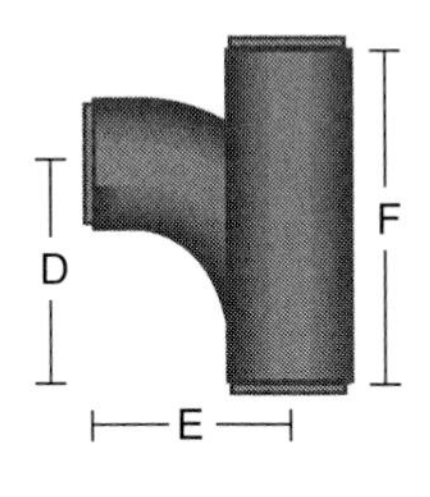

SANITARY TEE				
SIZE (IN.)	E	F	D	WEIGHT (LB.)
1 1/2 × 1 1/2	4 1/4	6 1/2	4 1/4	2.3
2 × 1 1/2	4 1/2	6 5/8	4 1/4	3.0
2 × 2	4 1/2	6 7/8	4 1/4	2.9
3 × 1 1/2	5	6	4 1/4	5.0
3 × 2	5	6 7/8	4 1/2	4.0
3	5	8	5	4.7
4 × 2	5 1/2	6 7/8	4 1/2	6.0
4 × 3	5 1/2	8	5	7.2
4	5 1/2	9 1/8	5 1/2	8.0
5 × 2	6 1/2	8 1/2	5	8.7
5 × 3	6	9 5/16	5 1/2	10.5
5 × 4	6	10 13/32	6	11.5
5	6 1/2	11 7/16	6 1/2	15.0
6 × 2	6 1/2	8 3/16	5	12.0
6 × 3	6 1/2	9 3/16	5 1/2	23.0
6 × 4	6 1/2	10 1/16	6	11.5
6 × 5	7	11 1/2	6 1/2	27.0
6	7	12 1/2	7	15.0
8 × 4	7 1/2	11 1/2	6 1/2	21.0
8 × 5	8	12 1/2	7	24.0
8 × 6	8	13 1/2	7 1/2	24.0
8	8	15 1/2	9 1/2	28.0

103F06B.EPS

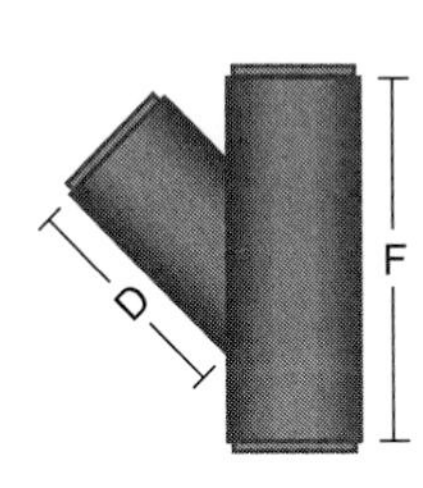

WYE			
SIZE (IN.)	D	F	WEIGHT (LB.)
1 1/2 × 1 1/2	4	6	2.2
2 × 2	4 5/8	6 5/8	2.6
3 × 1 1/2	4 5/8	6 5/8	3.0
3 × 2	5 5/16	6 5/8	3.7
3	5 3/4	8	4.9
4 × 2	6	6 5/8	5.5
4 × 3	6 1/2	8	7.2
4	7 1/16	9 1/2	8.8
5 × 2	7 1/2	8 1/16	7.2
5 × 3	8	9 11/16	10.5
5 × 4	8 1/2	11 3/16	12.0
5	9 1/2	12 5/8	14.5
6 × 2	8 1/4	8 5/16	9.5
6 × 3	8 3/4	9 3/4	11.0
6 × 4	9 1/4	11 3/16	13.0
6 × 5	10 1/4	12 1/2	16.0
6	10 3/4	14 1/16	18.0
8 × 3	9 7/8	10	19.0
8 × 4	10 3/8	11 7/16	22.0
8 × 5	11 3/8	12 7/8	25.0
8 × 6	11 13/16	14 13/16	28.0
8	13 3/8	17 1/8	38.5
10 × 4	11 11/16	12 5/8	37.0
10 × 6	13 1/8	15 7/16	46.0
10 × 8	14 11/16	18 3/8	52.0
10	16 1/2	21 1/2	72.0

103F06C.EPS

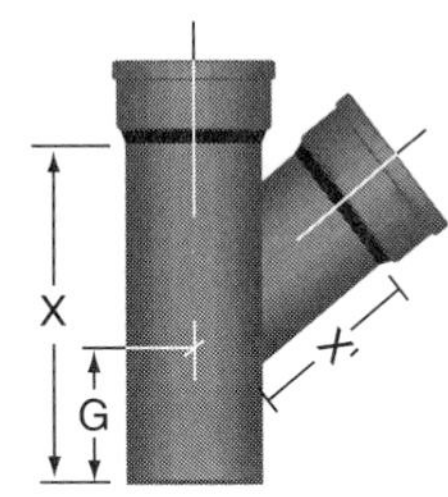

WYE					
				WEIGHT (LB.)	
SIZE (IN.)	G	X	X'	SV	XH
2 × 2	4	8	4	7	8
3 × 2	4 3/16	9	5	10	14
3 × 3	5	10 1/2	5 1/2	13	17
4 × 2	3 5/8	9	5 3/4	12	17
4 × 3	4 7/16	10 1/2	6 1/4	14	20
4 × 4	5 1/4	12	6 3/4	17	24
5 × 2	3 1/8	9	6 1/2	14	–
5 × 3	3 7/8	10 1/2	7	17	24
5 × 4	4 11/16	12	7 1/2	19	27
5 × 5	5 1/2	13 1/2	8	22	32
6 × 2	2 9/16	9	7 1/4	17	–
6 × 3	3 3/8	10 1/2	7 3/4	19	27
6 × 4	4 3/16	12	8 1/4	22	31
6 × 5	4 15/16	13 1/2	8 3/4	24	35
6 × 6	5 3/4	15	9 1/4	28	40
8 × 2	3 1/8	10 1/2	8 1/2	29	–
8 × 3	3 15/16	12	9	32	–
8 × 4	4 3/4	13 1/2	9 1/2	36	52
8 × 5	5 1/2	15	10	39	–
8 × 6	6 5/16	16 1/2	10 1/2	44	63
8 × 8	7 11/16	19 1/2	11 13/16	55	82
10 × 3	2 9/16	12	11	50	–
10 × 4	3 9/16	13 1/2	11 1/8	53	74
10 × 5	4 5/16	15	11 5/8	57	–
10 × 6	5 1/8	16 1/2	12 1/8	61	86
10 × 8	6 1/2	19 1/2	13 7/16	77	110
10 × 10	8	22 1/2	14 1/2	94	133
12 × 4	4 1/8	15	12 7/16	70	–
12 × 5	4 7/8	16 1/2	12 15/16	74	–
12 × 6	5 11/16	18	13 7/16	80	111
12 × 8	7 1/16	21	14 3/4	96	136
12 × 10	8 9/16	24	15 13/16	115	160
12 × 12	10 1/8	27	16 7/8	135	186
15 × 4	2 1/4	15	15	130	130
15 × 6	4	18	15 3/4	109	–
15 × 8	5 3/8	21	17 1/16	127	182
15 × 10	6 7/8	24	18 1/8	152	213
15 × 12	8 7/16	27	19 3/16	242	242
15 × 15	10 3/4	31 1/2	20 3/4	338	338

103F06D.EPS

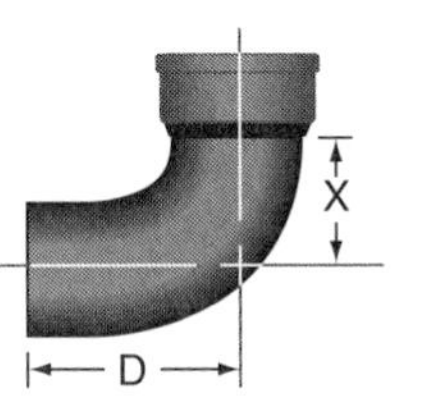

QUARTER BEND				
			WEIGHT (LB.)	
SIZE (IN.)	D	X	SV	XH
2	6	3 1/4	4	5
3	7	4	7	10
4	8	4 1/2	11	15
5	8 1/2	5	13	19
6	9	5 1/2	17	24
8	11 1/2	6 5/8	34	51
10	12 1/2	7 5/8	55	78
12	15	8 3/4	80	111
15	16 1/2	10 1/4	169	169

103F06E.EPS

Figure 6 ◆ Fitting dimensions for cast-iron pipe.

(A) ELLS

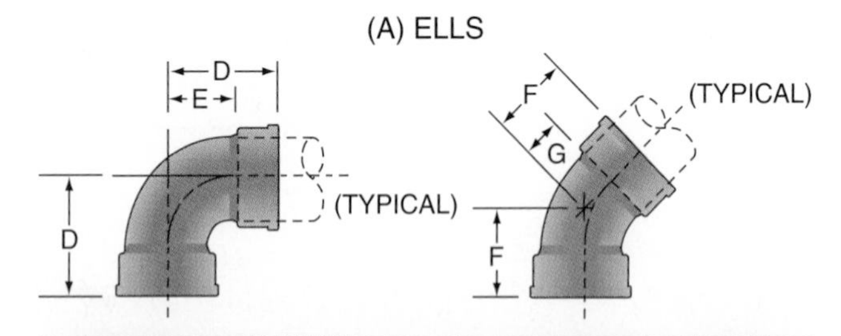

NOTES:
1. DIMENSIONS D AND F ARE CENTER TO END OF FITTING.
2. DIMENSIONS E AND G ARE CENTER OF FITTING TO END OF PIPE.
DIMENSIONS ARE IN INCHES

NOMINAL PIPE SIZE (IN.)	90 - DEG. ELBOWS			45 - DEG. ELBOWS		
	D	E	WT. 90° (LB.)	F	G	WT. 45° (LB.)
¼	1.88	1.56	0.13	1.02	0.70	0.09
⅜	1.94	1.62	0.17	1.20	0.89	0.11
½	2.25	1.94	0.25	1.38	1.06	0.17
¾	1.91	2.59	0.34	1.25	0.94	0.24
1	2.38	2.00	0.53	1.50	1.12	0.35
1¼	2.88	2.44	0.76	1.78	1.34	0.51
1½	3.50	2.94	0.96	2.19	1.62	0.67
2	4.28	3.72	1.81	2.53	1.97	1.21
2½	5.38	4.62	2.65	3.19	2.44	1.77
3	6.17	5.41	4.54	3.52	2.76	2.68
4	7.78	6.97	8.55	4.27	3.45	4.88
5	9.53	8.59	13.32	5.14	4.20	6.87
6	11.17	10.17	19.90	5.91	4.91	11.61
8	14.62	13.38	38.50	7.59	6.34	28.08
10	18.00	16.56		9.22	7.78	
12	21.25	19.69		10.70	9.14	

103F07A.EPS

(B) TEE

NOTES:
1. DIMENSION M IS THE CENTER TO END OF THE FITTING.
2. DIMENSION L IS THE CENTER OF FITTING TO END OF PIPE.
DIMENSIONS ARE IN INCHES

NOMINAL PIPE SIZE (IN.)	STRAIGHT TEES		
	M	L	WT. (LB.)
¼	0.69	0.38	0.13
⅜	0.91	0.59	0.17
½	1.09	0.78	0.23
¾	1.25	0.94	0.34
1	1.50	1.12	0.53
1¼	1.88	1.44	0.87
1½	2.25	1.69	1.21
2	2.50	1.94	1.87
2½	3.00	2.25	2.80
3	3.38	2.62	4.33
4	4.12	3.31	8.09
5	4.88	3.94	11.17
6	5.62	4.62	16.58
8	7.00	5.75	27.27
10	8.50	7.06	
12	10.00	8.43	

103F07B.EPS

Figure 7 ◆ Fitting dimensions for copper DWV pipe.

$$50 - (1\tfrac{1}{4} + 1\tfrac{1}{4})$$
$$50 - 2\tfrac{1}{2} = 47\tfrac{1}{2}$$

You need to cut a pipe that is 47½ inches long.

3.4.2 End-to-Center Measurement

For an end-to-center measurement, just subtract one fitting allowance since there is just one fitting. Use *Figure 4* to calculate the end-to-center measurement for 17½" of ¾" steel pipe.

First determine the fitting allowance or take-off by subtracting the thread-in from the center-to-face measurement:

$1\tfrac{3}{8}" - \tfrac{1}{2}" = \tfrac{11}{8}" - \tfrac{4}{8}" = \tfrac{7}{8}"$ is the fitting allowance.

Now subtract the fitting allowance from the end-to-center measurement.

$17\tfrac{1}{2}" - \tfrac{7}{8}" = 16\tfrac{12}{8}" - \tfrac{7}{8}" = 16\tfrac{5}{8}"$
of ¾" steel pipe is needed.

3.4.3 Face-to-Face Measurement

For a face-to-face measurement you would need to add the pipe makeup for each fitting. Use *Figure 4* to calculate the measurement for 6¾" of ½" steel pipe.

$$F - F = 6\tfrac{3}{4}"$$

Pipe makeup or thread-in = ½"

$6\tfrac{3}{4}" + \tfrac{1}{2}" + \tfrac{1}{2}" = 7\tfrac{3}{4}"$ of ½" steel pipe is needed.

Fitting Allowances

Keep in mind that fitting allowances and makeup vary for each type of pipe. PVC, cast-iron, and copper pipes each have different fitting allowances. The dimensions from the manufacturer are the actual fitting allowances or take-offs for PVC, copper, and hub-and-spigot cast iron pipe. No-hub cast-iron fittings do not have a makeup since there is no socket depth or thread-in as in other fittings. The fitting allowances for these pipes also differ completely from carbon steel threaded pipe.

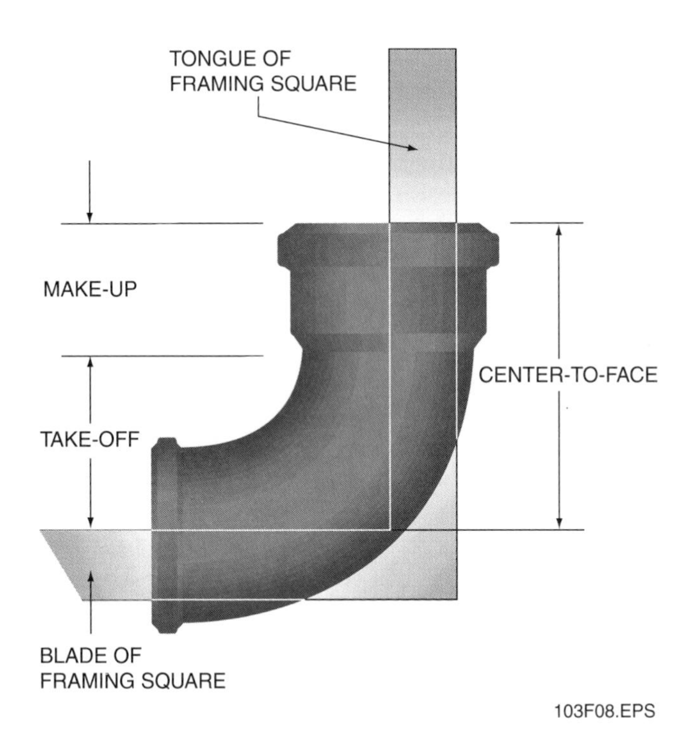

Figure 8 ◆ Measuring makeup with a framing square.

Review Questions

Section 3.0.0

1. The extension of the centerline of the pipe and centerline of the fitting intersect inside the fitting to create a(n) _____.
   a. centerline
   b. single line
   c. center point
   d. end-to-end dimension

2. The distance that a pipe screws into a fitting is called the _____.
   a. thread
   b. makeup
   c. center point
   d. center-to-face dimension

3. A nominal ½-inch pipe that is 10 inches long from the end to the center of an ell is abbreviated as _____.
   a. ½–10 E–C
   b. ½" C–Ell
   c. ½" 10" E–C ell
   d. 10" E–Ell C

4. A _____ can be used in the field to find approximate fitting take-off and pipe makeup.
   a. speed square
   b. framing square
   c. piping rule
   d. folding rule

5. A measurement of a length of pipe that includes the threaded ends is called a(n) _____ measurement.
   a. center-to-center
   b. face-to-face
   c. end-to-thread
   d. end-to-end

6. The end-to-end dimension required for a ⅜" threaded pipe with a center-to-center length of 15¾" is _____.
   a. ⅝"
   b. 14⅜"
   c. 14½"
   d. 15¾"

7. The end-to-end dimension required for a ¾" threaded pipe with a center-to-center length of 41⅞" is _____.
   a. 40⅛"
   b. 41⅞"
   c. ⅞"
   d. 41"

8. The end-to-end dimension required for a 1" threaded pipe with a center-to-center length of 42" is _____.
   a. 42"
   b. 40"

c. 1½"
d. 44"

9. The end-to-end dimension required for a 1½" threaded pipe with a face-to-face length of 18⅞" is _____.
   a. 21⅞"
   b. ½"
   c. 18⅞"
   d. 19⅞"

10. The end-to-end dimension required for a ½" threaded pipe with a face-to-center length of 29" is _____.
    a. 29⅝"
    b. ⅝"
    c. 29"
    d. 1⅛"

4.0.0 ◆ FIGURING 45-DEGREE OFFSETS

The 45-degree offset is very common in plumbing. It's used to get around an object such as a steel beam or other pipes. An offset occurs when a run of pipe changes direction. The term refers to the difference between the centerlines of two parallel pipes. After it changes direction, the new line remains parallel with the centerline of the original pipe.

4.1.0 45-Degree Right Triangle

A 45-degree offset creates a triangle with a 90-degree angle. Such a triangle is called a **right triangle** (see *Figure 9*).

The length of the pipe that runs between the fittings in an offset is called the **travel.** The distance between the centerlines of the two runs of pipe is called the offset. The remaining leg of the right triangle is called the **run.** (See *Figure 9.*)

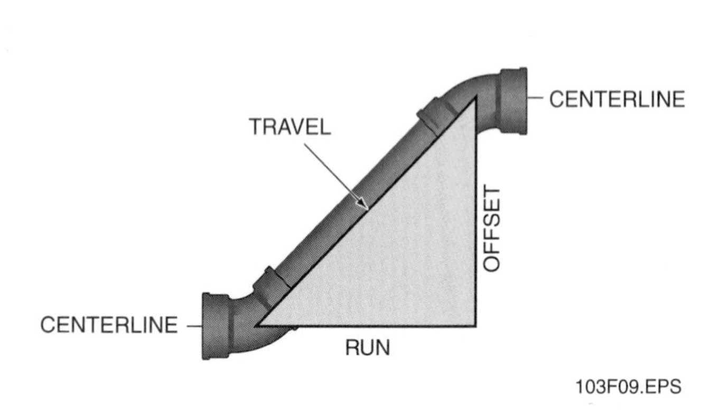

Figure 9 ◆ 45-degree offset.

4.2.0 Finding the Travel

In a 45-degree right triangle, the offset is always the same as the run because two legs of the triangle are equal. To find the travel, you can multiply the length of the run or the offset (since they are equal) times a **constant.**

A constant, as its name suggests, is a number that never changes. For a 45-degree right triangle, the constant is 1.414. Let's look at an example.

4.2.1 Study Example

Figure 10 shows a 45-degree offset. The offset and run are each 12 inches. Find the travel.

Multiply 12 inches by the constant 1.414 to get 16.97 inches. This is the center-to-center distance between the fittings. If this were 4-inch DWV copper pipe, the end-to-end length of the pipe would be 14¹³⁄₁₆ inches. To figure that out, subtract the fitting take-offs from the travel (see the copper fittings table in *Figure 7*). That gives you 14.81 inches, which you would then convert from a decimal to a fraction.

Figure 11 shows a typical installation that requires an offset. Assume that the pipe is being installed from the left. You must first determine where pipe A should end. Because you are using 45-degree ells, the run and offset must be equal. Therefore, if the offset is 18 inches, the run must also be 18 inches. Remember that the 18-inch dimension is taken from the center of the fitting, so you need to know the center-to-face dimension of the fitting as well as the thread engagement, if required.

To find the travel, multiply the offset by the constant 1.414.

$$18" \times 1.414 = 25.45" = 25\tfrac{7}{16}"$$

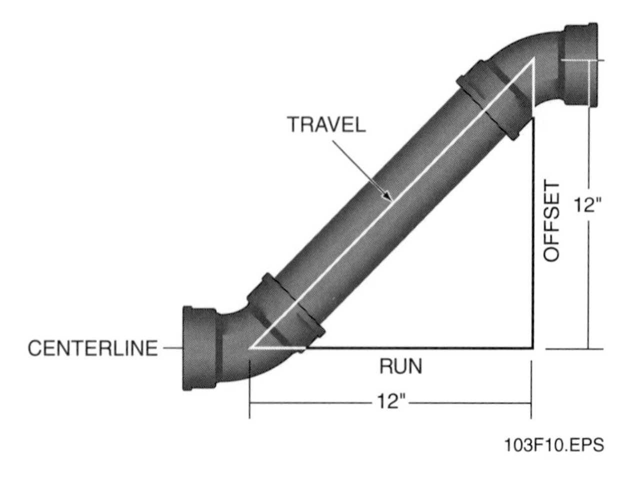

Figure 10 ◆ Figuring the travel.

4.3.0 Where Does the Constant Come From?

Piping offsets form right triangles, and there is a special relationship among the sides of a right triangle that can be expressed in a formula. This formula, called the **Pythagorean theorem,** says that in any right triangle, the longest leg (called the hypotenuse) squared is equal to the sum of the other two legs squared (see *Figure 10*). In piping terms, it looks like this:

$$\text{Travel}^2 = \text{Run}^2 + \text{Offset}^2$$

In practice, you can use the theory to check a corner to be sure it's square. To do this, measure from the corner 3 feet along one wall and place a mark at that point, then measure 4 feet along the other wall and mark that point. Then use a ruler to measure the straight-line distance between those two points. If that distance is 5 feet, the corner is square. You can see this is true by using the Pythagorean theorem as shown below.

$$3^2 + 4^2 = 5^2$$

$$9 + 16 = 25$$

Because the run and the offset are equal in a 45-degree right triangle, the constant (1.414) can be calculated by using 1 unit for the offset and 1 unit for the run:

$$\text{Travel}^2 = 1^2 + 1^2$$

or

$$\text{Travel}^2 = 2$$

To find the value of travel, we must take the square root of each side of the equation:

$$\sqrt{\text{Travel}^2} = \sqrt{2}$$

This yields the following:

$$\text{Travel} = \sqrt{2}$$

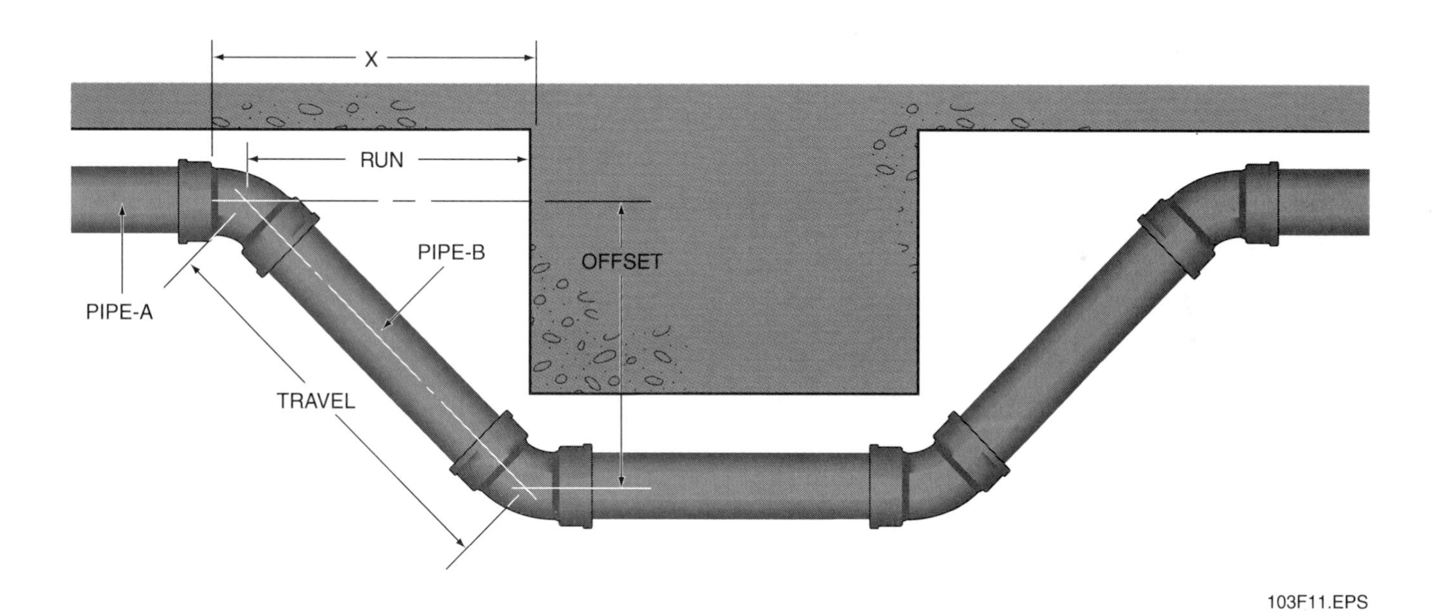

Figure 11 ◆ Calculating pipe offset.

ON THE LEVEL

Using the Constant 1.414

Remember that only 45-degree offsets can be figured with the constant 1.414. Other offsets are figured with different constants, but those offsets are not discussed in this module.

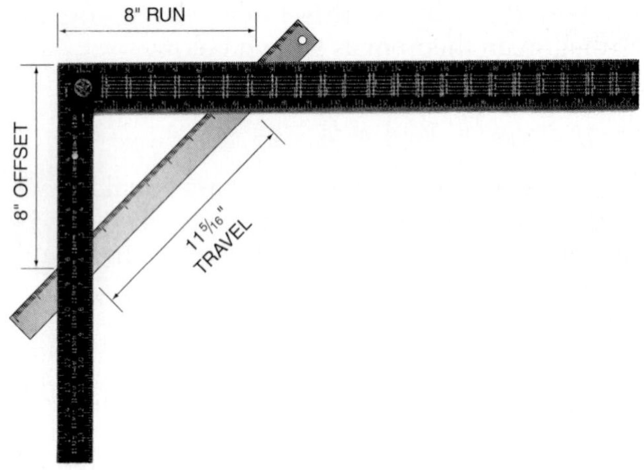

Figure 12 ◆ Finding travel with a framing square.

The square root of 2 is 1.414.

Because this equation uses one unit, the constant can be multiplied by whatever length you have for the offset or run to determine travel in a piping problem. Say you are using 45-degree elbows and need to calculate the travel with an offset of 36 inches.

$$\text{Travel} = 1.414 \times 36$$

$$\text{Travel} = 50.904 = 50\tfrac{15}{16}$$

4.4.0 Finding the Travel with a Framing Square

You may find yourself in the field without a calculator or without paper on which to figure the travel for offsets. If that happens, you can use a framing square and a wooden rule or tape measure to figure 45-degree offsets very accurately.

Because the offset and run are equal, the framing square can show an exact representation of the problem. *Figure 12* shows the method.

Lay the tape measure or rule at the offset dimension on both the blade and tongue of the square. Be sure to use *either* the outer numbers *or* inner numbers, but do not mix the two. Read the travel from your rule or tape. It's more accurate to **burn an inch** (deduct an inch from the measurement to account for the room required to grip the folding rule with your fingers) when doing this. If you hold on the inch mark, remember to deduct one inch from your reading.

4.4.1 Study Example

Use a framing square and tape measure or wooden rule to figure the travel for the following offsets.

1. 10 inches
2. 14 inches
3. 8 inches
4. 5½ inches
5. 11¾ inches

How would you figure a 33-inch offset with this method?

Review Questions

Section 4.0.0

1. The difference between the centerlines of two parallel pipes is called the _____.
   a. travel
   b. angle
   c. offset
   d. engagement
2. A right triangle that has two equal legs has _____ degrees for the other two angles.
   a. 90
   b. 30
   c. 60
   d. 45
3. The constant to use to calculate travel for any 45-degree angle is _____.
   a. 14.14
   b. 1,414
   c. 1.414
   d. 0.1414
4. The length of pipe that runs between the fittings in an offset is called the _____.
   a. travel
   b. constant
   c. theorem
   d. right triangle
5. If the offset and run of a 45-degree offset installation are both 11 inches, the approximate travel is _____.
   a. 15½ inches
   b. 121 square inches
   c. 63½ inches
   d. 11½ inches

Summary

Skilled, productive plumbers know mathematics. They "keep math in their toolbox" and treat it with the respect a precision instrument deserves. This module has introduced you to methods of measuring pipe and shown you how to find the cut length of pipe that runs between fittings. It has also introduced you to the basics of figuring simple offsets. With this groundwork, you can progress in your math skills and your trade skills. The more you know, the more you can take pride in your abilities.

PROFILE IN SUCCESS

Cary Mandeville

Instructor
PMA/CEFGA Apprentice Program
High Shoals, GA

Cary is from northeastern Georgia, near Atlanta. He lived in small towns and in the country while growing up. He worked for some time in a sewing plant in this large textile region of the United States. It wasn't until he went to Georgia Tech that he began working construction to help pay his tuition. He worked a range of crafts from framing to plumbing, but he recalls spending most of his time busting concrete and digging ditches. Since that time, Cary has been a company owner, partner, and instructor. Frustrated by the lack of qualified workers, he sold his interest in the company, went back to college for an education degree, and started an apprenticeship school.

How did you become interested in the construction industry?

During my first summer at Georgia Tech, I got a job working in steel construction. The job was helping me pay my way through school. I did some house framing as well before working in the plumbing trade. I basically began learning the trade by working with experienced plumbers. It was a slow process. Finally, a foreman recognized my interest and abilities and took me on as his helper. He began to show me systematically how to develop my plumbing skills. After that, I moved around and took my skills with me—each job paid more than the last.

After two or three years working for big companies, I went out on my own. I took mostly small jobs and figured out what to do as I went. A bunch of us got together and hired a plumbing inspector to teach us the plumbing code. The first time I tried for my license, I failed the test. I came back, studied harder, and passed it on the second try. I spent almost 15 years in a one- or two-man shop. In fact, I still have my first customers today. I have even done work for their children, who have since grown up, gotten married, and have places of their own.

In 1989, another plumber and I started a company. I had worked mostly in residential and service. My partner did commercial and industrial jobs. We bought a building and hired workers. Soon, we had 10 people working for us, but we couldn't find qualified plumbers. There was no schooling or training available at that time in Georgia. We tried doing in-house training with limited success. After seven years, my partner bought me out of the business. I went back and got my education degree and started an apprenticeship training school.

The school started with 35 or 40 apprentices sponsored by contractors and two instructors in two locations. Now we have over 90 apprentices and five instructors operating in four locations.

What do you think it takes to be a success in your trade?

Based on what I have seen with my own company and what I have learned in my discussions with the 35 companies involved with the apprenticeship school, the number one requirement for success is showing up for work. You would be surprised how many of the workers don't seem to think it is important to show up for work on time and prepared—with proper tools and equipment, proper clothing, and proper attitude. To be successful, you have to be mature and accept the responsibility of earning a living.

The next most important thing is practicing good communication skills. This includes notifying the boss or your supervisor if you have to miss a day of work for illness or an emergency. When you can't take basic steps like this, you are telling the employer that you can't be counted on. That will lose you job after job. The same is true if you have a problem on the job. You must learn how to communicate with your fellow workers. It is the key to success.

What are some of the things you do in your role of instructor?
I help trainees to develop their skills, aptitude, and attitude so they can be successful craftworkers. I work with them to foster an interest in how things work and how to translate book theory into everyday applications. One key skill is learning how to take what you learn and apply and adapt it to the endless variety of situations you encounter in plumbing jobs. It means taking a subject like math and realizing that you use it to design layouts, figure how to pipe different jobs, and get stuff done on the job. There is a place for all skill levels in this industry. It is my role to help each individual work to his or her potential and add value to his or her career.

What do you like most about your job?
My favorite moment is when I see the light go on—the "aha!" when *why* something works translates into understanding how to use it on the job. The best feeling is seeing a trainee understand a concept, learn a new skill, and develop the aptitudes and attitudes that will prove successful in a plumbing career. I think I am also going to enjoy when apprentices start graduating from the program. The school is still young right now. But already I have seen apprentices getting more responsibility from their employers. One guy has already become a junior foreman and got a company truck. Seeing that type of change in someone's life and work is what I like most.

What would you say to someone entering into the trades today?
You know, I still love plumbing. I may not be doing it every day now, but my love of the trade is still strong. Going into the trades today is different. When I started, it took me close to 10 years to learn my craft. Now, with apprenticeship training, a new person can cut that time in half. I would say, "Take advantage of all the training you can get." The better you are, the more money you will earn and the more opportunities will come your way.

There is tremendous flexibility in the crafts. You can work for a company and move up and maybe eventually start your own company. But success stories shouldn't create the impression that there is nothing to it, that craftwork comes easy. It takes years to develop craft skills. And it takes talent and skill to be a success. People entering the trades should realize that it takes about four to five years to become a journeyman. There is no overnight success—success comes from hard work and dedication to a craft.

GLOSSARY

Trade Terms Introduced in This Module

Back: Part of the fitting that is opposite to the side with an opening or face.

Blade: The longer of two sides of a framing square.

Burn an inch: Practice of deducting an inch from the reading (measurement) to account for the room required to grip the folding rule with your fingers.

Center: A point exactly halfway between two other points or surfaces.

Centerline: On a drawing, a line that shows the center of an object.

Center point: Point created where the centerline of the pipe and the centerline of the fitting meet within the fitting. The center point is used to determine the correct length of a pipe.

Constant: A number that never changes when used in an equation. For example, when finding the travel of a 45-degree right triangle, always use the constant 1.414.

Dimension lines: Lines on a drawing that indicate the measurements of an object.

Extension lines: Lines used on a drawing to indicate a dimension (measurement) away from the actual points of the dimension. This method is used when a drawing would be too crowded or cluttered if the dimension were shown within the two points.

Face: The open end of a fitting where a pipe is joined to the fitting, such as the opening of the inlet or either end.

Fitting allowance: The distance from the end of the pipe that goes into a fitting to the center of the fitting. (Also called fitting take-off.)

Framing square: Specialized square with tables and formulas imprinted on the blade for making quick calculations, such as finding travel and determining area and volume.

Offset: In measuring travel, the distance between two parallel pipes connected together with fittings such as 45-degree elbows. An offset is often used to go around an obstruction.

Pythagorean theorem: In any right triangle, the longest leg squared is equal to the sums of the other two legs squared. As applied to plumbing, the theorem is $\text{Travel}^2 = \text{Run}^2 + \text{Offset}^2$.

Remainder: The leftover amount in a division problem. For example, in the problem $34 \div 8$, 8 goes into 34 four times ($8 \times 4 = 32$) and 2 is left over, or, in other words, it is the remainder.

Right triangle: Any triangle with a 90-degree angle.

Run: In measuring travel, the horizontal distance from the center of a direction-changing fitting to the center of the next direction-changing fitting.

Schematic drawings: Simple, single-line representations (drawings) of pipe and fittings.

Thread makeup: The distance that a pipe screws into a fitting. (Also called thread engagement or thread-in.)

Throat: The part of the fitting where you thread in another pipe or fitting.

Tongue: The shorter of two sides of a framing square.

Travel: The measurement of the length of pipe that runs between the center of the fittings in an offset.

ANSWER KEY

Answers to Review Questions

Section 2.0.0

1. b
2. b
3. d
4. c
5. a
6. b
7. b
8. c
9. a
10. d
11. a
12. b
13. a
14. d
15. b
16. c
17. c
18. b
19. d
20. b
21. a
22. d
23. a
24. c
25. d

Section 3.0.0

1. c
2. b
3. c
4. b
5. d
6. c
7. a
8. b
9. d
10. a

Section 4.0.0

1. c
2. d
3. c
4. a
5. a

NCCER CRAFT TRAINING USER UPDATES

The NCCER makes every effort to keep these textbooks up-to-date and free of technical errors. We appreciate your help in this process. If you have an idea for improving this textbook, or if you find an error, a typographical mistake, or an inaccuracy in the NCCER's Craft Training textbooks, please write us, using this form or a photocopy. Be sure to include the exact module number, page number, a detailed description, and the correction, if applicable. Your input will be brought to the attention of the Technical Review Committee. Thank you for your assistance.

Instructors – If you found that additional materials were necessary in order to teach this module effectively, please let us know so that we may include them in the Equipment/Materials list in the Instructor's Guide.

Write: Curriculum Revision and Development Department
National Center for Construction Education and Research
P.O. Box 141104, Gainesville, FL 32614-1104

Fax: 352-334-0932

E-mail: curriculum@nccer.org

Craft ______ Module Name ______

Copyright Date ______ Module Number ______ Page Number(s) ______

Description

______

______

______

______

(Optional) Correction

______

______

______

(Optional) Your Name and Address

______

______

______

Module 02104-00

Introduction to Plumbing Drawings

Course Map

This course map shows all of the modules in the first level of the Plumbing Curriculum. The suggested training order begins at the bottom and proceeds up. Skill levels increase as you advance on the course map. The local Training Program Sponsor may adjust the training order.

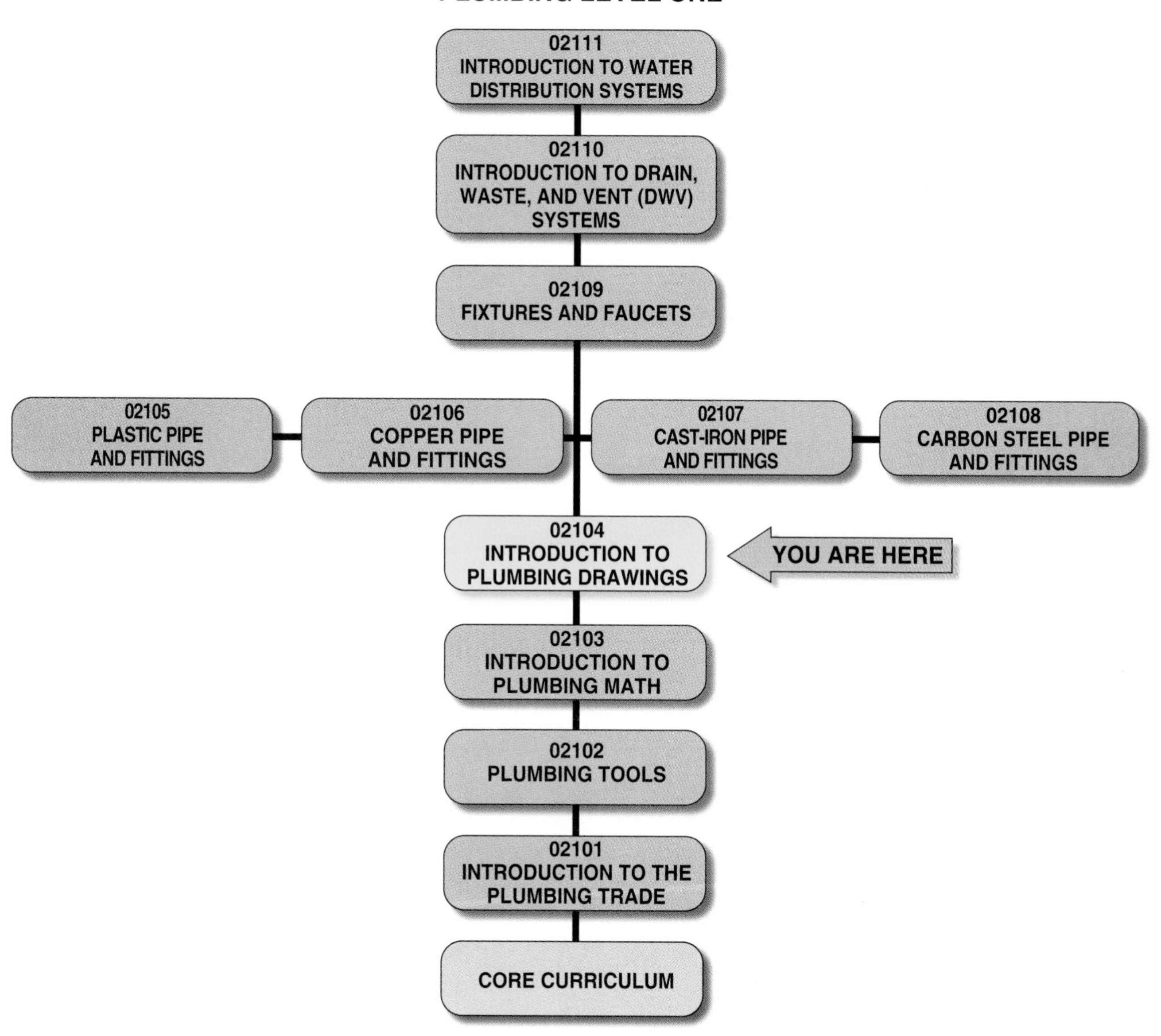

104CMAP.EPS

Copyright © 2000 National Center for Construction Education and Research, Gainesville, FL 32614-1104. All rights reserved. No part of this work may be reproduced in any form or by any means, including photocopying, without written permission of the publisher.

MODULE 02104 CONTENTS

Figures

Tables

MODULE 02104

Introduction to Plumbing Drawings

Objectives

When you finish this module, you will be able to:

1. Identify pictorial (isometric and oblique), schematic, and orthographic drawings, and discuss how different views are used to depict information about objects.
2. Identify the basic symbols used in schematic drawings of pipe assemblies.
3. Explain the types of drawings that may be included in a set of plumbing drawings and the relationship between the different drawings.
4. Interpret plumbing-related information from a set of plumbing drawings.
5. Convert plan view drawings to simple isometric drawings.
6. Use an architect's scale to draw lines to scale and to measure lines drawn to scale.
7. Discuss how local code requirements apply to certain drawings.

Prerequisites

Before you begin this module, it is recommended that you successfully complete task modules: Core Curriculum; Plumbing Level One, Modules 02101 through 02103.

Materials Required for Trainees

1. This module
2. Appropriate personal protective equipment
3. Sharpened pencil and paper
4. Architect's scale

1.0.0 ◆ INTRODUCTION

Before the first shovel of earth is moved, a building has already had a long life on paper. Every building, regardless of its size, begins life on the drawing board or at the computer terminal of a designer. These drawings, referred to by the broad term **blueprints,** provide the **details** of construction. The ability to read drawings is an important skill that each plumber must have. Developing this skill takes time and experience. This module introduces the types of drawings plumbers encounter and the **symbols** used on those drawings. The module describes the parts of a typical drawing and covers the basics of reading plumbing blueprints and drawings.

A complete set of plumbing plans contains both drawings and **specifications (specs).** The drawings show the size and shape of the structure. The specifications give information about the quality and type of components that must be installed. A plumber must refer to both the drawings and the specifications to get the necessary information to buy and install the proper materials.

The drawings that present the details of construction are often called blueprints. This term originally referred to architects' drawings that appeared as a white line drawing on a solid blue background. Today's blueprints are likely to be reproduced on white paper, usually with blue lines, but sometimes with black or dark reddish lines. A drawing with dark reddish lines is called a **sepia** (drawing).

The drawings of the future are likely to be produced through **computer-aided drafting (CAD),** which many architectural firms and contractors are already using. These drawings have their own look. The biggest advantage of CAD is that it is easy to make changes in the drawings.

In this module you will learn about the components of a complete set of residential drawings, as an example of architectural drawings. Then you will learn how the individual parts are shown on plumbing drawings. And finally, you will learn how those drawings are made and how to read them.

2.0.0 ◆ COMPONENTS OF CONSTRUCTION DRAWINGS

A complete set of construction drawings typically will contain a **plot plan,** a **foundation plan, floor plans, elevation drawings,** details or section drawings, **electrical drawings, HVAC** (heating, ventilation, and air-conditioning) **drawings,** and a set of **plumbing drawings.**

Each drawing contains a **title block,** which gives such facts as the owners and the address of the property, the date of the drawing, how many pages are included in the full set of drawings (for example, "Page 1 of 8," meaning that there are eight pages in the full set and this is the first page), and the **scale** of the drawing. The scale indicates the size relationship between the drawing and the completed building. You will learn more about scale later in this module.

Some of the drawings also include **dimensions,** which show actual distances; specifications, which describe the quality of materials to be installed and other facts related to the installation; and **notes** about special requirements for the job. These points are covered later in this module.

2.1.0 Plot Plan

The plot plan shows the location of the building on the property and shows where the water and sewer mains are installed (see *Appendix A*).

Building codes may say that nothing can be built on certain parts of the land. For example, **setback** requirements give the minimum distance that must be maintained between the street and the structure. **Easements** specify places where utilities can be installed. Codes may set a minimum width for **side yards** to provide access to rear yards, to reduce the possibility of a fire jumping from one building to another, or to promote ventilation.

In addition to locating the building, the plot plan will usually show the location of the utilities. Different symbols are used to indicate the water, storm sewer, sanitary sewer, and gas mains.

2.2.0 Foundation Plan

The foundation plan gives the size and shape of the foundation. It also shows the location of footings for steel posts that support the first floor; all drainage, waste, and vent (DWV) piping that must be installed below the concrete floor; and the water supply piping that must be installed below the first floor. A foundation plan may also indicate the positions of floor drains and other piping that must be installed beneath the basement floor.

2.3.0 Floor Plan

The floor plan shows the layout of rooms on each level of the house and the shape and size of each room (see *Appendix A*). You will use floor plan drawings to locate the position of water supply and waste piping.

It is standard practice to install the cold-water supply at the right of each fixture. Having the cold water controlled by the right-hand faucet and the hot controlled by the left-hand faucet prevents burns because people are familiar with this setup.

2.4.0 Elevations

Elevation drawings show the finished exterior appearance of the building. You can use these drawings to find information about where exterior hose bibbs go, how vent stacks go through the roof, and how storm drainage is supposed to look.

2.5.0 Details

Details are enlarged section drawings of important structural elements or of cabinets, staircases, and other special features of the building.

2.6.0 Electrical Drawings

Electrical drawings are sometimes superimposed on the floor plan. These drawings show the location of outlets, switches, and electrical fixtures.

2.7.0 HVAC Drawings

HVAC drawings show the location of the furnace and air-conditioning equipment, as well as the location of ducts and registers, or piping and radiators.

2.8.0 Plumbing Drawings

Plumbing drawings show the piping system, with the location of fixtures and pipe runs and the size and type of pipe to be installed.

The water supply and DWV systems are shown on **isometric drawings,** which sometimes are included in the plumbing drawings. (Isometric drawings are covered later in this module.) These drawings show the diameter of each run of pipe and the fittings required to make the installation. You can guess the length of each run of pipe by looking at the measurements on the floor plans, but you will not know the exact length of pipe until you are actually installing it.

Review Questions

Section 2.0.0

1. The ______ plan shows the location of the sewer and water mains and the location of the building on the property.
   a. foundation
   b. plot
   c. elevation
   d. plumbing

2. For information on vent stacks that go through the roof, you would refer to the _________ plan.
   a. details
   b. HVAC
   c. floor
   d. elevation

3. A plumber should consult ______ plans to make sure there are no conflicts with furnace and air-conditioning equipment.
   a. floor
   b. HVAC
   c. foundation
   d. elevation

4. To identify where floor drains and DWV pipes will be installed beneath the basement floor, refer to the ________ plan.
   a. floor
   b. details
   c. foundation
   d. plot

5. _________ leave space for utility companies to install their utilities.
   a. Setbacks
   b. Easements
   c. Details
   d. Side yards

3.0.0 ◆ READING PLUMBING DRAWINGS

To read plumbing drawings, you need to understand scale, dimensioning, specifications, and notes. You also need to be aware of local plumbing codes. Knowing these things will help you make sense of the drawings.

3.1.0 Scale

Blueprints and other types of construction drawings are drawn to scale. This simply means that the actual building, room, or object being shown is reduced in proportion from full size to a smaller size. Because the reduction is proportional, you can measure the drawing and determine the actual size of the object.

The advantage of drawing to scale is that each of the parts is proportional to the physical object itself. Obviously, a wall that measures 8 feet high and 14 feet wide cannot be drawn full size. Using scale to draw the wall allows the architect, designer, or plumber to represent the wall in the same proportions as it exists but smaller.

The instrument that makes it possible both to draw and measure scaled drawings is the **architect's scale** (see *Figure 1*). The architect's scale is a ruler. It may be either flat or triangular in shape. If it is flat, it contains four different scales; if it is triangular, it contains 10 scales. *Table 1* shows the 10 scales on a triangular architect's scale and also the full size at the top.

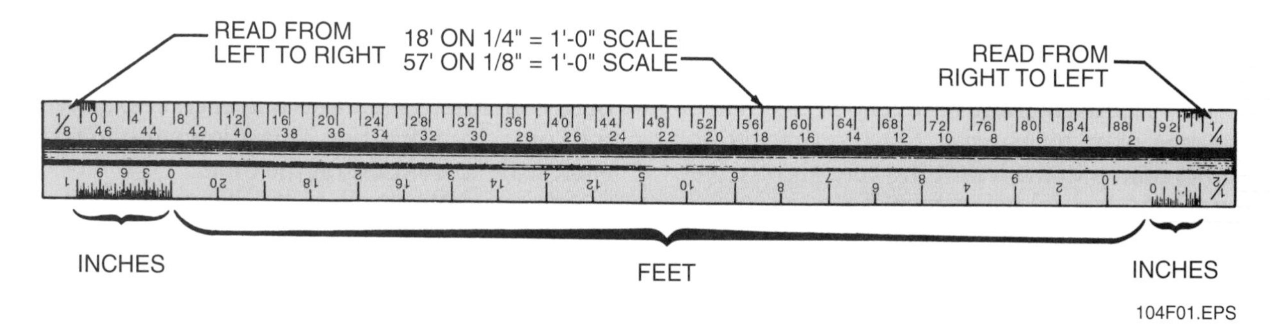

Figure 1 ◆ Architect's scale.

Table 1 The Scales and Amount of Reduction

Designation	Scale	Reduction
16	12" = 1'	Full-size
3	3" = 1'	One-quarter size
1½	1½" = 1'	One-eighth size
1	1" = 1'	One-twelfth size
¾	¾" = 1'	One-sixteenth size
½	½" = 1'	One twenty-fourth size
⅜	⅜" = 1'	One thirty-second size
¼	¼" = 1'	One forty-eighth size
3/16	3/16" = 1'	One sixty-fourth size
⅛	⅛" = 1'	One ninety-sixth size
3/32	3/32" = 1'	One one hundred twenty-eighth size

Table 2 Value of Each Division

Designation	Number of Divisions	Value
3	96	⅛"
1½	48	½"
1	48	½"
¾	24	½"
½	24	½"
⅜	12	1"
¼	12	1"
3/16	12	1"
⅛	6	2"
3/32	6	2"

Notice that all scales are given in reference to 1 foot.

The architect's scale can be read either from left to right or from right to left, depending on which scale you are reading. For example, you would read the ¼ scale from right to left and the ⅛ scale from left to right.

Each scale is set up the same way. First, there is the designation of the scale itself, such as ¼ or 3/32. The designation is followed by a series of lines. These lines represent exactly 1 foot as it is drawn with this scale.

Notice that the number 0 is found at one end of the 1-foot representation on every scale. This is where you start measuring. On the ¼ scale, for example, marks to the left of the zero signify feet, and marks to the right signify inches.

Each scale has a different number of marks to the right of zero. Not all of them signify inches. *Table 2* shows the value of these marks for each scale.

Figure 2 shows a close-up view of the ¼ scale and what each of the marks means.

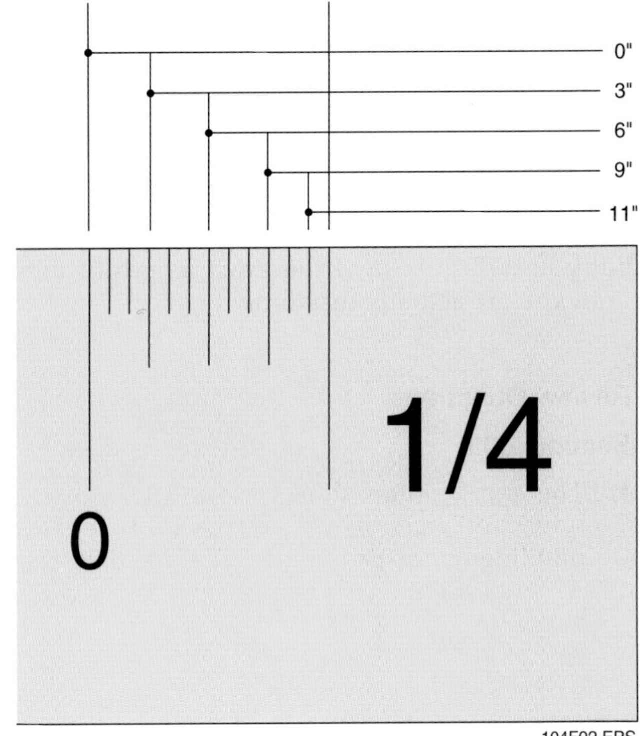

Figure 2 ◆ The ¼ scale.

As you look at the architect's scale, you may be confused by the numbers. Look at the ¼ scale in *Figure 1*. As you move to the left you will see a line of numbers beginning with 0 and ending with 46, as well as a line of numbers beginning with 92 and ending with 0. Remember, you read the ¼ scale from right to left. Only the numbers from 0 to 46 are used on the ¼ scale. The other numbers are for the ⅛ scale, which is read from left to right.

The system is the same for every face of the architect's scale. Turn your scale until you see the 1 and ½ scales. Reading the ½ scale from left to right, you see a row of numbers beginning with 0 and ending with 20. Reading the 1 scale from right to left, you see a row of numbers beginning with 0 and ending with 10. You will see the same system on the 3/32 and 3/16 scales, the 1½ and 3 scales, and the ⅜ and ¾ scales.

You may have noticed that the scales are paired on the architect's scale so that each is exactly one-half or twice the other. This allows the marks for each scale to be used on the other.

For example, look at the ¼ scale in *Figure 1*. Between 0 and 2 feet on this scale, there are four divisions. (The shorter lines along the top belong to the ⅛ scale, but you also use them when you read the ¼ scale.) Each mark along the ¼ scale represents 6 inches in the length of an actual object.

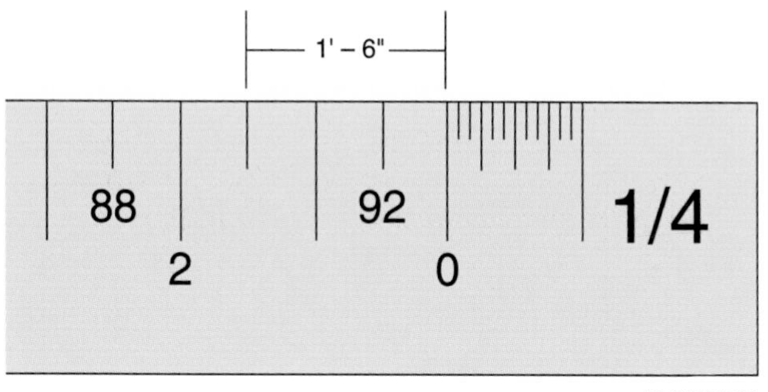

Figure 3 ◆ 1 foot, 6 inches mark on the ¼ scale.

Figure 4 ◆ 1 foot, 10 inches on the ¼ scale.

Thus, a line drawn from 0 to the mark before the 2 would be 18 inches, or 1 foot, 6 inches, long (see *Figure 3*).

Figure 4 shows how 1 foot, 10 inches would be represented on the ¼ scale.

As you have learned, you can usually find the scale of a drawing on the title block. Typical scales for plumbing drawings are ⅛" = 1' or ¼" = 1'. If the title block shows a scale of 1' = ¼", this means that for every ¼ inch on the drawing, the real object takes up 1 foot. The drawing is 1/48 the size of the real object. Scale makes drawings convenient to handle.

3.2.0 Dimensioning

Dimensions show actual distances. To dimension a distance on a drawing, you may need to use **extension lines.** Extension lines establish the limits of the measurement, as *Figure 5* shows. Arrows, dots, or slashes may be used to indicate the limits of the dimension. It is important to notice where the dimension is taken from.

Some dimensions are shown from end to end. Others are shown from centerlines.

3.3.0 Notes

Some of the information on a set of drawings is in the form of notes. From these notes you can learn about the materials you will need and special installation requirements. Learn to look for the *Notes* section on drawings (see *Figure 6*).

3.4.0 Specifications

Specifications, or specs, describe the quality of the materials to be installed. Also, specs may say that you must follow certain practices in installing and testing the plumbing system. A basic residential plan like the one in the sample drawings for this module may not contain a separate set of specs. In this case, you would follow the local plumbing code for the type and quality of materials and installation practices that must

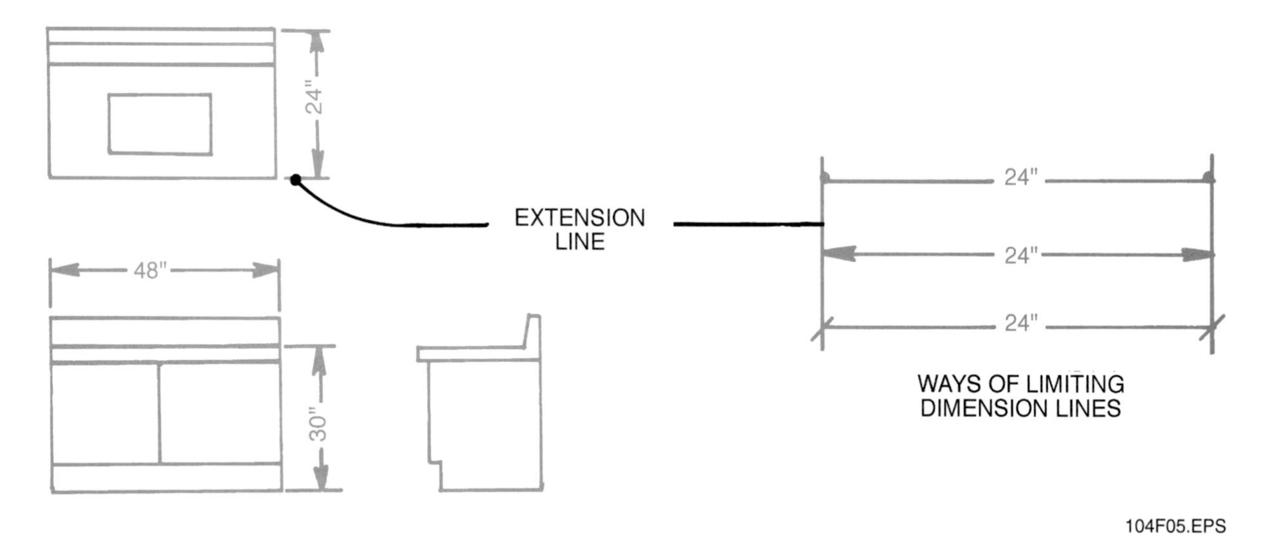

Figure 5 ◆ Extension lines.

NOTES BY SYMBOL "○"

① CAP EXISTING INLET AND OUTLETS TO EXISTING ACID DILUTION BASIN BELOW FLOOR. REWORK EXISTING ACID WASTE LINE TO CONNECT TO NEW 4" ACID WASTE FROM ABOVE.

② REWORK EXISTING 3" AW TO CONNECT TO NEW 4" AW DOWN.

③ REFER TO ARCH. DRAWINGS FOR FURRINGS.

④ 2½" CW, 1½" HW–UP TO 1ST FLOOR CLG. ¾" HWR DN. TO BELOW FLOOR RE: ARCH. DRAWINGS FOR FURRING.

⑤ SLEEVE STRUCTURAL WALL IN CRAWL SPACE. REFER TO STRUCTURAL DRAWINGS.

⑥ 2" WASTE & 4" AW DOWN IN WALL/CHASE. COORDINATE WITH HVAC PIPING.

⑦ REFER TO ARCHITECTURAL DRAWINGS FOR WALL FURRING.

104F06.EPS

Figure 6 ◆ Sample notes.

be used. Many times you will find that the plans and specifications for a residence will not be complete. Plumbing drawings may not be included beyond the location of fixtures on the floor plan. When this happens, it is the plumber's responsibility to design the piping system so that it meets all local requirements. You need much more knowledge about plumbing codes and sizing of pipe before you can be expected to undertake this responsibility.

3.5.0 Role of Plumbing Codes

For now, though, you should be aware that the local plumbing code governs the installation of all plumbing systems in the community. The basic purpose of the code is to protect the health and safety of the people who use the system by preventing improper installations. This code is the law in the community, and it is enforceable through the courts. The code sets the minimum quality that is allowed for all plumbing installations. The plans and specifications for a particular building may call for installation that is of higher quality than that required by the code, which is fine. But any plan or specification that calls for a plumbing installation that is of lower quality is in violation of the code and should not be followed.

Review Questions

Section 3.0.0

1. To _________ a drawing means to show the actual building, room, or object reduced in proportion from full size to a smaller size.
   a. dimension
   b. spec
   c. isometric
   d. scale

2. The instrument used to draw and measure scaled drawings is a(n) ______________.
   a. dimension scale
   b. architect's scale
   c. orthographic scale
   d. plumber's scale

3. The number ______ is found at one end of the 1' representation on every scale and is the point where you start measuring.
   a. 0
   b. 1
   c. ¼
   d. 2

4. To check for certain practices that must be used in installing and testing a plumbing system, a plumber refers to the ________.
   a. title block
   b. scale
   c. dimensions
   d. specifications

5. Use _______ to establish the limits of a measurement on a drawing.
   a. specifications
   b. scales
   c. extension lines
   d. notes

4.0.0 ◆ TYPES OF DRAWINGS

To communicate ideas without confusion, the construction industry has developed several techniques to show the size and shape of objects accurately. Plumbers work with many different types of drawings. For example, plumbers must be able to find the size and dimensions of fittings from manufacturers' catalogs. They must also be able to interpret details about assemblies. Fixture drawings are an important part of the trade.

Even though each type of drawing may use a different technique to convey information, there are many similarities.

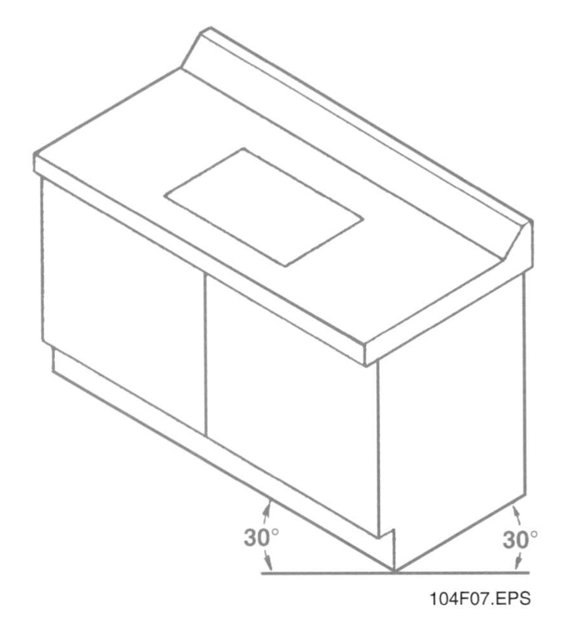

Figure 7 ◆ Isometric drawing.

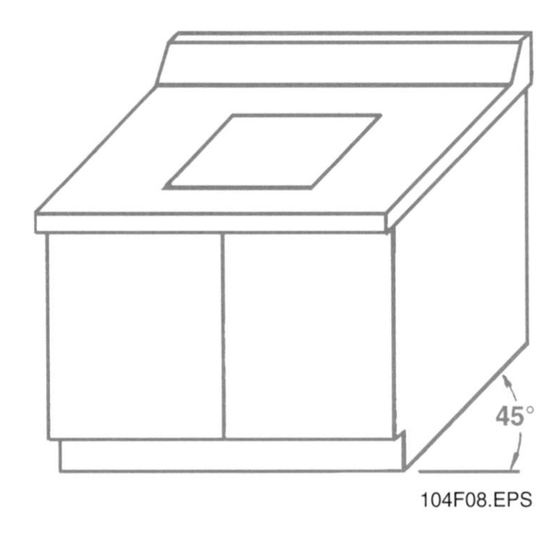

Figure 8 ◆ Oblique drawing.

4.1.0 Pictorial Drawings (Isometric and Oblique)

Pictorial drawings give the impression of three dimensions. The advantage of pictorial drawings is that they give a picture that is very similar to the actual object. *Figure 7* is a type of pictorial drawing known as an isometric drawing. As a plumber, you will encounter isometric drawings often, and they are discussed in greater detail in an *On the Level* feature in this module. Isometrics create an illusion of three dimensions because vertical lines are drawn vertically but horizontal lines are projected at 30 degrees and seem to go back into the horizon.

Figure 8 shows another type of pictorial drawing called an **oblique drawing.** Oblique drawings create the illusion of depth by using lines that are drawn at 45 or 60 degrees to the horizontal.

Figures 7 and *8* are similar in that the sides and top of the object are not actually rectangular. The corners are not drawn at 90-degree angles even though the object itself has 90-degree corners.

4.2.0 Schematic Drawings

Most plumbing sketches are **single-line drawings** or **schematic drawings.** In a single-line drawing, the line represents the centerline of the pipe, so these drawings can be used to represent pipe of any diameter. Compare the isometric representation of a stack in *Figure 11* with the schematic drawing in *Figure 12.* (This drawing has standard symbols, which will be discussed later.)

4.3.0 Orthographic Drawings

To communicate information about the exact size and shape of an object, a designer often uses **orthographic drawings,** which show dimensions that are proportional to the actual physical dimensions. In orthographic drawings, the designer draws lines that are scaled-down representations of real dimensions. Every 12 inches, for example, may be represented by ¼ inch on the drawing.

An orthographic drawing (see *Figure 13*) is made by viewing the object from one or more of six possible directions and recording what the object looks like in each of the views.

You can create more views of the object by looking at it from other angles, as shown in *Figure 14.* Although you can get as many as six different views, generally you will not need more than three to show the object completely.

To make orthographic drawings understandable, designers often draw the views in the proper relationship to one another. Notice, in *Figure 15,* how the views from different angles are related to the finished drawing.

Figure 16 shows both a pictorial drawing and an orthographic drawing of the same object. Study and compare these drawings carefully. To read an orthographic drawing, you have to learn to visualize what the finished object looks like from the views given. This will require a lot of effort and practice.

Orthographic drawings show an object's shape, its size, and the relationship between its parts. Architects and designers may also include

Isometric Drawings

An isometric drawing is a pictorial drawing in which all vertical lines are shown vertically and all horizontal lines are projected at a 30-degree angle. As a plumber, you must be able to make isometric sketches of piping assemblies and to read isometric drawings of piping assemblies, even very complex ones.

You will learn the basics of making simple isometric sketches.

To make isometric sketches, you need to understand how the illusion of three dimensions is created on a flat piece of paper.

The simple box shown in *Figure 9* may be the easiest way to begin thinking about this process. Note that vertical edges on the box remain vertical in the sketch.

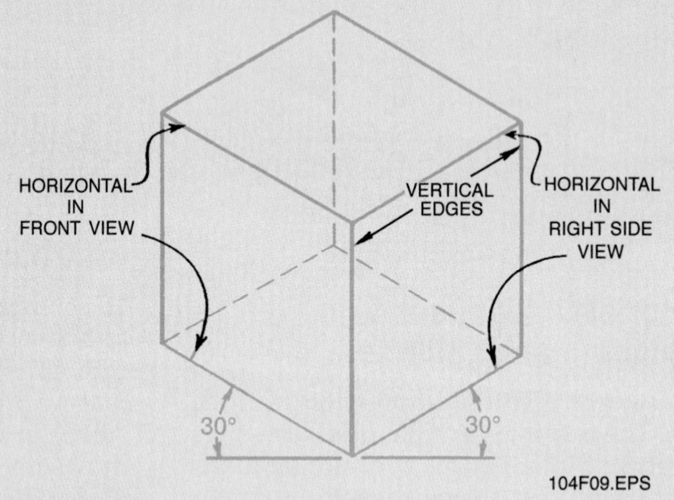

Figure 9 ◆ Simple box.

Lines that are horizontal in the front view are drawn 30 degrees from horizontal, or approximately in the 10 o'clock position. Lines that are horizontal in the right-side view are drawn at 30 degrees from horizontal, in approximately the 2 o'clock position. The dashed lines show the hidden back and bottom edges.

Try sketching several different boxes. Some should be cubes. Others should be long and thin or tall and narrow. Does each of the sketches produce an illusion of three dimensions?

The isometric box is a good way to begin illustrating piping systems because it relates directly to the shape of most buildings and rooms. The piping system typically needs to fit within or near the walls.

Sketches can be made without including the walls as part of the sketch. However, the plumber must know where the walls are located and position the fitting to make the required bends in the piping system. *Figure 10* shows the steps you would use for making the sketch shown in *Figure 11.*

Dimensions on isometric drawings are usually not to scale. Distortion is permitted on these drawings so that they can show three dimensions, and this distortion makes scaling difficult.

Riser diagrams are more complex isometric drawings of all the plumbing in the building (see *Appendix A*). Plumbers use them to do material **take-offs.** These diagrams identify the types of fittings so that the plumber can order what's required—like a shopping list.

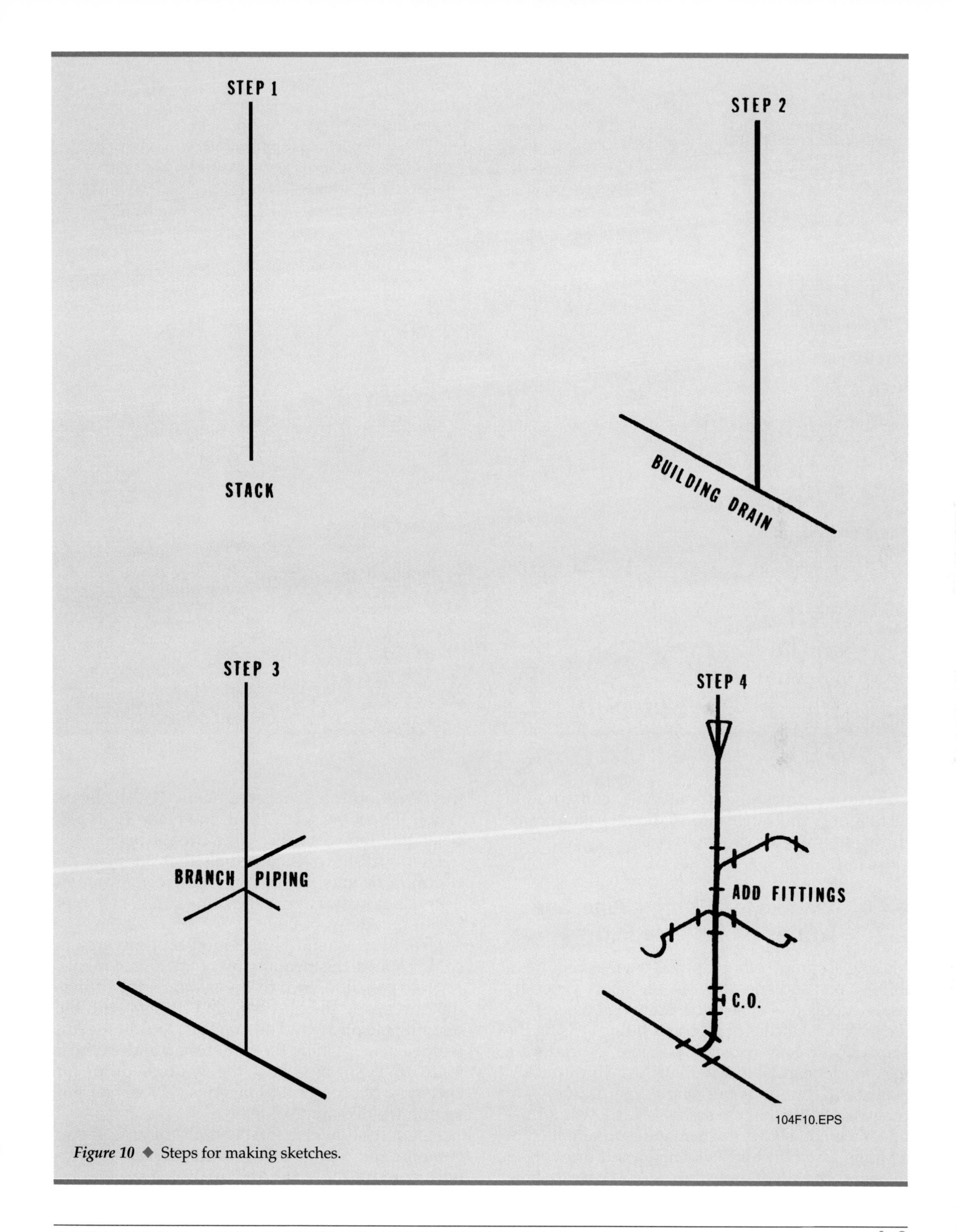

Figure 10 ◆ Steps for making sketches.

The riser diagram shows vertical and horizontal piping along with sizes and a riser number that refers back to the full set of plumbing drawings. Diagrams typically include a domestic water diagram and a DWV diagram.

To convert plan view drawings to isometric, remember that any horizontal lines from the plan view convert to an angled line (30 degrees) on an isometric drawing. A full circle drawn at the end of an angled line indicates that the line continues up from that end of the horizontal line. A half-circle indicates that the pipe continues downward from the end of the horizontal line. These lines refer to vertical lines on the plan view.

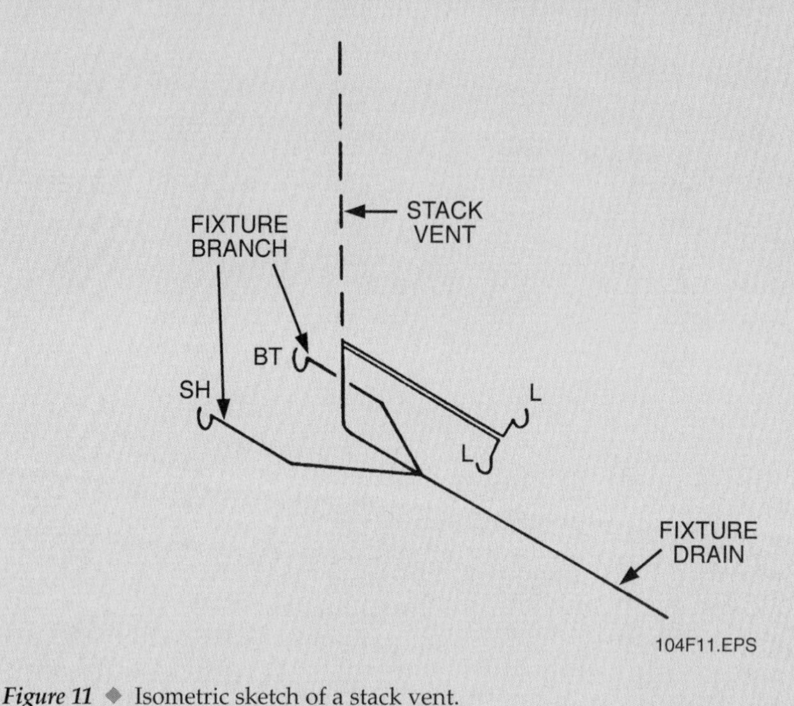

Figure 11 ◆ Isometric sketch of a stack vent.

descriptive notes about materials, construction techniques, and other information. Orthographic drawings are often drawn to scale.

4.4.0 Symbols for Fittings, Pipe, and Fixtures on Schematic Drawings

Making accurate orthographic drawings requires a lot of work and special tools. Plumbers generally use symbols and schematic drawings when they are making sketches of pipe and fittings. Using the symbols correctly makes it possible to produce a very understandable schematic drawing quickly. Schematic drawings are single-line drawings of pipe assemblies that use symbols to show information. *Figure 17* shows a schematic representation of a piping assembly and an orthographic drawing of the same thing. Note how the symbols vary to show the direction the 90-degree elbow (L, ell) faces. When the elbow faces front, the circle is closed; when the elbow faces back, the circle is open.

Figure 18 shows the symbols that are used for common fittings. However, architects and engineers do not always use the same symbols to represent the same objects.

Usually, a legend is provided on drawings to communicate the meaning of the various symbols.

Note the difference in the symbols for fittings that are facing the observer and fittings with the opening facing away. For example, look at the difference in tees where the outlet is up and where it is down. Also, note how the symbols differ for each type of pipe joint (flanged, screwed, bell and spigot, welded, and soldered).

Figure 19 shows typical symbols for pipe. If you combine the pipe symbols with the fitting symbols, you can show almost any piping system.

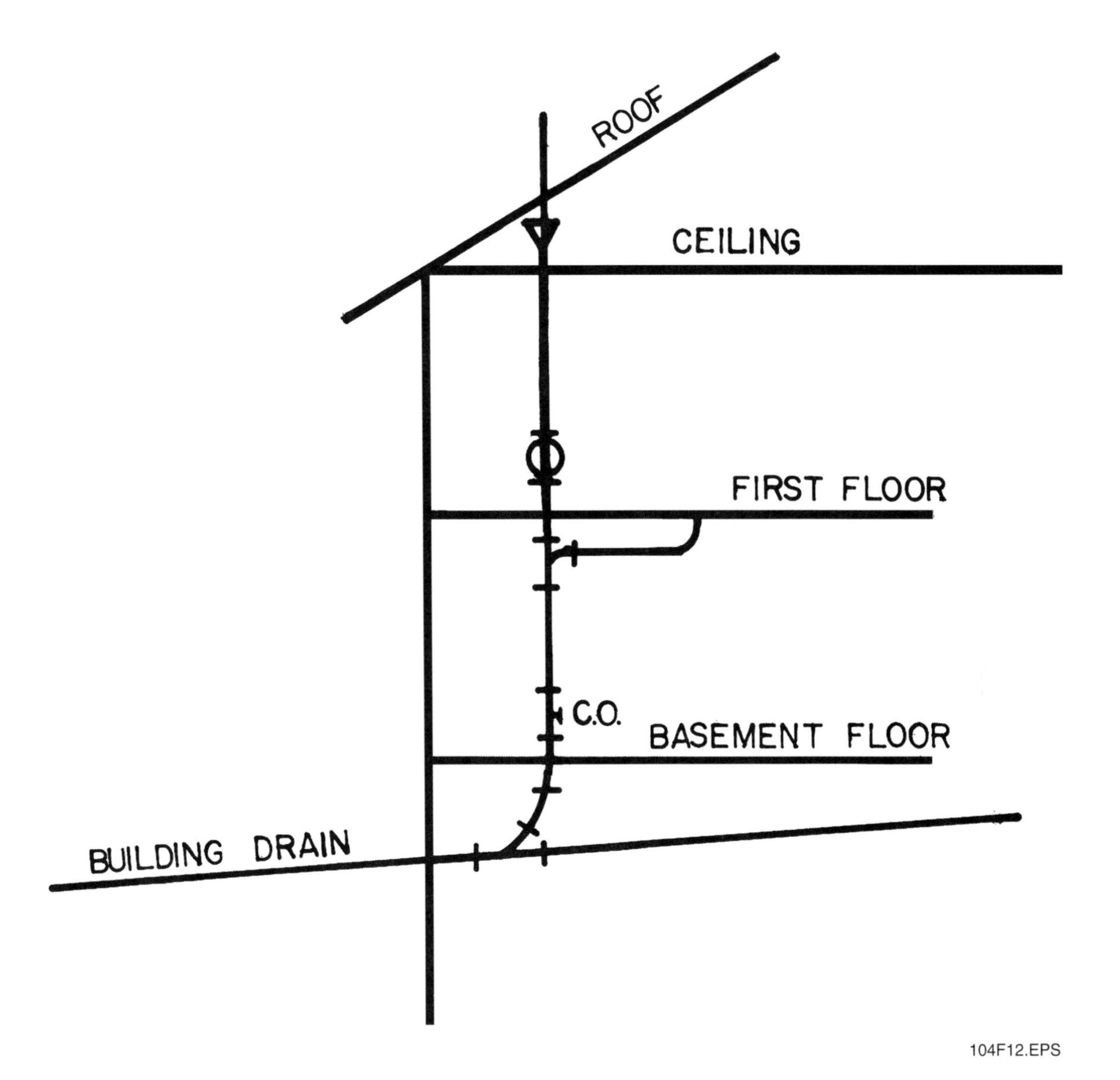

Figure 12 ◆ Schematic sketch of a piping stack.

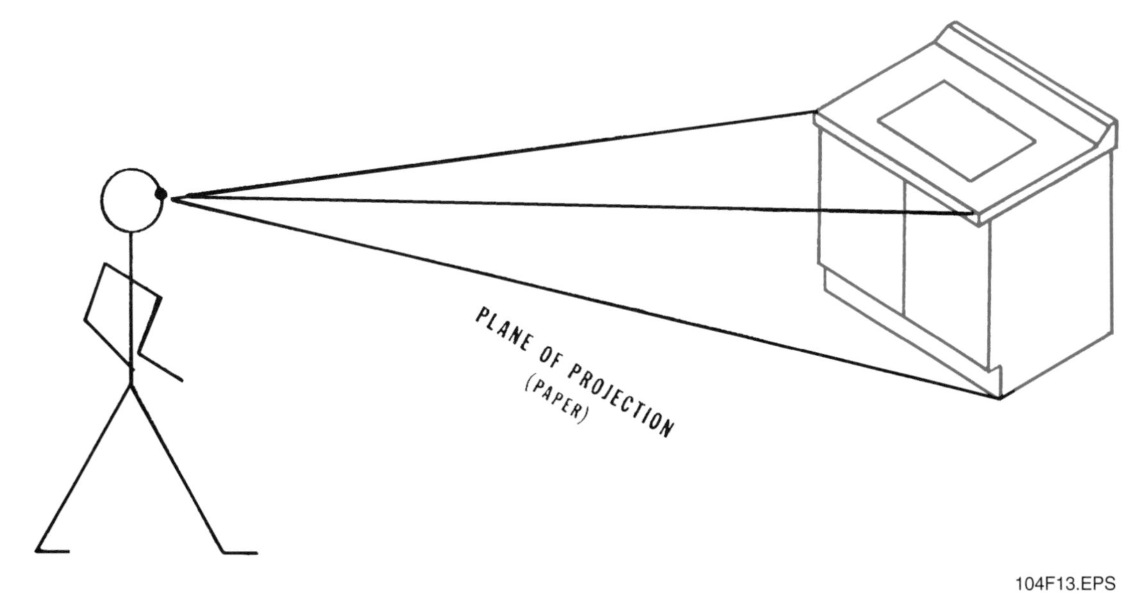

Figure 13 ◆ Orthographic projection.

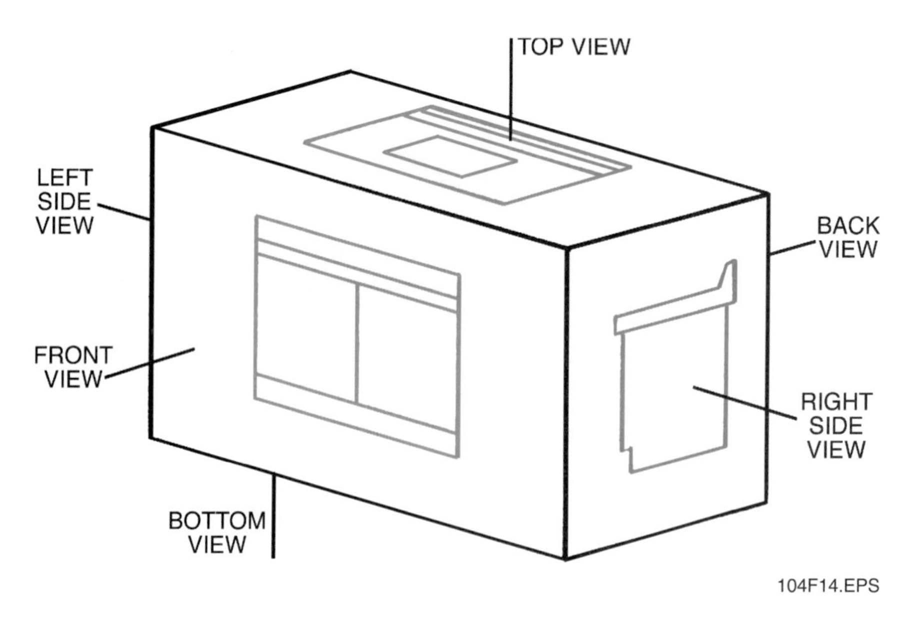

Figure 14 ◆ Additional planes for orthographic projection.

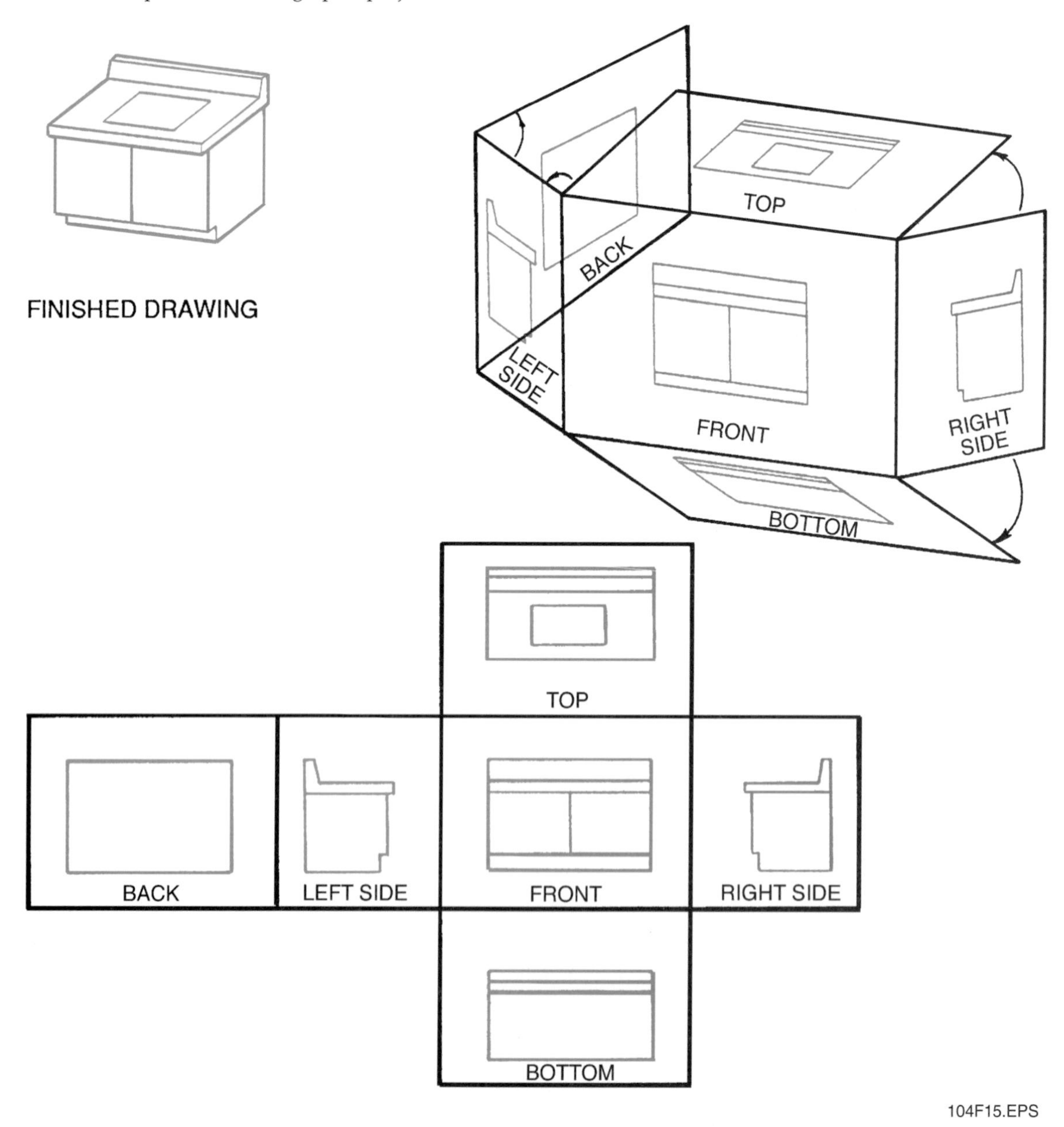

Figure 15 ◆ Relationship of orthographic views.

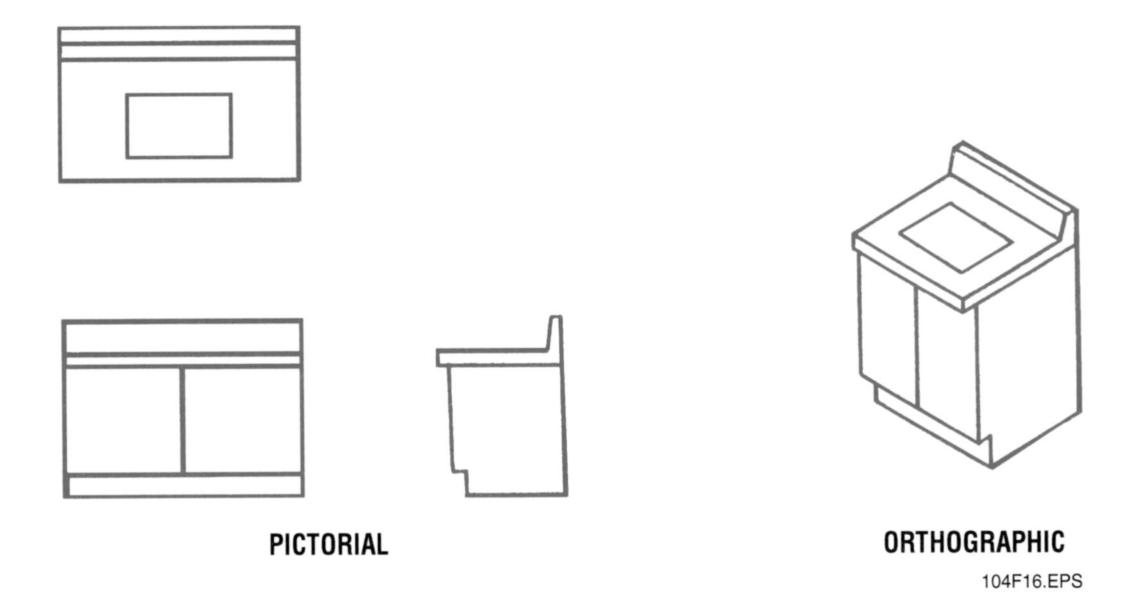

Figure 16 ◆ Orthographic and pictorial views.

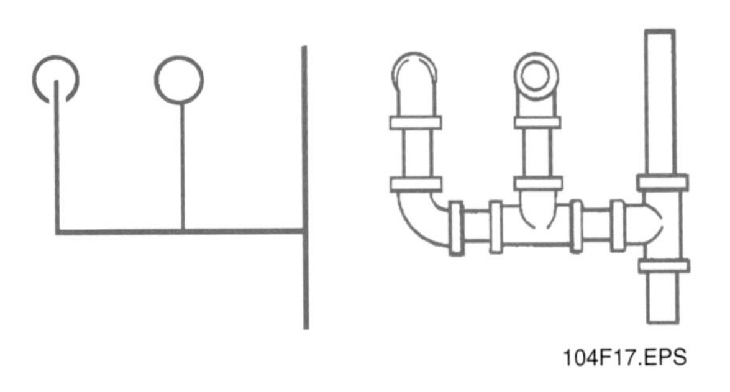

Figure 17 ◆ Schematic and orthographic drawing.

Figure 20 shows the symbols for plumbing fixtures. These symbols are useful to show what fixtures are connected to various runs of pipe and to help you understand blueprints that you will see on the job.

4.5.0 Types of Plumbing Drawings

As a plumber, you will see different types of drawings. Study each drawing carefully and figure out what information it contains and how it shows that information.

4.5.1 Catalog Drawings

Manufacturers of pipe and fittings publish catalogs that contain information about their products. Many of them publish *dimensional catalogs* that show each of their fittings along with overall sizes and, in some cases, weights. Plumbers look at these **catalog drawings** to determine such factors as overall dimensions, actual inside dimensions, fitting depth, and laying length of the fittings.

Figure 21 shows a plastic DWV system P trap with a solvent-welded joint.

The method used to show the dimensions of the fitting is typical. The drawing uses capital letters (A, B, etc.) that are keyed to a table. Information about size is given in the table.This method is efficient because information can be given for different sizes of fittings.

Figure 22 uses a similar technique to give information about cast-iron wye. Note that the table gives information for 39 separate fittings. By checking the various combinations of pipe sizes (G, X, and X′), a plumber can find a combination that will work for the installation.

SV stands for service weight and *XH* for extra heavy pipes. Service weight is the type of pipe leading to the main at the street connection.

4.5.2 Fixture Drawings

Plumbers often look at **fixture drawings** to determine whether a certain fixture will fit in the available space. *Figure 23* is a drawing of a typical lavatory. Two views are required to show the size and shape of this lavatory. The dimensions for roughing-in the water supply pipes and the drains

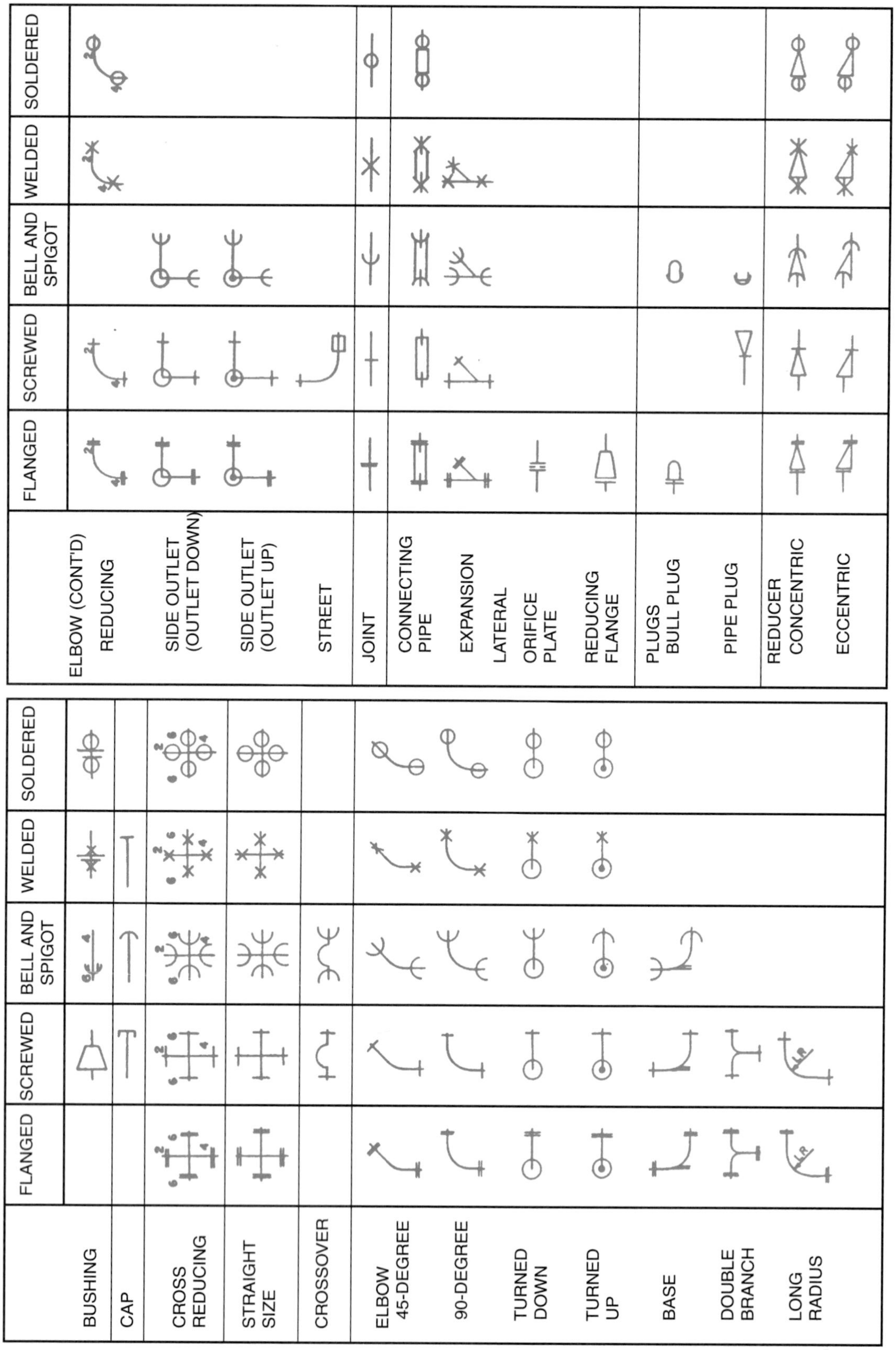

	FLANGED	SCREWED	BELL AND SPIGOT	WELDED	SOLDERED
BUSHING					
CAP					
CROSS REDUCING					
STRAIGHT SIZE					
CROSSOVER					
ELBOW 45-DEGREE					
90-DEGREE					
TURNED DOWN					
TURNED UP					
BASE					
DOUBLE BRANCH					
LONG RADIUS					

	FLANGED	SCREWED	BELL AND SPIGOT	WELDED	SOLDERED
ELBOW (CONT'D) REDUCING					
SIDE OUTLET (OUTLET DOWN)					
SIDE OUTLET (OUTLET UP)					
STREET					
JOINT					
CONNECTING PIPE					
EXPANSION					
LATERAL					
ORIFICE PLATE					
REDUCING FLANGE					
PLUGS BULL PLUG					
PIPE PLUG					
REDUCER CONCENTRIC					
ECCENTRIC					

104F18A.EPS

Figure 18 (1 of 3) ◆ Fitting symbols.

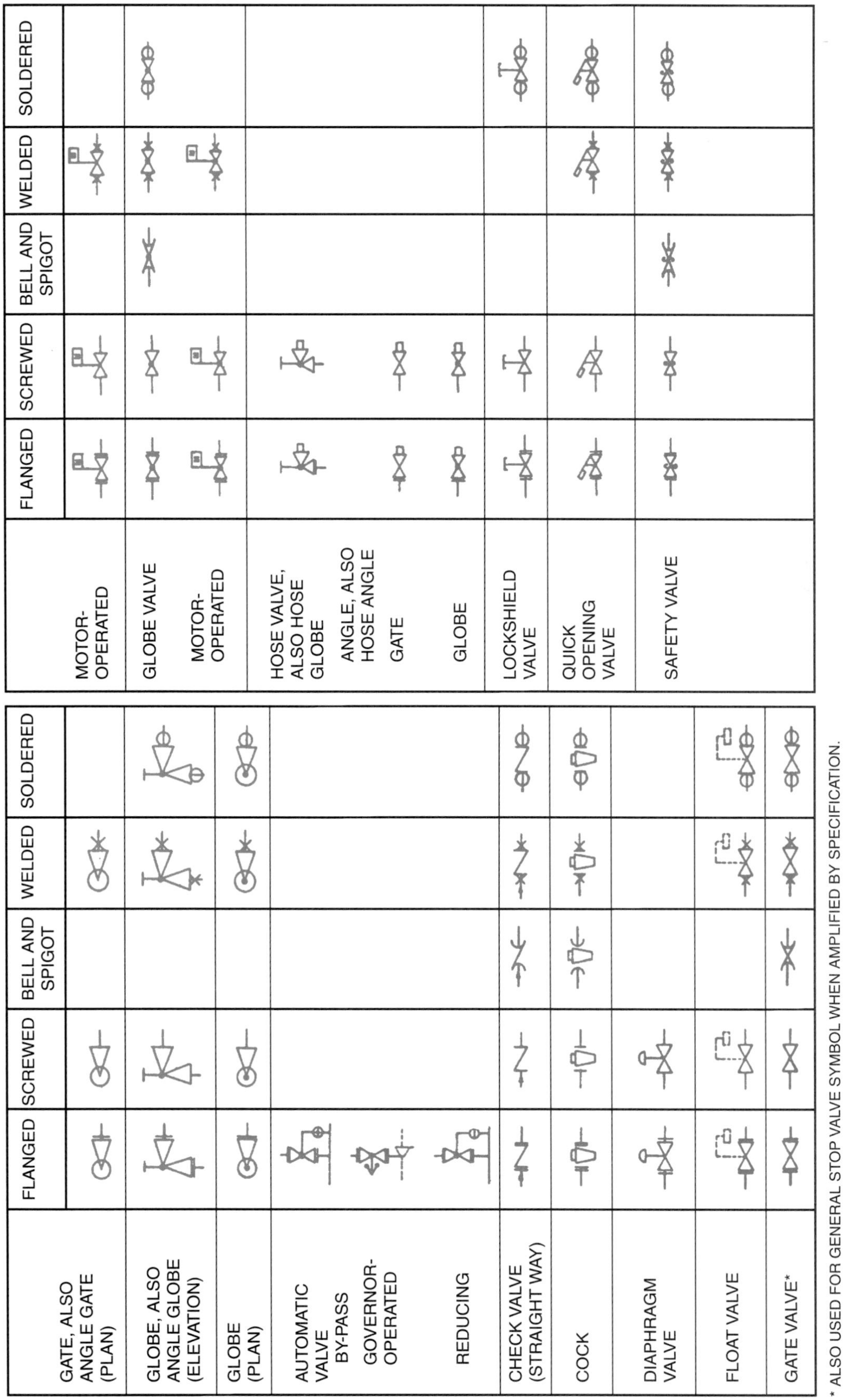

	FLANGED	SCREWED	BELL AND SPIGOT	WELDED	SOLDERED
GATE, ALSO ANGLE GATE (PLAN)					
GLOBE, ALSO ANGLE GLOBE (ELEVATION)					
GLOBE (PLAN)					
AUTOMATIC VALVE BY-PASS					
GOVERNOR-OPERATED					
REDUCING					
CHECK VALVE (STRAIGHT WAY)					
COCK					
DIAPHRAGM VALVE					
FLOAT VALVE					
GATE VALVE*					

* ALSO USED FOR GENERAL STOP VALVE SYMBOL WHEN AMPLIFIED BY SPECIFICATION.

	FLANGED	SCREWED	BELL AND SPIGOT	WELDED	SOLDERED
MOTOR-OPERATED					
GLOBE VALVE					
MOTOR-OPERATED					
HOSE VALVE, ALSO HOSE GLOBE					
ANGLE, ALSO HOSE ANGLE					
GATE					
GLOBE					
LOCKSHIELD VALVE					
QUICK OPENING VALVE					
SAFETY VALVE					

104F18B.EPS

Figure 18 (2 of 3) ◆ Fitting symbols.

	FLANGED	SCREWED	BELL AND SPIGOT	WELDED	SOLDERED
SLEEVE					
TEE STRAIGHT SIZE					
(OUTLET UP)					
(OUTLET DOWN)					
DOUBLE SWEEP					
REDUCING					
SINGLE SWEEP					
SIDE OUTLET (OUTLET DOWN)					
SIDE OUTLET (OUTLET UP)					
UNION					
ANGLE VALVE CHECK, ALSO ANGLE CHECK					
GATE, ALSO ANGLE GATE (ELEVATION)					

104F18C.EPS

Figure 18 (3 of 3) ◆ Fitting symbols.

vary depending on the type of drain installed, as shown in the table below the drawing. Look at this drawing and answer the following questions:

1. What is the diameter of the opening that must be cut out of the countertop?
2. What is the overall depth of the lavatory?
3. If a pop-up drain is used, how far from the floor should the drain pipe be located?
4. How far apart must the cold and hot water connections be on the faucet that is installed in this lavatory?

Did You Know?

Drawings like the one in *Figure 23* are also called (1) fixture rough-in sheets, (2) manufacturers' fixture spec sheets, and (3) cut sheets.

Similar drawings are used to show other types of fixtures. *Figure 24* shows a bathtub. The top view is typical of views shown in orthographic drawings, but the side and end views are drawn as section views. Section views show the internal shape of objects. Also, part of the end view has been eliminated and the side view moved to the left of its normal location. Study this drawing carefully and then answer the following questions:

1. In how many different sizes is this bathtub manufactured?
2. What is the overall length of the smallest bathtub available?
3. How high above the subfloor will the side of the bathtub be?
4. Will an opening need to be cut in the subfloor to permit installation of the bathtub?
5. How far above the subfloor will the stub-out for the shower head be located?

Figure 25 shows a rough-in drawing for a typical water closet (toilet). Notice how it shows the overall dimensions and the location of the water supply piping.

4.5.3 Assembly Drawings

Assembly drawings show how to assemble complex objects. They also show the relationship of the individual parts to the whole object.

Figure 26 shows a pop-up assembly for a lavatory with all its parts. Note how the parts are shown so that you can easily visualize how the pop-up is put together.

Figure 27 is a more complicated kind of assembly drawing than *Figure 26.* Each part is numbered to correspond to a number in the parts list. Note that both a handle version and a push-button version of the flush valve are shown on the same drawing. Also, note that two types of control stops may be installed. With this kind of drawing, a plumber should be able to disassemble and correctly reassemble any plumbing product.

Figure 28 shows an assembly drawing of a kitchen faucet. With this drawing, a plumber should be able to identify needed repair parts and make the necessary replacements.

WASTE WATER

DRAIN OR WASTE–ABOVE GRADE	————
DRAIN OR WASTE–BELOW GRADE	— — — -
VENT	- - - - - - - -
COMBINATION WASTE AND VENT	—— CWV ——
ACID WASTE	—— AW ——
ACID VENT	- - - AV - - -
INDIRECT DRAIN	—— D ——
STORM DRAIN	—— SD ——
SEWER-CAST IRON	S-CI
SEWER-CLAY TILE BELL & SPIGOT	S-CT
DRAIN-CLAY TILE BELL & SPIGOT	————

OTHER PIPING

GAS-LOW PRESSURE	—— G —— G ——
GAS-MEDIUM PRESSURE	—— MG ——
GAS-HIGH PRESSURE	—— HG ——
COMPRESSED AIR	—— A ——
VACUUM	—— V ——
VACUUM CLEANING	—— VC ——
OXYGEN	—— O ——
LIQUID OXYGEN	—— LOX ——

104F19.EPS

Figure 19 ◆ Pipe symbols.

Did You Know?

The symbol C_L stands for centerline.

4.5.4 Cutaway Drawings

For many products, a **cutaway,** or section, drawing is enough to show how a product is constructed. The gate valve shown in *Figure 29* is a good example of this kind of drawing. Note that each part is numbered to correspond to the parts list. Also, note that the dimensions of many sizes of this valve are given in the specifications table. This method eliminates the need to make a separate drawing for each size valve.

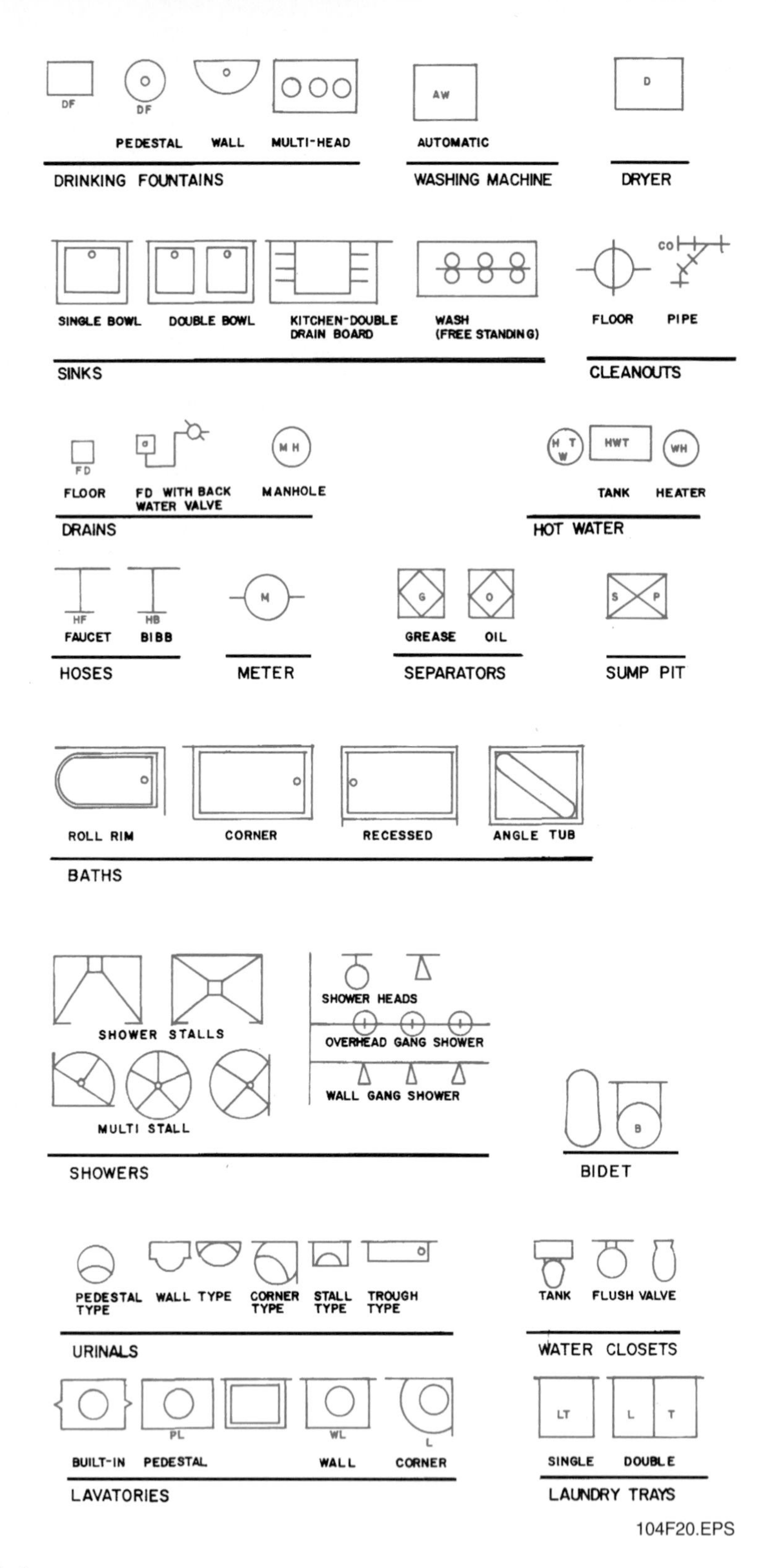

Figure 20 ◆ Fixture symbols.

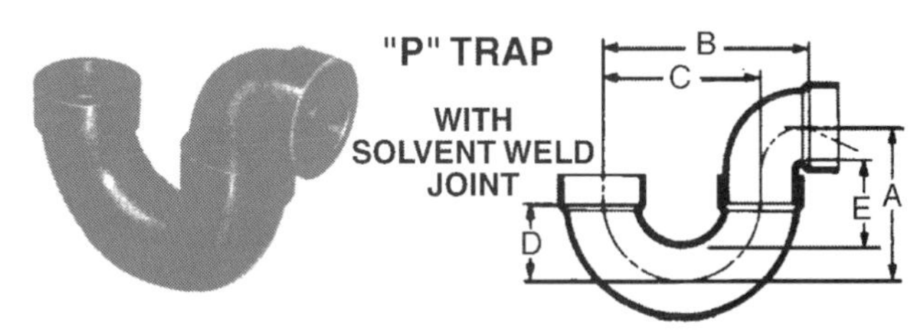

SIZE	DESCRIPTION H x H	A	B	C	D	E
1 1/2		3 3/4	5 7/32	4	1 13/16	2 1/8
2		4 9/16	7 1/2	5 1/4	2 1/4	2 1/2
3		6 5/16	8 7/16	6 5/16	2 11/16	3 1/4
4		7 7/8	10 13/16	8 1/8	3 1/2	3 7/8

104F21.TIF

Figure 21 ◆ Catalog drawing of a P trap.

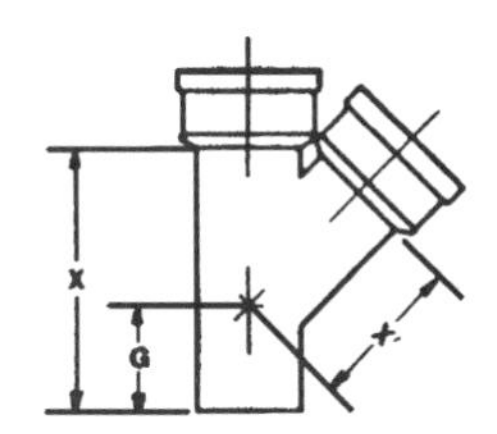

Wye					
				Weights	
Size	G	X	X′	SV	XH
2 × 2	4	8	4	7	8
3 × 2	4 3/16	9	5	10	14
3 × 3	5	10 1/2	5 1/2	13	17
4 × 2	3 5/8	9	5 3/4	12	17
4 × 3	4 7/16	10 1/2	6 1/4	14	20
4 × 4	5 1/4	12	6 3/4	17	24
5 × 2	3 1/8	9	6 1/2	14	20
5 × 3	3 7/8	10 1/2	7	17	24
5 × 4	4 11/16	12	7 1/2	19	27
5 × 5	5 1/2	13 1/2	8	22	32
6 × 2	2 9/16	9	7 1/4	17	23
6 × 3	3 3/8	10 1/2	7 3/4	19	27
6 × 4	4 3/16	12	8 1/4	22	31
6 × 5	4 15/16	13 1/2	8 3/4	24	35
6 × 6	5 3/4	15	9 1/4	28	40
8 × 2	3 1/8	10 1/2	8 1/2	29	42
8 × 3	3 15/16	12	9	32	47
8 × 4	4 3/4	13 1/2	9 1/2	36	52
8 × 5	5 1/2	15	10	39	57
8 × 6	6 5/16	16 1/2	10 1/2	44	63
8 × 8	7 11/16	19 1/2	11 13/16	55	82
10 × 3	2 9/16	12	11	50	71
10 × 4	3 9/16	13 1/2	11 1/8	53	74
10 × 5	4 5/16	15	11 5/8	57	80
10 × 6	5 1/8	16 1/2	12 1/8	61	86
10 × 8	6 1/2	19 1/2	13 7/16	77	110
10 × 10	8	22 1/2	14 1/2	94	133
12 × 4	4 1/8	15	12 7/16	70	97
12 × 5	4 7/8	16 1/2	12 15/16	74	104
12 × 6	5 11/16	18	13 7/16	80	111
12 × 8	7 1/16	21	14 3/4	96	136
12 × 10	8 9/16	24	15 13/16	115	160
12 × 12	10 1/8	27	16 7/8	135	186
15 × 4	2 1/4	15	15	130	130
15 × 6	4	18	15 3/4	109	152
15 × 8	5 3/8	21	17 1/16	127	182
15 × 10	6 7/8	24	18 1/8	152	213
15 × 12	8 7/16	27	19 3/16	170	242
15 × 15	10 3/4	31 1/2	20 3/4	204	290

104F22.EPS

Figure 22 ◆ Catalog drawing of a wye.

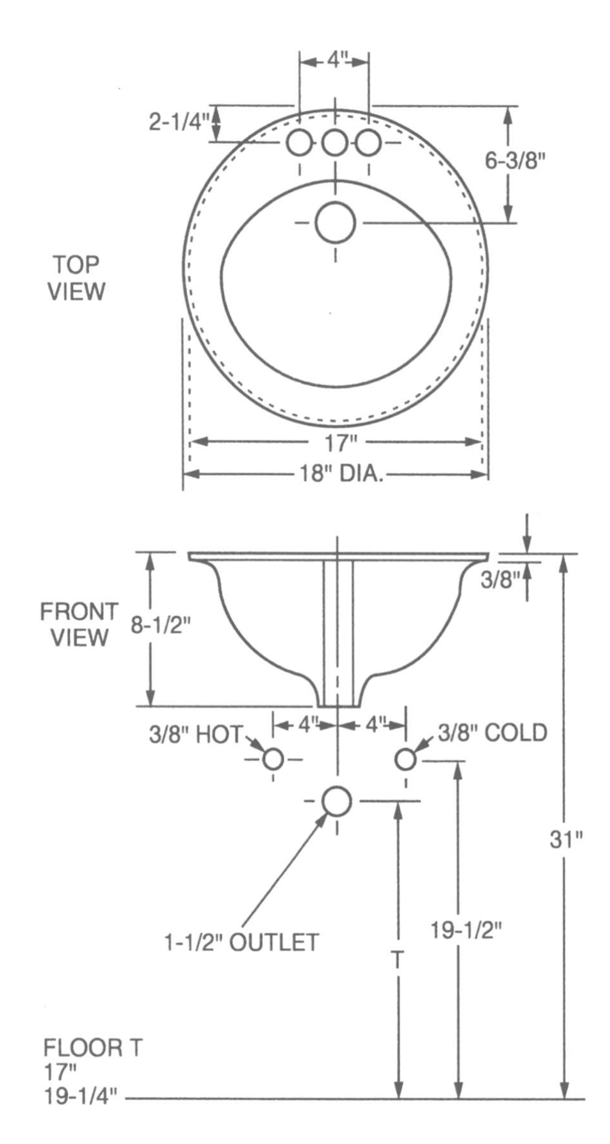

104F23.EPS

Figure 23 ◆ Lavatory fixture drawing.

4.5.5 Sketches

Being able to communicate with other workers is a skill every plumber must have. For example, you must be able to tell another plumber how pipe is to be installed or how the building structure must be modified to install plumbing. Sketches are often made on whatever surface is convenient—scraps of paper, wood, the floor, or the wall. These sketches can communicate very effectively.

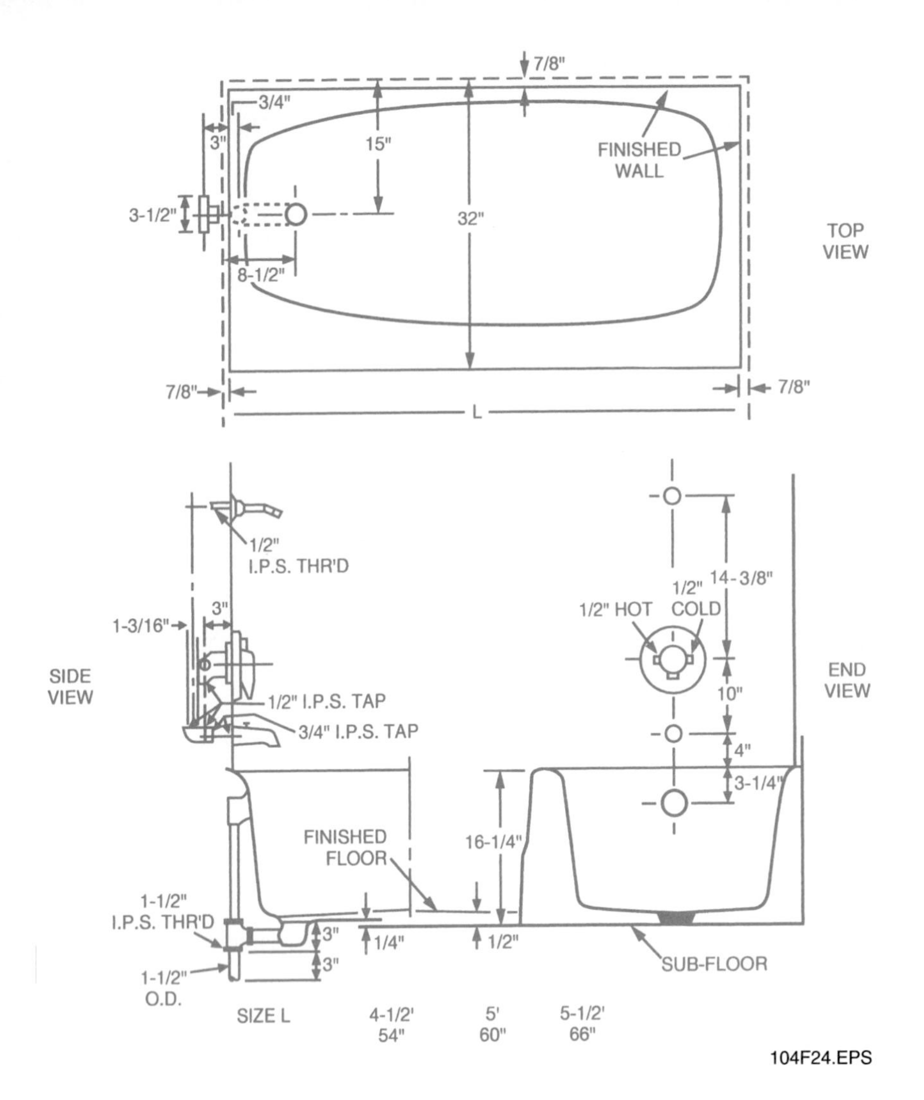

Figure 24 ◆ Bathtub fixture drawing.

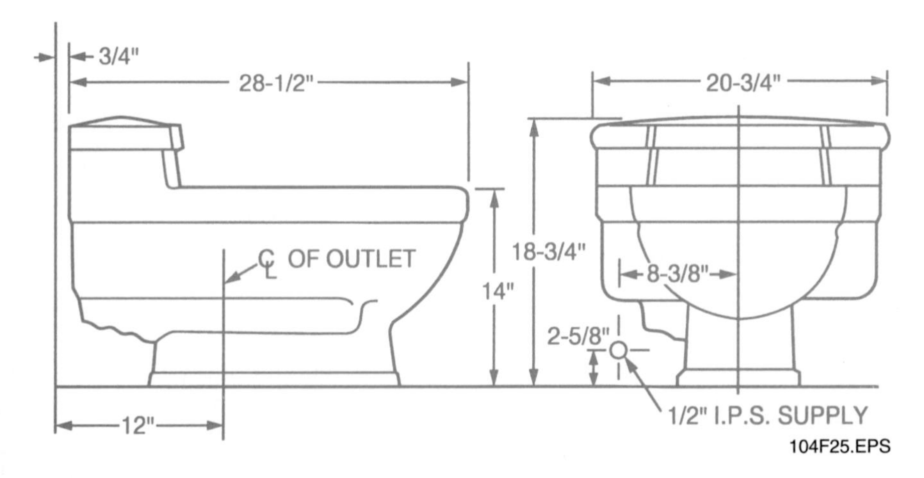

Figure 25 ◆ Water closet fixture drawing.

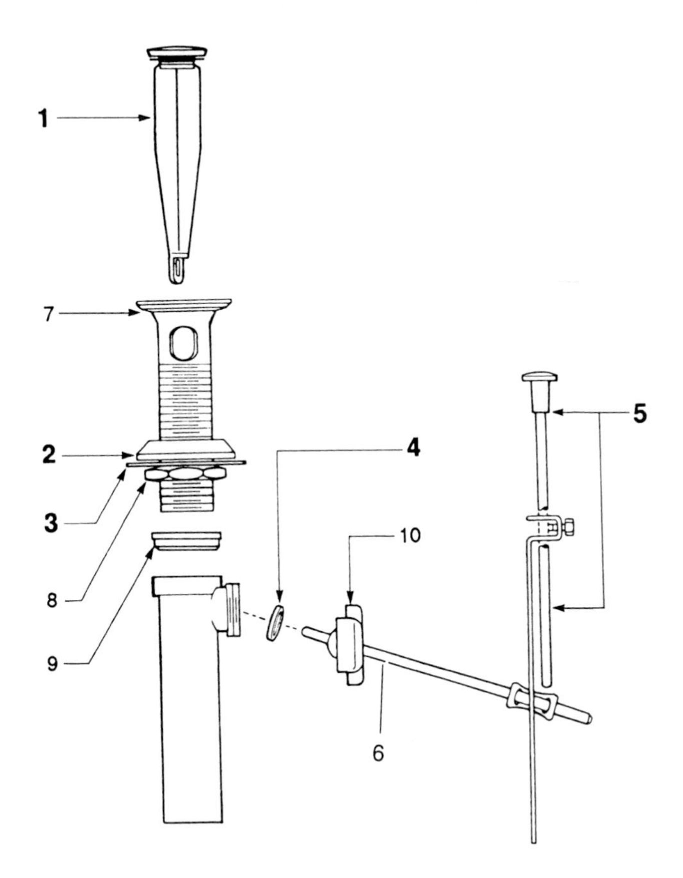

MOEN

METALLIC/NON-METALLIC LAVATORY WASTE ASSEMBLIES MODELS 96497 & 96107

ITEM	CAT. NO.	MFG.	DESCRIPTION
1	52403 ✔	12314	**Waste Plug Assy.**
2	52102 ✔	11987	**Gasket**
3	58689	11990	**Washer**
4	52249 ✔	12312	**Pivot Seat**
5	58579 ✔	11986	Rod & Knob (bulk)
	58700 ✔		Rod & Knob (carded)
6	51910✔	11985	**Pivot Rod**
10	58675 ✔	11998	Pivot Nut
PARTS AVAILABLE THROUGH SPECIAL ORDER			
7	----	11984	Waste Seat Kit
8	----	12313	Mounting Nut
9	----	11997	Seal

104F26.EPS

Figure 26 ◆ PO (plugged opening) pop-up assembly drawing.

Review Questions

Section 4.0.0

1. A(n) _______drawing gives the impression of three dimensions and provides a picture that visually resembles the physical object.
   a. orthographic
   b. schematic
   c. isometric
   d. linear

2. A(n) ____________ drawing is made by looking at the object from one or more of six possible angles.
   a. projection
   b. orthographic
   c. oblique
   d. pictorial

3. Using ________ correctly makes it possible to produce a very understandable schematic drawing quickly.
   a. symbols
   b. 30-degree angles
   c. assembly drawings
   d. isometrics

4. Plumbers look at _________ drawings to determine such factors as overall dimensions, actual inside dimensions, fitting depth, and laying length of the fittings.
   a. schematic
   b. isometric
   c. cutaway
   d. catalog

5. A(n) __________ drawing shows the relationship of the individual parts to the object as a whole.
   a. oblique
   b. isometric
   c. assembly
   d. schematic

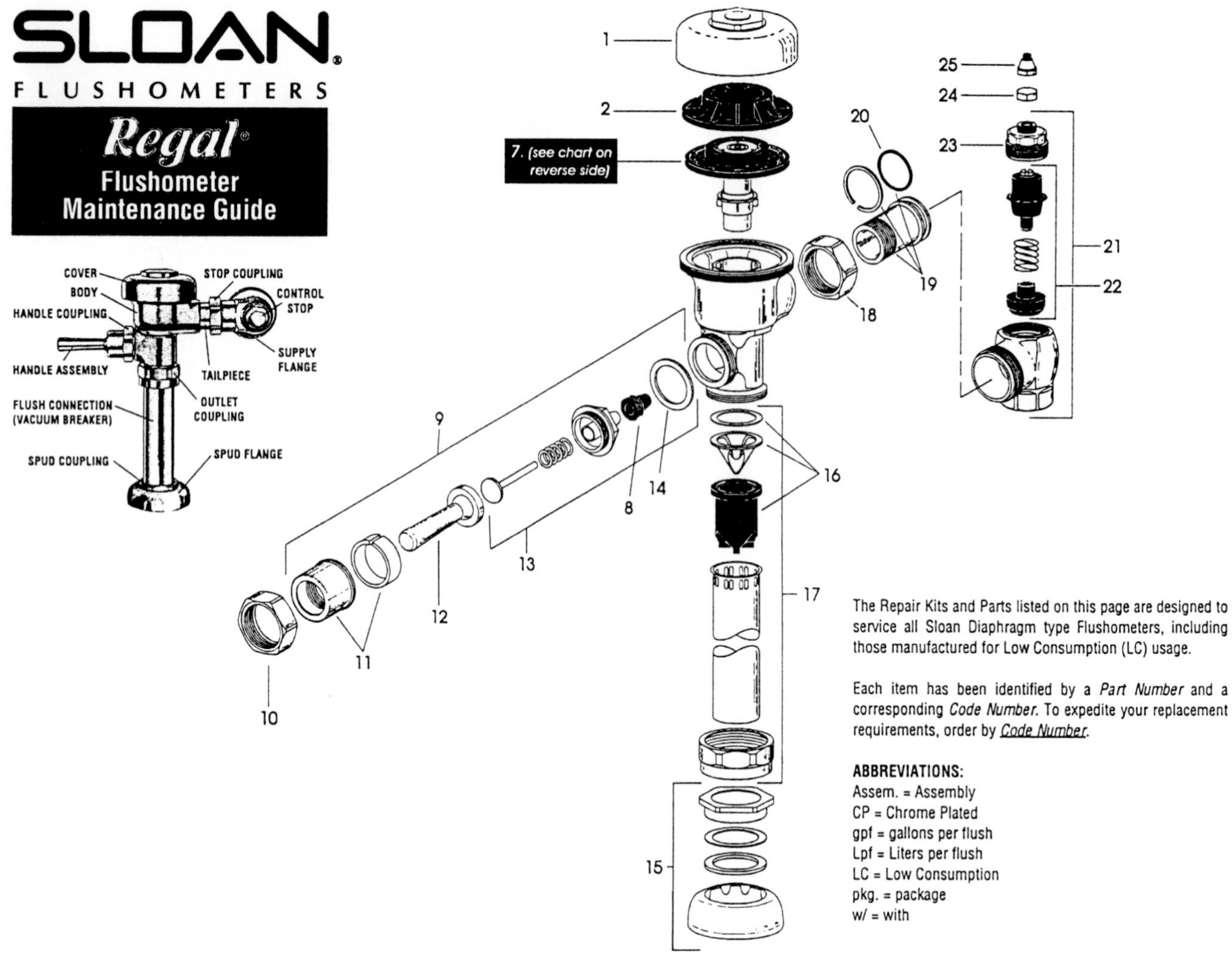

Figure 27 ◆ Assembly drawing of a flush valve.

Summary

This module has introduced plumbing drawings and suggested how a plumber can get information from them. It is important to remember that it takes many different types of drawings, along with the specifications for the project, to get a complete picture of the plumbing requirements. Pictorial drawings show depth. Orthographic drawings show objects as if you were looking at them head-on from certain angles. The ability to read and understand different kinds of drawings is an essential plumbing skill, along with the ability to make orthographic or isometric sketches and to draw plumbing assemblies. This module has introduced the basics of reading and interpreting drawings. It is a building block for future study. Blueprint reading, like all other plumbing skills, must be practiced and studied throughout your career.

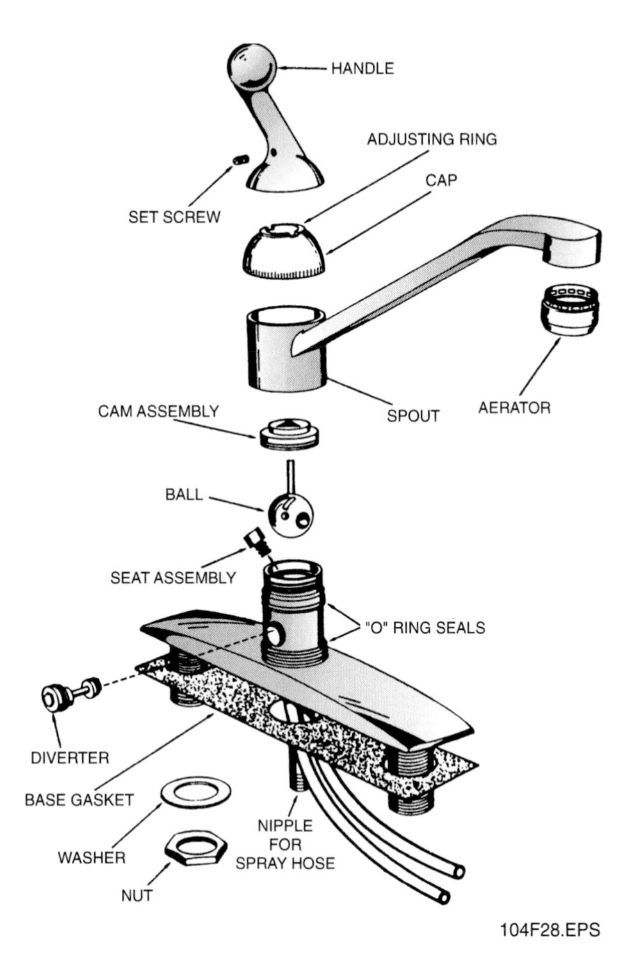

Figure 28 ◆ Assembly drawing of a kitchen faucet.

Parts and Materials

Item	Description	Material
1	Body	ASTM B-62
2	Handwheel	Malleable Iron
3	Handwheel Nut	ASTM B-16
4	Identification Plate	Brass
*5	Stem	ASTM B-16, B-62
*6	Packing Nut	ASTM B-16, B-62
7	Packing	Teflon-Asbestos
*8	Stuffing Box	ASTM B-16, B-62
9	Centerpiece	ASTM B-62
*10	Disc	ASTM B-62

**B-62 used in larger sizes*

Specifications

Valve Size	Cat. No.	Dim. A	Dim. B	Dim. C	Weight Pounds
3/8	50812	4 1/8	1 5/8	2 1/16	.8
1/2	50813	4 1/4	2 3/16	2 1/16	.9
3/4	50814	4 7/16	2 7/8	2 1/2	1.4
1	50815	5 7/16	3 3/8	2 1/2	2.1
1 1/4	50816	6 1/4	3 7/16	2 1/2	2.9
1 1/2	50817	7 1/16	3 3/4	3	3.8
2	50818	8 1/4	4 3/8	3 1/2	6.3
2 1/2	51537	9 3/16	5 3/8	4 1/8	10.9
3	51538	11 1/8	6 1/4	5 1/4	16.4

All Dimensions in Inches

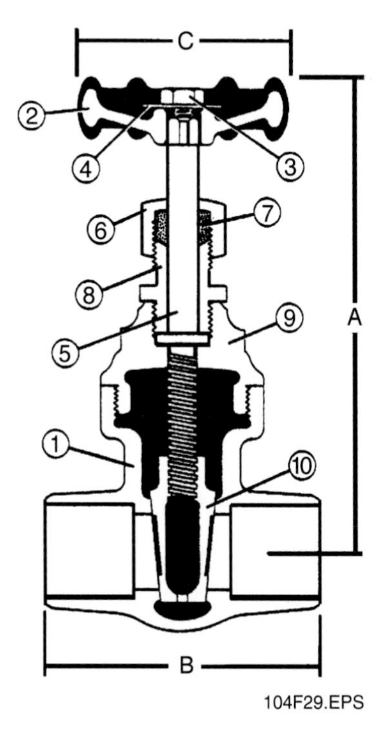

Figure 29 ◆ Cutaway drawing of a gate valve.

PROFILE IN SUCCESS

James Lee

Plumbing
Instructor, Atlantic Technical Center
Coconut Creek, FL

While still a teenager, James, a native of Panama City, Florida, worked as a foreman in his father's plumbing business, supervising two to three employees. He wasn't convinced that he liked plumbing growing up, but returned to the trade after college. He worked for both mechanical and plumbing companies before starting his teaching career.

How did you become interested in the construction industry?

I attended Tom Henney Technical Center during high school. Then I attended Florida A&M where I got my bachelor's degree in political science. I was aspiring to be an attorney during those days, although I later decided that was not for me. Following college graduation, I went back to work for my dad, enrolled in the apprenticeship-training program, and got my journeyman's license.

In 1986, I moved to Ft. Lauderdale, where I joined the union. I worked as a foreman for two plumbing companies and a mechanical contractor. I oversaw mechanical work, plumbing installation and service, and even pipefitting work for a shipbuilding company.

I decided that I did not want to be in the field every day, so in 1987 I applied for an instructor's position at Atlantic Technical Center and have been here ever since. Our program is a feeder for apprenticeship-training programs. Our students are part of the SkillsUSA VICA (Vocational Industrial Club of America) program that precedes the National Association of Plumbing, Heating, and Cooling Contractors (NAPHCC) apprenticeship-training program.

Someday I hope to join a U.S. embassy or Peace Corps effort to teach plumbing in less-developed countries. It is a fundamental trade with many facets. I think it would be a great challenge.

What do you think it takes to be a success in your trade?

To be successful, you have to learn 100 percent of the theory and 100 percent of the practical skills. Most important, however, are your skills. You have to be someone a company can depend on. If you are dependable and willing to work, you will find success in the trades.

I also recommend apprenticeship training to get a leg up. A program of this type prepares you with both the practical skills and the theoretical knowledge you need. But, if you don't stick it out, you can't succeed. To be a success in plumbing—or any trade—you have to decide to make it through the apprenticeship phase and get your journeyman's license. That is the basis for success.

What are some of the things you do in your role of instructor?

I teach VICA students the theory and practical skills of plumbing. Our students perform most of the maintenance jobs on campus as well as some installation jobs. Several of my students have won state and regional VICA skills contests. Basically, I help these students prepare to enter a formal apprenticeship-training program where they will work for a company during the day and go to class one or two nights a week. Besides teaching students how to work with plumbing tools and equipment, I also teach them how to become dependable, valuable employees.

As a plumber, I did a lot of service warranty work, troubleshooting leaks inside walls and underneath slabs. My biggest challenge was dealing with unhappy and disgruntled customers. Dealing tactfully and courteously with customers is a critical job skill.

What do you like most about your job?

When I was a plumber, I liked the challenge of the different situations I faced each day. No two plumbing jobs are exactly the same. You have to be able to understand and apply what you know to each new circumstance. You don't just learn a set of skills. You learn how to adapt theory and skills to individual challenges. It's a great learning experience every day.

As an instructor, what I like most is the success of my former students. Lots of my students keep in touch with me and let me know how they have made it in their trade. Their success is my greatest reward.

What would you say to someone entering into the trades today?

When I am interviewing students who want to join the VICA program, I stress the many facets of the plumbing trade. I tell students to concentrate on learning new technologies. Keeping up with an industry that changes constantly is a challenge. I also emphasize that plumbing is hard work. You have to be willing to keep at it and to persevere. Preparation and opportunity make success.

I tell anyone interested in the trades, "The only time success comes before work is in the dictionary."

GLOSSARY

Trade Terms Introduced in This Module

Architect's scale: A kind of ruler that uses smaller units to represent feet, such as ½ inch or ¼ inch to represent 1 foot. The units are used so that all building measurements are in proportion to the actual measurements but on a smaller scale for the drawings.

Assembly drawing: A drawing that shows how to assemble a complex product. It also shows the relationship of the individual parts to the object as a whole.

Blueprint: A construction drawing. The name comes from the original method of producing these drawings, which resulted in a blue background with white lines.

Catalog drawings: Drawings of plumbing fixtures that are found in manufacturers' catalogs.

Computer-aided drafting (CAD): A sophisticated design program used on computers. Drawings can be created in two or three dimensions.

Cutaway drawing: Section drawing that shows the construction elements of a particular part of the building or of a fixture.

Details: Sections of construction drawings that are enlarged to make them clearer.

Dimensions: The measurements of an object.

Easement: A designated right-of-way, such as the access guaranteed to utility companies for repair of utilities that are located on, or cross over, private land.

Electrical drawing: Drawing that shows the location of outlets, switches, and electrical fixtures. May be superimposed on the floor plan.

Elevation drawing: The drawing of a structure showing a side, front, or back view.

Extension line: A line used on a drawing to locate a dimension (measurement) away from the actual points of the dimension. This method is used when a drawing would be too crowded or cluttered if the dimension were shown within the two points.

Fixture drawing: A drawing that shows the components of a fixture in detail.

Floor plan: A construction drawing of a building looking down at the floor from above (bird's-eye view). The plan shows at least the outline of the wall locations and lengths to scale.

Foundation plan: A construction drawing showing the placement and dimensions of a building foundation.

HVAC drawing: (Heating, ventilation, and air-conditioning)—A construction drawing that shows the placement of the furnace and air-conditioning equipment, and the location of ducts and registers or pipes and radiators.

Isometric drawing: A pictorial drawing that creates the feel of a three-dimensional object. All horizontal lines are projected at a 30-degree angle.

Notes: Brief comments, instructions, or information on a construction drawing or other document.

Oblique drawing: A pictorial drawing that shows the shape of an object. It shows the front of the object with the body of the object at a slight angle.

Orthographic drawings (projections): Construction drawings that show straight-on views of the different sides of an object. Used for elevation drawings.

Pictorial drawing: A drawing that shows a three-dimensional view of an object.

Plot plan: A plan view drawing of a structure that includes the dimensions of the building site, the location of the structure in relation to the property boundaries, elevation of key points, existing and finish contour lines, utility services, and compass directions.

Plumbing drawings: Construction drawings that show the location of fixtures and pipe runs, and give the size and type of pipe to be installed.

Scale: The relationship of the dimensions on a drawing to the actual dimensions of the structure; for example, in a ¼ inch scale, ¼ inch represents 1 foot.

Schematic drawing: A single-line drawing of a plumbing system or electrical wiring routing or circuit.

Sepia (drawing): A print or construction drawing with dark reddish-brown lines on a light background. Used to make blueline prints.

Setback: The distance a local code requires a building to be set back from the street.

Side yards: Space along the sides of a structure that provides access to rear yards, reduces the possibility of fire jumping from one building to the next, and promotes ventilation around the structure.

Single-line drawing: A plumbing drawing that uses a single line to represent the centerline of a pipe. Single-line drawings can be used to represent a pipe of any diameter.

Specifications (specs): Written requirements included with the drawings or blueprints of a construction project. They provide more details or descriptions of the technical standards that must be met during construction.

Symbol: A mark or drawing used to indicate a specific object, material, class, or entity. A legend shows the symbols used on a drawing and their meanings.

Take-off: The process by which detailed lists are compiled, based on drawings and specifications, of all the material and equipment necessary to construct a project (also called material take-off).

Title block: A block on the front of a construction drawing or blueprint, located in the lower right corner, that provides identification and revision information about the drawing.

APPENDIX A

Plumbing Drawings

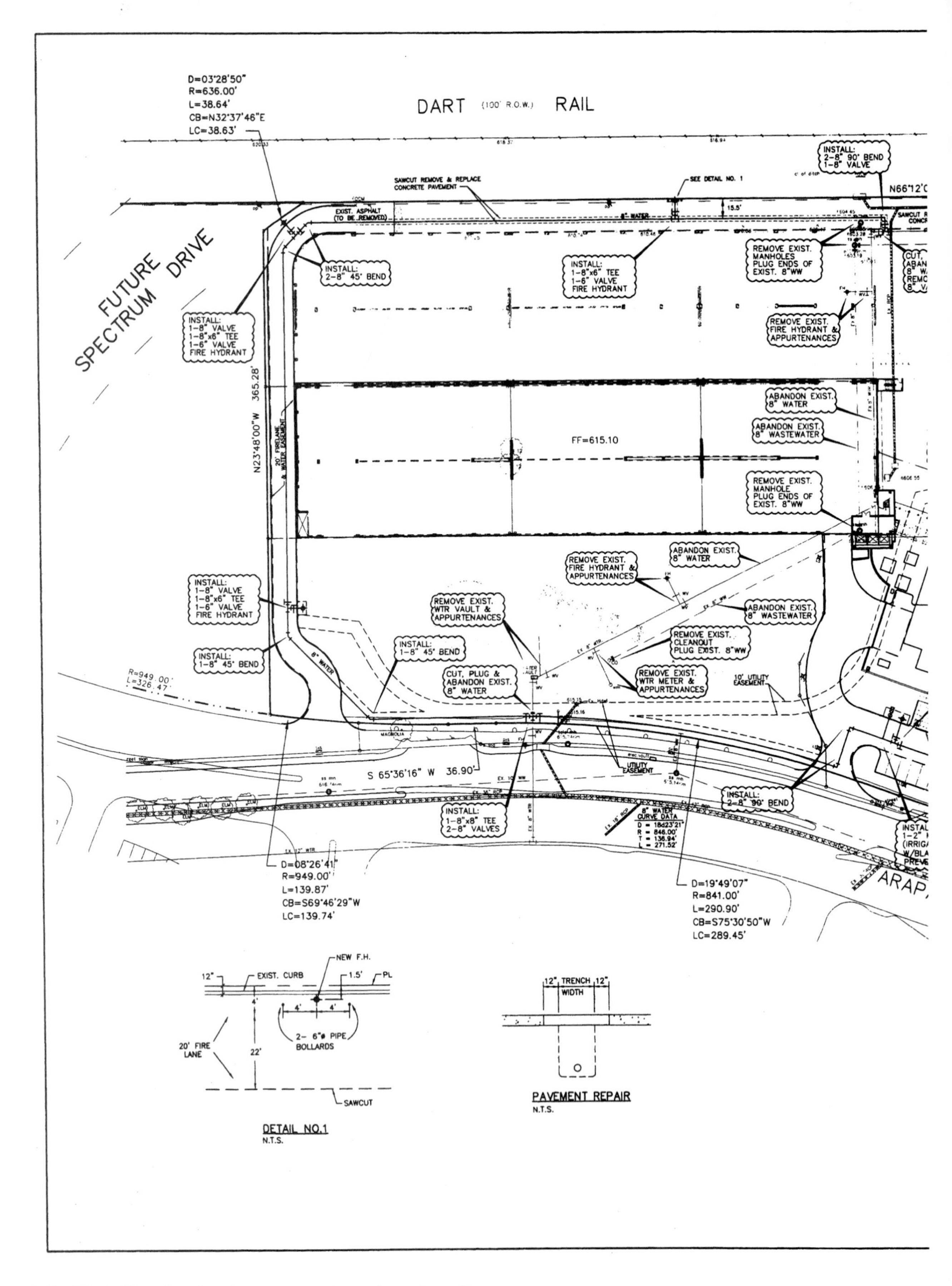

Figure 1 (1 of 2) ◆ Plot plan showing sewer, water main, and gas lines.

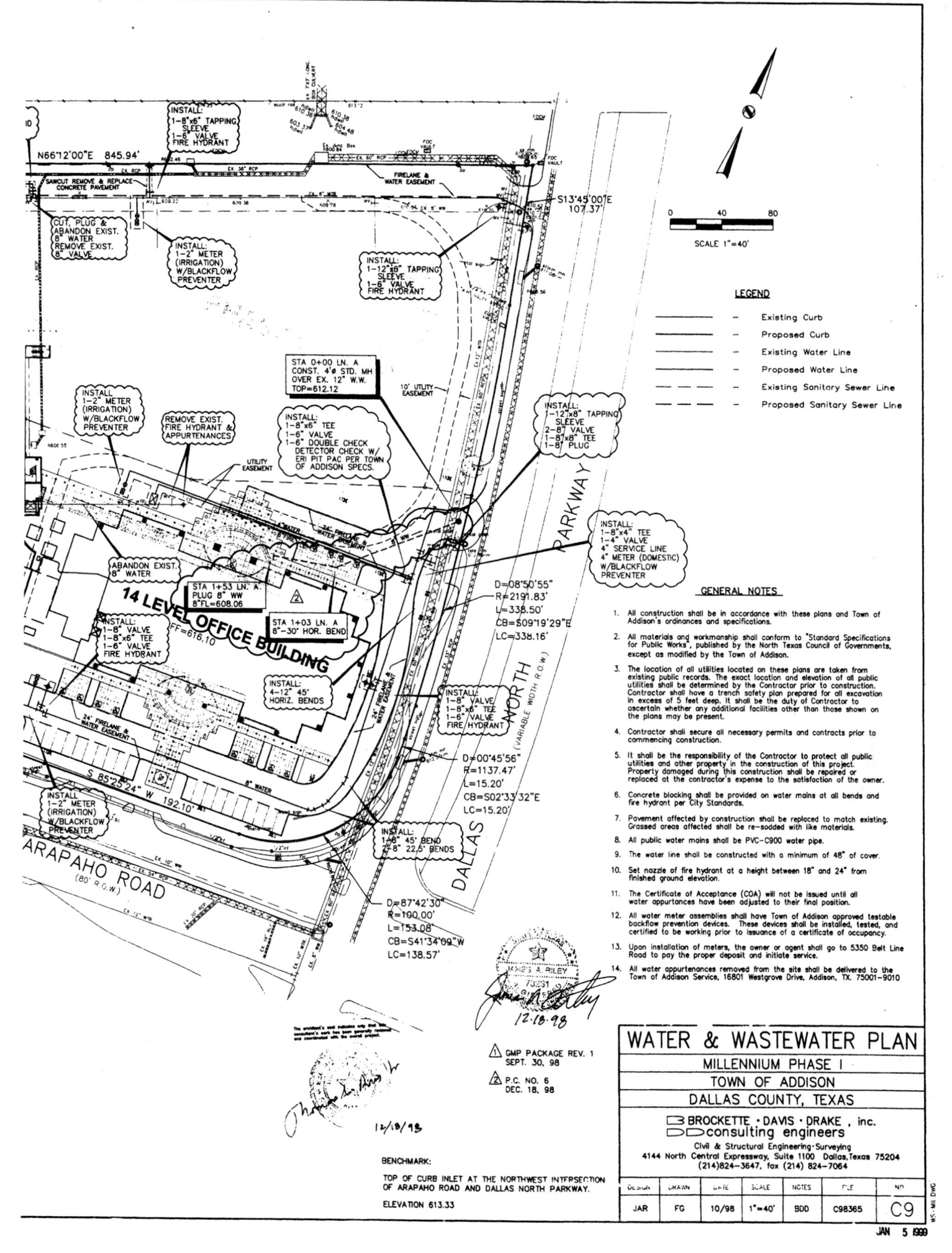

Figure 1 (2 of 2) ◆ Plot plan showing sewer, water main, and gas lines.

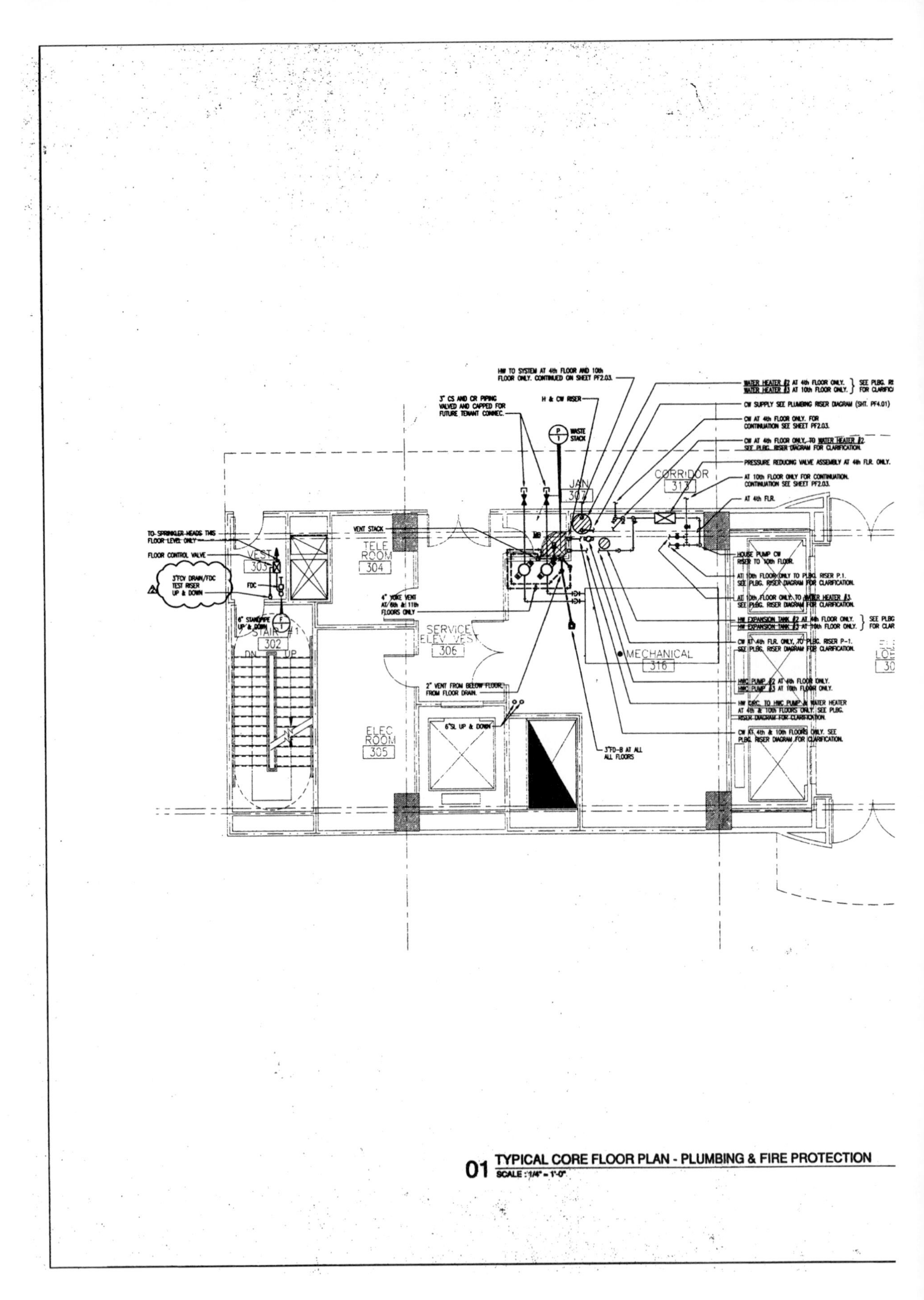

Figure 2 (1 of 2) ◆ Floor plan: plumbing.

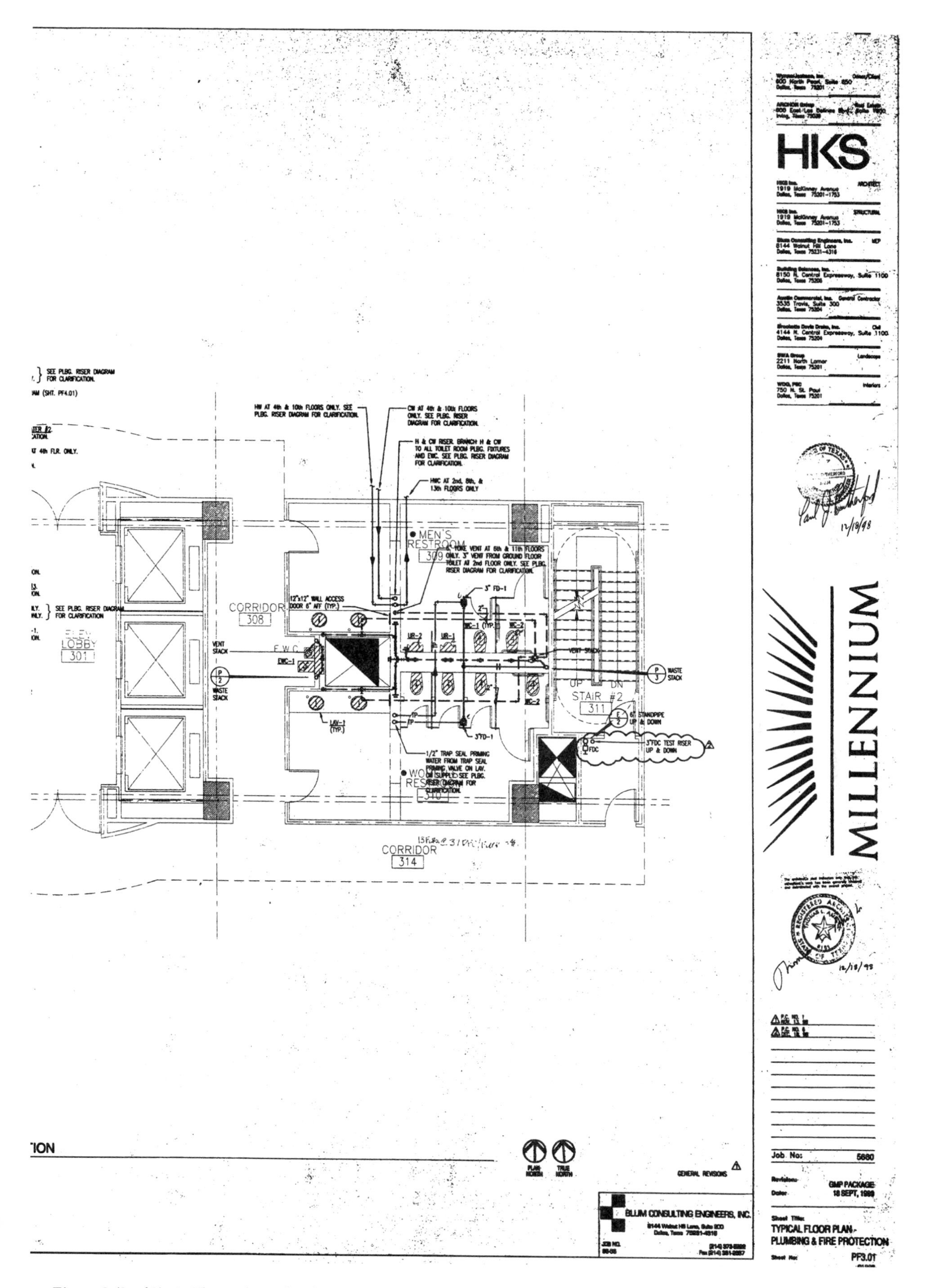

Figure 2 (2 of 2) ◆ Floor plan: plumbing.

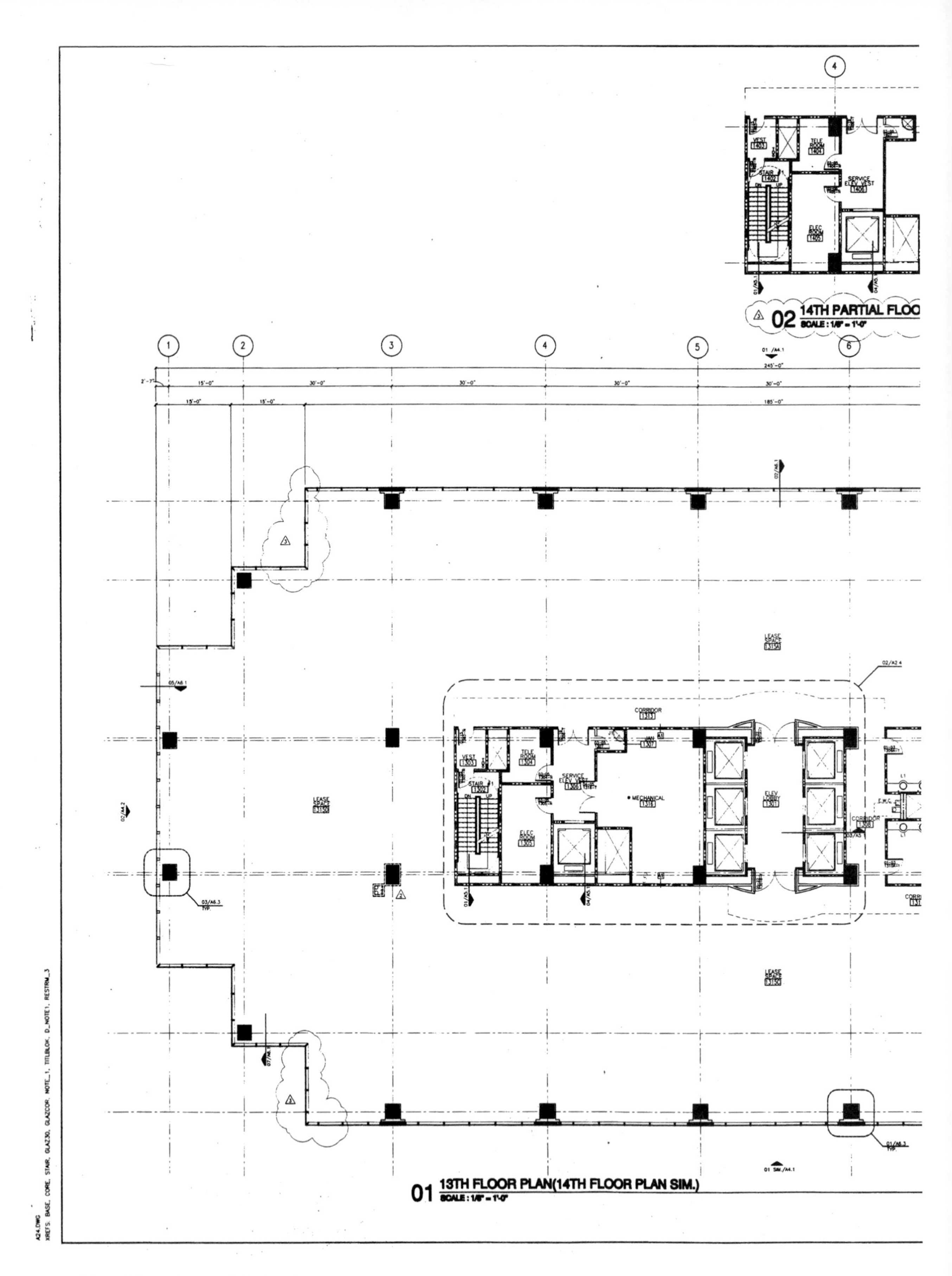

Figure 3 (1 of 2) ◆ Floor plan: architectural.

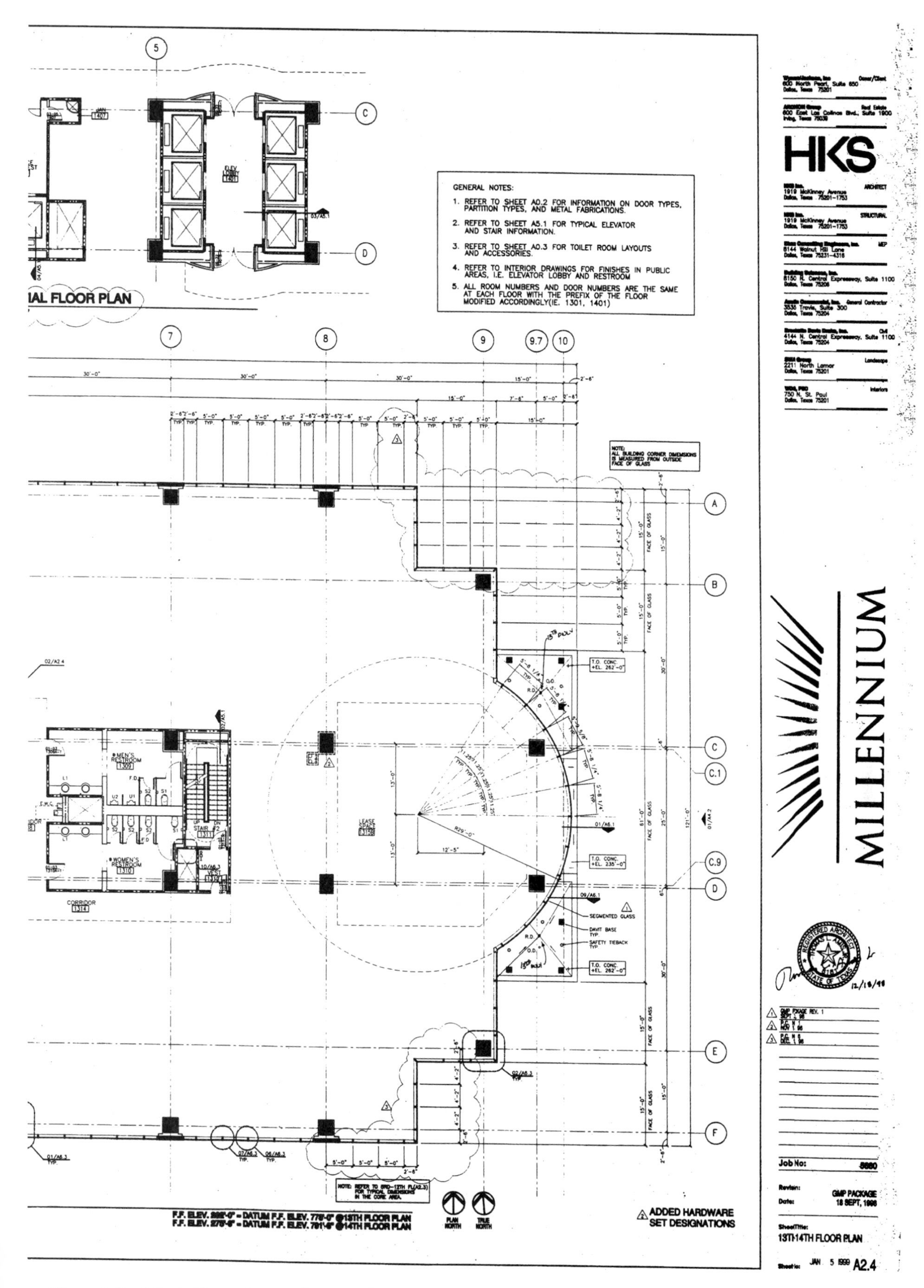

Figure 3 (2 of 2) ◆ Floor plan: architectural.

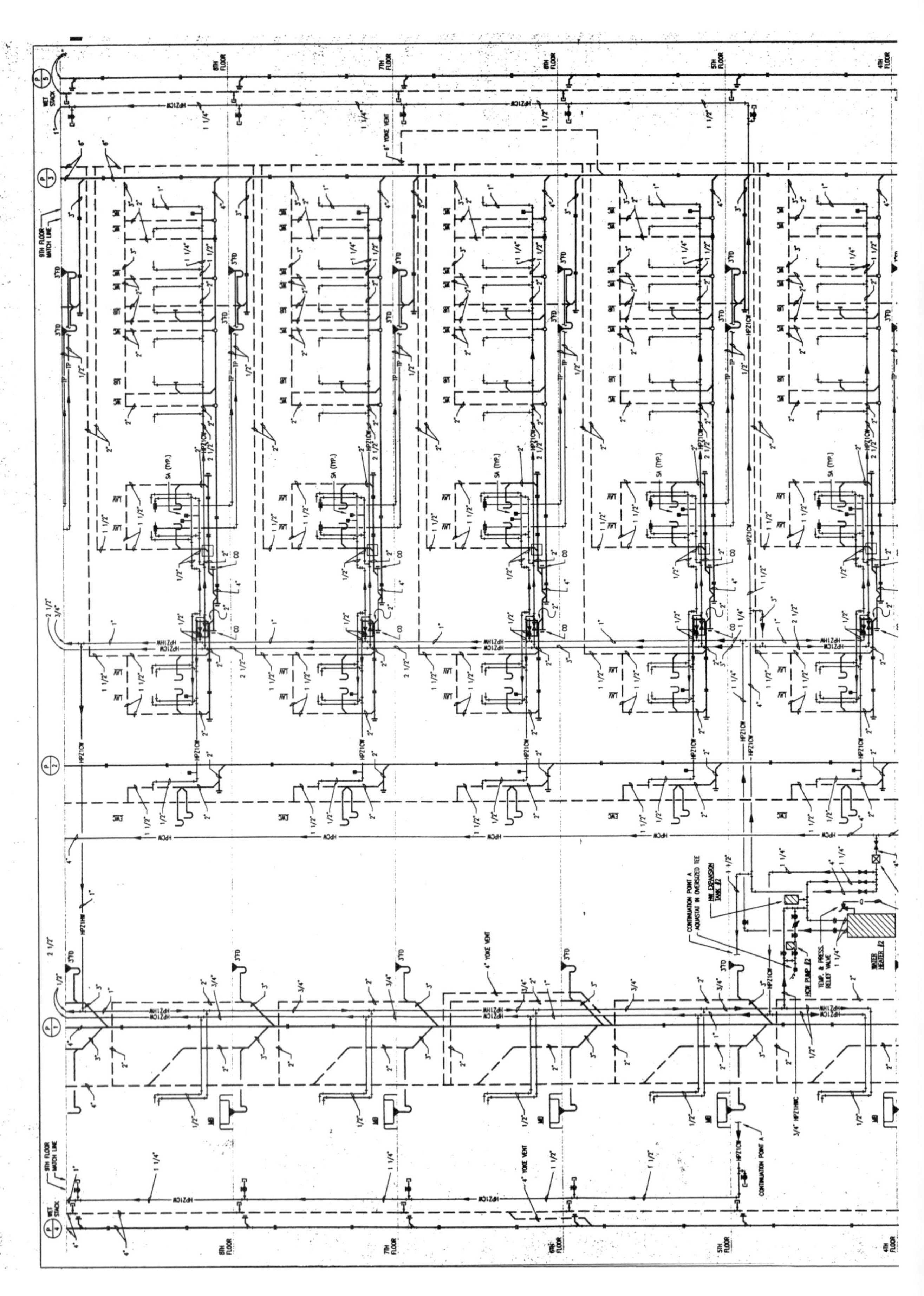

Figure 4 (1 of 2) ◆ Riser diagram with material take-off.

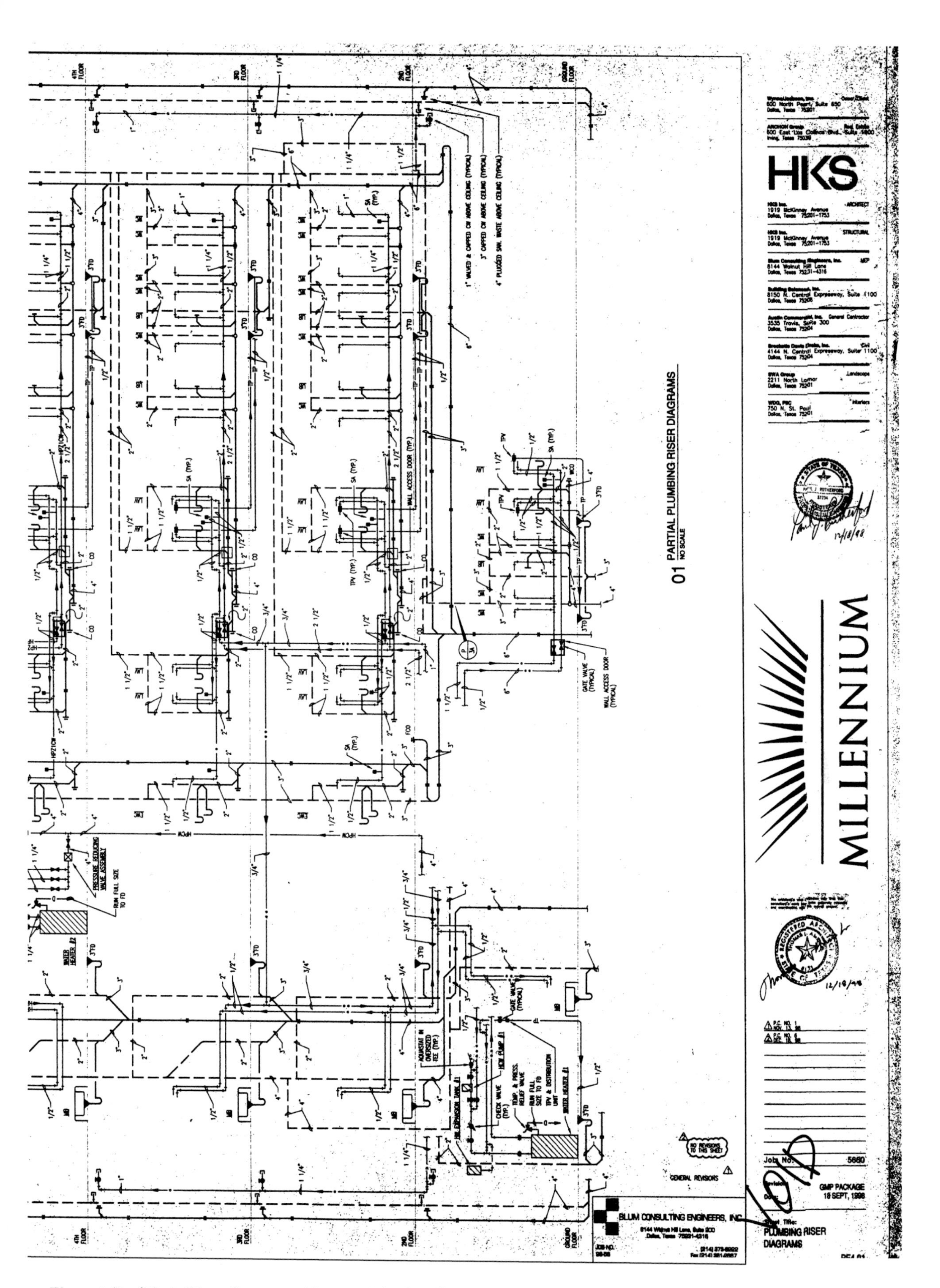

Figure 4 (2 of 2) ◆ Riser diagram with material take-off.

ANSWER KEY

Answers to Review Questions

Section 2.0.0

1. b
2. d
3. b
4. c
5. b

Section 3.0.0

1. d
2. b
3. a
4. d
5. c

Section 4.0.0

1. c
2. b
3. a
4. d
5. c

References

Blueprint Reading for Plumbers, 1989 Edition. J. Guest, B. D'Arcangelo, and B. D'Arcangelo. Albany, NY, 1989: Delmar Publications, Inc.

NCCER CRAFT TRAINING USER UPDATES

The NCCER makes every effort to keep these textbooks up-to-date and free of technical errors. We appreciate your help in this process. If you have an idea for improving this textbook, or if you find an error, a typographical mistake, or an inaccuracy in the NCCER's Craft Training textbooks, please write us, using this form or a photocopy. Be sure to include the exact module number, page number, a detailed description, and the correction, if applicable. Your input will be brought to the attention of the Technical Review Committee. Thank you for your assistance.

Instructors – If you found that additional materials were necessary in order to teach this module effectively, please let us know so that we may include them in the Equipment/Materials list in the Instructor's Guide.

Write: Curriculum Revision and Development Department
National Center for Construction Education and Research
P.O. Box 141104, Gainesville, FL 32614-1104

Fax: 352-334-0932

E-mail: curriculum@nccer.org

Craft ______________________ Module Name ______________________

Copyright Date ______________ Module Number ______________ Page Number(s) ______________

Description

____

____

____

(Optional) Correction

____

____

(Optional) Your Name and Address

____

____

Module 02105-00

Plastic Pipe and Fittings

Course Map

This course map shows all of the modules in the first level of the Plumbing curriculum. The suggested training order begins at the bottom and proceeds up. Skill levels increase as you advance on the course map. The local Training Program Sponsor may adjust the training order.

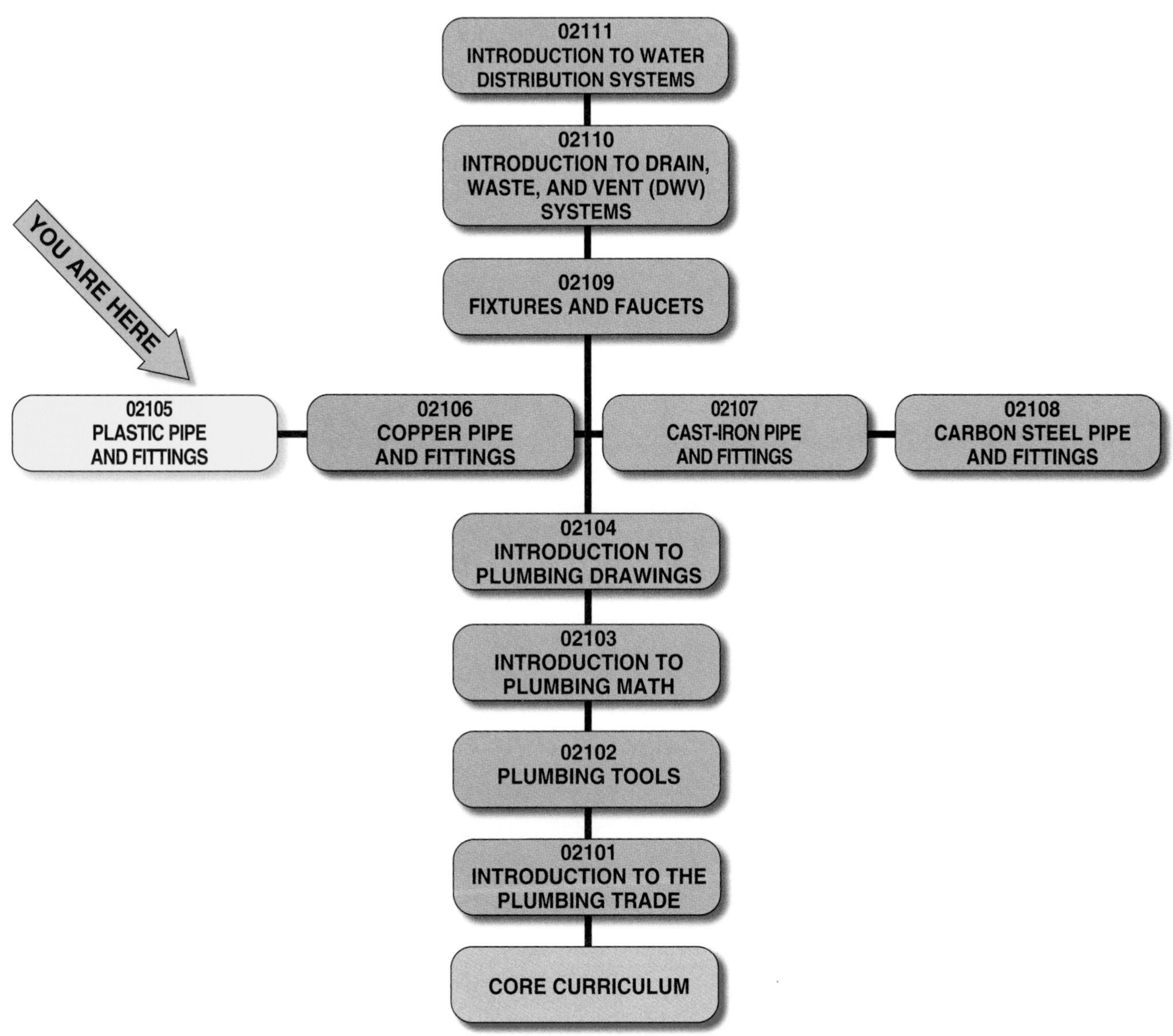

Copyright © 2000 National Center for Construction Education and Research, Gainesville, FL 32614-1104. All rights reserved. No part of this work may be reproduced in any form or by any means, including photocopying, without written permission of the publisher.

MODULE 02105 CONTENTS

Figures

Tables

MODULE 02105

Plastic Pipe and Fittings

Objectives

When you finish this module, you will be able to:

1. Identify the common types of materials and schedules of plastic piping.
2. Identify the common types of fittings and valves used with plastic piping.
3. Identify and determine the kinds of hangers and supports needed for plastic piping.
4. Identify the various techniques used in hanging and supporting plastic piping.
5. Demonstrate the ability to properly measure, cut, and join plastic piping.
6. Follow basic safety precautions for the installation, operation, and maintenance of plastic tubing.
7. Identify the hazards and safety precautions associated with plastic piping.

Prerequisites

Before you begin this module, it is recommended that you successfully complete task modules: Core Curriculum; Plumbing Level One, Modules 02101 through 02104.

Materials Required for Trainees

1. This module
2. Appropriate personal protective equipment
3. Sharpened pencil and paper

1.0.0 ◆ INTRODUCTION

Plastic has revolutionized plumbing, a trade that was based for hundreds of years on metal pipe and fittings. The use of durable plastic pipe and fittings, which can be **solvent welded** by chemicals, has changed the way plumbers work.

Did You Know?

Although plastic materials became popular in the 1950s and 1960s, plastics date back more than a century to the introduction of celluloid in 1870. The 1930s and 1940s saw the introduction of Lucite, Plexiglas, and nylon. **PVC (polyvinyl chloride),** the first plastic used for plumbing purposes, was under development in the 1930s.

Plastics contain polymers—long chains of molecules that we can mold, cast, or force out through a die into desired shapes. We can heat and shape resins in the form of pellets, powders, and solutions in various combinations to give the finished product the desired properties. Strong, durable, and **chemically inert,** plastics are ideal for many plumbing uses.

Plastic has some drawbacks: Temperature changes cause it to expand and contract (more often and more than metal), and it can be flammable and give off fumes when it burns. Plumbers use plastic today in a wide variety of ways, including drain, waste, and vent (DWV) piping; water distribution; chemical waste systems; and fuel gas piping.

ON THE LEVEL

Restrictions on Plastics

Consult your local codes for restrictions on the use of plastics in commercial fire-rated buildings and in multistory applications.

2.0.0 ◆ MATERIALS

Several kinds of plastic pipe are available today, each with different applications and installation requirements (see *Table 1*).

Table 1 Plastic Pipe and Fittings

Pipe	Uses
ABS	DWV sanitary systems, corrosive waste
PVC	DWV sanitary systems, cold water service
CPVC	Hot and cold water distribution
PE	Cold water service, gas service
PEX	Hot and cold water distribution

2.1.0 ABS

ABS (acrylonitrile-butadiene-styrene) pipe and fittings are made from a thermoplastic resin (see *Figure 1*). It is the standard material for many types of DWV systems. Schedule 40 ABS pipe has the same wall thickness as standard Schedule 40 steel pipe. (See discussion of Schedule 40 in the following sections.) ABS pipe is available in diameters ranging from 1½ inches to 6 inches. ABS pipe wall comes in **solid wall** or **cellular core wall** construction, which are interchangeable in plumbing applications.

Easy to handle and easy to install, ABS is light; a 3-inch-diameter, 10-foot-long section weighs less than 10 pounds. It performs well at extreme temperatures because ABS absorbs heat and cold slowly—an important feature for a system that handles both hot and cold wastes.

ABS is highly resistant to household chemicals. In tests, it showed no effect from such common products as Tide™ detergent, Clorox™ bleach, Drano™, Liquid Plumber™, and Sani-Flush™. Sewage treatment plants use ABS because it stands up to the highly corrosive and abrasive liquids commonly found in such systems.

ABS is strong and lasts a long time. In 1959, ABS pipe was used in an experimental residence. Twenty-five years later, an independent research firm dug up and analyzed a section of the pipe and found no evidence of rot, rust, or corrosion. The pipe also withstands earth loads, slab foundations, and high surface loads without collapse.

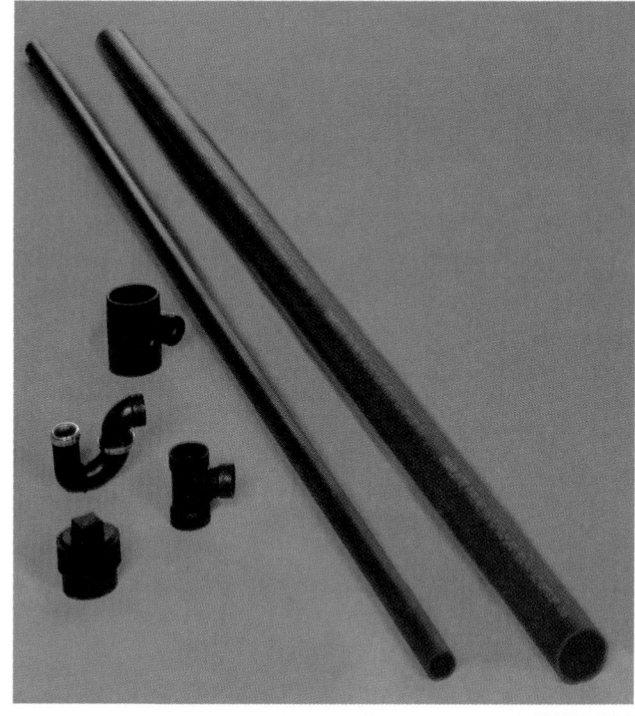

105F01.EPS

Figure 1 ◆ ABS pipe and fittings.

Solvent cement that joins ABS pipe and fittings temporarily softens the joining surface. This process eliminates the need for torches or lead pots, which helps prevent the risk of fire during plumbing installations.

2.2.0 PVC

PVC (polyvinyl chloride) is a rigid pipe with high-impact strength that is manufactured from a thermoplastic material (see *Figure 2*). The material has an indefinite life span under most conditions. PVC is frequently used in cold water systems. It is also used to transport many chemicals because of

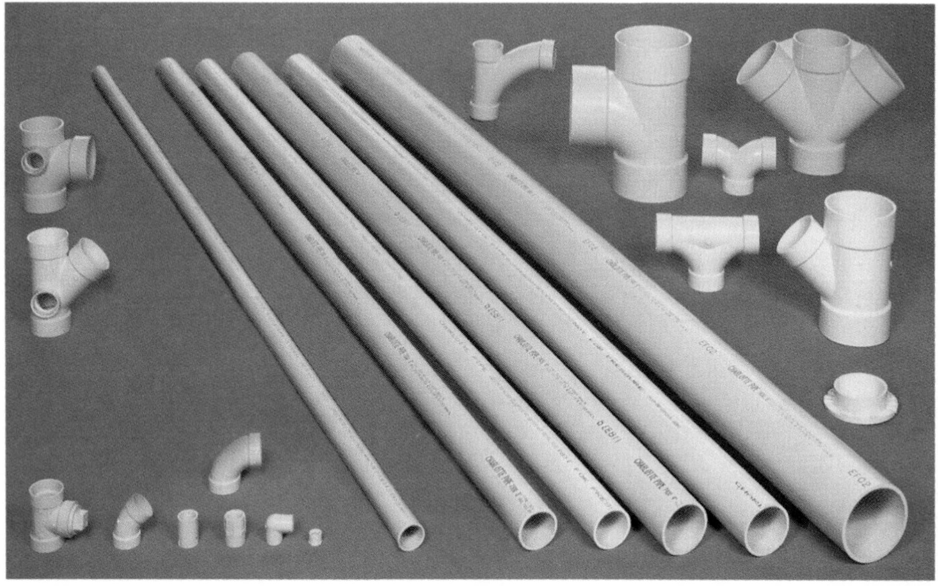

105F02.EPS

Figure 2 ◆ PVC pipe and fittings.

its chemical-resistant properties. PVC is available in solid wall and cellular core wall construction.

Solid core PVC pipe can be used in high-pressure systems but only to carry low-temperature water. It must be protected from sunlight because ultraviolet rays degrade the thermoplastic materials. It is lightweight, is easy to handle and install, has joint flexibility that handles ground movement without leaking, and has long life with no maintenance as long as it is protected from sunlight.

PVC pipe can be joined with solvent cement to form a solvent weld. Solvent weld joints, however, are rigid and not well suited for underground applications in large pipe sizes. The larger pipe is relatively stiff and will not flex easily if the soil shrinks or swells. The other basic joining process for PVC pipe is the bell and spigot, which is covered later in this module. The bell and spigot pipe has a bell on one end with an internal **elastomer** seal. The spigot (straight end) of the next pipe is then fitted into the bell end to form a fluid-tight joint.

PVC also can be threaded and joined to steel pipe with a **transition fitting.** The threading process lowers the allowable working pressure of the system by 50 percent because so much material is removed from the pipe for the threads.

2.3.0 CPVC

CPVC (chlorinated polyvinyl chloride) pipe and fittings are made from an engineered vinyl polymer (see *Figure 3*). It is used worldwide in hot and cold water distribution systems. Improvements made to its parent polymer, polyvinyl chloride, added high-temperature performance and improved impact resistance to this material. CPVC is produced in standard copper tube sizes (CTS) from ½ inch to 2 inches, with a full line of fittings. **Bushings** are available to connect CTS CPVC pipe to CPVC Schedule 40 and 80 pipe for systems requiring piping diameters larger than 2 inches.

CPVC is acceptable under many model codes for use inside the building. Its molecular structure practically eliminates condensation in the summer and heat loss in the winter, decreasing the likelihood of costly drip damage to walls or structure. CPVC pipe's smooth, friction-free interior surfaces result in lower pressure loss and higher flow rates, and provide less opportunity for bacteria growth. CPVC does not break down in the presence of aggressive (chemically reactive) water. Unlike metal pipes, it does not rust, pit, or scale.

CPVC is lightweight and easy to install. Recent improvements to CPVC have made the pipe stronger and more durable during installation. In laboratory tests down to 20°F, CPVC pipe withstood a hammer drop that substantially damaged copper piping under the same conditions. This is a clear advantage for plumbers working in cold-weather states.

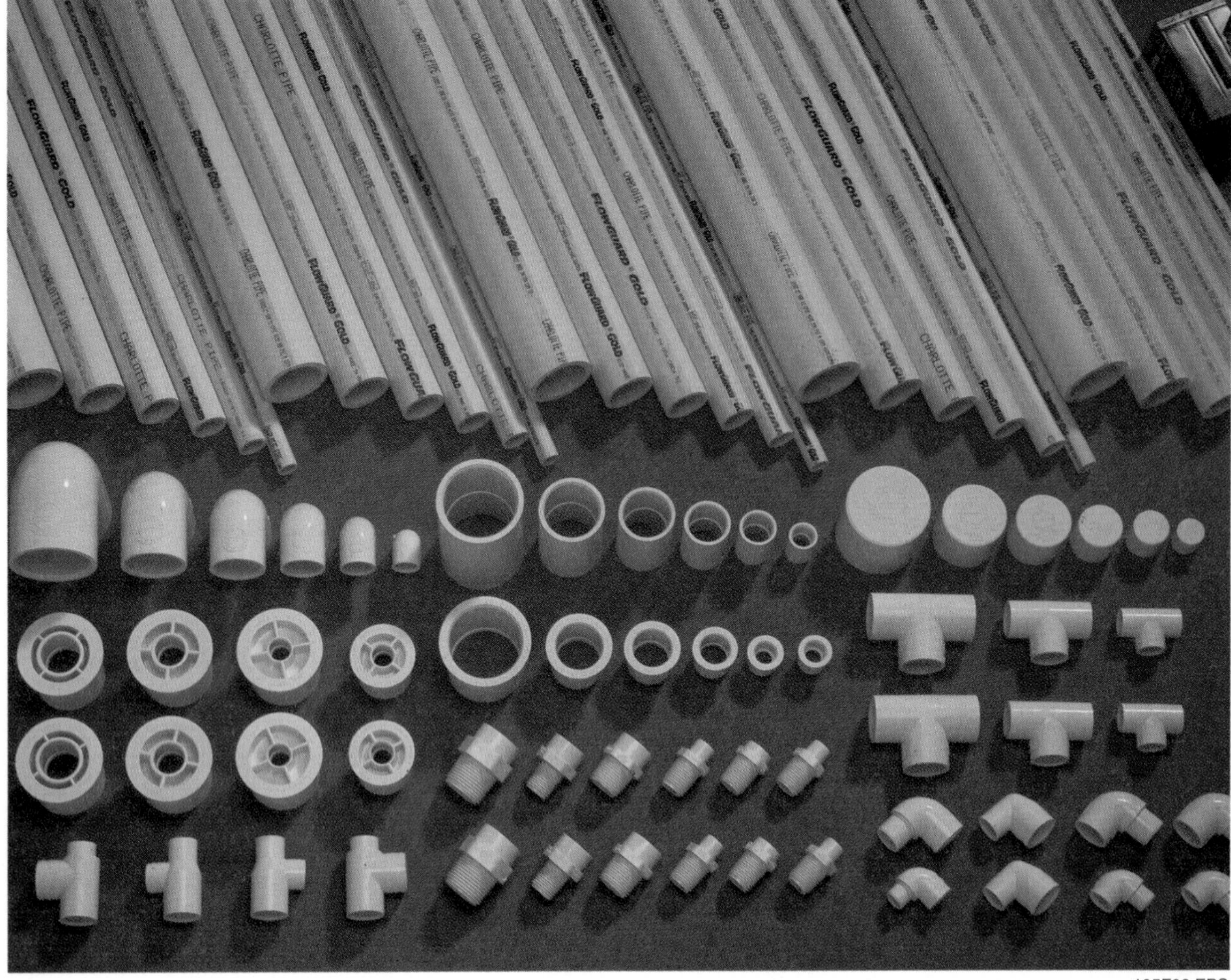

105F03.EPS

Figure 3 ◆ CPVC pipe.

Plumbers join CPVC by solvent welding. The solvent cement is applied to the pipe end and the inside of the fitting end, temporarily softening the materials. The cement cures quickly and fuses the joint together.

2.4.0 PE

PE (polyethylene) is a thermosetting plastic. It is commonly used as tubing because of its strength, flexibility, and chemical-resistant properties (see *Figure 4*). PE is also corrosion resistant, which makes it ideal for transporting chemical compounds. It will not deteriorate when exposed to the ultraviolet rays of the sun. PE piping can be installed outdoors without a protective coating. PE is used for cold water and underground gas service lines (outside buildings).

Because it is resistant to chemicals, PE is not joined by solvent welding. It can be joined using heat fusion or with mechanical joints and clamps. PE joined by fusion is similar to a weld on steel—the materials of the joined parts merge so they are indistinguishable from each other. This process gives the joint the same positive characteristics as the pipe itself.

Did You Know?

PE is the most inert of all pipe, meaning that it is unlikely to react with other substances. That is why it is used in hospital tubing and in soda fountains.

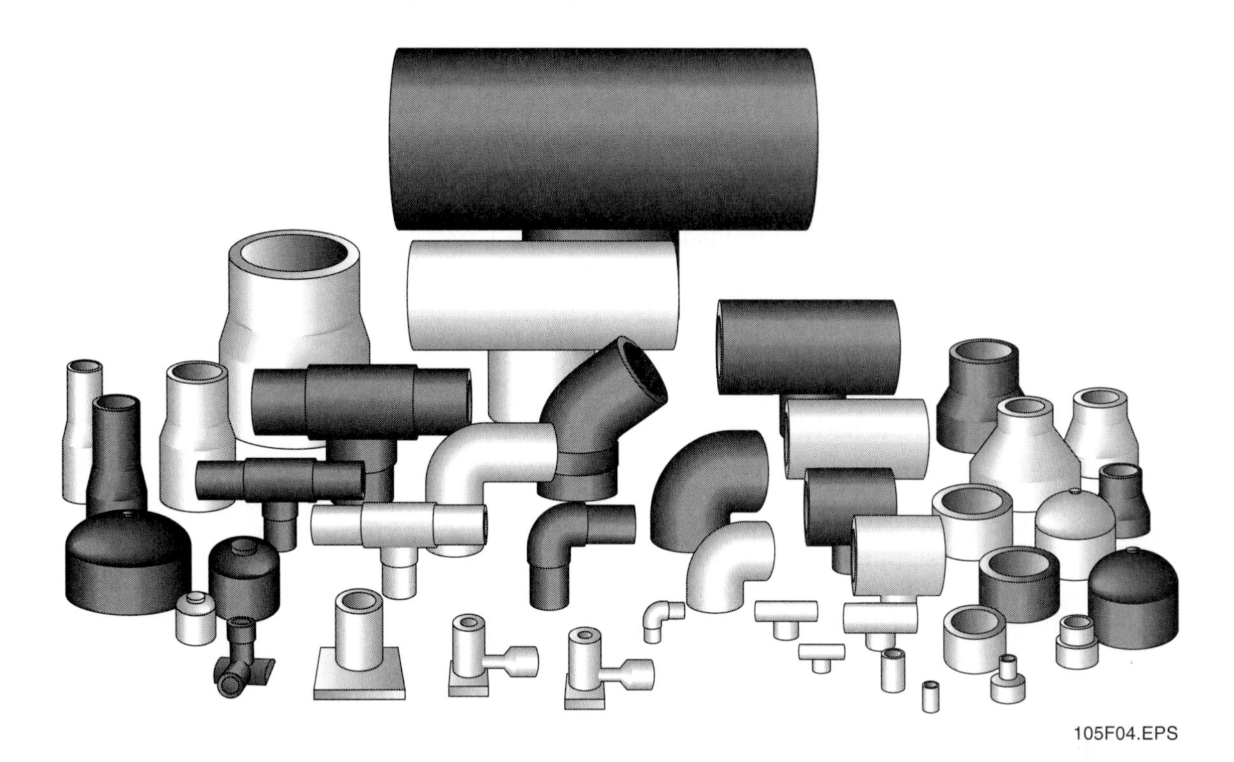

Figure 4 ◆ PE tubing.

2.5.0 PEX

PEX (cross-linked polyethylene) tubing is formed when high-density polyethylene is subjected to heat and high pressure (see *Figure 5*). PEX resists high temperatures, pressure, and chemicals. It is suitable for potable (safe to drink) hot and cold water systems, hydronic (circulating water or steam) radiant floor systems, baseboard and radiator connections, and snow-melt systems.

PEX tubing is joined using a system: Each connection requires a PEX ring and specific type of fitting. To connect the PEX tubing, place a PEX ring at the cut end of the tubing. Then use an expander tool to enlarge the tubing and ring together. Quickly slide the expanded tube and ring over the appropriate fitting, making sure that the tubing reaches the top of the fitting. Within a few seconds, the tubing and ring shrink back to size and hold the fitting tightly.

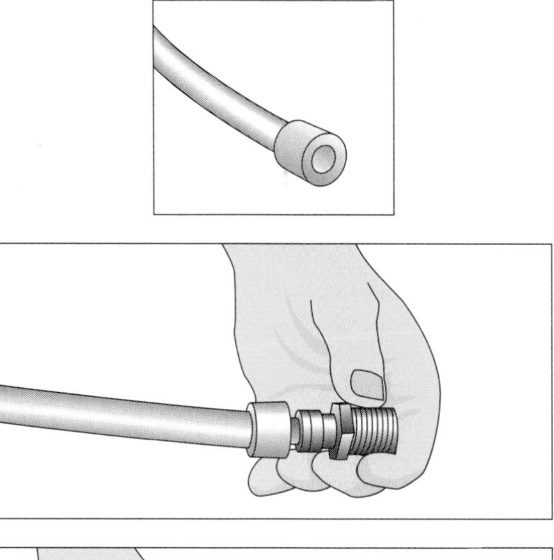

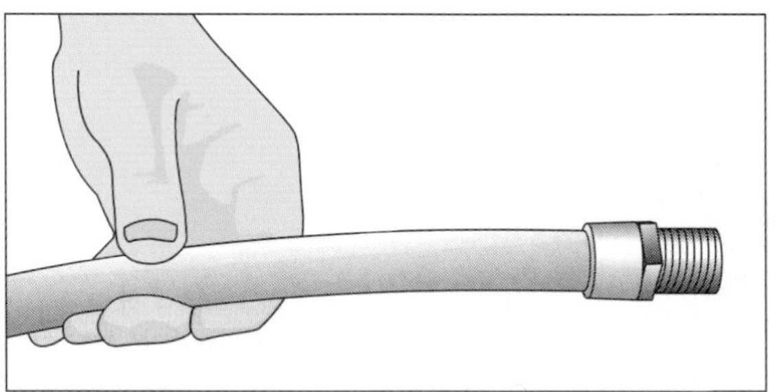

Figure 5 ◆ PEX tubing.

2.6.0 Solvent Cements

Solvent cements vary, depending on the materials to be solvent welded. Each type of plastic has its own cement, and the cement may vary based on the size of the pipe and fittings. Be careful to choose the right cement for the job. Ask your supervisor if you are unsure which cement to choose.

2.7.0 SDRs and Schedules

Plastic pipe is manufactured with various wall thicknesses, commonly called schedules or **SDRs** (size dimension ratios).

SDR relates wall thickness to the diameter of the pipe, using a set ratio of wall thickness to diameter. This ratio is designated by a number, such as SDR 35. The larger the SDR number, the thinner the wall versus the diameter. So, for instance, SDR 17 pipe is thicker than SDR 35 pipe. This ratio is maintained so that as the pipe diameter gets bigger, the wall gets thicker.

In an effort to simplify and standardize the use of plastic pipe and fittings, the manufacturers designated two SDR ratios as Schedule 40 (the most commonly used) and Schedule 80 (thicker wall). Schedule 40 is designated in most plumbing codes as the minimum for use in or under buildings.

2.8.0 Labeling (Markings)

Plastic pipe and fittings must be labeled and approved for use in plumbing systems. The label on a plastic pipe will include:

- Nominal pipe diameter–copper tube size (CTS) or iron pipe size (IPS)
- Schedule or SDR
- Type of material–PVC, CPVC, and so on
- **Pressure rating**–usually expressed as pounds per square inch **(psi)** at a certain temperature
- Relevant standard for the pipe–for example, ASTM D1785
- Listing body or laboratory–for example, the National Sanitation Foundation (NSF) or the Uniform Plumbing Code (UPC)—that certifies the manufacture of the pipe
- Manufacturer
- Country of origin

This label ensures that a recognized third party (a listing body or laboratory) has tested the material and approved it for the use intended (see *Figure 6*). Each solvent cement used on the installation is also listed and labeled for its specific use.

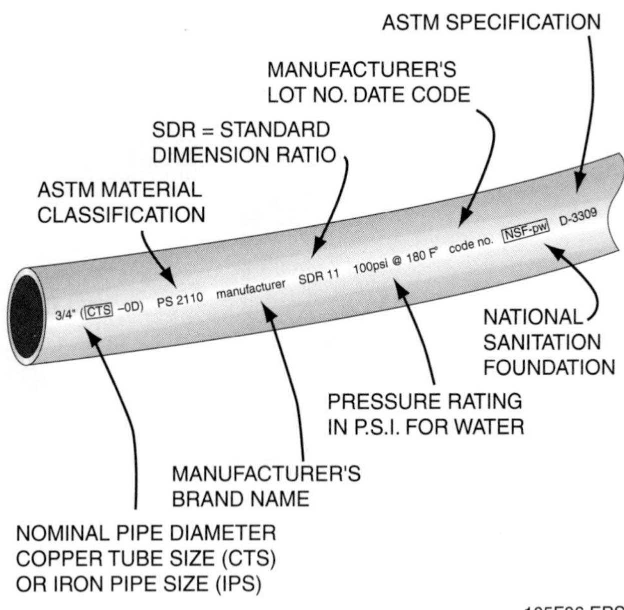

Figure 6 ◆ Typical pipe label.

Did You Know?

The American Society for Testing and Materials (ASTM) was organized in 1898 and has grown into one of the largest voluntary standards development systems in the world.

ASTM is a not-for-profit organization that provides a forum for producers, users, consumers, and representatives of government and academia to meet on common ground and write standards for materials, products, systems, and services. ASTM publishes standard test methods, specifications, practices, guides, classifications, and terminology. ASTM's standards development activities cover metals, paints, plastics, textiles, petroleum, construction, energy, the environment, consumer products, medical services and devices, computerized systems, electronics, and many other areas.

More than 10,000 ASTM standards are published each year in the 72 volumes of the *Annual Book of ASTM Standards.*

ON THE LEVEL

Choosing the Correct Material

Carefully choose material to be used for different plumbing purposes, such as PVC, which can be used for both DWV and water piping. The PVC is the same color in both cases (white), but it is tested and labeled differently for each use. If you use the wrong type, as indicated by its label, your local inspecting authority could reject the installation.

Review Questions

Section 2.0.0

1. A _____ temporarily softens the pipe and the fitting materials, allowing them to be fitted together and fused.
   a. solvent weld
   b. bushing
   c. solvent cement
   d. polymer

2. _____ has high-temperature performance and improved impact resistance compared to PVC.
   a. CPVC
   b. PB
   c. ABS
   d. PE

3. _____ pipe is widely used in hot and cold water distribution systems.
   a. ABS
   b. PB
   c. PVC
   d. CPVC

4. _____ tubing cannot be joined by solvent welding because it resists chemicals.
   a. PB
   b. PE
   c. ABS
   d. PVC

5. _____ on different types of plastic pipe is standardized to provide the plumber with information about the materials being used.
   a. Sizing
   b. Labeling
   c. Coloring
   d. Threading

3.0.0 ◆ COMMON FITTINGS

A pipe's use determines what kinds of fittings and joints are used. The following sections discuss the fittings used in water supply and DWV systems.

3.1.0 Water Supply Fittings

Pressure-type fittings for use with water supply are short turn (radius) with ledges or shelves. The short turn (radius) describes the curve or bend of a fitting that changes direction. As the radius increases, the change in flow direction becomes smoother. This is particularly important for waste drain systems where the smoother flow keeps organic waste from being deposited along the pipeline.

The most common types of water supply fittings include (see *Figure 7*):

- *Union*–to add a branch line in an existing system
- *Reducer*–to connect pipes of different sizes
- *Elbow*–to change the direction of rigid pipe by either 90 degrees or 45 degrees
- *Tee*–to provide an opening to connect a branch pipe at 90 degrees to the main pipe run
- *Coupling*–to join two lengths of the same pipe size when making a straight run
- *Cap*–to plug water outlets when testing the system or to create an air chamber to eliminate water hammer
- *Plug*–to close openings in other fittings or to seal the end of a pipe
- *Manifold*–to run several water supply lines from the main supply to different fixtures

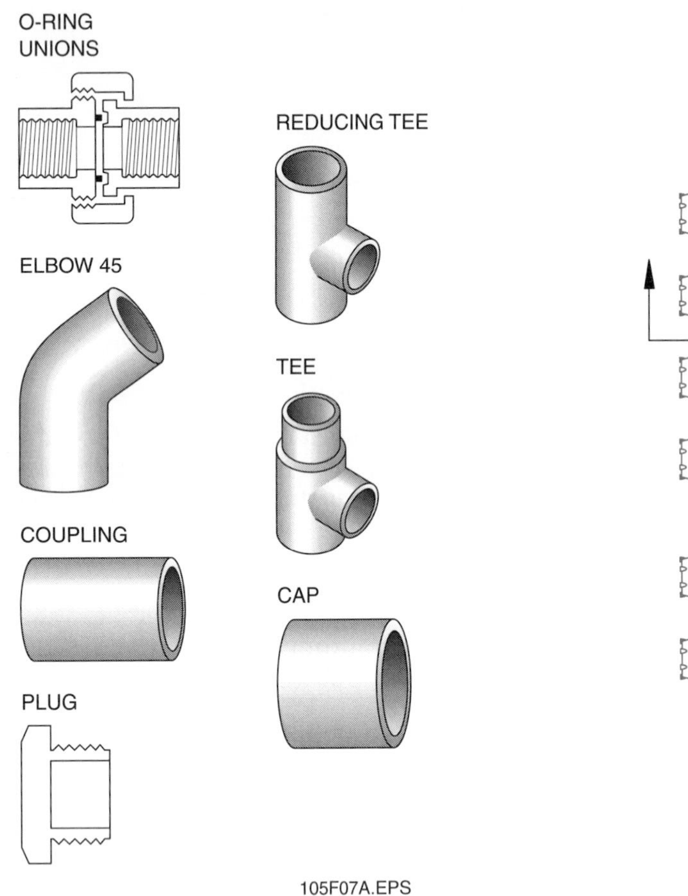

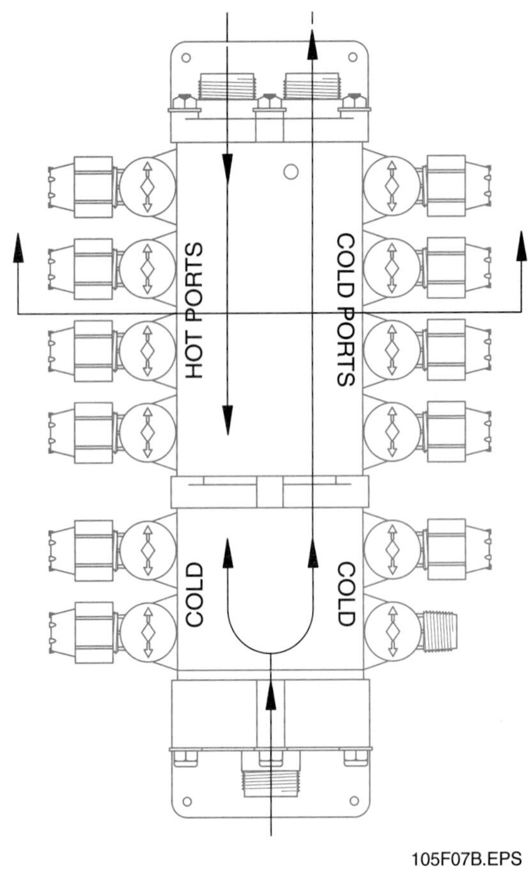

Figure 7 ◆ Water supply fittings.

3.2.0 DWV Fittings

Drainage fittings have smooth interior passages with no ledges. Drainage fittings have a longer radius that makes directional changes smoother and less likely to collect solids.

The most common types of DWV fittings include (see *Figure 8*):

- *Tee*–to make 90-degree turns; changes directions in a drain pipe from horizontal to vertical
- *Wye (Y)*–to make 45-degree turns; used to connect horizontal waste pipes and branch drains to the building main drain
- *Bend*–to change direction of either horizontal or vertical drainage and vent pipe; classified by degree of turn and radius of the curve (e.g., 1/16 bend = 22.5 degrees, 1/8 bend = 45 degrees, 1/4 bend = 90 degrees)
- *Coupling*–to join two lengths of pipe
- *Closet bend*–to receive waste from the water closet
- *Closet flange*–to connect the closet bend to the water closet

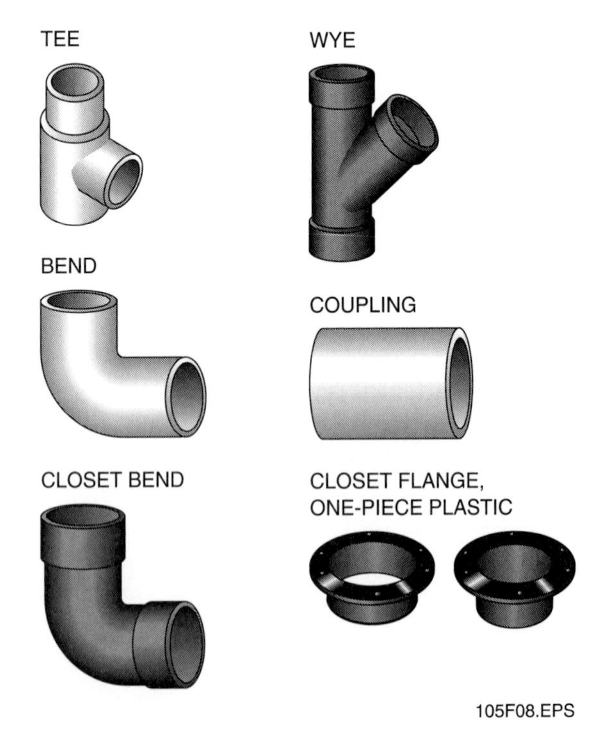

Figure 8 ◆ DWV fittings.

3.3.0 Joining Fittings and Pipes

Typically, water supply and DWV fittings are joined using the same installation techniques. Among the installation methods are solvent weld, ring-tight gasket, and heat fusion.

Solvent-weld fittings have sockets that the pipe fits into (see *Figure 9*). Often these are an **interference fit**—that is, the fit gets tighter as the pipe is pushed into the socket.

Ring-tight gasket fittings have a rubber O-ring or gasket in the socket; the pipe must be beveled at the leading edge to allow it to pass the gasket that forms the seal (see *Figure 10*).

Fusion fittings may have a butt, not a socket. This butt has the same **outside diameter (OD)** and **inside diameter (ID)** as the pipe. On these fittings, the ends of the fitting and pipe are heated and then pressed together to form a joint (see *Figure 11*).

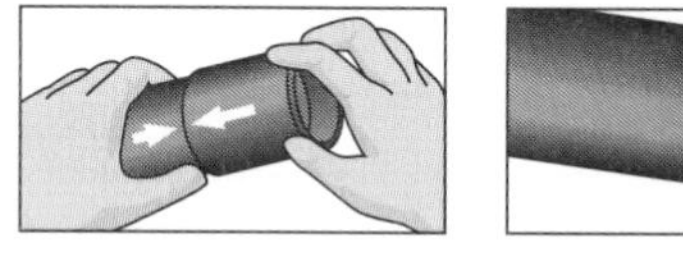

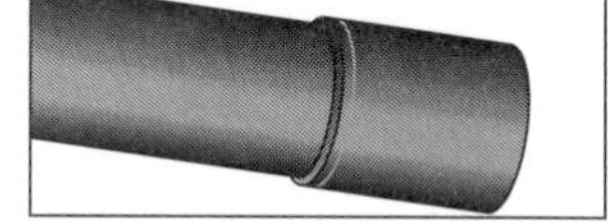

105F09.EPS

Figure 9 ◆ Solvent-welded fitting.

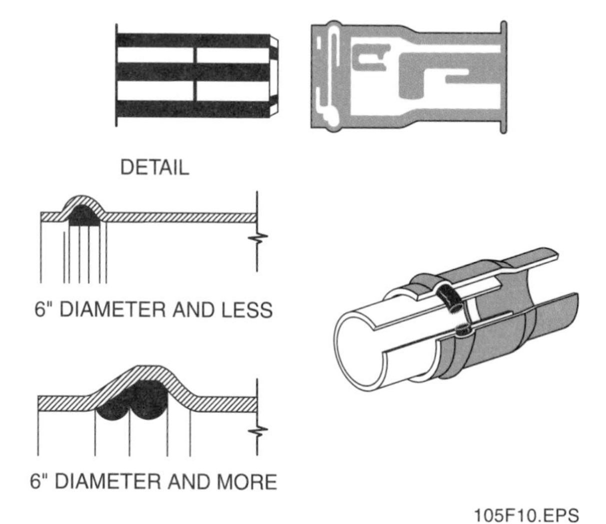

105F10.EPS

Figure 10 ◆ Ring-tight gasket fitting.

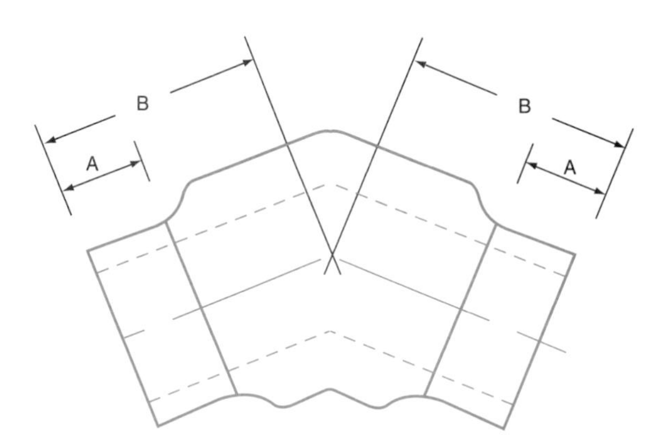

45 DEGREE ELBOW BUTT FUSION MOLDED

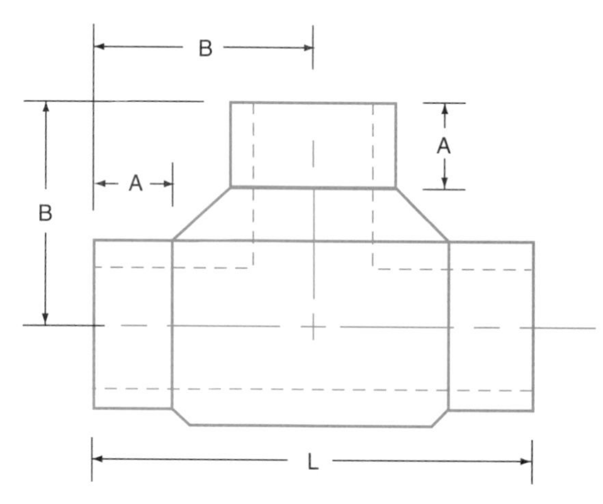

TEE BUTT FUSION MOLDED

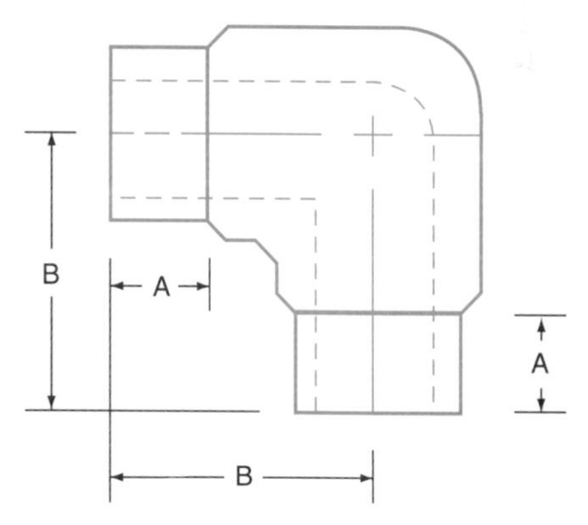

90 DEGREE ELBOW BUTT FUSION MOLDED

105F11.EPS

Figure 11 ◆ Fusion fitting.

Joining Pipe

Before you cement a joint, make sure the pipe and fitting are free of dust, dirt, water, and oil. Solvent cement acts fast. It is important to move quickly and efficiently when you are joining pipe using solvent welding.

Review Questions

Section 3.0.0

1. In DWV systems, _____ fittings are used to change the direction of rigid pipe by either 90 degrees or 45 degrees.
   a. elbow
   b. coupling
   c. manifold
   d. tee

2. A wye fitting changes the direction of pipe by ____ degrees.
   a. 30
   b. 40
   c. 45
   d. 22

3. _____ join two lengths of the same pipe size when you are making a straight run.
   a. Couplings
   b. Manifolds
   c. Unions
   d. Reducers

4. _____ gasket fittings use a rubber O-ring or gasket in the socket to make sure the joint is leakproof.
   a. Fusion
   b. Solvent-weld
   c. Ring-tight
   d. Union

5. _____ fittings have the same inside and outside diameter as the pipe.
   a. Solvent-weld
   b. Union
   c. Fusion
   d. Ring-tight

4.0.0 ◆ HANGERS AND SUPPORTS

Plastic pipe is supported in different ways, and some materials may need more support than others. The type of support depends on the material, its size, whether it is in the horizontal or vertical position, and its use.

Did You Know?

Local codes often include spacing requirements for different types and sizes of plastic pipe. See *Table 2* for a typical local code.

The makeup of the hangers and support system will depend on the specifications, the local code, and the materials being used. Hangers and anchors should be strong enough to support the weight of the pipe and its contents, maintain its alignment, and not sag. Piping in the ground should be laid on a firm bed for its entire length. Plastic piping requires special considerations because it expands and contracts much more than other kinds of piping.

4.1.0 Hangers

Plumbing installations planned by architects and engineering consultants include plans and specifications that completely describe the proposed system. Specifications are based on codes or ordinances and must be followed. A specification for pipe hangers, for example, may read, "All piping shall be supported with hangers spaced not more than 10 feet apart (on center). Hangers shall be the malleable iron split-ring type and shall be as manufactured by XYZ Hangers, Inc., or other approved vendor."

Table 2 Typical Local Code

1994 UPC (IAPMO)			
Materials	**Type of Joints**	**Horizontal**	**Vertical**
Cast Iron Hub and Spigot	Lead and Oakum	5 feet (1.5 m). except may be 10 feet (3.0 m) where 10 foot (3.0 m) lengths are installed[1, 2, 3]	Base and each floor not to exceed 15 feet (4.6 m)
	Compression Gasket	Every other joint, unless over 4 feet (1.2 m), then support each joint[1, 2, 3]	Base and each floor not to exceed 15 feet (4.6 m)
Cast Iron Hubless	Shielded Coupling	Every other joint, unless over 4 feet (1.2 m), then support each joint[1, 2, 3, 4]	Base and each floor not to exceed 15 feet (4.6 m)
Copper Tube and Pipe	Soldered, Brazed, or Welded	1½ inch (38.1 mm) and smaller, 6 feet (1.8 m), 2 inch (50.8 mm) and larger, 10 feet (3.0 m)	Each floor, not to exceed 10 feet (3.0 m)[5]
Steel and Brass Pipe for Water or DWV	Threaded or Welded	¾ inch (19.1 mm) and smaller, 10 feet (3.0 m), 1 inch (25.4 mm) and larger, 12 feet (3.7 m)	Every other floor, not to exceed 25 feet (7.6 m)[5]
Steel, Brass, and Thread Copper Pipe for Gas	Threaded or Welded	½ inch (12.7 mm), 6 feet (1.8 m), ¾ (19.1 mm) and 1 inch (25.4 mm), 8 feet (2.4 m), 1¼ inch (31.8 mm) and larger, 10 feet (3.0 m)	½ inch (12.7 mm), 6 feet (1.8 m), ¾ (19.1 mm) and 1 inch (25.4 mm), 8 feet (2.4 m), 1¼ inch (31.8 mm) and larger, every floor level
Schedule 40 PVC and ABS DWV	Solvent Cemented	All sizes, 4 feet (1.2 m) Allow for expansion every 30 feet (9.1 m)[3, 6]	Base and each floor. Provide mid-story guides. Provide for expansion every 30 feet (9.1 m)[6]
CPVC	Solvent Cemented	1 inch (25.4 mm) and smaller, 3 feet (0.9 m), 1¼ inch (31.8 mm) and larger, 4 feet (1.2 m)	Base and each floor. Provide mid-story guides[6]
Lead	Wiped or Burned	Continuous support	Not to exceed 4 feet (1.2 m)
Copper	Mechanical	In accordance with standards acceptable to the Administrative Authority	
Steel and Brass	Mechanical	In accordance with standards acceptable to the Administrative Authority	

[1]Support adjacent to joint, not to exceed eighteen (18) inches (0.5 m).
[2]Brace at not more than forty (40) foot (12 m) intervals to prevent horizontal movement.
[3]Support at each horizontal branch connection.
[4]Hangers shall not be placed on the coupling.
[5]Vertical water lines may be supported in accordance with recognized engineering principles with regard to expansion and contraction, when first approved by the Administrative Authority.
[6]See the appropriate IAPMO Installation Standard for expansion and other special requirements.

Hangers (see *Figure 12*) are used for horizontal support of pipes and piping. The main purpose of hangers and brackets is to keep the piping in alignment and to prevent it from bending or distorting, but they can also be used as a vibration isolator. Horizontal hangers can be attached to wooden structures with lag screws or large nails. If you do not expect vibration problems, you may use commonly accepted plumbing practices. If vibration may be a problem, the specifications should tell you what materials to use.

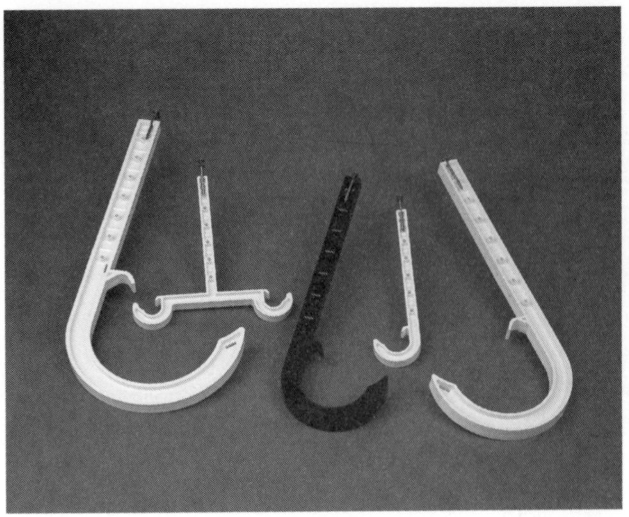

105F12A.EPS

(A) J-HOOKS

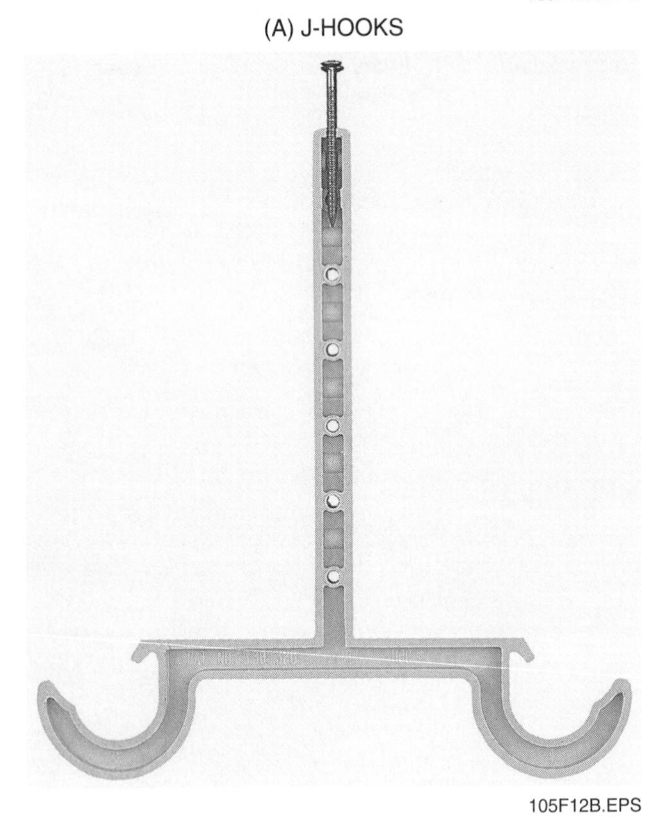

105F12B.EPS

(B) DOUBLE-J HOOKS

105F12C.EPS

(C) LOCKING TUBE STRAP

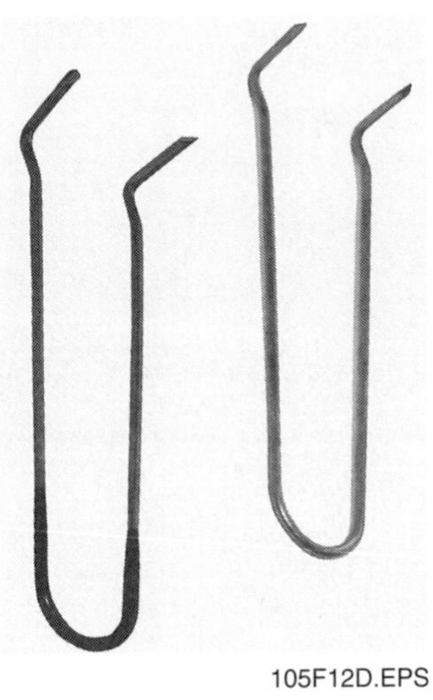

105F12D.EPS

(D) PIPE HOOKS

Figure 12 ◆ Hangers.

4.2.0 Fastening

Beam clamps or C-clamps (see *Figure 13*) are used to fasten pipes to beams and other metal structures. Other horizontal support clamps and brackets are shown in *Figure 14* (see page 5–14). Vertical hangers (see *Figure 15* on page 5–15) or pipe risers consist of a friction clamp that can be attached to structural site components to support the vertical load of the pipe. Special fasteners are used to attach hangers to masonry, concrete, or steel.

105F13A.EPS

(A) SUSPENSION CLAMPS

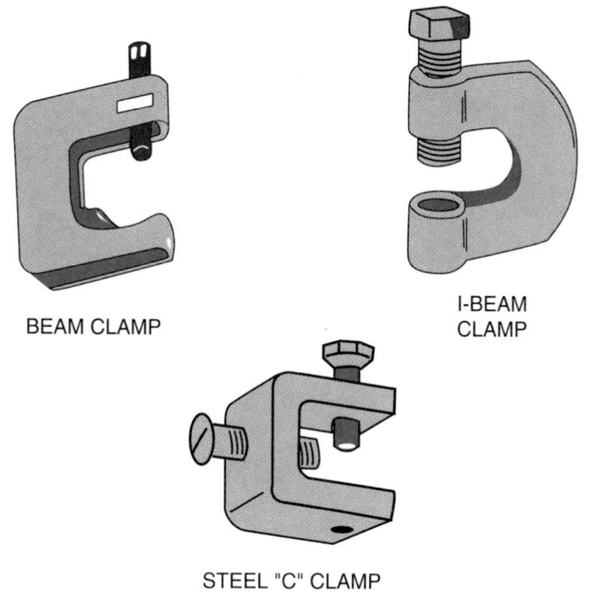

105F13B.EPS

Figure 13 ◆ Beam clamps, C-clamps, and suspension clamps.

4.3.0 Supports

PVC and ABS pipe should be supported with pipe hangers (see *Figure 12*). Avoid hangers that may cut or squeeze pipe and tight clamps or straps that prevent pipe from moving or expanding.

Support piping at intervals of not more than 4 feet, and at branches, at changes of direction, and when using large fittings. Although supports should provide free movement, they must prevent lateral runs from moving up, which could create a reverse grade on branch piping and back up the system. Size any holes made for pipe through framing members to allow for free movement.

Use supports for vertical piping at each floor level or as required by the design. Mid-story guides can provide greater stability for vertical pipes that run up through the building.

105F14A.EPS

(A) PLUMBER'S TAPE

105F14C.EPS

(C) SUPPORT BRACKET IN PLACE

105F14B.EPS

(B) PIPE SUPPORT

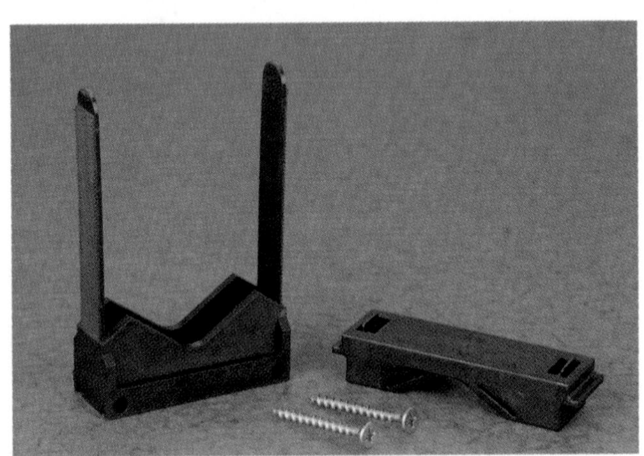

105F14D.EPS

(D) SUPPORT BRACKET

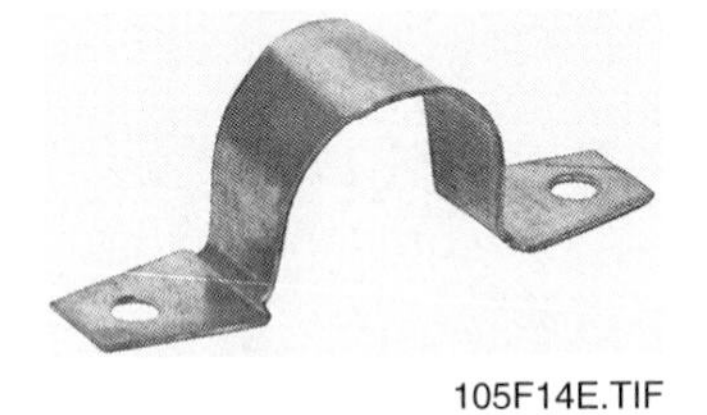

105F14E.TIF

(E) TWO-HOLE STRAP

Figure 14 ◆ Support clamps and brackets.

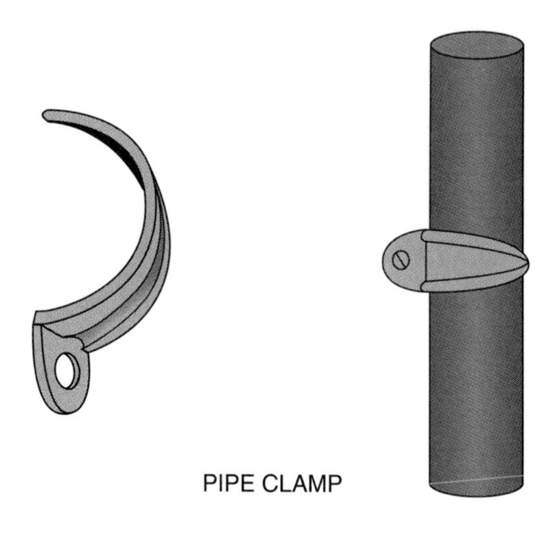

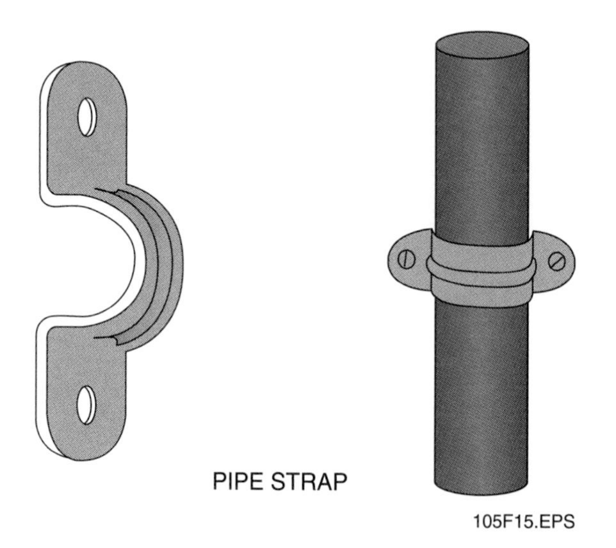

Figure 15 ◆ Vertical hangers.

5.0.0 ◆ MEASURING, CUTTING, AND JOINING

Techniques for measuring, cutting, and joining vary depending on the materials you are using and the function of the pipe (DWV or gas, for instance). The following sections cover ABS, PVC, CPVC, PEX, and PE pipe used in DWV and water systems.

5.1.0 Measuring

It is important to plan ahead when you are installing ABS and PVC pipe and fittings in DWV systems. These systems have built-in pitch or fall (which you will learn more about in *Introduction to DWV Systems*), and they must be laid out accurately, with pipes cut to exact lengths. Mistakes cannot be fixed later with heat or hammers.

When you are measuring any pipe, be sure to allow for depth of joints. Take measurements to the full depth of the socket, not with pipe partly inserted into the socket.

Because the solvent cement sets up quickly, you must dry fit the installation, then mark alignment for fittings before you make the joint.

5.2.0 Cutting

Plastic pipe requires a square cut for good joint integrity. Cutting tubing as squarely as possible provides the best bonding area within a joint. If there is any indication of damage or cracking at the tubing end, cut off at least 2 inches beyond any visible crack.

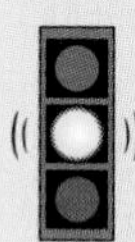

CAUTION

Follow all manufacturer-recommended precautions when cutting or sawing pipe or when using any flame, heat, or power tools.

5.2.1 Cutting PVC and ABS Pipe

You can cut PVC and ABS pipe with appropriate pipe cutters, a handsaw, or a power saw equipped with a carbide tip or abrasive blade. Plastic pipe-cutter wheels are available to fit standard cutters. You can also use lightweight, quick-adjusting cutters designed exclusively for plastic piping.

To make sure you get a square cut, use a power saw on large jobs and a miter box on small jobs

Use Sharp Wheels

Do not use pipe cutters with dull wheels—especially wheels that have been used to cut metal—because they will exert too much pressure, causing larger shoulders and burrs that you will need to remove.

Mark Carefully

To be sure of proper alignment in the final assembly, carefully mark for position any fittings that will be rolled or otherwise aligned.

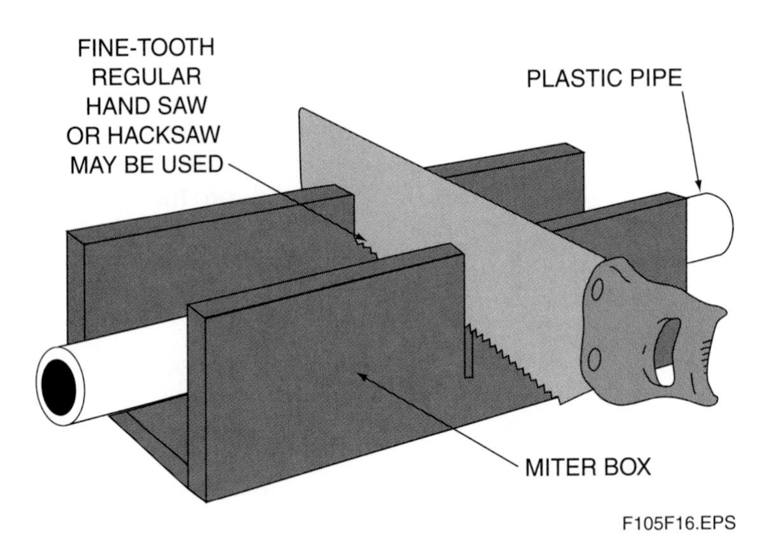

Figure 16 ◆ Using a miter box on small jobs.

(see *Figure 16*). If these are not available, **scribe** the pipe and cut to the mark.

After cutting the pipe, ream it inside and **chamfer** the edge to remove burrs, shoulders, and ragged spots.

5.2.2 Cutting CPVC and PVC Tubing

You can easily cut CPVC and PVC tubing with a wheel-type plastic tubing cutter, a hacksaw, or other fine-toothed hand or power saws. The use of ratchet cutters is permitted as long as blades are sharpened regularly. When you are using a saw, use a miter box to ensure a square cut.

With CPVC and PVC tubing, burrs and filings can prevent proper contact between tube and fitting during assembly, so they should be removed from the outside and inside of the tubing. A chamfering tool is preferred, but you can also use a pocketknife or a file. A slight bevel on the end of the tubing will ease entry of the tubing into the fitting socket and lessen the chances of pushing solvent cement to the bottom of the joint.

5.3.0 Joining

Plastic pipe and fittings are joined with solvent cement that temporarily softens the joining surfaces. This brief softening period lets you seat the pipe into the socket's interference fit. The softened surfaces then fuse together, and joint strength develops as the solvents evaporate. The resulting joint is stronger than the pipe itself. Because the cement hardens fast, it is important to move quickly and efficiently when joining pipe (see *Figure 17*).

CAUTION

Use protective eyewear and gloves during any solvent-weld procedure and consult the Material Safety Data Sheet (MSDS) on the cements being used. An MSDS should be available to anyone on the job site.

To join plastic tubing, follow these steps:

Step 1 Before you cement a joint, use a clean, dry cloth to make sure that both the pipe and the fittings are free of dust, dirt, water, and oil. Also check the dry fit of the tubing and fitting.

Step 2 Using the correct applicator, apply the primer. While the surfaces are still wet with primer, apply the cement. Use the applicator that comes with the cement or use an ordinary bristle paintbrush. For fast application, the width of the applicator should be at least half the diameter of the pipe.

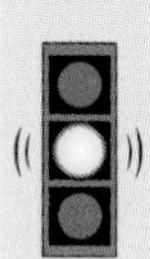

CAUTION

Avoid unnecessary skin or eye contact with primers and cements or soldering materials. If contact occurs, wash immediately.

Step 3 Hold the joint together until you have a tight set. A proper joint normally shows a bead around its entire perimeter.

Step 4 After setting, wipe the excess cement from the pipe.

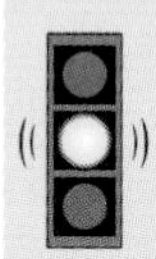

CAUTION

Make sure to apply primers and cements or soldering materials in a well-ventilated area.

Curing time depends on weather, application technique, the cement being used, and the degree of interference fit between pipe and fitting.

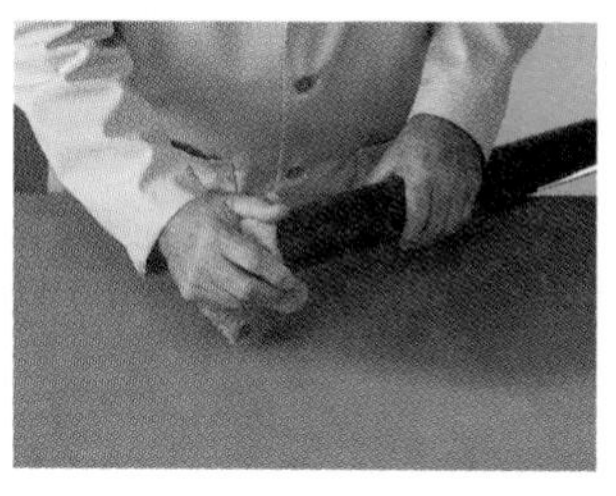

105F17A.TIF

(A)

105F17B.TIF

(B)

105F17C.TIF

(C)

105F17D.TIF

(D)

105F17E.TIF

(E)

105F17F.TIF

(F)

105F17G.TIF

(G)

105F17H.TIF

(H)

Figure 17 ◆ Joining plastic pipe.

5.3.1 Joining CPVC and PVC Pipe and Fittings

When you join CPVC and PVC pipe and fittings (see *Figure 18*), the tubing should make contact with the socket wall one-third to two-thirds of the way into the fitting socket. At this stage, tubing should not go to the bottom of the socket. Use only CPVC or PVC cement or an all-purpose cement conforming to ASTM F-493 standards, or the joint may fail.

To join CPVC or PVC pipe and fittings, follow these steps:

Step 1 Clean the bell interior and the spigot area of all dirt thoroughly with a rag or brush.

Step 2 Clean the surfaces to be joined by lightly scouring the ends of the pipe with emery paper, then wiping with a clean cloth. Do this to ensure that the primer will soften the plastic before you apply solvent cement.

Step 3 Cut the pipes to the lengths you need, then test fit the pipes and fittings. The pipes should fit tightly against the bottom of the hubbed sockets on the fittings.

Step 4 Mark the pipe and fitting with a felt-tipped pen to show the proper position for alignment. Also mark the depth of the fitting sockets on the pipes to make sure that you fit them back in completely when joining.

Step 5 Apply the primer to the surfaces being joined. The primer will soften the plastic in preparation for the solvent-weld process.

Step 6 While the primer is still wet, apply a heavy, even coat of cement to the pipe end. Use the same applicator without additional cement to apply a thin coat inside the fitting socket. Too much cement can cause clogged waterways. Do not allow excess cement to puddle in the fitting and pipe assembly.

Step 7 Immediately insert the tubing into the fitting socket, rotating the tube one-quarter to one-half turn while inserting. This motion ensures an even distribution of cement in the joint. Properly align the fitting.

Step 8 Hold the assembly for about 10 seconds, allowing the joint to set up. An even bead of cement should be visible around the joint. If this bead does not appear all the way around the socket edge, you may not have applied enough cement. In this case, remake the joint to avoid the possibility of leaks. Wipe excess cement from the tubing and fitting surfaces.

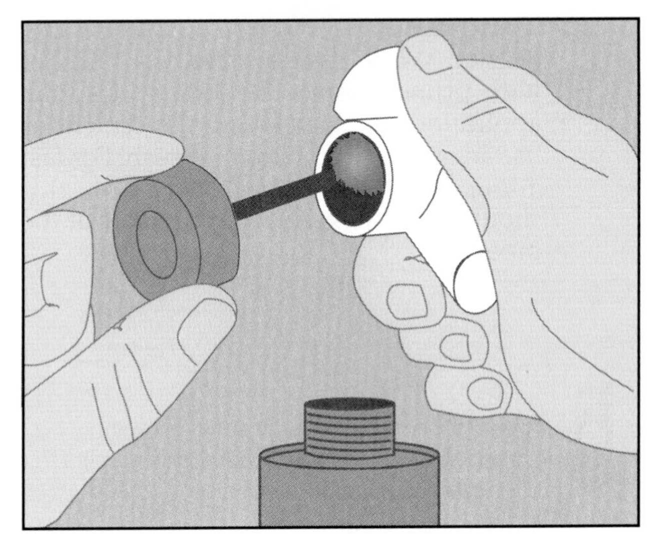

105F18.EPS

Figure 18 ◆ Joining CPVC and PVC pipe and fittings.

Purple versus Clear Primer

Two types of plastic pipe primer are available. One of them has purple dye. Inspectors will look specifically for the purple stain as proof that the pipes were primed before the solvent cement was applied. If you use clear primer, there will be no telltale stain.

In some cases, with permission, you may use a clear primer on trim-out work.

To install PVC gravity sewer pipe, follow these steps (see *Figure 19*):

Step 1 Fold the gasket into a heart shape with the nose or rounded part of its cross section facing out of the mouth of the bell.

Step 2 Insert the gasket into the bell and work it into its groove until it is smooth and free from waves. You may have to snap the gasket, or wet it with clean water or a wet rag, to make sure it goes in place completely.

Step 3 After the gasket has been placed in the bell, thoroughly coat its exposed surface with lubricant. Then apply the lubricant to the entire surface of the spigot end up to the memory mark. Make especially sure that the tapered portion of the spigot is thoroughly coated. After lubrication, the pipe is ready to be joined.

Step 4 Line up the spigot with the bell and insert it straight into the bell. The spigot end of the pipe has a mark to indicate the proper depth of insertion. This mark will be about flush with the end of the bell when the joint is fully assembled. The memory mark must never be more than ⅜ inch from the end of the bell after assembly.

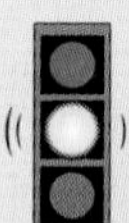

CAUTION

When you are installing a ring-tight PVC gasket, you must assemble the pipe either by hand or by using a bar or block. Never swing or stab the pipe to join it.

105F19.EPS

Figure 19 ◆ Installing PVC gravity sewer pipe.

The manufacturer's instructions will show minimum cure times for different size tubes at different temperatures before you can pressure-test the joint. Solvent cement and cure times also depend on relative humidity. Cure time is shorter for drier environments, smaller sizes, and higher temperatures. The use of primer or the presence of hot water increases the cure time required for pressure testing.

Use special care when assembling CPVC and PVC systems in extremely low temperatures (below 40°F) or extremely high temperatures (above 100°F). In extremely hot environments, make sure both surfaces to be joined are still wet with cement when you put them together.

5.3.2 Joining PEX Tubing

When joining PEX tubing, first square-cut the tubing perpendicular to the length of the tubing, using a cutter designed for plastic tubing. Remove all excess material or burrs that might affect the fitting connection. Slide a PEX ring over the end of the tube, and extend it no more than 1⁄16 inch (see *Figure 20*).

Separate the handles of a separator tool and insert the tool's expansion head into the end of the tubing until it stops. Be sure you have the correct size expander head in the tool. Place the free handle of the tool against your hip, or place one hand on each handle when necessary. Fully separate the handles and bring them together. Repeat this process until the tubing and ring are snug against the shoulder on the expansion head. Before the final expansion, withdraw the tubing from the tool and rotate the tool one-eighth turn. This prevents the tool from forming ridges in the tubing.

Expand the tubing one final time. Immediately remove the tool and slide the tubing over the fitting until the tubing reaches the stop on the fitting. Hold the fitting in place for two or three seconds until the tubing shrinks onto the fitting so that it holds the fitting firmly. The tubing and PEX ring must be seated against the shoulder of the fitting for a proper connection.

Good connections have the PEX ring snug against the stop on the fitting shoulder. If there is more than 1⁄16 inch between the ring and the fitting, square-cut the tubing 2 inches away from the fitting and make another connection using a new PEX ring.

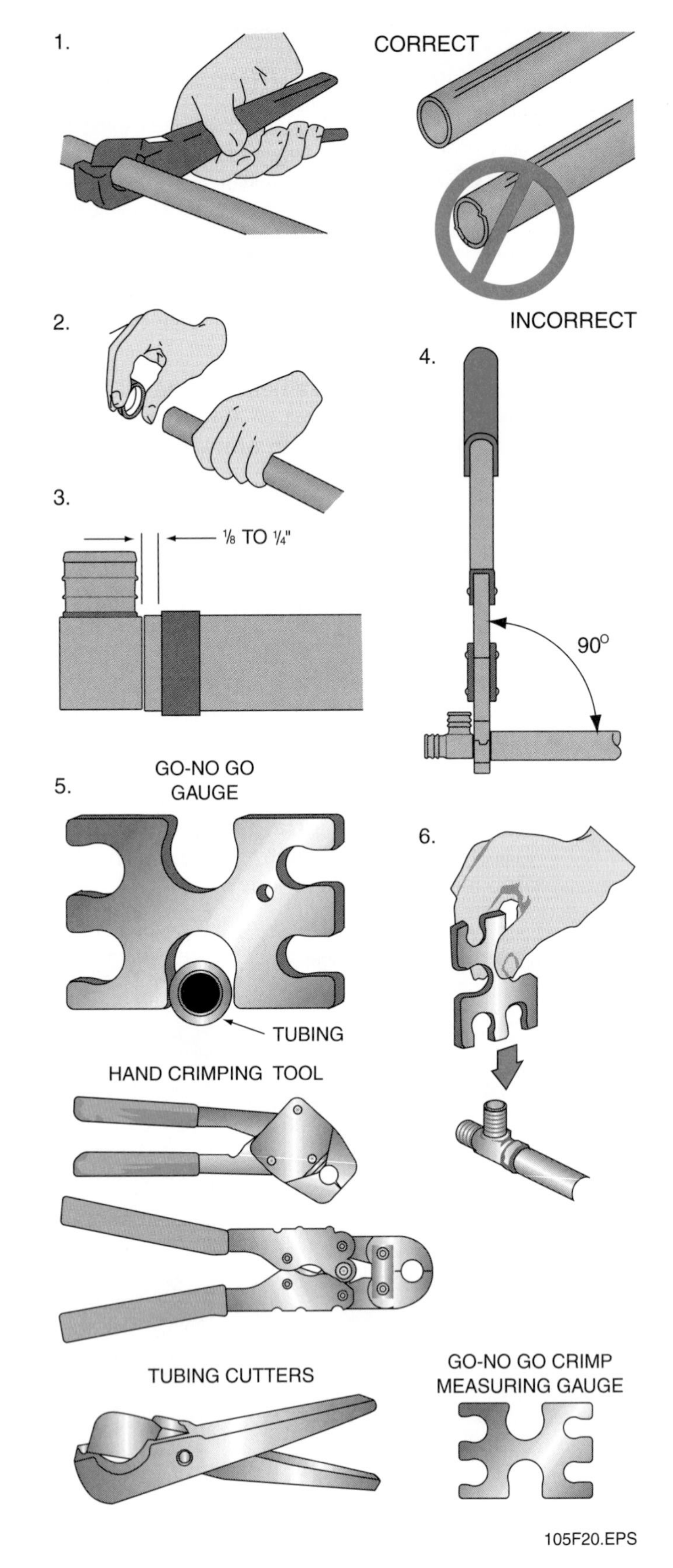

Figure 20 ◆ Joining PEX tubing.

5.3.3 Joining PE Tubing

PE fusion (used in gas transmission systems) often involves the use of expensive tools (see *Figure 21*) and requires training and certification. Manufacturers of PE products and joining equipment often provide this training and certification free of charge. New techniques that involve **compression collars** for joining PE are now becoming popular because less training is required for their use.

5.4.0 Pressure Testing

Once an installation is completed and cured, the systems should be **hydrostatically pressure-tested** in accordance with local code requirements. This means the system should be filled with water and all air should be bled from the highest and farthest points in the run.

If a leak is found, the joint must be cut out and discarded. A new section can be installed using couplings.

CAUTION

After testing a connection, thoroughly flush the system for at least 10 minutes to remove any remaining trace amounts of solvent cement.

During subfreezing temperatures, water should be blown out of the lines after testing to avoid possible damage from freezing.

CAUTION

Use air testing only when hydrostatic testing is not practical. Use extreme caution, because air under pressure is explosive. For air testing, you must notify all site personnel of the test, you must use protective eyewear, and you must take precautions to prevent impact damage to the system during the test.

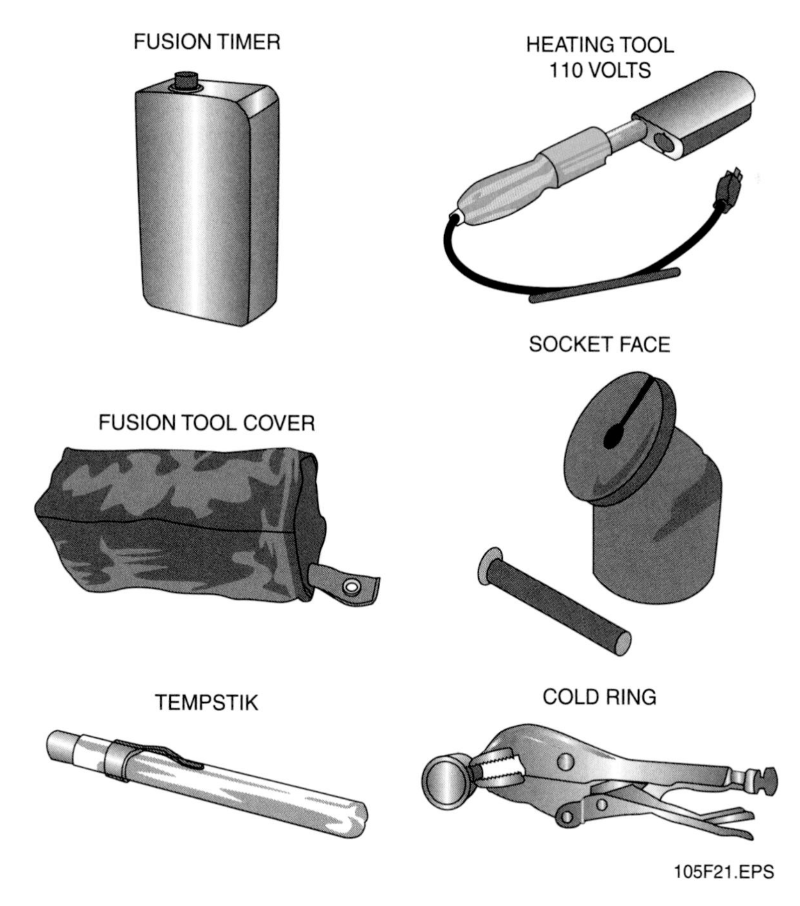

Figure 21 ◆ PE fusion tools.

Review Questions

Sections 4.0.0–5.0.0

1. _____ are used as vertical support for pipes and tubing.
   a. Beam clamps
   b. C-clamps
   c. Pipe risers
   d. Special fasteners

2. Place piping supports at intervals of no more than _____ feet or as specified by local code.
   a. 8
   b. 4
   c. 6
   d. 12

3. When you are measuring any pipe that needs to be cut and joined, be sure to calculate the depth of the _____.
   a. weld
   b. fixture
   c. joint
   d. square

4. To make sure that the pipes join correctly, make a clean, _____ cut using a power saw or miter box.
   a. beveled
   b. angled
   c. smooth
   d. square

5. Be sure there is proper _____ when you are applying primers and cements or soldering materials.
   a. ventilation
   b. adhesion
   c. connection
   d. humidity

Summary

The use of plastics in the plumbing trade has greatly increased the effectiveness and safety of indoor and outdoor plumbing systems. The introduction of the different types of plastic pipe and tubing means that today's plumbers must be knowledgeable about new techniques, tools, and applications. Plastic may seem like the answer to all plumbing problems, but exactly which type of plastic you use makes a big difference. You must learn the special techniques for cutting and joining a variety of plastic pipe types. Recognizing the common types of materials used on a job site is a basic skill.

Using plastic pipe eliminates one of the most common hazards associated with the plumbing industry—fire. When plumbers used to join pipe using molten lead, accidents involving spilled lead and tipped-over heating furnaces were a constant concern. But plastic pipe has its own hazards. Fumes from solvent chemicals are harmful; solvents should be used only in well-ventilated areas. You must be familiar with the types of safety hazards each variety of pipe and chemical solvent presents. It is your responsibility to know how to protect yourself on the job site.

Developing technology affects plumbers more now than ever. Manufacturers are constantly researching and producing new products. To be competitive in the modern plumbing industry, you must keep up with innovations and improvements in installation techniques and materials. Plumbing provides a lifelong learning opportunity.

PROFILE IN SUCCESS

Charles Owenby

Plumbing Instructor
JF Ingram Technical College
Datsville, AL

Charles was born in West Virginia and raised in Montgomery, Alabama. He watched his older brother, a plumber, leave for work early each day and return home early. That piqued his interest in plumbing. He got a job working on the utility crew of a company putting in subdivisions and has been a plumber ever since. Today, Charles still loves the plumbing profession, as demonstrated by his devotion to teaching. He even teaches at local vocational schools in his spare time.

How did you become interested in the construction industry?

I started working in 1967 with a company building subdivisions. After a bit, I moved over to the new construction part of the company and became a registered apprentice. I went to school one night a week and learned gas, water, and drainage piping installation, and I kept moving up.

In 1969, I joined the Seabees, a branch of the Navy that's a mobile construction battalion. I continued in plumbing while doing my military service. We worked like a large construction company. Steel workers, masons, carpenters, site developers, and plumbers all worked on building bases, landing strips, and recreation centers and performed camp maintenance. I worked in plumbing, HVAC, water treatment, and sewage treatment. For a time, I was assigned to Puerto Rico where we provided the outlying islands with freshwater systems.

In 1971, after my discharge, I went back to my previous job. I took my journeyman's exam and was then qualified to run plumbing jobs—working with apprentices. I also joined the Navy Reserve and served 28 years. During Desert Shield and Desert Storm, I was assigned to the Reservist Brigade staff as the training chief for 26 battalions. I was in charge of reviewing their annual training plans.

In the meantime, I passed my master plumber exam that qualified me to perform contracts and bid on jobs. Alabama law requires a company to have at least one master plumber when bidding on contract work.

In 1981, I got my bachelor's degree at Troy State University, which provides overseas military training, and became an instructor at JF Ingram Technical College. In 1993, I received my master's degree in counseling.

Meanwhile, as a reservist, I switched to the environmental unit of the Seabees. We conducted environmental audits, performed hazardous waste planning and control, monitored underground storage tanks, and helped with compliance issues related to government regulations.

What do you think it takes to be a success in your trade?

Plumbing is an art and a science. Too many people think of plumbing as just fixing leaking faucets and repairing toilets. Nothing could be farther from the truth. To be a success as a plumber, you have to be flexible enough to adapt to ever-changing situations on different jobs—no two jobs are exactly alike. Some companies may specialize in one aspect of plumbing—water supply piping, DWV piping, maintenance and repair—but you have to know something about each facet of the trade. One sure route to success is to find a mentor—someone you can work with who will help you learn and grow in your career.

You must be prepared to work your way up through the skill levels required to be a great plumber. Apprentices don't make decisions; they have to do what they are told to do. Journeymen make decisions. Master plumbers run jobs and figure prominently in a company's achievements. Your work attitude, willingness to learn, and your work ethic are all important to your success.

Patience is a virtue. Never try to hurry through a job. Lack of proper planning or not understanding what is needed can cause big problems. As the old saying goes, "If you don't have time to do it right the first time, when are you going to have the time to correct it?"

What are some of the things you do in your role of instructor?
I teach courses in plumbing using the National Center's curriculum. We emphasize safety, blueprint reading, job ethics, and technical skills and techniques. In Alabama, all apprentices must register with the state to be allowed to work in plumbing and to use the tools. After about two years as an apprentice, students can take the journeyman's exam.

I emphasize safety on many different levels. As an instructor, I am responsible for equipping my students with the knowledge of how to work safely with tools and equipment and how to protect themselves in the work environment. I also stress the effect that their work has on the safety of others. For example, working on medical gas installations requires many specialized skills. A wrong connection can cause serious injury or death. Of course, I also remind students to know and adhere to state and local codes.

I also work to correct the stereotype of plumbers. This is a great profession with a wonderful potential for making a living. People who are willing to dedicate themselves to learning the skills, knowledge, and art of plumbing will be successful.

What do you like most about your job?
I enjoy sharing the profession with students. I like informing people about the art and science that plumbing requires. I don't think enough people understand the art of plumbing—it is often covered up by wallboard. The scientific element of plumbing is equally as challenging and fascinating.

I also love the profession. I could talk for days about what I like most about plumbing.

What would you say to someone entering into the trades today?
Learning a trade is a process. After almost 34 years in plumbing, I am still striving and still learning. It's the challenge that makes it all so interesting. Plumbing is a good profession that will allow you to make a good living. The market is growing and there are lots of jobs available. I tell people entering the trades today, "There are so many opportunities."

GLOSSARY

Trade Terms Introduced in This Module

ABS (acrylonitrile-butadiene-styrene): Plastic pipe and fittings used extensively in drain, waste, and vent (DWV) systems.

Bushing: A usually removable cylindrical lining for an opening, such as for a mechanical part, used to limit the size of the opening, resist abrasion, or serve as a guide.

Cellular core wall: Plastic pipe wall that is low-density, lightweight plastic containing entrained (trapped) air.

Chamfer: To bevel the edge of construction material to a 45-degree angle.

Chemically inert: Does not react with other substances or materials.

Compression collar: A piece of hardware that uses compression force to connect sections of polyethylene piping.

CPVC (chlorinated polyvinyl chloride): Plastic pipe and fittings used extensively in hot and cold water distribution systems.

Elastomer: Any of various elastic substances resembling rubber, such as polyvinyl elastomers.

Fusion fitting: A fitting with a butt that has the same outside diameter and inside diameter as the pipe.

Hydrostatically pressure-tested: Filled with water and with all air bled from the highest and farthest points in the run.

Inside diameter (ID): The distance between the inner walls of a pipe; the standard measure of tubing used in heating and plumbing.

Interference fit: Fit that tightens as the pipe is pushed into the socket.

Outside diameter (OD): The distance between the outer walls of a pipe.

PB (polybutylene): Formerly used for plumbing pipe; it is no longer used but is still found in some residences.

PE (polyethylene): Flexible plastic pipe, tubing, and fittings that do not deteriorate when exposed to sunlight.

PEX (cross-linked polyethylene): Tubing and fittings made with heat and high pressure. It has high-temperature, pressure, and chemical resistance.

Pressure rating: The pressure at which a component or system may be operated continuously.

psi: A measurement of pressure in terms of pounds per square inch.

PVC (polyvinyl chloride): Plastic pipe and fittings used for cold water distribution and for industrial water and chemicals.

Ring-tight gasket fitting: Fitting with a rubber O-ring or gasket in the socket.

Scribe: To mark material with a sharp-pointed instrument to indicate the cutting line.

SDR: Size dimension ratio; used to express the size of a pipe or tube.

Solid wall: Plastic pipe wall that does not contain trapped air.

Solvent-weld: To join two pipes using a solvent cement that softens the material's surface.

Transition fitting: A special fitting that allows threaded plastic pipe to be connected to steel pipe.

ANSWER KEY

Answers to Review Questions

Section 2.0.0

1. c
2. a
3. d
4. b
5. b

Section 3.0.0

1. a
2. c
3. a
4. c
5. c

Sections 4.0.0–5.0.0

1. c
2. b
3. c
4. d
5. a

NCCER CRAFT TRAINING USER UPDATES

The NCCER makes every effort to keep these textbooks up-to-date and free of technical errors. We appreciate your help in this process. If you have an idea for improving this textbook, or if you find an error, a typographical mistake, or an inaccuracy in the NCCER's Craft Training textbooks, please write us, using this form or a photocopy. Be sure to include the exact module number, page number, a detailed description, and the correction, if applicable. Your input will be brought to the attention of the Technical Review Committee. Thank you for your assistance.

Instructors – If you found that additional materials were necessary in order to teach this module effectively, please let us know so that we may include them in the Equipment/Materials list in the Instructor's Guide.

Write: Curriculum Revision and Development Department
National Center for Construction Education and Research
P.O. Box 141104, Gainesville, FL 32614-1104

Fax: 352-334-0932

E-mail: curriculum@nccer.org

Craft ________________ Module Name ________________________________

Copyright Date ________ Module Number ________ Page Number(s) ________

Description

____

____

____

____

(Optional) Correction

____

____

____

(Optional) Your Name and Address

____

____

____

Module 02106-00

Copper Pipe and Fittings

Course Map

This course map shows all of the modules in the first level of the Plumbing curriculum. The suggested training order begins at the bottom and proceeds up. Skill levels increase as you advance on the course map. The local Training Program Sponsor may adjust the training order.

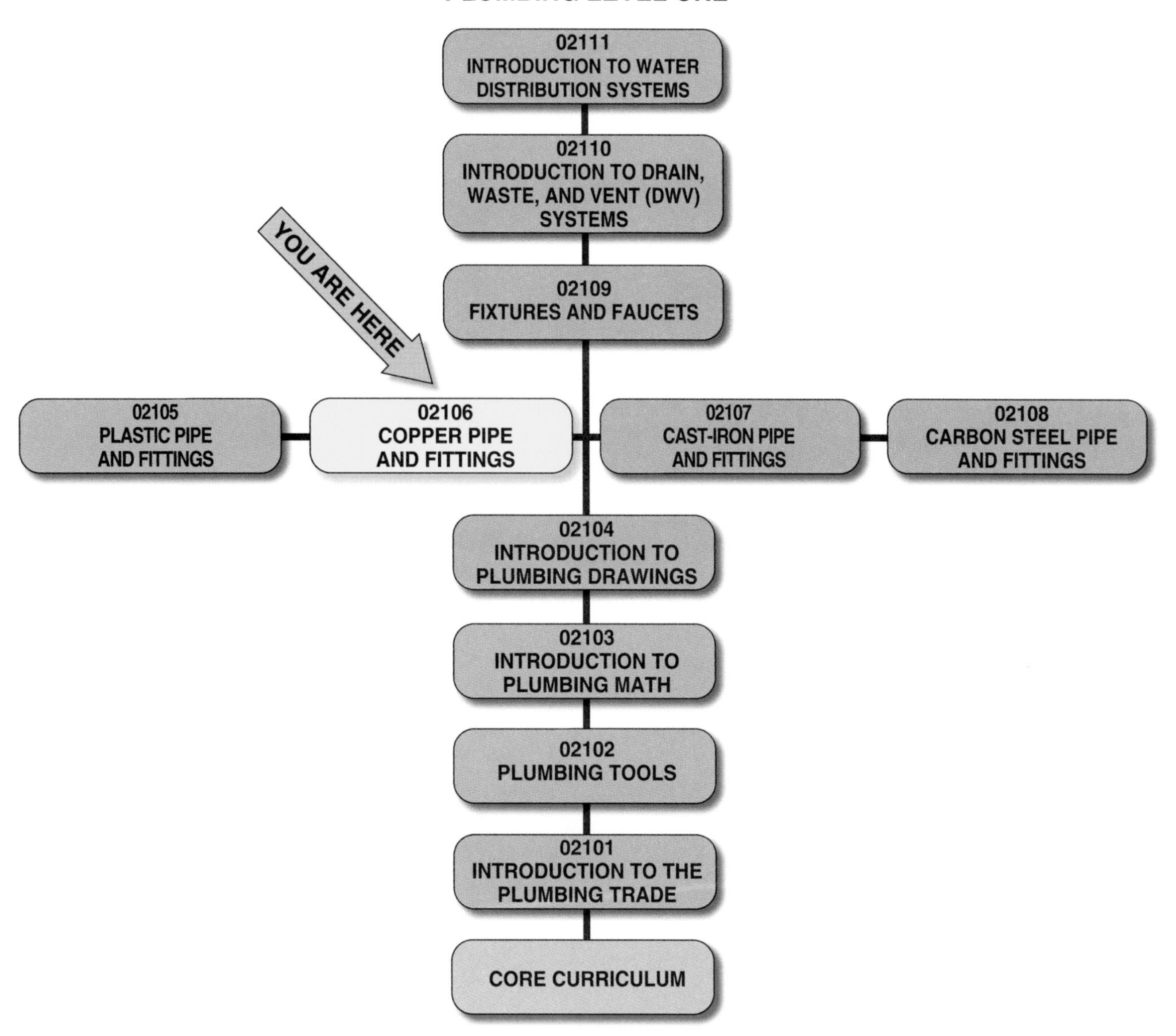

106CMAP.EPS

Copyright © 2000 National Center for Construction Education and Research, Gainesville, FL 32614-1104. All rights reserved. No part of this work may be reproduced in any form or by any means, including photocopying, without written permission of the publisher.

MODULE 02106 CONTENTS

Figures

Table

MODULE 02106

Copper Pipe and Fittings

Objectives

When you finish this module, you will be able to:

1. Identify the common types of materials and schedules used with copper piping.
2. Identify the common types of fittings and valves used with copper piping.
3. Identify the techniques used in hanging and supporting copper piping.
4. Demonstrate the ability to properly measure, ream, cut, and join copper piping.
5. Identify the hazards and safety precautions associated with copper piping.

Prerequisites

Before you begin this module, it is recommended that you successfully complete task modules: Core Curriculum; Plumbing Level One Modules 02101 through 02105.

Materials Required for Trainees

1. This module
2. Appropriate personal protective equipment
3. Sharpened pencil and paper

1.0.0 ◆ INTRODUCTION

Copper pipe and fittings are used daily in piping systems. They have a wide range of uses—for hot and cold water supply; for drain, waste, and vent (DWV) systems; for fuel gas supplies; and for transporting refrigerant in residential and smaller commercial air-conditioning systems, although it was much more expensive than traditional ferrous (iron) pipe and fittings. As more copper sources were discovered, the price dropped, and now copper pipe and fittings are common.

Copper is a mineral that is mined out of the ground. The copper is then melted and molded into various sizes, lengths, and angles. Copper was first used in plumbing in the early 1800s, but it was not widely used then.

2.0.0 ◆ MATERIALS

There are seven types of copper **tubing:** K, L, M, DWV, **ACR** (air-conditioning and refrigeration), Medical Gas, and G. Each type represents a series of sizes with different wall thicknesses. The tubing can be drawn (hard) or **annealed copper** (soft). All types of tube are available in a hard form. Hard forms come in 12- to 20-foot lengths and in sizes ranging from ¼ inch to 12 inches. Types K, L, ACR, and G also are available in soft coils in 40- to 100-foot lengths and in sizes ranging from ⅛ inch to 2 inches.

2.1.0 Sizing

Types K, L, M, DWV, and Medical Gas use nominal (standard) sizing. This means that the **outside diameter (OD)** is always ⅛ inch larger than the **nominal size** (see *Figure 1*). For example, the outside diameter of a ¾-inch nominal type M tube measures ⅞ of an inch. This allows the same fittings to be used with the different wall thicknesses

Did You Know?

Copper is one of our most plentiful metals. It has been in use for thousands of years. A piece of copper pipe used by the Egyptians more than 5,000 years ago is still in good condition. When the first Europeans arrived in the New World, they found the Native Americans using copper for jewelry and decoration. Much of this copper was from the region around Lake Superior, which separates northern Michigan and Canada.

One of the first copper mines in North America was established in 1664 in Massachusetts. During the mid-1800s, new deposits of copper were found in Michigan. Later, when the miners went West in search of gold, they uncovered some of the richest veins of copper in the United States.

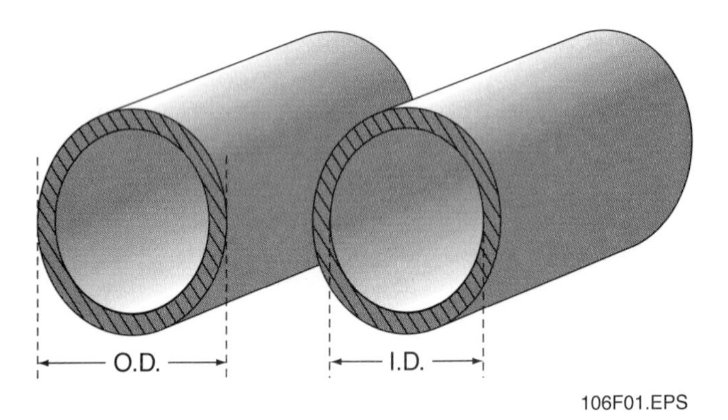

Figure 1 ◆ Sizing.

and **inside diameters (ID)** of the different types of copper tubes. The inside diameter, on the other hand, is usually close to the stated size. A ½-inch ID means that there is approximately ½ inch between the inside walls of the pipe.

Types ACR and G use actual outside diameter sizing. A ⅞-inch outside diameter ACR copper tube is actually the same outside diameter as ¾-inch K, L, or M copper tube. This means that the same fittings also could be used with these pipes.

Strength, formability (how easily the pipe can be bent), and other mechanical factors are considered in choosing the type of copper tube to be used. Plumbing and mechanical codes govern what types may be used. Plumbers must always consult local plumbing and mechanical codes.

Following are examples of types of applications that show which type of copper tube can be used successfully:

- *Underground water service*–Type M hard for straight lengths joined with fittings or Type L soft where coils are more convenient.
- *Water distribution systems*–Type M for both above and below ground.
- *Chilled water mains*–Type M for all sizes. Type DWV, where approved, may be used for sizes of 1¼ inch and larger; however, joints must be made with solder-joint pressure fittings.
- *Drainage, waste, and vent systems*–Type DWV for aboveground and belowground waste, soil, and vent lines; roof and building drains; and sewers. DWV copper tubing is often used as fixture drains, such as for the drain pipe from a sink, in cast iron DWV systems. DWV tubing is thinner than Type M tubing.
- *Heating and solar heating*–For radiant panel and water heating and for snow-melting systems, use Type L soft where the coils are formed in place or prefabricated. Use Type M where straight lengths are used. Type M of all sizes is used for water heating and low-pressure steam. Type DWV, where approved, may be used for sizes of 1¼ inch or larger; however, joints must be made with solder-joint pressure fittings. Type L can be used for condensate return lines.
- *Fuel oil, liquefied petroleum, and natural gas services*–Follow local codes for copper tube.
- *Nonflammable medical gas systems*–Medical Gas tube types K or L that have been cleaned for oxygen service.
- *ACR systems*–Types L or ACR may be used in air-conditioning and refrigeration systems. Type M is used on air-conditioning condensate drains.
- *Ground source heat pump systems*–Types L or ACR where the ground coils are formed in place or prefabricated may be used in these systems.
- *Fire sprinkler systems*–Type M hard is recommended. Where bending is required, Types K or L are recommended.

Drawn copper tubing, which is widely used in commercial refrigeration and air-conditioning systems, comes in straight lengths of 20 feet and in sizes from ⅜-inch OD to 4-inch OD. The lengths are filled with nitrogen and plugged at each end to maintain a clean, moisture-free internal condition. This tubing is intended for use

Did You Know?

U.S. industries use close to three million tons of copper each year. About two-thirds of the copper comes from the mining process; the rest comes from recycled copper. Discarded objects, dismantled buildings, and worn-out machinery are sources of "secondary" copper. This scrap copper can be melted down and reused. Many recycling stations pay for scrap copper.

with formed fittings to make the necessary bends or changes in direction. It is more self-supporting than annealed copper tubing (**ACR tubing**); therefore, it needs fewer supports.

2.2.0 Labeling (Markings)

The labels on copper pipe contain a lot of useful information. Types K, L, M, G, DWV, and Medical Gas must be permanently incised (marked) to show the tube type, the name or trademark of the manufacturer, and the country of origin. This information is also printed on the hard tubes in a color that identifies the tube type.

The labeling on most soft copper is stamped into the product. *Table 1* shows the seven types of copper tubing, the corresponding color code, and the common application for which the tubing is used. *Always check local codes.*

Table 1 Copper Tubing Color Codes

Type	Color Code	Application
K	Green	Underground and interior service
L	Blue	Above-ground service
M	Red	Above-ground water supply and DWV
DWV	Yellow	Above-ground DWV piping
ACR	Blue	Air-conditioning, refrigeration, natural gas, liquefied petroleum (LP) gas
Medical Gas	K–Green L– Blue	Medical gas
G	Yellow	Natural gas, liquefied petroleum (LP)

Review Questions

Section 2.0.0

1. When copper tubing is described as being annealed, the tubing is _____.
   a. soft
   b. hard
   c. incised
   d. color marked
2. _____ copper tubing comes in 12- to 20-foot lengths.
   a. DWV
   b. Hard
   c. ACR
   d. Soft
3. Types K, L, M, DWV, and Medical Gas copper tubing are measured using _____.
   a. inside diameter sizing
   b. outside diameter sizing
   c. nominal sizing
   d. copper tube sizing
4. _____ copper tubing is often used as fixture drains.
   a. Type M
   b. DWV
   c. Medical gas
   d. Type L
5. Type _____ copper tubing is red.
   a. M
   b. K
   c. DWV
   d. L

3.0.0 ◆ COMMON FITTINGS AND VALVES

The types of fittings and valves used with copper piping depend on how the piping is used. Use as few fittings as possible. Fewer fittings mean fewer chances for leaks and **pressure drops.**

3.1.0 Water Fittings and Valves

Common copper fittings for use with water supply systems (see *Figure 2*) include the following:

- *90-degree ell and 45-degree ell*–an elbow used to change the direction of the pipeline.
- *Drop ear ell*–an elbow that allows you to attach the pipeline to the building frame; frequently used at the last joint before the pipe comes through the wall to be attached to a fixture.

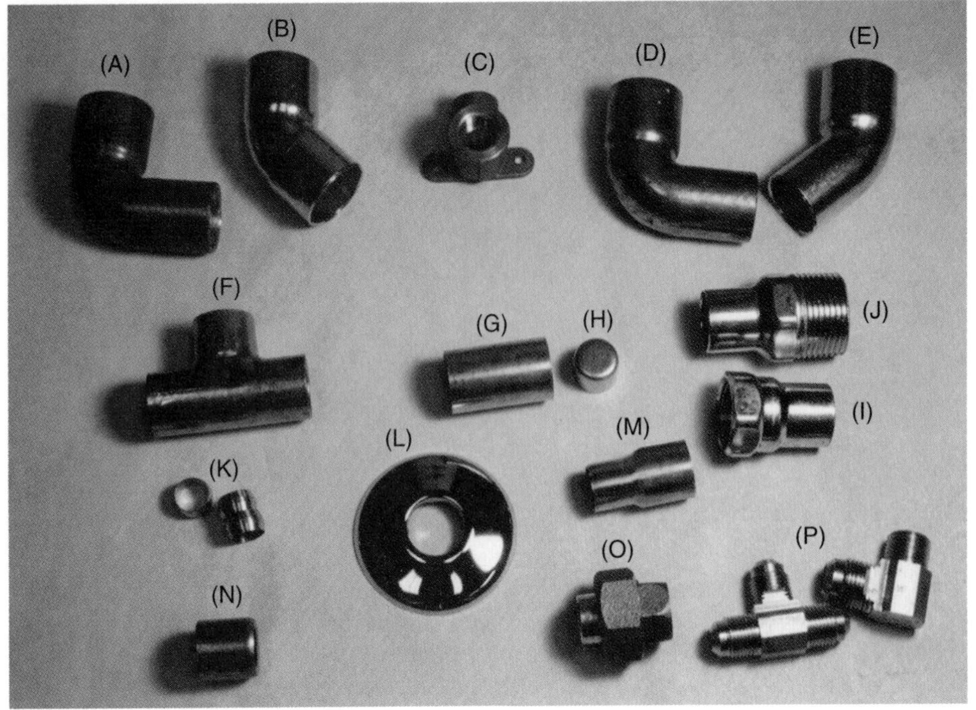

106F02A-P.EPS

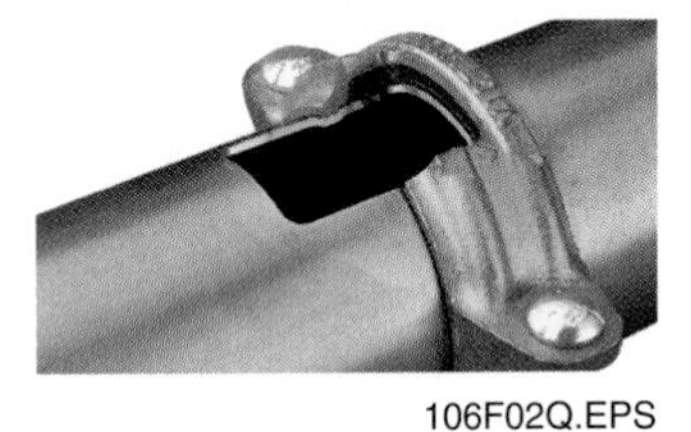

106F02Q.EPS

(Q)

Figure 2 ◆ Fittings for copper tubing. (A) 90-degree ell. (B) 45-degree ell. (C) Drop-ear ell. (D) Street 90. (E) Street 45. (F) Tee. (G) Coupling. (H) Cap. (I) Female adapter. (J) Male adapter. (K) Sweat-to-compression adapter. (L) Flange. (M) Reducer Coupling. (N) Bushing. (O) Sweat union. (P) Flare fittings. (Q) Grooved copper fittings.

- *Street 90*–an elbow fitting that has a run with male threads.
- *Street 45*–an elbow fitting with male threads.
- *Tee*–used to make branches at 90-degree angles to the main pipe. Reducing tees are used to make branches at 90-degree angles from the main pipe to smaller outlet pipes.
- *Coupling*–used to connect lengths of pipe on straight runs.
- *Cap*–has internal threads and is used to close the end of a threaded pipe.
- *Female and male adapters*–fittings **soldered** onto the end of a length of copper tubing to provide a threaded end for attaching to another threaded pipe.
- *Sweat-to-compression adapter*–used to adapt a soldered copper tube to a **compression joint** by means of a **ferrule.**
- *Sweat flange*–used to adapt soldered copper tube to iron pipe flange.
- *Reducer coupling and reducer bushing*–used to connect pipe-to-pipe or pipe-to-fitting with different sizes of pipe.
- *Sweat union*–soldered to copper tubes, which can then be joined by the male and female half unions by a threaded shell nut.

Did You Know?

Two kinds of solder fittings are available with copper tubing. The first is a wrought copper fitting, which is made from copper tubing that is shaped into different types of fittings. Wrought copper fittings are generally lightweight, are smooth on the outside, and have thin walls. The second type of fitting used with copper tubing is a cast solder fitting. This type of fitting is made using a mold. The first cast fittings had holes in the sockets to put solder in. Today, the heated copper is poured into the mold and allowed to cool. Cast solder fittings have a rough exterior and come in a wide variety of shapes. They are heavier than wrought solder fittings.

- *Flare fittings*–fittings with a flared end that can be joined with a male cone-shaped tubing end or union.
- *Grooved copper*–a mechanical coupling material for rigidly connecting copper tubing that has been **roll-grooved.**

ON THE LEVEL

Plumbing Codes

Plumbing codes have been developed to protect the water supply and our health and safety. Most cities, counties, and states have adopted model codes that are based on suggested national standards. These model codes may be subject to local interpretation and usually reflect local conditions. For example, an area that often has earthquakes or floods will have special requirements to deal with these conditions. The specifications for each job should detail these requirements. Installers take responsibility for code violation corrections, which can be expensive. While standards are guidelines to improve performance or reliability, codes and ordinances are mandatory, and plumbers must follow these rules.

If tee fittings are not properly installed they can cause a condition known as **bullheading,** in which the outlet opening on the branch is larger than the opening on the straight run. This causes turbulence, which adds to the pressure drop caused by liquid moving from a smaller pipe to a larger pipe. Bullheading turbulence may also cause **hammering** or banging in the line. If more than one tee is installed in the line, a straight piece of pipe with a length between tees of 10 times the pipe's diameter is recommended to reduce turbulence. For example, a pipe that is four inches in diameter should have 40 inches of pipe between each tee (10 inches × 4 inches = 40 inches).

Solder, or sweat, fittings are special fittings made of copper or brass that are used for soldering or **brazing** copper pipe. Soldering is a type of heat bonding; copper pipe is joined when a soft filler metal is melted in the joint between the two pipes. Solder fittings are made slightly larger than the pipes to be joined, leaving only enough room for solder to flow into the joint. Adapter fittings allow copper tubing to be joined with threaded pipe on one end while the other end is soldered.

Flare fittings used in refrigeration and air-conditioning consist of a variety of elbows, tees, and unions. They are **drop-forged** brass and are accurately machined to form the 45-degree flare face. The fittings used are based on the size of the tubing. Flare nuts (see *Figure* 2) are hexagon-shaped (have six sides). An adjustable, or open-end, wrench is used with these fittings.

Grooved copper fittings (see *Figure* 2) are made for connecting copper tubing in sizes from 2 to 6 inches. These fittings have grooved ends, so they can be installed using a wrench. This eliminates the need for soldering, sweating, or brazing.

Common valves for water lines (see *Figure* 3) include the following:

- *Gate*–has a wedge-shaped or tapered metal disc that fits into a smooth-ground surface or seat with the same shape, which allows a straight-line flow with little obstruction. It is a good choice for lines that will remain either completely open or closed most of the time.
- *Globe*–has a partition between the inlet and outlet that blocks the flow through the valve. It uses a replaceable or fiber washer or metal disc. Because this valve is reliable and easy to repair, it is often used in water supply lines inside buildings.
- *Ball*–has a ball with a hole bored through its diameter, mounted on a spindle. When the valve is closed, the hole is at 90 degrees to the valve body so no flow can take place. When the valve is turned a quarter turn or opened completely, water flows through the hole. This valve is commonly used at the inlets and outlets of heat exchangers in HVAC (heating, ventilating, and air-conditioning) cooling systems and in systems where quick shutoffs may be necessary for in-line maintenance.
- *Compression, stop, and waste*–is opened or closed by raising or lowering a horizontal disc using a threaded stem. An elastomeric (rubberized) washer on the end of the stem seals the valve seat, closing off water flow. This valve is most commonly used for water faucets.
- *Check*–prevents reverse flow in a plumbing system. It permits free flow in one direction but automatically closes if liquid begins to flow the other way. This valve is commonly used on domestic and irrigation wells.
- *Stop*–controls flow of liquids or gases between a building and supply source; also called a ground-key valve.

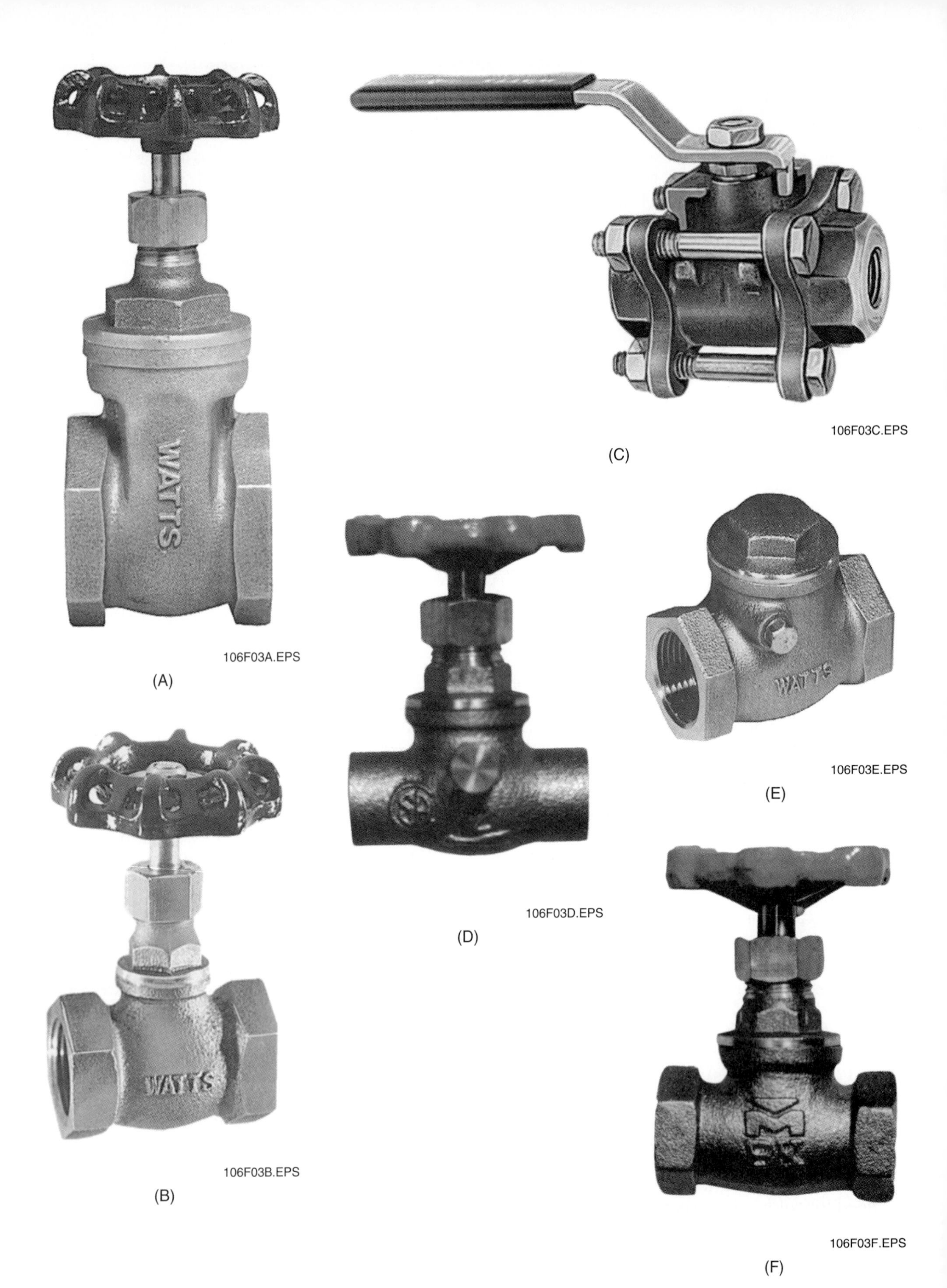

Figure 3 ◆ Common valves for water lines. (A) Gate valve. (B) Globe valve. (C) Ball valve. (D) Compression, stop, and waste valve. (E) Check valve. (F) Stop valve.

Did You Know?

To solder copper fittings, you must heat the fitting until the soldering paste starts to melt. This melting paste looks like beads of sweat on the pipe, which led to the name "sweat fittings."

3.2.0 DWV Fittings and Valves

Common fittings used in DWV systems are elbows (90 degrees, 45 degrees, 22½ degrees), male adapters, ferrule adapters, sanitary tees, cleanout tees, reducing tees, wyes, and reducers. Common copper valves for DWV systems are butterfly valves.

Review Questions

Section 3.0.0

1. _____ are used to change the direction of a pipeline.
   a. Adapters
   b. Couplings
   c. Elbows
   d. Caps
2. A _____ is used to adapt soldered copper pipe to an iron pipe flange.
   a. male or female adapter
   b. flare fitting
   c. sweat-to-compression adapter
   d. sweat flange
3. Because it is reliable and easy to repair, a _____ valve is often used in water supply lines inside buildings.
   a. globe
   b. gate
   c. ball
   d. compression
4. The valve most often used in water faucets is the _____ valve.
   a. ball
   b. compression stop and waste
   c. globe
   d. gate
5. _____ are used to make branches in a pipeline.
   a. Tees
   b. Adapters
   c. Couplings
   d. Flanges

4.0.0 ◆ HANGERS AND PIPE SUPPORTS

The plumber must install and support all pipes so that both the pipe and its joints remain leakproof. Improper support can cause the piping system to sag. This causes stress on the pipe and fittings and, over time, increases the chance of breaks or leaks in the piping system. Without proper support the drainage pipe can shift from its proper angle and form traps. These traps fill with liquid and solid wastes that block the pipeline. Copper pipe is supported in a variety of ways and with various sizes of hangers and supports.

4.1.0 Types of Hangers and Pipe Supports

Which hanger to use depends on the job **specifications.** Specifications are determined by considering a number of factors:

- The combined weight of the pipe fittings and valves
- The maximum weight of the contents that might be carried in the pipe
- What material (wood, concrete, steel) the hanger will be attached to
- The distance from the anchor point to the pipe
- The potential for corrosion between the hanger and the pipes, fittings, or valves
- The expansion and contraction of the piping system
- The vibration of equipment attached to the pipe system

Different types of hangers and supports are designed for holding and supporting pipe in either a horizontal or vertical position. They are manufactured in various materials including carbon steel, malleable iron, cast iron, and various plastics. They are available with different finishes, including copper plate, black, galvanized, chrome, and brass. To reduce corrosion, hangers and supports should be made of the same material as the pipe. If another material is used, the pipe must be shielded.

Hangers and supports can be placed into three major categories:

- Pipe attachments
- Connectors
- Structural attachments

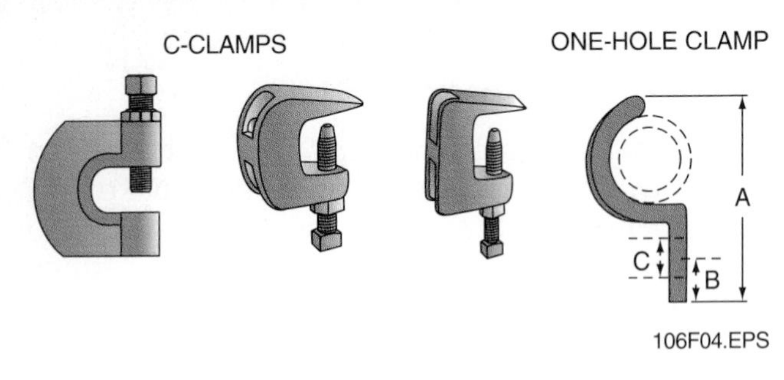

Figure 4 ◆ Pipe attachments.

Figure 5 ◆ Hangers for horizontal and vertical support.

4.1.1 Pipe Attachments

A pipe attachment is the part of the hanger that touches or connects directly to the pipe. It may be designed for either heavy duty or light duty, for covered (insulated) pipe or plain pipe. Examples of pipe attachments are shown in *Figure 4.*

Hangers are used for horizontal or vertical support of pipes and piping. The main purpose of hangers and brackets is to keep the piping in alignment and to prevent it from bending or distorting. Several styles of hangers are shown in *Figure 5.*

Piping can be supported on wood-frame construction with several styles of pipe attachments. These include pipe hooks, J hooks, tube straps, perforated band irons, and pipe clamps (see *Figure 6*). In these examples, the pipe attachment, connector, and structural attachment are all one unit.

Pipe hangers are used mainly to support pipe, but they can also be used as vibration isolators (see *Figure 7*). If you do not expect any vibration

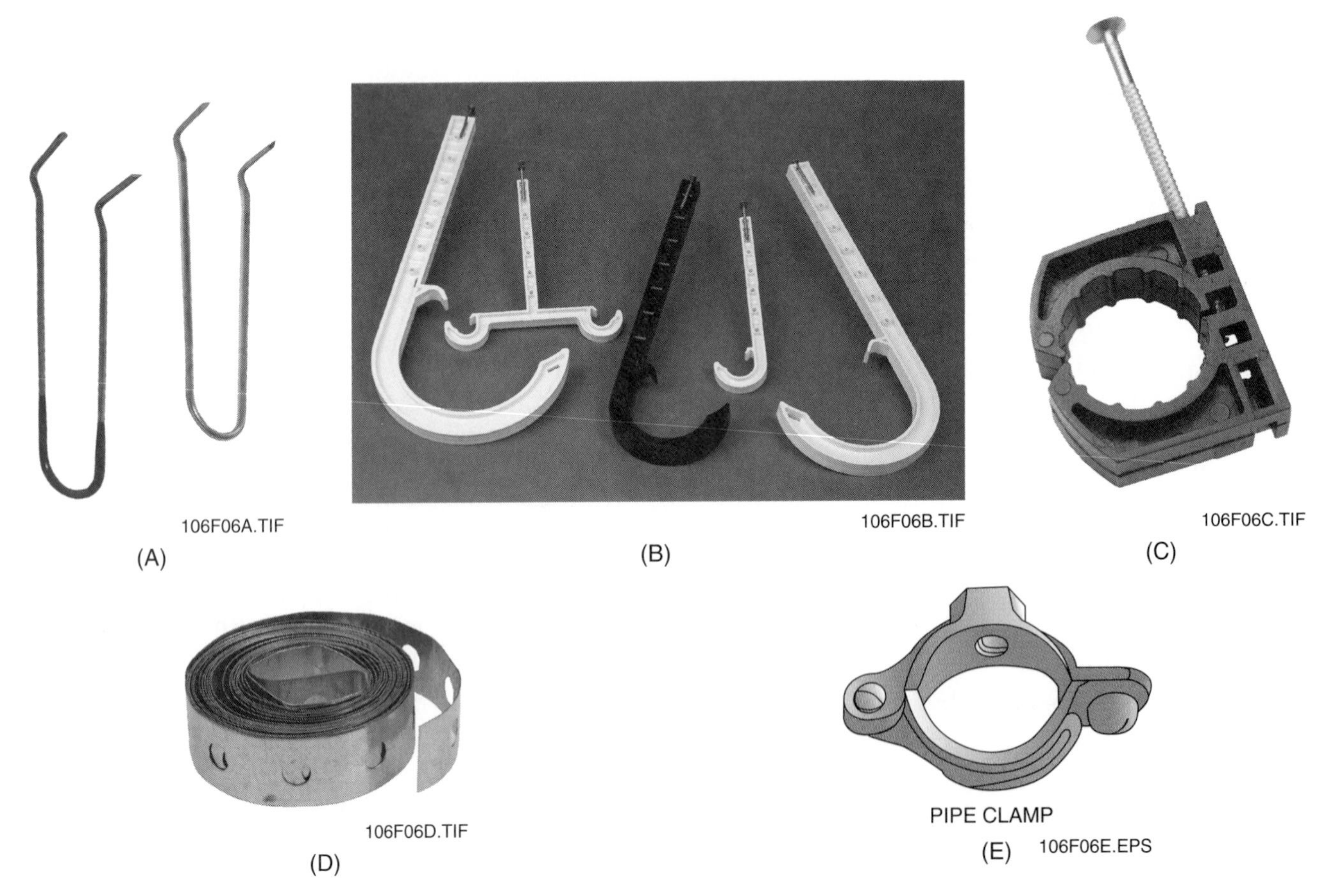

Figure 6 ◆ Hangers for installation on wood-frame construction. (A) Pipe hooks. (B) J hooks. (C) Locking tube strap. (D) Perforated band irons. (E) Pipe clamps.

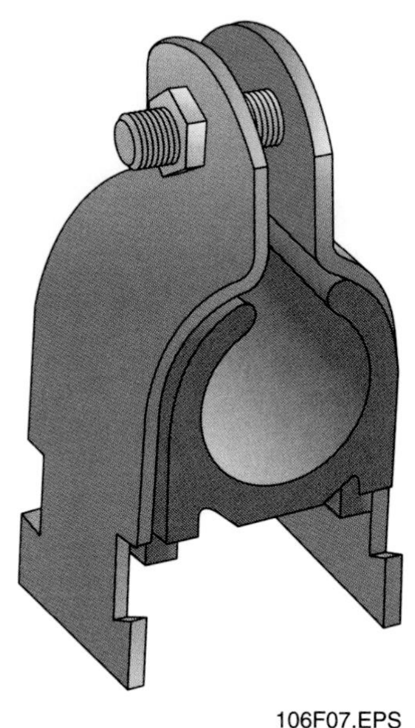

Figure 7 ◆ Strut VibraClamp™.

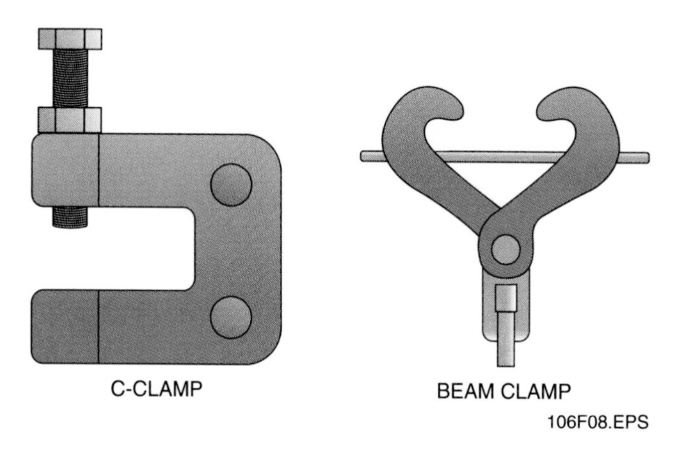

Figure 8 ◆ Clamps.

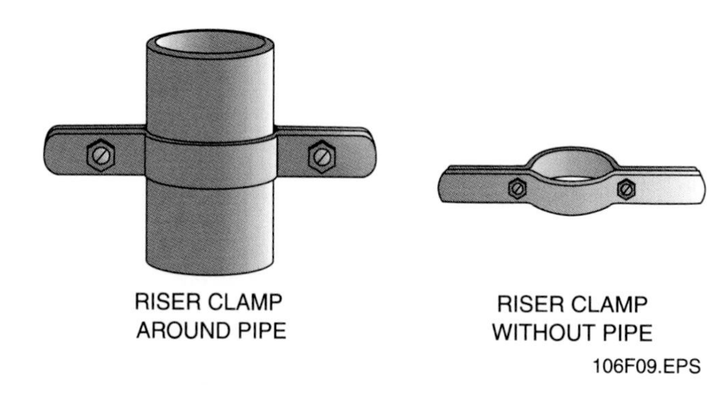

Figure 9 ◆ Vertical clamps.

problems, use commonly accepted plumbing practices. If vibration may be a problem, the specifications should tell you what materials to use.

For fastening pipes to beams and other metal structures, one-hole clamps, steel brackets, beam clamps, or C-clamps (see *Figure 8*) are used. Vertical hangers (see *Figure 9*) or pipe risers consist of a friction clamp that can be attached to structural site components to support the vertical load of the pipe. Special fasteners are used to attach hangers to masonry, concrete, or steel.

Other pipe attachments include universal pipe clamps and standard 1⅝-inch or 1½-inch channels (see *Figure 10*). The notched steel clamps are inserted by twisting them into position along the slotted side of the channel. The pipes can be aligned as close to one another as the couplings allow.

4.1.2 Connectors

The connector section of the hanger is the part that links the pipe attachment to the structural attachment. These connectors can be divided into two groups: rods and bolts, and other rod attachments.

The rod attachments include eye sockets, extension pieces, rod couplings, reducing rod couplings, hanger adjusters, turnbuckles, **clevises,** and eye rods. Some of these connectors are shown in *Figure 11*.

4.1.3 Structural Attachments

Structural attachments are used to anchor the pipe hanger assembly securely to the structure. Structural attachments include threaded drop-in anchors, plastic or lead mollies, toggles, C-beam clamps, pound-in nail anchors, wedge anchors, rapid rods, and strut channels.

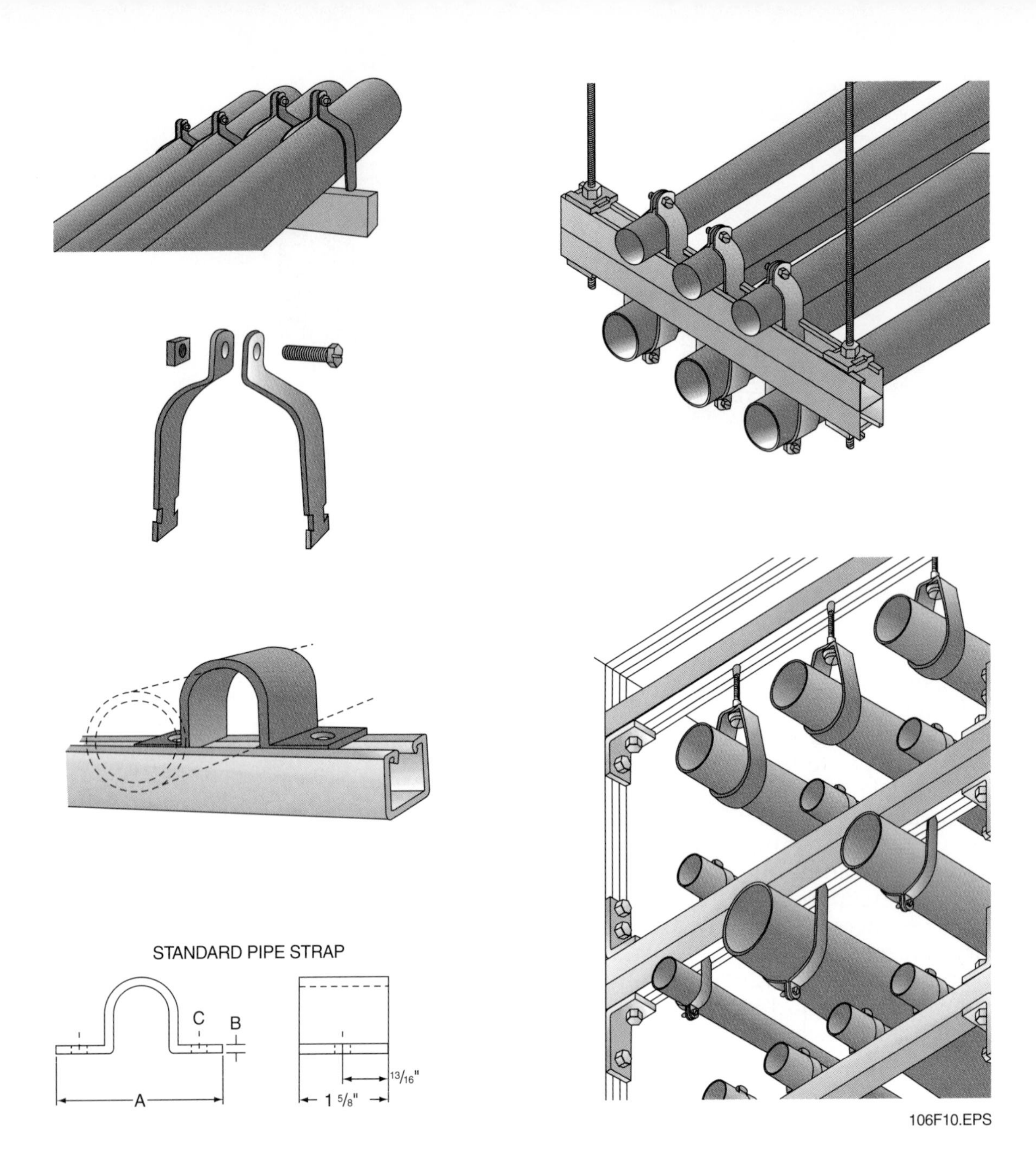

Figure 10 ◆ Universal pipe clamps and channels.

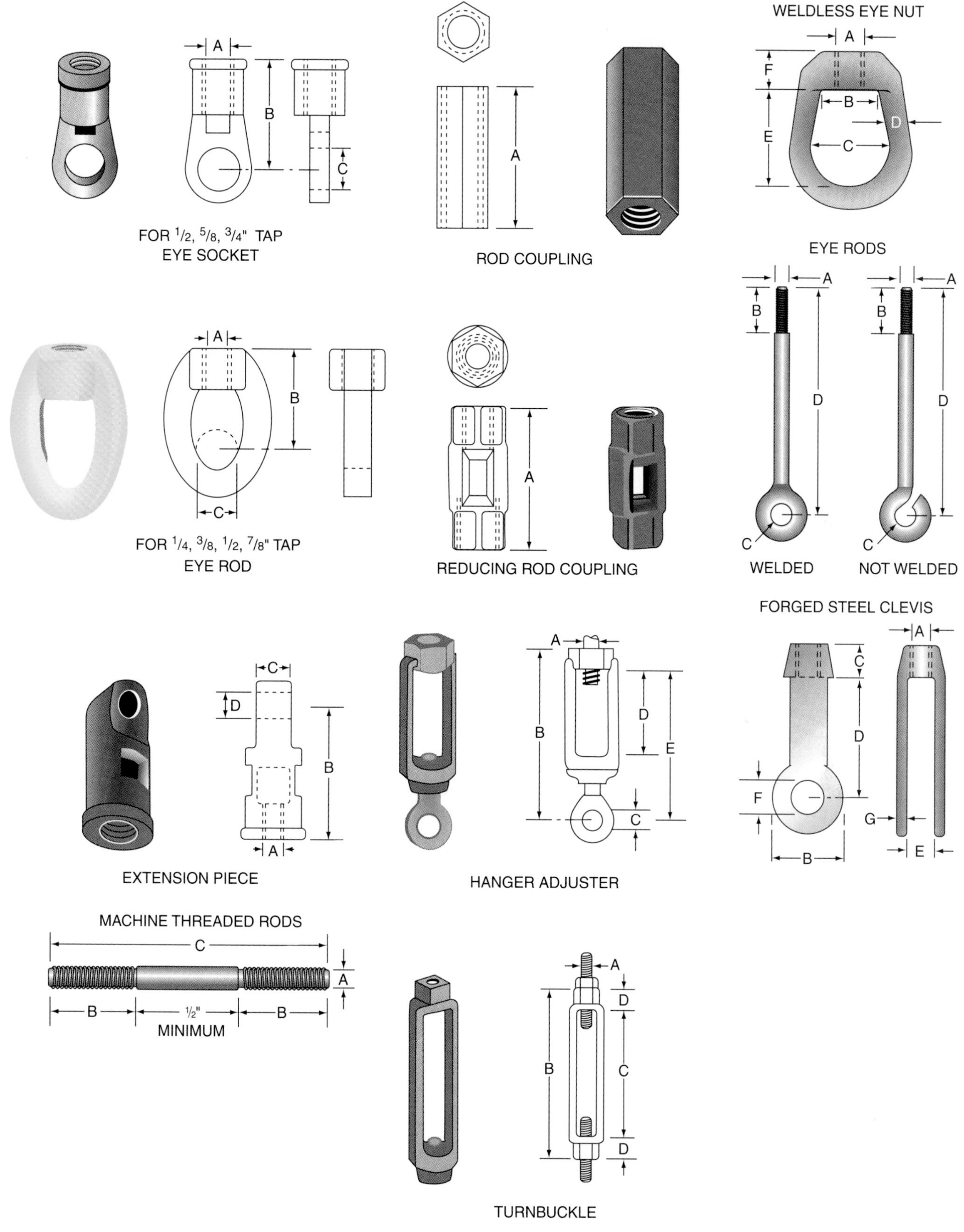

Figure 11 ◆ Rod attachment connectors.

4.2.0 Installing Hangers and Supports

What you use to install hangers and supports will vary (see *Figure 12*), depending on the item and what it is being attached to.

- Threaded drop-in anchors come in various sizes and are most commonly used in concrete ceilings, walls, or floors. Different sizes of anchors are rated for different weights or spread between anchors.
- Toggles are used for any hollow wall or ceiling support, including drywall and block.
- Drive nail anchors are used in concrete, brick, and stone.
- Rapid Rod™ hangers screw into wood. Various lengths of all-thread rod can be cut to a specified height. Common sizes of all-thread rod are ⅜ inch to ½ inch.
- C-clamps or beam clamps normally hook to iron or to a beam. These include set-bolts that must be tightened to ensure that they will not come loose.
- Channel (strut) comes in various lengths, widths, and heights and can be bolted or welded together. Various floor, wall, and overhead strut brackets are available.

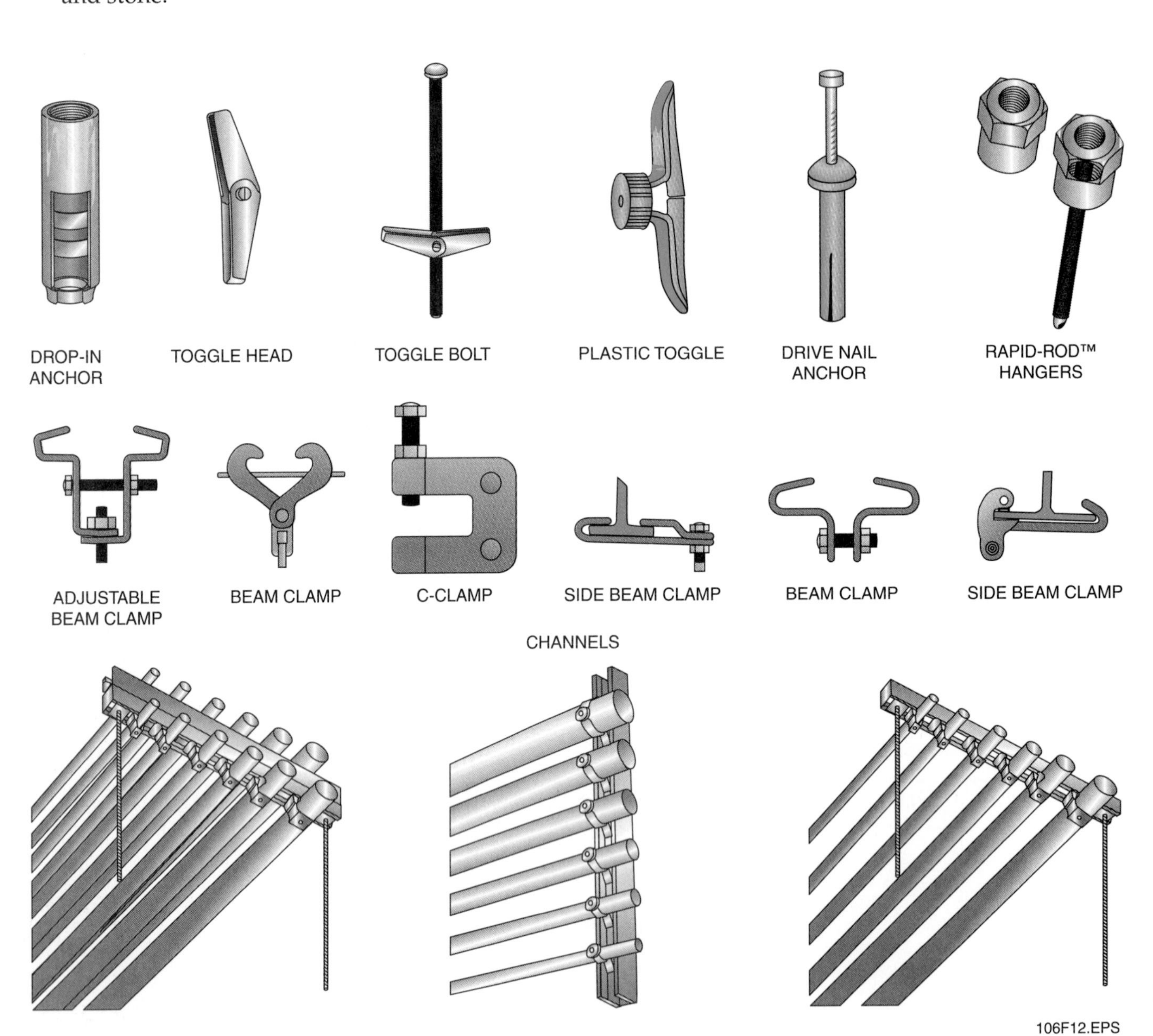

Figure 12 ◆ Installation hardware.

Review Questions

Section 4.0.0

1. Pipe hangers are used primarily to support pipe, but they can also be used as _____.
   a. pressure checks
   b. channel struts
   c. vibration isolators
   d. insulation support

2. _____ completely describe the proposed system, including the quality of the materials to be used and the work required.
   a. Specifications
   b. Schedules
   c. Blueprints
   d. Attachments

3. The principal purpose of hangers and brackets is to keep the piping in _____ and to prevent it from bending or distorting.
   a. isolation
   b. traction
   c. alignment
   d. joint

4. _____ can be attached to structural site components to support the vertical load of the pipe.
   a. J hooks
   b. Eye sockets
   c. Wedge anchors
   d. Pipe risers

5. To secure hangers and connectors to hollow wall or ceiling support, including drywall and block, use _____ as the attachments.
   a. anchors
   b. toggles
   c. rod screws
   d. C-clamps

5.0.0 ◆ MEASURING, CUTTING, REAMING, BENDING, AND JOINING

Some measuring, cutting, reaming, bending, and joining techniques are related specifically to copper. Different techniques may be used depending on what type of copper piping you're using and the function of the pipe.

5.1.0 Measuring

There are several types of measuring (see *Figure 13*):

- *End-to-end*–Measure the full length of the pipe, including both threaded ends.
- *End-to-center*–Use for pipe that has a fitting screwed on one end only; pipe length is equal to the measurement *minus* the end-to-center dimension of the fitting *plus* the length of thread engagement.
- *Center-to-center*–Use with a length of pipe that has fittings screwed onto both ends; pipe length is equal to the measurement *minus* the sum of the end-to-center dimensions of the fittings *plus* twice the length of the thread engagement.
- *End-to-face*–Use for pipe that has a fitting screwed on one end only; pipe length is equal to the measurement *plus* the length of the thread engagement.
- *Face-to-face*–Use for same situation as center-to-center measurement; pipe length is equal to the measurement *plus* twice the length of the thread engagement.
- *End-to-back*–Use for pipe that has a fitting screwed on one end only; pipe length is equal to the measurement *plus* the length of the screwed-on fitting, *plus* the length of the thread engagement.
- *Center-to-face*–Use with a length of pipe that has fittings screwed onto both ends; pipe length is equal to the measurement from the center of one of the screwed-on fittings to the face of the opposite fitting, *plus* twice the length of the thread engagement.
- *Face-to-back*–Use with a length of pipe that has fittings screwed onto both ends; pipe length is equal to the measurement, *plus* the distance from the face to the back of one screwed-on fitting, *plus* twice the length of the thread engagement.

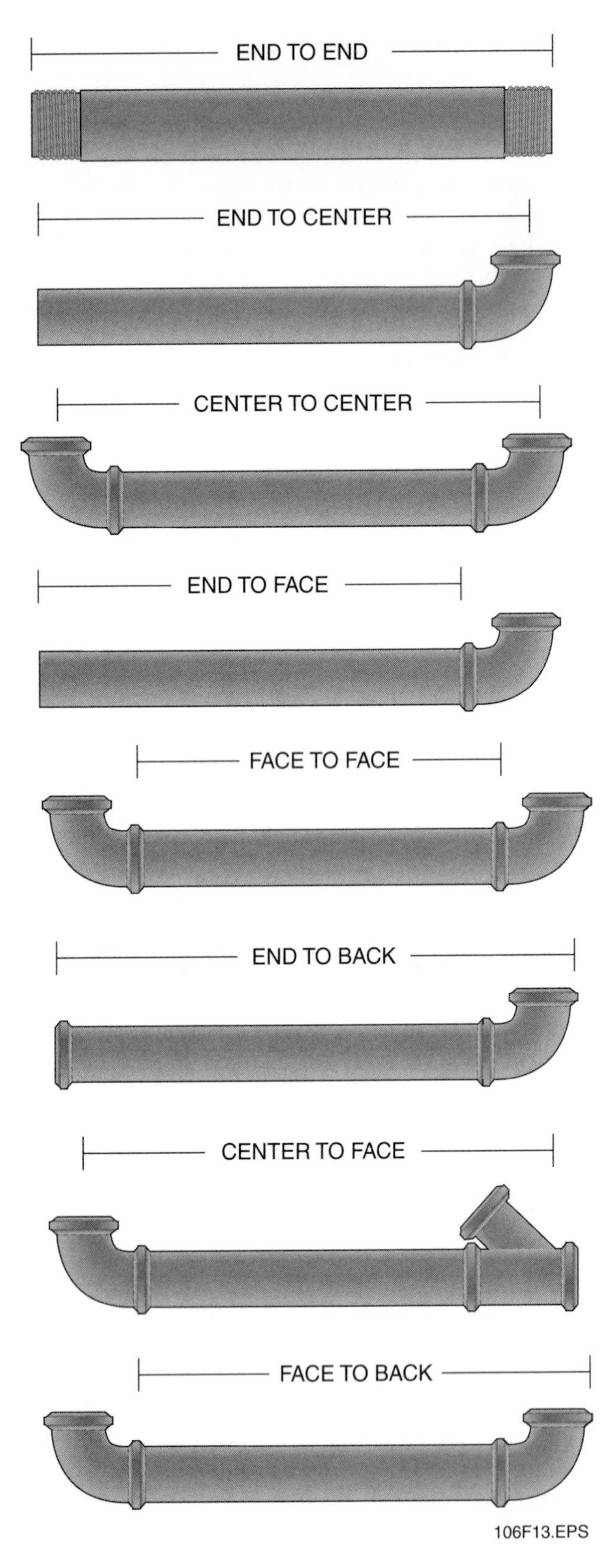

Figure 13 ◆ Measuring pipe.

5.2.0 Cutting

You can cut copper tubing with a handheld tube cutter, a hacksaw and sawing fixture, or a midget cutter. The tube cutter is preferred because it makes a cleaner joint and leaves no metal particles (see *Figure 14*). You must use a tube cutter that is the right size for the copper being cut and make sure that the proper cutting wheel is in place.

Inside cutters are used for trimming extended ends of installed water closet bowl and shower waste lines below the level of the flange (see *Figure 15*).

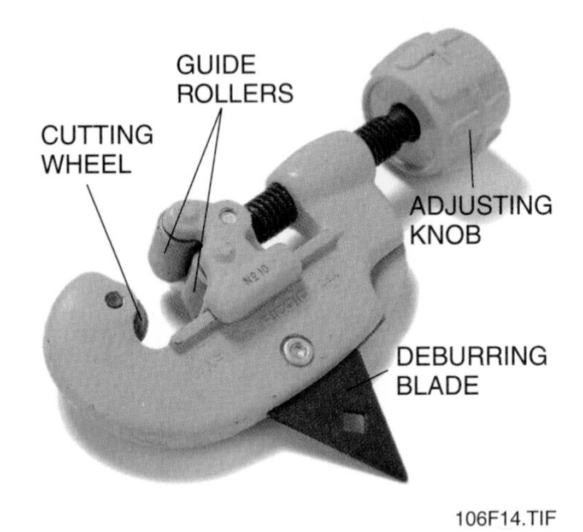

Figure 14 ◆ Handheld tube cutter.

Figure 15 ◆ Internal tube cutter.

To use the handheld tube cutter (see *Figure 16*), place it on the tube at the point where you want to cut. Tighten the knob, forcing the cutting wheel against the tube. Make the cut by rotating the cutter around the tube under constant pressure. After the cut is made, use the built-in deburring blade to remove any burrs from inside the tube. A variety of models are available, and the cutting ranges vary from ⅛-inch OD to 4⅛-inch OD.

To cut larger-size hard-drawn tubing, you can use a hacksaw and sawing fixture. This fixture helps square the ends and allows more accurate cuts. The blade of the hacksaw should have at least 32 teeth per inch. Avoid getting saw cuttings inside the tubing. File the end to produce a smooth surface, and carefully clean the inside of the tubing with a cloth.

Portable power tools with abrasive wheels to cut, clean, and buff the tubing can be used for cutting large quantities of pipe.

5.3.0 Reaming

You must ream all cut tube ends to the full inside diameter of the tube. Reaming removes the small burr (rough, inside edge) created when you cut the pipe. Burrs left on tubing can cause the pipe to corrode. A tube that is reamed correctly provides a smooth inner surface for better flow. You must also remove burrs on the outside of the tube to ensure a good fit. Tools used to ream tube ends include the reaming blade on the tube cutter, files (round or half-round), a pocket knife, and a deburring tool. An example of a deburring blade on a tube cutter can be seen in *Figure 14*.

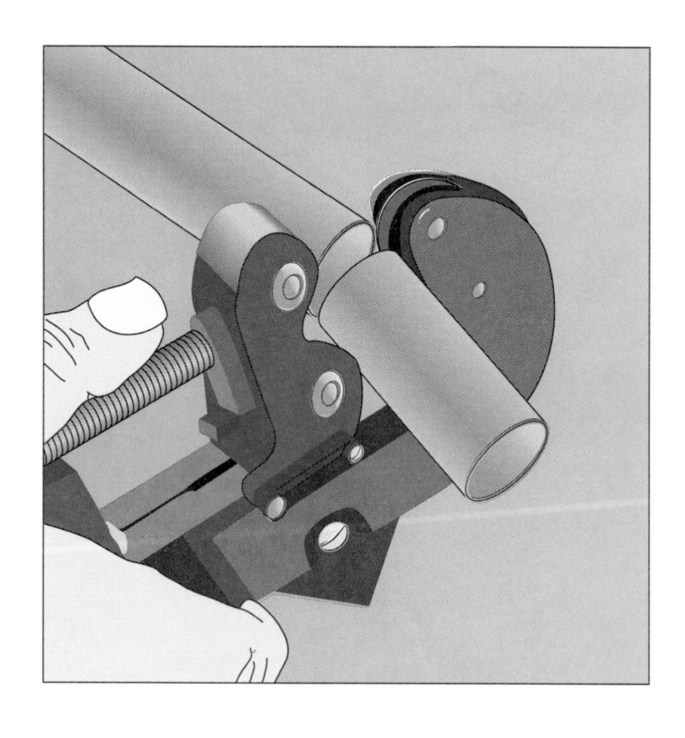

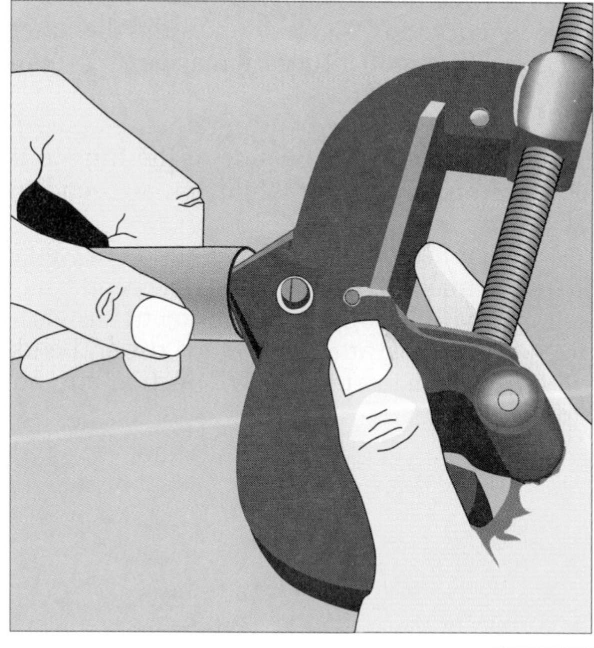

106F16.EPS

Figure 16 ◆ Using a handheld tube cutter.

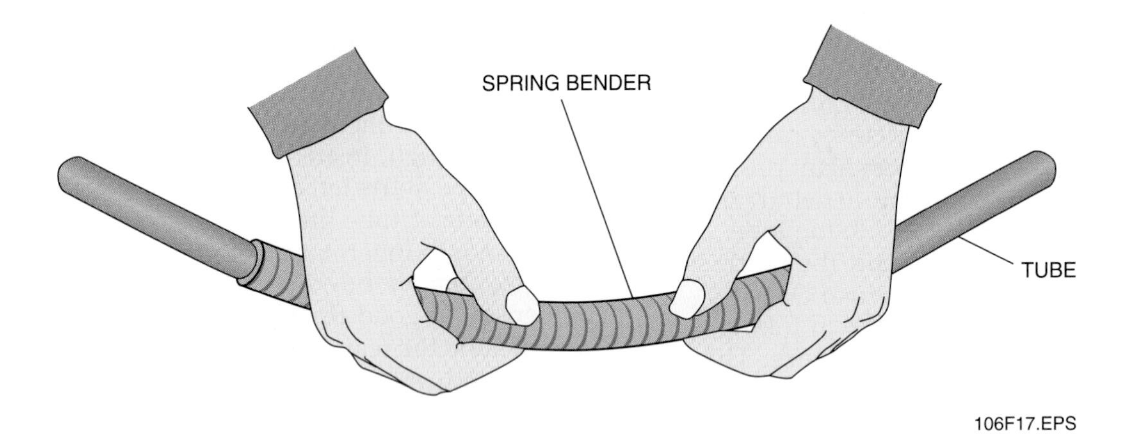

Figure 17 ◆ Tube-bending spring.

5.4.0 Bending

When you are using soft tubing, it is best to bend rather than cut it. Smaller-diameter tubing can easily be bent by hand. Take care not to flatten the tube with a bend that is too sharp. For large tubing, the minimum bending radius to which a tube may be curved is up to 10 times the diameter of the tube. For smaller tubes, it may be up to 5 times the tube's diameter.

You can use tube-bending springs (see *Figure 17*) placed on the outside of the tube to prevent it from collapsing while you are bending it by hand.

You can use tube-bending equipment (see *Figure 18*) to make accurate and reliable bends in both soft and hard tubing. This equipment has various sizes of forming attachments for use in bending tubing with diameters up to ⅞ inch at any angle up to 180 degrees.

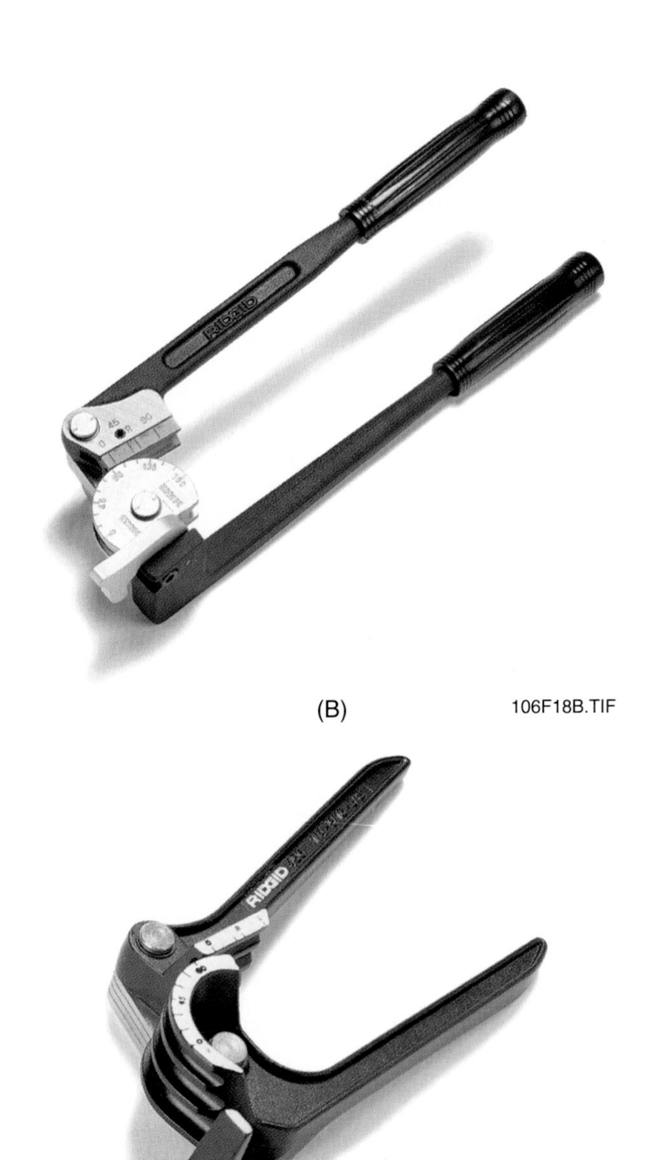

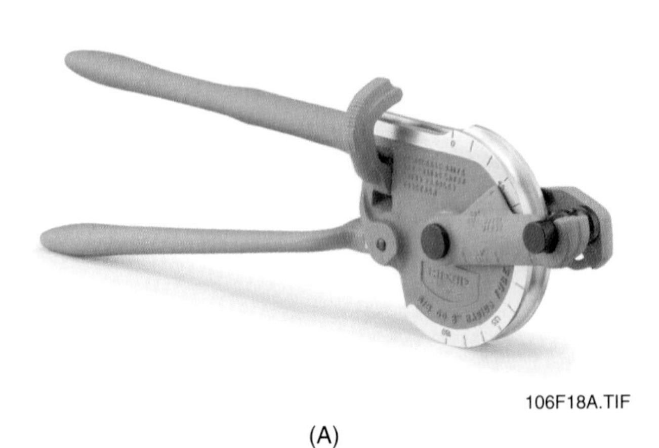

Figure 18 ◆ Tube-bending equipment. (A) Geared ratchet level-type tube bender. (B) Lever bender. (C) Lever bender for three sizes of soft tubing.

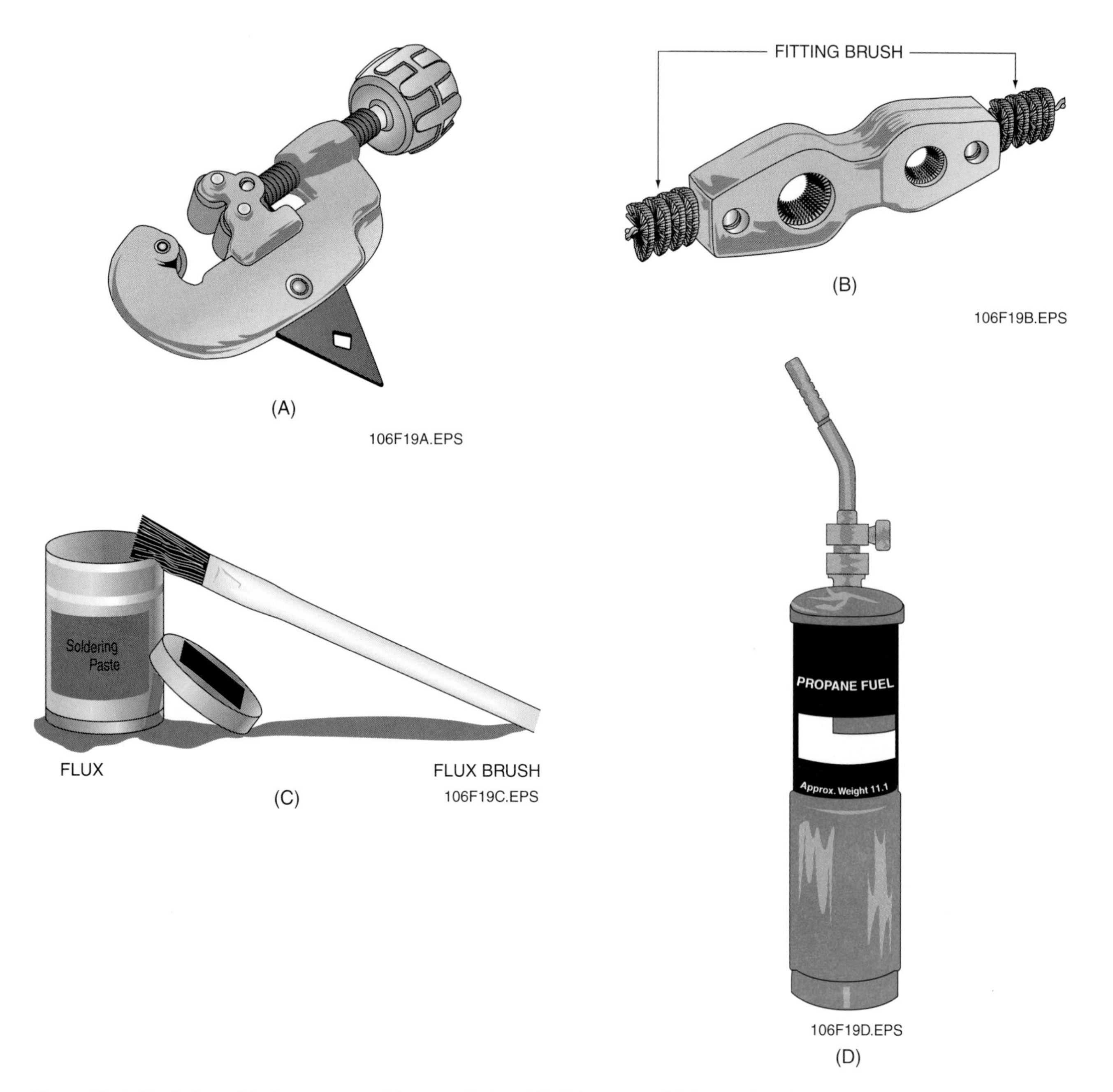

Figure 19 ◆ Tools for soldering copper tubing to a fitting. (A) Tube cutter. (B) Fitting brush. (C) Flux and flux brush. (D) Soldering torch.

5.5.0 Joining

Copper can be joined in one of four ways: a **sweat joint**, a compression joint, a **flare joint**, or a **swaged joint**.

A sweat joint is made by measuring, cutting, reaming, cleaning, and applying **flux** to the copper piping, then adding heat and soldering to a certain temperature until the solder flows into the joint. To solder copper tubing to fittings, you will use a tube cutter, fitting brush, acid or flux brush, and soldering torch (see *Figure 19*).

See *Figure 20* for an illustration of how to solder. The heat can be generated by electricity or various kinds of gases. Sweat joints are used with hard or soft copper with nominal sizes from ⅛ inch to 4 inches.

To solder copper tubing to a fitting, follow these steps:

Step 1 Measure and mark tubing to length, including the fitting allowance for the portion of the tube that will extend inside fittings.

Step 2 Use a tube cutter to cut the copper tubing.

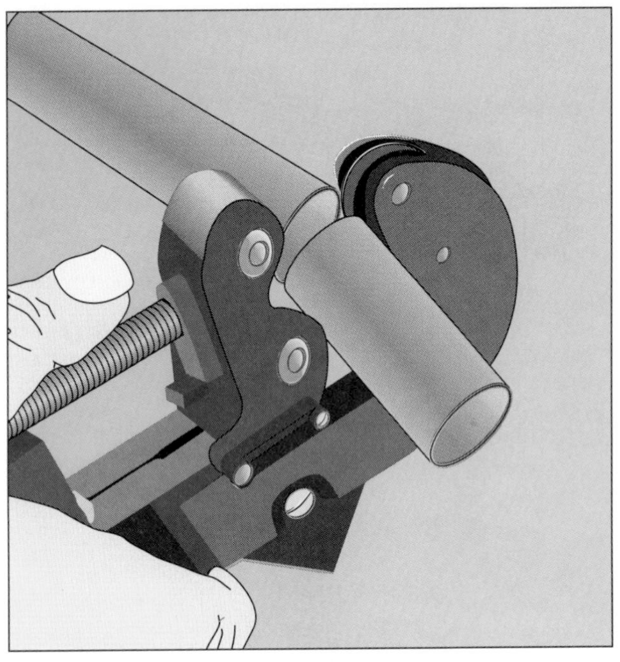

MEASURING, MARKING, AND CUTTING

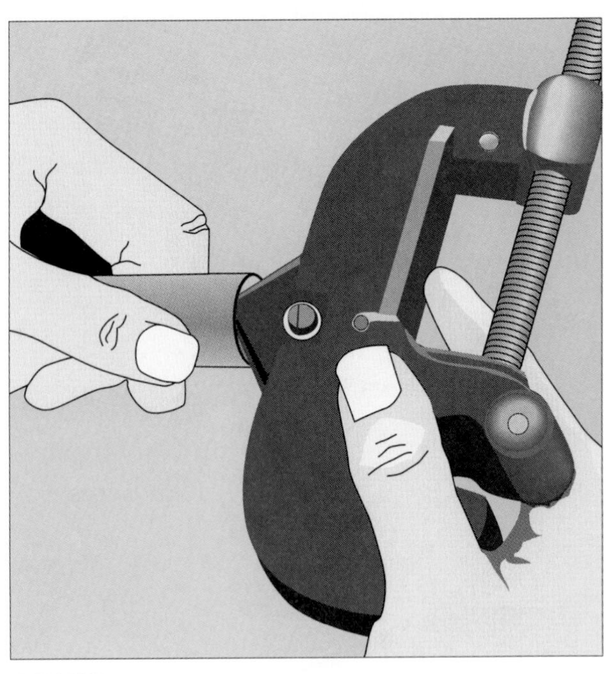

REAMING

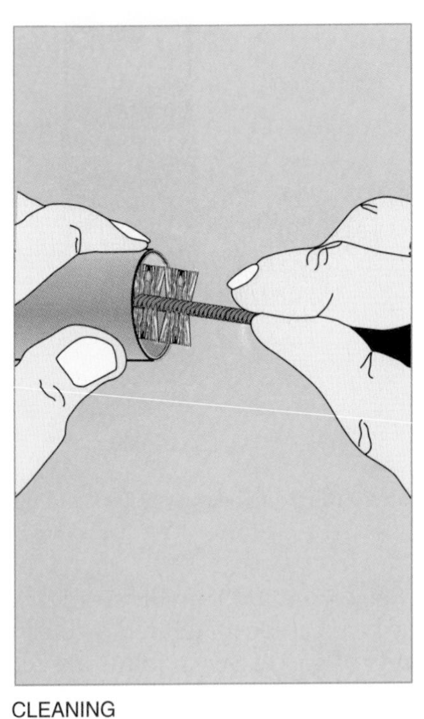

CLEANING

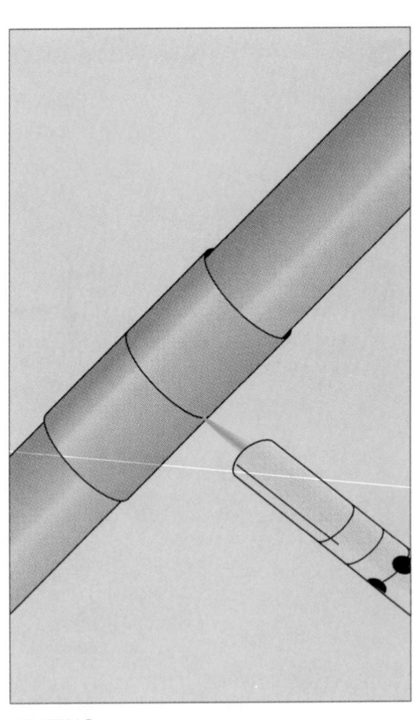

HEATING

106F20A.EPS

Figure 20 (1 of 2) ◆ Soldering.

Step 3 Use a reaming tool on the end of the tube cutter to remove the burr from the inside of the tubing.

Step 4 Clean the inside of each fitting by scouring it with a fitting brush (small wire brush).

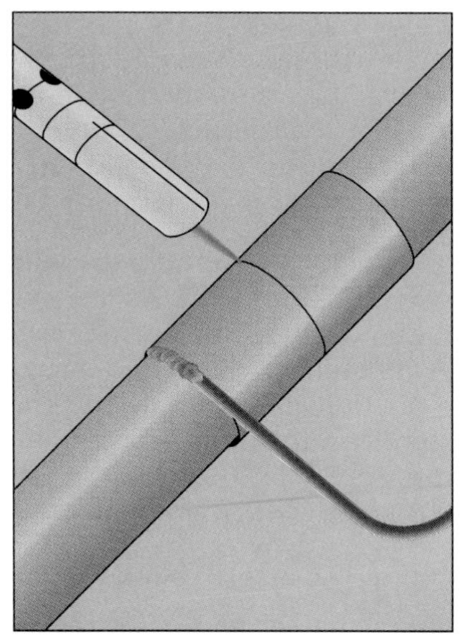
MELTING THE SOLDER

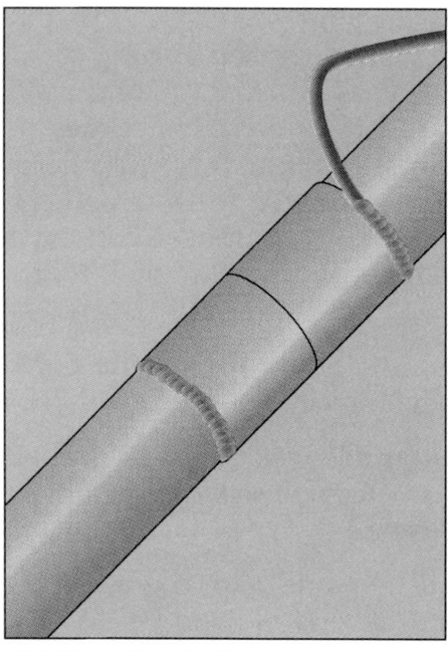
FILLING THE JOINTS

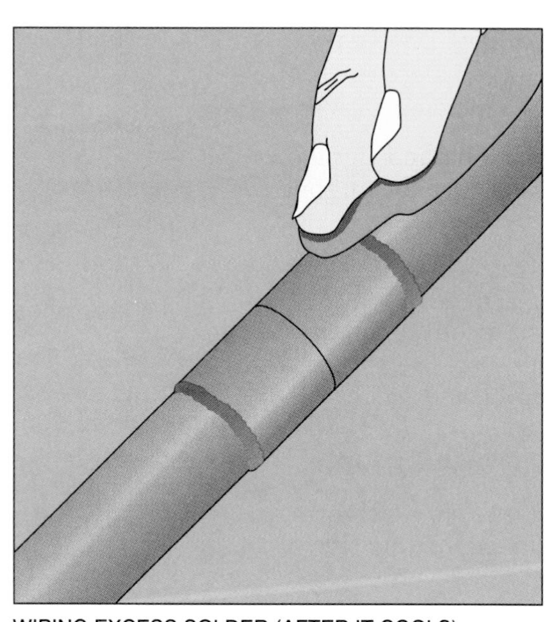
WIPING EXCESS SOLDER (AFTER IT COOLS)

106F20B.EPS

Figure 20 (2 of 2) ◆ Soldering.

Step 5 Apply a thin layer of soldering paste or flux to the end of each tube using a flux brush (small, stiff brush).

Step 6 Push the tubing into the fitting and turn it a few times to spread the flux evenly.

Step 7 Hold the tip of a soldering torch flame against the fitting until the soldering paste starts to sizzle. Then move the flame around to the other side of the fitting to heat it evenly. Some plumbers use a special soldering torch fired from a 20-pound tank of gas called a B-tank.

Step 8 When the flux starts to bubble, remove the torch and touch the wire solder to the point where the pipe enters the fitting. If the fitting and tubing are sufficiently hot, the solder will melt and be drawn into the joint. Once a line of solder shows completely around the joint, the connection is filled with solder.

A compression joint (see *Figure 21*) is a mechanical joint that is made by measuring, cutting, and reaming the pipes and using compression fittings. This kind of joint is sometimes used for joining refrigerant tubing. It is a popular method because it uses a threaded fitting and takes less time than making flared or heat-bonded connections. To create a compression joint, follow these steps:

Step 1 Measure, cut, and ream the pipe. Make sure the tube is cut square and deburred.

Step 2 Slip the nut over the tube, and slide the ring on the tube with the teeth facing the tube's end.

Step 3 Install the cone with the convex surface toward the end of the tube. To be sure the fitting goes together completely, make sure that ¼ inch of the tube extends beyond the cone when working with a ½-inch tube, and that ½ inch of the tube extends beyond the cone when using a ¾-inch tube.

Step 4 Push the nut onto the fitting and tighten. When the fitting squeaks, turn the nut one more full turn.

A flare joint is made by measuring, cutting, and reaming the pipes and using a flare fitting. This kind of joint is commonly used to join soft copper tubing from ¼ inch to 2 inches, because soft copper fittings should be leakproof and easily dismantled with the right tools.

Two kinds of flare fittings are popular:

- The single-thickness flare (see *Figure 22*) forms a 45-degree cone that fits up against the face of a flare fitting.
- A double-thickness flare (see *Figure 23*) is preferable with larger-size tubing when a single flare may be weak, such as under excessive pressure or expansion. Also, double-thickness flare connections can be taken apart and reassembled more easily without damage.

For both single- and double-thickness flares, a special flaring tool (see *Figure 24*) is used to expand the end of the tube outward into the shape of a cone, or flare.

In one operation, the single-thickness flare is formed, and then the lip is folded back into itself and compressed.

To make a flare joint, follow these steps:

Step 1 Cut tubing to the desired length.

Step 2 Slip the flare nut onto the cut tubing.

Step 3 Use the flaring tool to flare the tubing's ends so the fit is perfect.

Step 4 Slide the nuts onto each end of the flare tubing, then gently bend or shape the tubing by hand on the male thread of each fitting.

Step 5 Use a smooth-jaw adjustable wrench to tighten until the fitting is snug.

Step 6 Test the joint for leaks.

A swaged joint (see *Figure 25*) is created with a swaging tool, which is similar to a flaring tool. The block of the tool holds the tube, and the punch is forced into the end of the tubing. Swaging takes more time, but when it is properly done and soldered, it produces very secure soldered joints, reducing leak hazards.

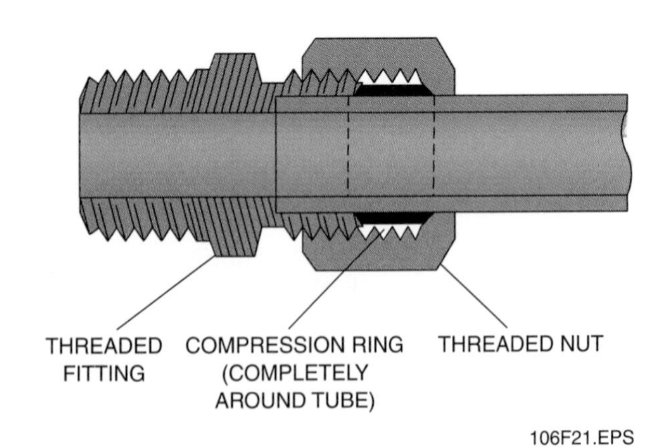

Figure 21 ◆ Compression joint.

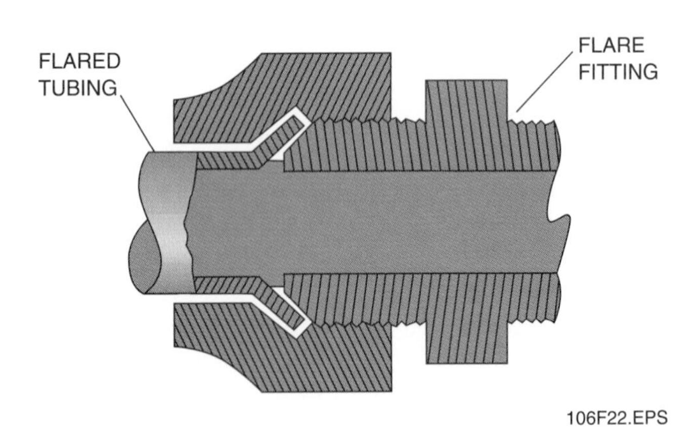

Figure 22 ◆ Single-thickness (angle) flare.

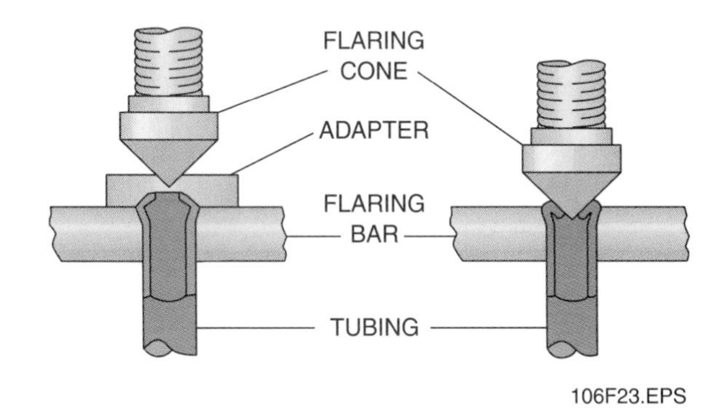

Figure 23 ◆ Double-thickness flare.

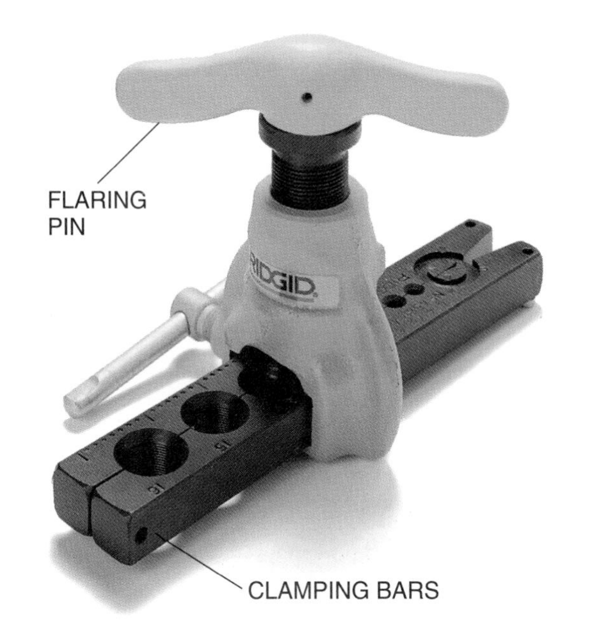

Figure 24 ◆ Flaring tool.

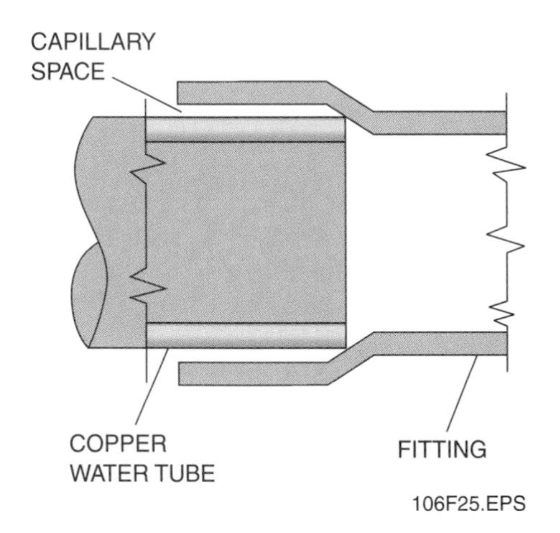

Figure 25 ◆ Swaged joint.

6.0.0 ◆ SAFETY

Always practice safety precautions to protect yourself and your co-workers from injury and to prevent equipment damage. Here are some general precautions to take when working with copper, and some precautions specific to the installation process:

- Do not try to do any work that you have not been specifically trained to do. Always work under the direct supervision of your instructor or foreman. For example, do not operate valves unless you know exactly what the result will be. Before you work independently on a system, you must know the temperature and pressure conditions that exist at every point in the system, and how those conditions can be affected by malfunctions or by changes in valve positions.
- Do not disconnect piping in a system that is under pressure; this action can cause serious injury.
- Wear goggles and gloves when you work with refrigerants. Do not work with refrigerants in poorly vented areas. Also, be aware that some refrigerants become toxic when they are exposed to an open flame.
- Always shut off electrical power unless it is absolutely necessary to work with it on. When you shut off electrical power, always use approved lockout/tagout procedures to avoid electrical shock. Check with your immediate supervisor before proceeding whenever electrical power is applied.

Insulating

Under certain temperature and humidity conditions, condensation will form on cold-water piping and may drip into equipment or occupied areas. Similarly, heat can escape from hot-water piping. To prevent these conditions, some piping is insulated.

Insulation is a material that prevents the transfer of heat. Cork, glass fibers, mineral wool, and polyurethane foams are examples of insulating materials. Insulation should be fire-resistant, moisture-resistant, and vermin-proof.

If the insulation cannot be installed before the tubing is connected, it must be split lengthwise to fit onto the pipe. Split seams and connecting seams must then be sealed (see *Figure 26*). Insulation should not be stretched, because its effectiveness will be reduced.

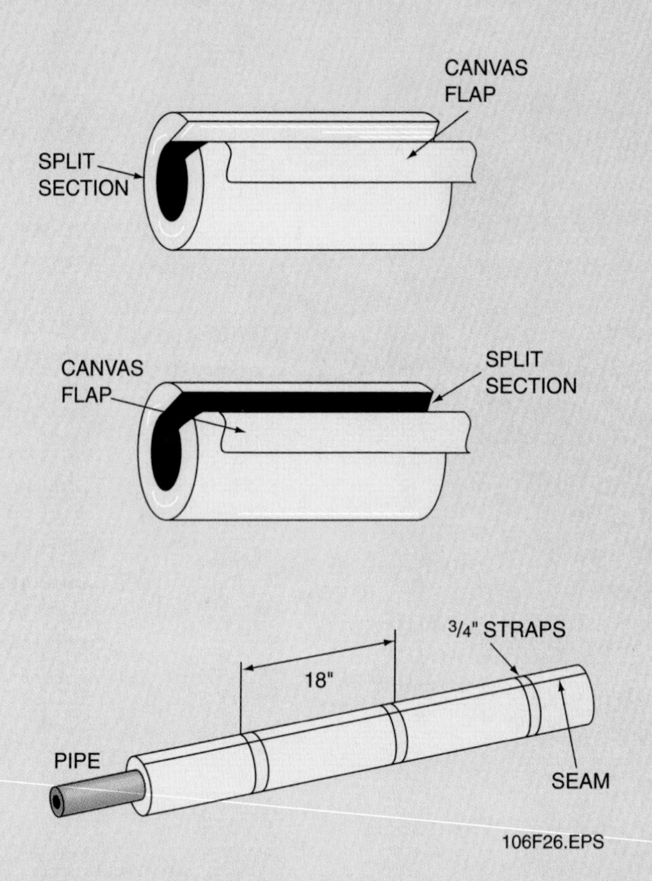

Figure 26 ◆ Installing insulation.

Some pipes are always insulated, and others are insulated only under certain conditions. Local building codes and job specifications will describe insulation requirements. Which pipes to insulate and under what conditions will be covered in detail in a later module.

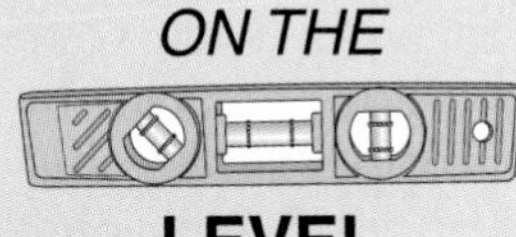

Testing

A completed installation must be inspected to make sure that the work has been done according to the job specifications and applicable codes. Then the system must be leak tested to be sure that all connections are secure.

The plumber is required to supply all the equipment needed for testing. This includes test plugs, caps, plugs, and a source of compressed air (if air testing is required). Water testing can be done with test plugs and a hose. To test, close all outlets to the piping system with test plugs, caps, and plugs. Fill the system with water or air under pressure and look for leaks.

Water testing the DWV system requires a minimum of 10 feet of water head. To do this, fill the vent stack completely. Once the DWV piping is filled, inspect the whole piping system for leaks. This inspection is required before any of the system can be covered. The plumbing inspector will also be checking for cross connections, defective or inferior materials, and poor work.

To water test the water supply system, close all outlets and fill the system with water from the main. Inspect the system for leaks or other potential problems. Where codes specify water testing at pressures higher than those available from the normal water supply, the plumber can use a hydraulic test pump to produce the required pressure (see *Figure 27*).

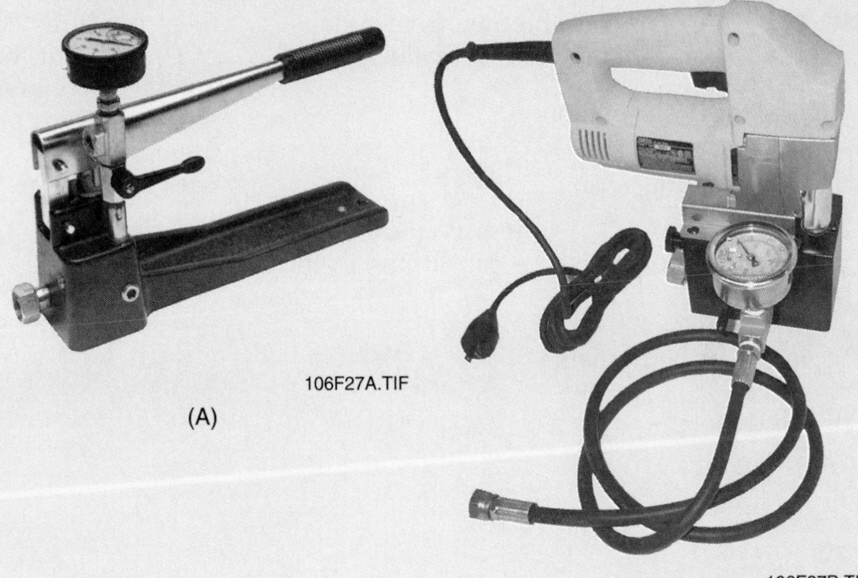

Figure 27 ◆ Hydraulic test pumps. (A) Hand-operated. (B) Portable.

Air testing is similar to water testing, except the piping is filled with compressed air. A pressure gauge at the test plug allows the inspector or plumber to determine if any pressure is lost somewhere in the piping. Soapsuds are useful for detecting leaks. The suds are brushed onto joints and any escaping air forms bubbles.

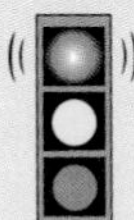

WARNING!

Oxygen, acetylene, or other gases should never be used to pressure-test a refrigerant system. Oxygen will cause an explosion if it comes into contact with refrigerant oil. Acetylene is highly flammable. The only gas other than refrigerant that should be introduced into a system is nitrogen.

Review Questions

Sections 5.0.0–6.0.0

1. A(n) _____ joint is commonly used to join soft copper tubing from ¼ inch to 2 inches.
   a. compression
   b. flare
   c. sweat
   d. angle

2. A _____ joint is a mechanical joint that is made by measuring, cutting, and reaming.
   a. swaged
   b. sweat
   c. compression
   d. heated

3. A(n) _____ measure is the full length of the pipe, including both threaded ends.
   a. face-to-face
   b. end-to-end
   c. center-to-center
   d. face-to-end

4. A _____ helps prevent the tube from collapsing while you are bending it by hand.
   a. tube-reaming tool
   b. hand-held burring blade
   c. tube-shielding cover
   d. tube-bending spring

5. Do not operate _____ unless you know the temperature and pressure conditions in the system.
   a. power tools
   b. soldering equipment
   c. valves
   d. hand tools

Summary

A plumber must be able to work with several kinds of piping, including copper. You have learned about soft and hard copper tubing and its uses. You have also learned about the various fittings and valves that plumbers use to join copper tubing for water distribution systems and DWV systems. You now know that piping must be properly supported and insulated, and you know the different kinds of hangers and supports that you can use. The requirements for pipe hangers and insulation are usually specified in the job specifications or by local building codes.

Copper can be joined in four ways: a sweat joint, a compression joint, a flare joint, and a swaged joint. This module has introduced you to all four methods of joining copper pipe, as well as how to measure, cut, ream, and bend pipe that you are going to join.

A completed piping installation must be inspected and tested. At this point, you will not be doing the testing, but you should be familiar with the procedures.

Throughout this module you have read about special safety practices to prevent injury to yourself and others or damage to equipment. Follow these practices as you go into the field and work with copper pipe and fittings.

PROFILE IN SUCCESS

Michael McConnaha

Plumbing
Field Foreman, Tank and Piping Contractors
Carmel, IN

Michael was born and raised in Lebanon, Indiana. He left Lebanon for a couple of years, but decided there wasn't a nicer place to live and went back to stay. A 15-year construction veteran, he has done everything from HVAC to cabinetry to plumbing. Currently, he is a field foreman for Tank and Piping Contractors and teaches apprenticeship training as well.

How did you become interested in the construction industry?
During high school I worked in a wood shop because I always liked working with my hands. I went to college for about year and a half on a football scholarship, but felt I was wasting my time. Even though my parents are college graduates, I didn't want to continue in college. At a career night presentation, I learned about a technical school for HVAC training. I liked the sound of it and enrolled in their one-year intensive training program. I started working full-time at a cabinet shop while going to school five days a week.

After graduating, I tried working in the HVAC trade, but the construction industry was down. I wanted to make more money, so I decided to go back to cabinetry. At that age, money is more important than planning for a career.

Shortly after that, two friends in the plumbing trade told me they were looking for plumbers and asked if I would be interested. They knew about my construction experience and figured I could learn plumbing skills quickly. About this time, I had started to think seriously about having a career, and not just bounce from job to job, so I signed on with them. I worked with a journeyman and also enrolled in the ABC apprenticeship-training program that uses NCCER's Standardized Craft Training.

My mentor is a third-generation plumber who is the plumbing superintendent at Tank and Piping. He knows a lot and has been a great help to me. I've passed my journeyman's exam and am currently running jobs as the plumbing field foreman on different sites. I took my master plumber exam this year and expect the results shortly.

What do you think it takes to be a success in your trade?
First, I would say to be a success at any trade—not just plumbing—you have to be willing to work. Any career takes devotion and hard work, and you have to be willing to learn lots of different skills. There is so much to plumbing—commercial, residential, industrial. You have so many opportunities to develop different areas such as process piping; hazardous waste systems; water distribution and supply; drainage, waste, and vent systems; maintenance and repair—and the list goes on. It's better not to settle on any one aspect at the beginning.

I recommend apprenticeship-training programs because you get exposed to the many different facets of plumbing. You learn skills that you may not use today but that could be valuable in a future job.

What are some of the things you do in your role of field foreman?
At Tank and Piping, we are working foremen. That means that in addition to supervising the apprentices and helpers, we actually help with the installation and whatever other work is going on. I review the job blueprints and take-offs to plan the layout and execution of the different jobs. That's my favorite part. By understanding the blueprints, I can figure out what is going where and how. From there I schedule the workflow.

We do many different types of jobs. We just finished a truck stop project that involved installing showers, restrooms, a restaurant, a coffee shop, and more. A few weeks ago we installed the gas piping for a commercial machine shop. The variety of jobs is a great part of the challenge.

I also teach a four-hour apprenticeship training class one night a week. It's a lot of work, but very satisfying.

What do you like most about your job?
I like layout and design. My favorite job is taking the blueprints and figuring out how to get the job done. The best challenge is working on mechanical rooms. In a mechanical room, all the piping comes from all over the structure into this one small room. The pipes hook up to machines such as a boiler, water treatment and softener, and compressors. The art of it comes in when you have all those pipes laid out, and the room looks great. That's the best.

What would you say to someone entering the trades today?
Give it time. Expect ups and downs. Working in the trades is like any other job—you may not like it every single day. But it's fun and less restrictive than working in an office. And it's rewarding—both for the money and the pride you get from accomplishing so much. What I would say to someone entering the trades is, "Don't give up too early." You may have a rotten week near the beginning, but know that it will get better. Stick it out.

As an apprentice, you'll need to learn to work with the journeyman overseeing the job. Some want to stand over you every minute; others will tell you what needs to be done and leave you to it. You can learn from both types. The most important thing to remember is if you don't understand, *ask!* The key to success is learning to do a job right. That is true even if you make mistakes. Mistakes are a great learning experience.

GLOSSARY

Trade Terms Introduced in This Module

ACR: Air-conditioning and refrigeration system.

ACR tubing: Annealed copper tubing that is manufactured specifically for use in air-conditioning and refrigeration work.

Annealed copper: Soft copper tubing produced through the annealing process of heating a material and slowly cooling it to relieve internal stress. This process reduces brittleness and increases toughness.

Brazing: Hard soldering; a way to join metal using an alloy with a melting point higher than that of solder, but lower than that of the metals being joined.

Bullheading: Caused when the branch pipe of a tee is larger than the main line (for example, 1 × 1 × 2) and the flow of liquid hits the back wall of the tee. A bullhead is a tee fixture in which the main line is smaller than the branch.

Clevis: An iron (or link in a chain) bent into the form of a horseshoe, stirrup, or letter U, with holes in the ends to receive a bolt or pin.

Compression joint: A method of connection in which tightening a threaded nut squeezes a compression ring to seal the joint.

Drop-forged: A product made when heated metal is pounded or shaped between dies with a drop hammer or press.

Ferrule: A tube or metallic sleeve, fitted with a screwed cap to the side of a pipe to provide access for maintenance.

Flare joint: A fitting in which one end of each tube to be joined is flared outward using a special tool. The flared tube ends mate with the threaded flare fitting and are secured to the fitting with flare nuts.

Flux: A substance that helps fuse metals and prevents the metal from oxidizing, which would damage the metal.

Hammering: Hydraulic shock caused by rapid velocity changes in liquids passing through tubes or pipes, usually as a result of fast-acting valves interrupting the flow of water. The resulting back pressure hits the elbow and tees with a powerful force, sometimes banging them into nearby structural members.

Inside diameter (ID): The distance between the inner walls of a pipe. The standard measure for tubing used in heating and plumbing applications.

Insulation: A substance that retards the flow of heat.

Nominal size: Approximate measurement in inches of the inside diameter (ID) of pipe for most copper pipes. However, nominal size of ACR tubing is based on the outside diameter (OD).

Outside diameter (OD): The distance between the outer walls of a pipe. Used as the standard measure for ACR tubing.

Pressure drop: A decrease in pressure from one point to another caused by friction losses in a fluid system, such as a water system.

Roll-grooved: Process of making copper piping that is compatible with grooved copper fittings.

Solder: An alloy, usually having a lead or tin base, which is used to join metals by fusion.

Specification: A document that describes the quality of the materials and work required. Specifications are the source for the quality of tubing, fixtures, hangers, and so forth to be used on a project.

Swaged joint: A pipe joint in which the diameter of one of the pipes to be joined is expanded using a special tool. The other pipe then fits inside the swaged pipe.

Sweat joint: A pipe joint made by applying solder to the joint and heating it until it flows into the joint.

Tubing: (1) A hollow cylinder, which may be metallic or nonmetallic, used for conveying fluids. The tubing wall varies in thickness depending on the type of material, such as plastic, steel, or copper. (2) A conduit used to protect runs of electrical wire.

ANSWER KEY

Answers to Review Questions

Section 2.0.0
1. a
2. b
3. c
4. b
5. a

Section 3.0.0
1. c
2. d
3. a
4. b
5. a

Section 4.0.0
1. c
2. a
3. c
4. d
5. b

Sections 5.0.0–6.0.0
1. b
2. c
3. b
4. d
5. c

Additional Resources

For advanced study of topics covered in this task module, the following books are suggested:

The Copper Tube Handbook. New York: Copper Development Association, 1995.

Modern Plumbing. E. Keith Blankenbaker. South Holland, IL: The Goodheart-Willcox Company, Inc., 1997.

Modern Refrigeration and Air Conditioning. Andrew D. Althouse, Carl H. Turnquist, and Alfred F. Bracciano. South Holland, IL: The Goodheart-Willcox Company, Inc., 1996.

Pipefitter's Handbook, Third Edition. Forrest R. Lindsey. New York: Industrial Press, Inc., 1967.

Refrigeration and Air-Conditioning Technology, Second Edition. William C. Whitman, William M. Johnson, and Bill Whitman. Albany, NY: Delmar Publishers, Inc., 1995.

NCCER CRAFT TRAINING USER UPDATES

The NCCER makes every effort to keep these textbooks up-to-date and free of technical errors. We appreciate your help in this process. If you have an idea for improving this textbook, or if you find an error, a typographical mistake, or an inaccuracy in the NCCER's Craft Training textbooks, please write us, using this form or a photocopy. Be sure to include the exact module number, page number, a detailed description, and the correction, if applicable. Your input will be brought to the attention of the Technical Review Committee. Thank you for your assistance.

Instructors – If you found that additional materials were necessary in order to teach this module effectively, please let us know so that we may include them in the Equipment/Materials list in the Instructor's Guide.

Write: Curriculum Revision and Development Department
National Center for Construction Education and Research
P.O. Box 141104, Gainesville, FL 32614-1104

Fax: 352-334-0932

E-mail: curriculum@nccer.org

Craft ______ Module Name ______

Copyright Date ______ Module Number ______ Page Number(s) ______

Description ______

(Optional) Correction ______

(Optional) Your Name and Address ______

Module 02107-00

Cast-Iron Pipe and Fittings

Course Map

This course map shows all of the modules in the first level of the Plumbing curriculum. The suggested training order begins at the bottom and proceeds up. Skill levels increase as you advance on the course map. The local Training Program Sponsor may adjust the training order.

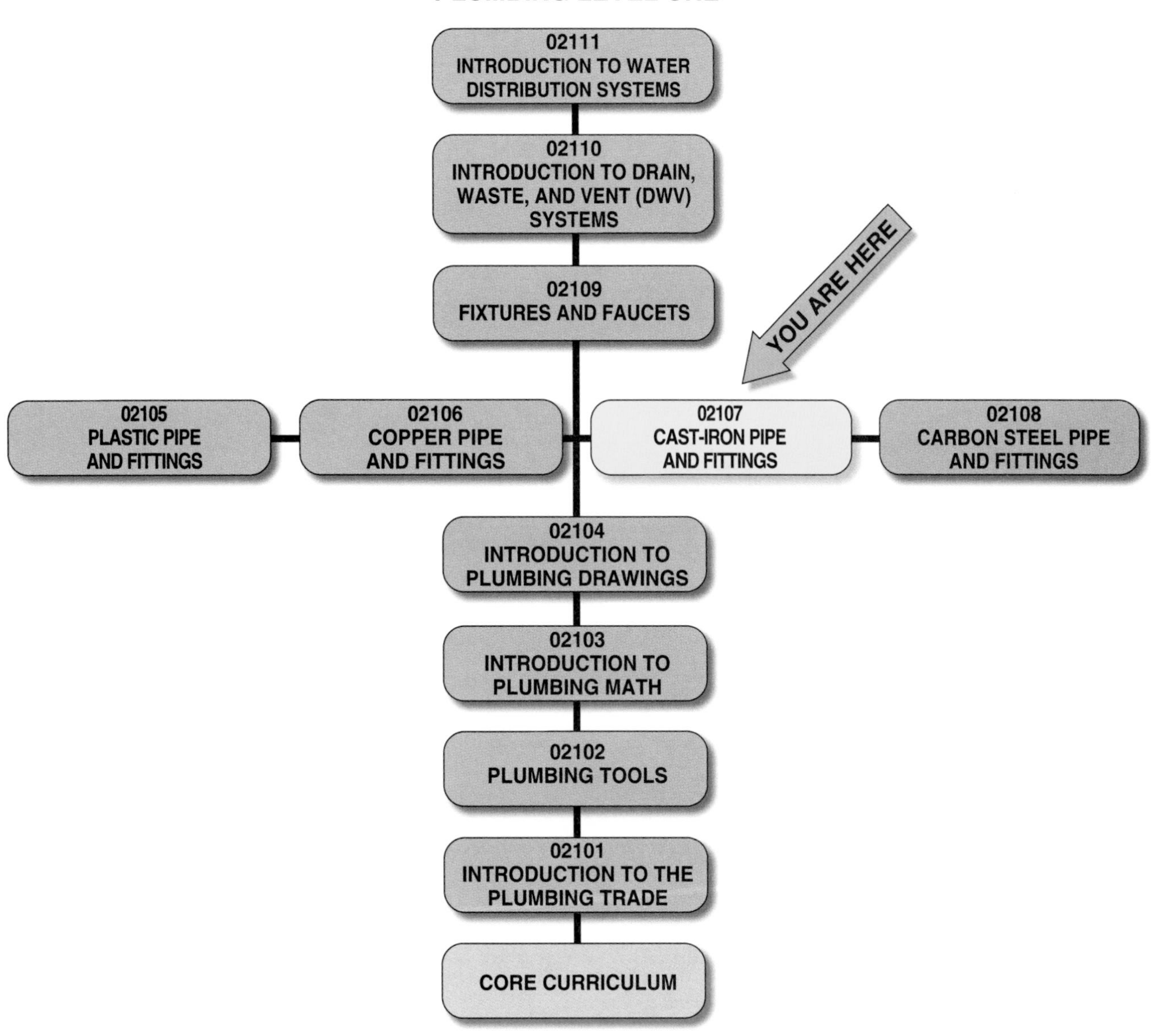

Copyright © 2000 National Center for Construction Education and Research, Gainesville, FL 32614-1104. All rights reserved. No part of this work may be reproduced in any form or by any means, including photocopying, without written permission of the publisher.

MODULE 02107 CONTENTS

Figures

Table

MODULE 02107

Cast-Iron Pipe and Fittings

Objectives

When you finish this module, you will be able to:

1. Identify the common types of materials and schedules used with cast-iron piping.
2. Identify the common types of fittings used with cast-iron piping.
3. Identify the various techniques used in handling and supporting cast-iron piping.
4. Demonstrate the ability to properly measure, cut, and join cast-iron piping.
5. Identify the hazards and safety precautions associated with cast-iron piping.

Prerequisites

Before you begin this module, it is recommended that you successfully complete task modules: Core Curriculum; Plumbing Level One, Modules 02101 through 02106.

Materials Required for Trainees

1. This module
2. Appropriate personal protective equipment
3. Sharpened pencil and paper

1.0.0 ◆ INTRODUCTION

Cast-iron pipe, often called soil pipe, is used for drain, waste, and vent (DWV) systems in residential, commercial, and industrial plumbing. Cast iron is a strong and durable piping material that resists loads, corrosion, and abrasion. It also can withstand extreme temperature changes.

This module introduces **hub-and-spigot cast-iron pipe** and fittings and **no-hub cast-iron pipe** and fittings. It discusses sizes, weights, joining procedures, and supporting techniques.

2.0.0 ◆ MATERIALS

Hub-and-spigot cast-iron pipe is shown in *Figure 1*. Notice the double-hubbed section of pipe. This pipe can be cut into two single-hub pipes.

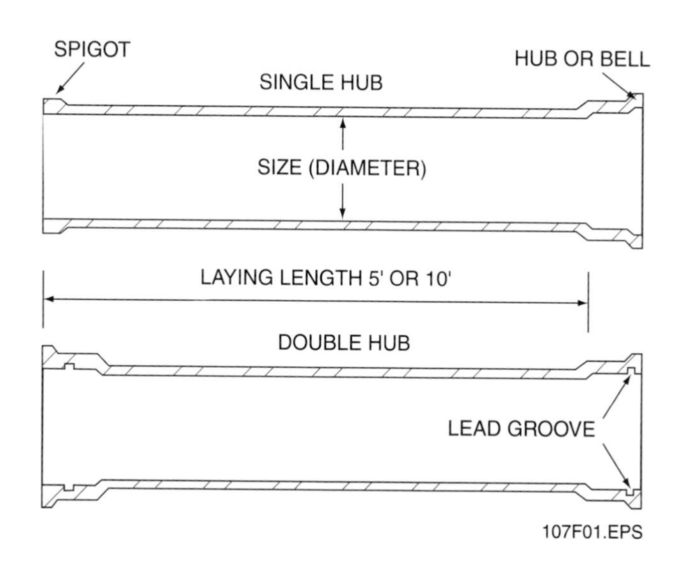

Figure 1 ◆ Hub-and-spigot cast-iron pipe.

Did You Know?

Iron is one of the oldest known metals. Prehistoric people used iron from meteorites to make tools, weapons, and other objects. The Iron Age began about 1200 B.C.E. in the Near East. It was marked by refined iron tools and weapons. Iron was the backbone of the Western world's growing technology until the 1800s.

Iron was the main industrial metal until the Industrial Revolution. In the mid-1800s, an economical way of making steel (a mixture of iron and carbon) was discovered. Steel soon replaced iron as the basic material for making tools and equipment for industry.

Rock with high concentrations of iron is called *iron ore.* This ore is mined from beneath the earth's surface. To convert iron ore into iron products, the ore is refined to remove impurities. Refining iron ore takes extreme heat and a reducing agent to draw oxygen out of the iron ore. Coke is a grayish-black substance composed of mostly carbon. When the coke is burned, it gives off intense heat. The burning coke draws the oxygen out of the iron ore. Limestone is also mixed in and used to remove impurities from the ore. The limestone soaks up unwanted minerals such as sulfur and phosphorus. The result is refined iron that can be used to make iron products.

Cast iron is melted iron poured into molds or shapes. Any type of iron formed by casting or using a mold is called cast iron.

Cast-iron pipe was first introduced in the United States at the beginning of the 1800s, when it was used to replace rotting wooden piping in water supply and gaslight systems in Pennsylvania. Since that time, cast-iron pipe has been used extensively in the plumbing industry.

Different kinds of iron are used to make cast-iron pipe and fittings. Gray cast iron, with a controlled carbon content, has been used for many years for water main piping and sanitary waste piping. The disadvantage of gray cast iron is its brittleness. In 1955, **ductile iron** pipe was introduced and has since become the industry standard for modern water and wastewater systems. The addition of magnesium or cerium (minerals) to molten gray iron causes the graphite flakes to form globular nodules. This change makes ductile iron less brittle than gray iron. Ductile iron is used for cast-iron pipe and fittings.

No-hub cast-iron pipe is shown in *Figure 2.* As the name suggests, this kind of pipe does not have a hub at either end. It is joined with a **compression joint.** No-hub cast-iron pipe and fittings are used in all areas of plumbing where cast iron is recommended. Although it is a relatively new method for connecting cast-iron pipe, no-hub is very popular. This method was not meant to replace hub-and-spigot cast-iron pipe. No-hub pipe was produced to provide the plumber with an easier method of joining cast iron in areas where space is limited.

2.1.0 Sizes

Hub-and-spigot cast-iron pipe is manufactured in different lengths and diameters. Hub-and-spigot is made in lengths of 5 feet or 10 feet and in diameters of 2 inches, 3 inches, 4 inches, 5 inches, 6 inches, 8 inches, 10 inches, 12 inches, and 15 inches. The smaller diameters are used in residential construction. Many types of fittings are not manufactured for pipe sizes larger than 8 inches. Generally, when sizes above 15 inches are required under a building, ductile iron pipe is specified.

The double-hub pipe, as shown in *Figure 1,* is available in 30-inch lengths. This pipe can be cut into two short single-hub pipes, which can be useful for a job that requires many joints.

Cast-iron pipe is classified by its weight. Two weights of pipe are manufactured: **service weight** (lightweight) and extra-heavy (only available for hub-and-spigot). Service weight is the most commonly used pipe. However, if a structure is subject

107F02.EPS

Figure 2 ◆ No-hub cast-iron pipe.

Ductile Iron and Water Lines

Ductile iron is a type of cast iron in which magnesium is added to molten gray iron to reform the graphite flakes into nodular shapes. It reduces the brittleness of the cast iron and therefore holds up under the weight of the building and soil.

Most large water lines are made with ductile iron. A standard **soil pipe cutter** will not cut through ductile iron. You must use a demolition saw to cut ductile iron water pipes.

to vibration, settling, or excessive corrosion, extra-heavy pipe and fittings are recommended. Service weight is marked SV; the term *service weight* indicates the type of pipe, not its actual weight, which can vary. Extra-heavy is marked XH.

No-hub cast-iron pipe is also manufactured in lengths of 5 feet or 10 feet. It is available in diameters of 1½ inches, 2 inches, 3 inches, 4 inches, 5 inches, 6 inches, 8 inches, and 10 inches. No-hub and hub-and-spigot pipe are made to generally the same thickness.

2.2.0 Labeling

All cast-iron pipe and fittings with the C trademark (see *Figure 3*) are manufactured according to the standards set by the Cast Iron Soil Pipe Institute. According to these standards, all pipe must be measured before distribution from the manufacturer. All pipe and fittings must be labeled with the manufacturer's name or trademark, with the Institute's trademark, and with the correct diameter of the pipe or fitting (see *Figure 3*). This label is located about 1½ inches from the end of the pipe. *Figure 4* shows a typical fitting label.

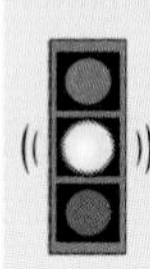

CAUTION

If the no-hub plumbing materials are not labeled correctly, they might not have been manufactured according to the standards described above. In most cases, plumbing codes do not permit unlabeled connections to be installed.

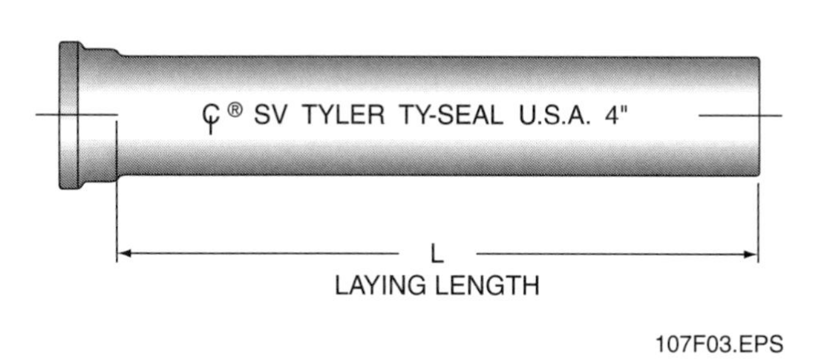

Figure 3 ◆ Pipe label.

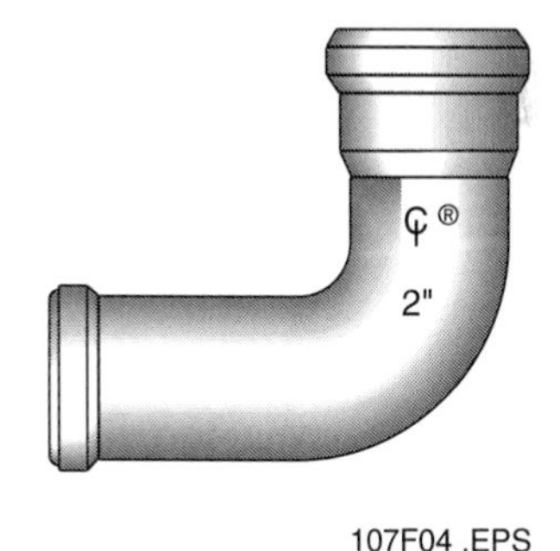

Figure 4 ◆ Fitting label.

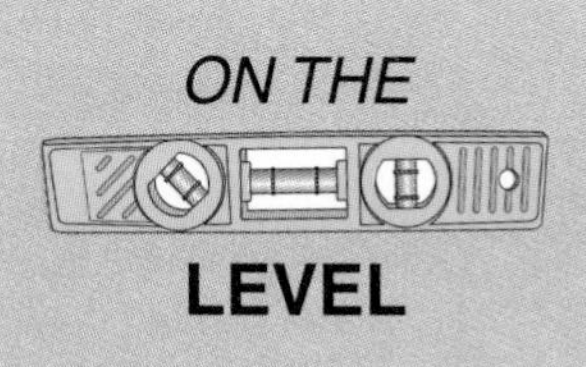

Dura-Iron

Dura-Iron is a high-silicon cast-iron hub-and-spigot pipe used for its resistance to highly corrosive waste. Chemical factories and photography labs use this type of pipe. Be careful when you are working with Dura-Iron—the fittings are expensive and break easily. This type of hub-and-spigot pipe is joined using lead and oakum.

3.0.0 ◆ COMMON FITTINGS AND VALVES

Almost any kind of pipe assembly is possible with the wide variety of pipe fittings made today. This section identifies the common types of cast-iron fittings and the more important dimensions of each.

Information on the common fittings available for hub-and-spigot cast-iron pipe and for no-hub pipe is available from manufacturers.

3.1.0 Bends

Bends are fittings that allow piping to change direction. The name given to each bend indicates the fractional part of a full circle that the bend turns the run of pipe. For example, a 1/8 bend turns the pipe 1/8 of 360 degrees, or 45 degrees. Bends are available in most standard pipe sizes, the smallest being 1½ inches in diameter. *Figure 5* shows the basic types of bends.

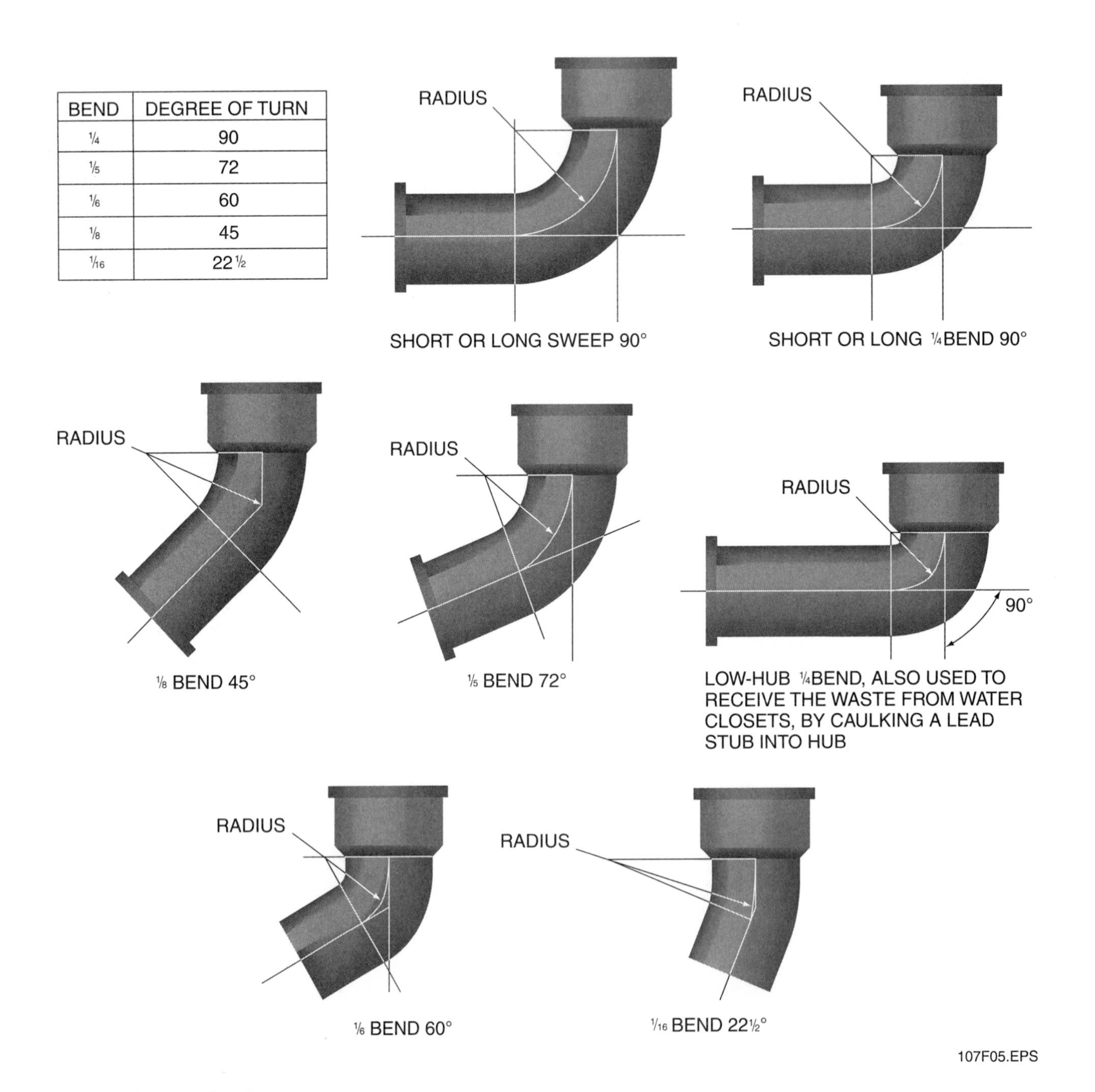

BEND	DEGREE OF TURN
1/4	90
1/5	72
1/6	60
1/8	45
1/16	22 1/2

Figure 5 ◆ Common bends.

Long bends are similar to standard bends, except that in long bends, one leg of the fitting is longer than the other (see *Figure 6*). Long bends typically are dimensioned with two numbers. The first number indicates the diameter of the bend; the second number is the length of the longer leg of the fitting. For example, a 4 × 8, ¼ bend is 4 inches in diameter, and its longer leg is 8 inches in length.

Long bends are expensive and should be used only at the base of plumbing stacks.

Sweeps are similar to ¼ bends except that the radius of the curve is much greater (see *Figure 5*). This allows for smoother flow of wastes through the piping system. There are two types of sweeps: long sweeps and short sweeps. Long sweeps and short sweeps differ in two important ways: the radius of the curve made by the fittings and the laying length.

A number of bends have side or **heel inlets.** Small vents or drains can be connected to these openings. Fittings with side inlets are called either right-hand or left-hand fittings, depending on which side the opening appears. To determine if the fitting is right- or left-hand, place the fitting with the inlet end facing you and the outlet end down, as shown in *Figure 7*.

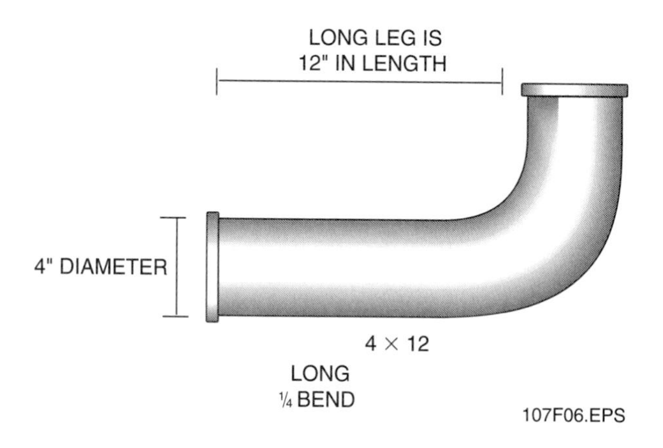

Figure 6 ◆ Long bend.

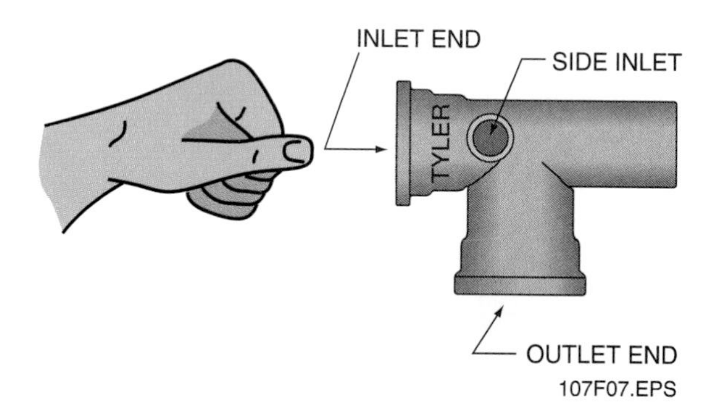

Figure 7 ◆ Identifying the hand of an inlet.

Closet bends connect the water closet to the main drainage piping (see *Figure 8*). Both the inlet and outlet ends of the fitting are manufactured with break-off grooves to permit easy cutting. Closet bends with left-hand or right-hand side openings can be selected. They can be regular, offset 1 inch, offset 2 inches, or reducing.

The **closet flange** is used to connect the water closet to the closet bend (see *Figure 9*).

3.2.0 Branches

Wyes and **double wyes** are used to join two or three drains into one pipe (see *Figure 10*). Wyes connect branches to the main stack at 45-degree angles. Sometimes the branch fitting has a threaded or tapped opening in either the main or branch hub. A cleaned plug (used to close the fitting or pipe) or a threaded pipe from a plumbing fixture can be connected to this type of opening. An inverted wye can be used in the venting system.

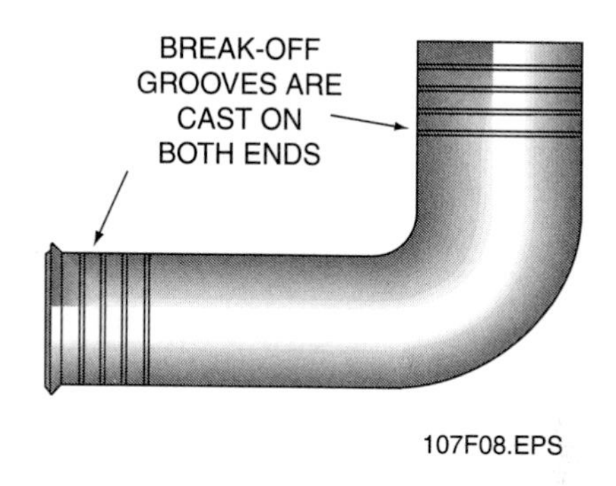

Figure 8 ◆ Closet bend.

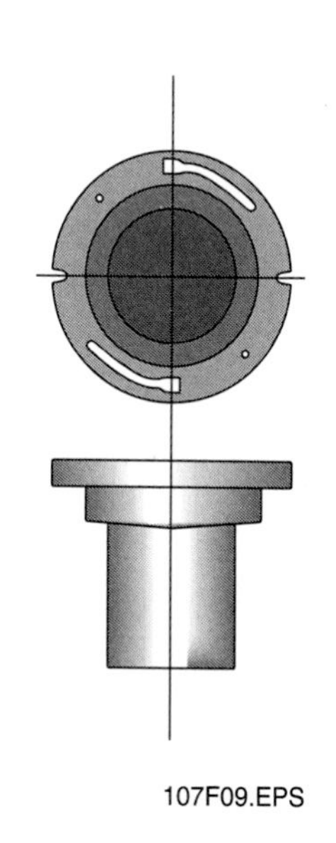

Figure 9 ◆ Closet flange.

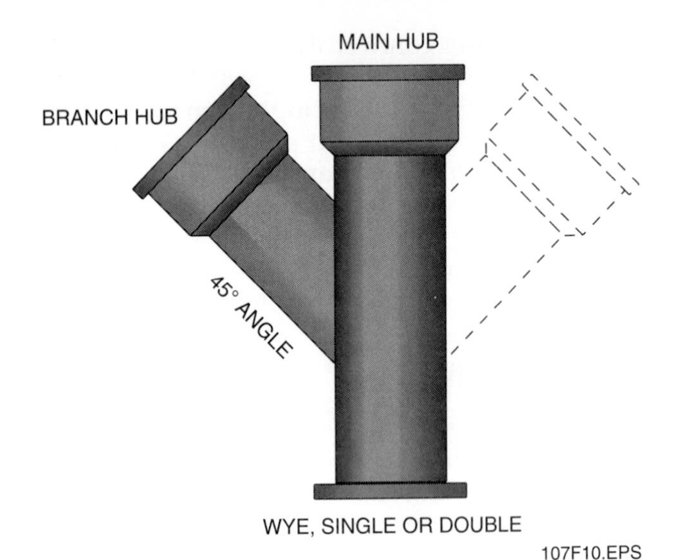

Figure 10 ◆ Wye.

The **sanitary tee** is used to connect pipe at right-angle intersections (see *Figure 11*). The **double sanitary tee** is sometimes called a **sanitary cross.** Tees can also be used on vent lines. They are available with threaded or tapped openings. The dimensions are specified to indicate the diameter of the openings. If all openings are the same diameter, only one dimension is given. When the branch openings are smaller than the straight-through openings, the fitting is dimensioned with two numbers, such as 4 × 3. The first number is the diameter of the openings on the straight-through run; the second number is the diameter of the branch openings.

Sanitary tees also are produced with tapped or threaded side openings. These fittings are designed as right-hand or left-hand. You can identify a branch fitting by placing the branch toward yourself and the outlet end of the run lower than the branch.

Figure 11 ◆ Tees.

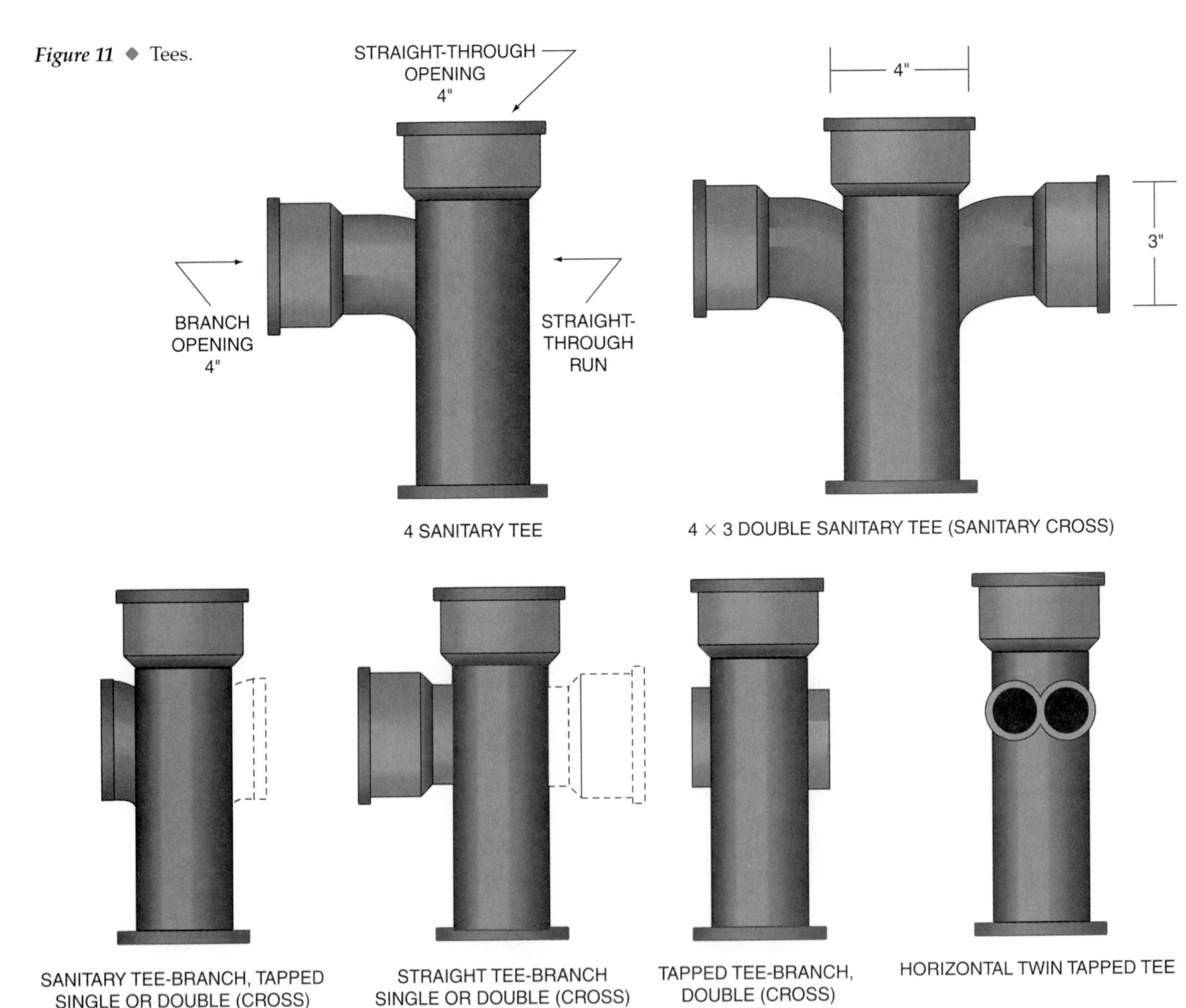

3.3.0 Increasers and Reducers

An **increaser** or **reducer** is used to connect one size pipe to another (see *Figure 12*). For example, you may need a larger-sized pipe above the heated area of a building to keep frost buildup from closing the opening in the vent stack. You would use an increaser to connect a larger pipe to the top of the stack.

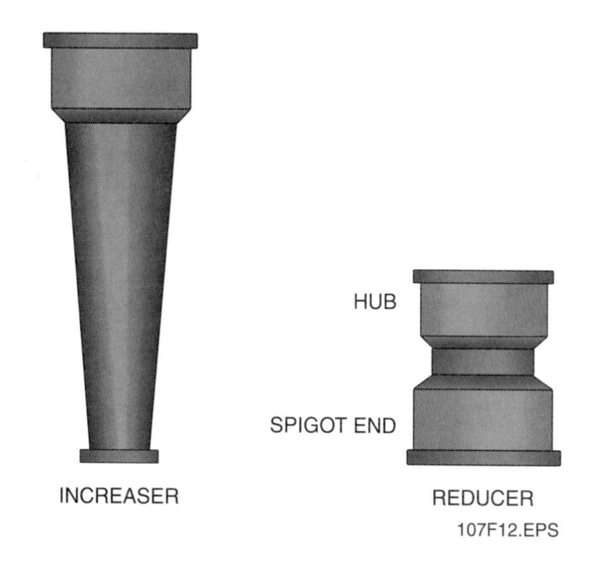

Figure 12 ◆ Increaser and reducer.

3.4.0 Traps

Traps are installed between the drainage line and fixtures or drains (see *Figure 13*). They are designed to hold a water seal. This prevents sewer gas from escaping into the building. Many traps are designed with cleanout plugs that can be removed for cleaning out the trap.

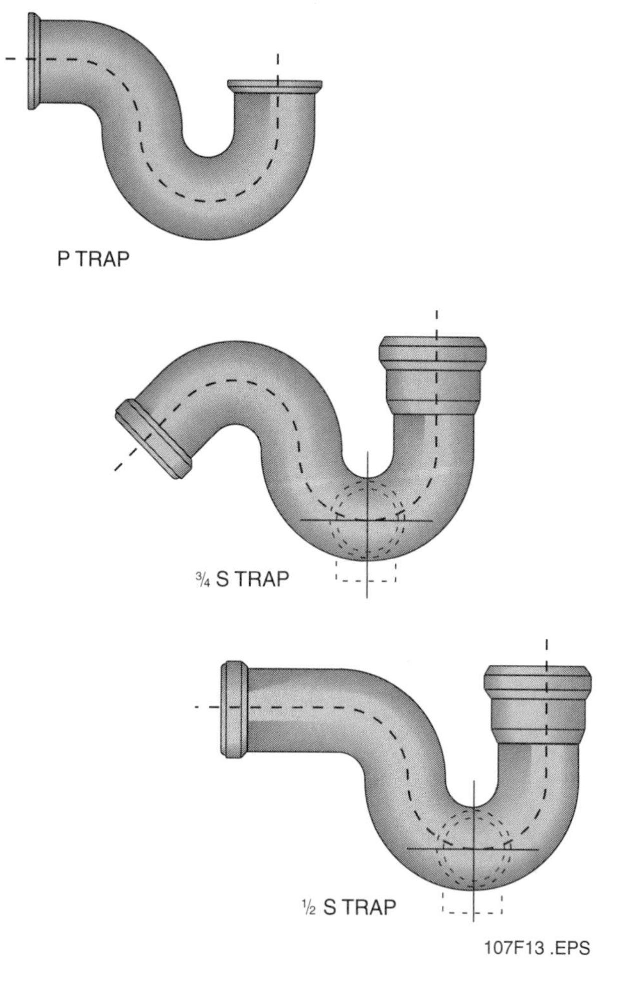

Figure 13 ◆ Traps.

3.5.0 Backwater Valve

The **backwater valve** is a specially designed cast-iron **check valve** for drainage pipe systems (see *Figure 14*). The valve is available for either hub-and-spigot (service weight) or no-hub pipe connections. A check valve prevents fluid from going backwards in the pipe system. When backwater valves are installed underground, they are usually in a valve box or vault, so a plumber can get to them easily.

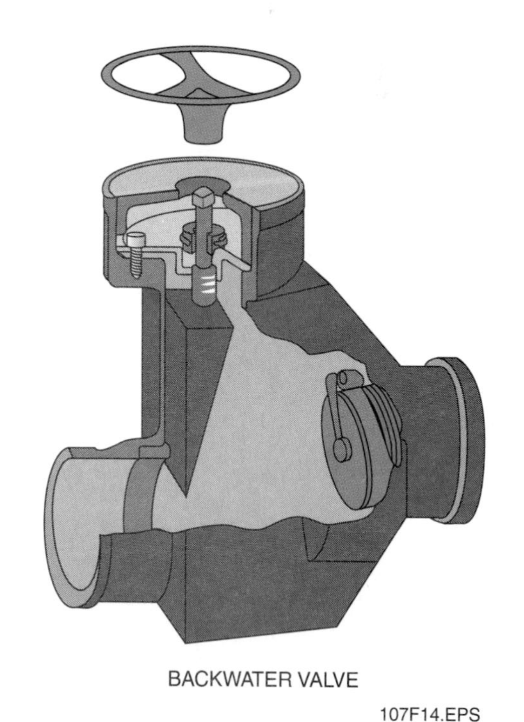

Figure 14 ◆ Backwater valve.

> **ON THE LEVEL**
>
> ### *Traps*
>
> The trap is a simple, elegant solution to a plumbing problem that basically separates our living space from our sewage system. The water seal in a trap prevents gas and fumes from entering the building through the fixtures. Traps also act as a barrier to vermin that may enter a building's living space through the plumbing piping. Many underdeveloped nations do not use traps in their waste systems.

Review Questions

Sections 2.0.0–3.0.0

1. _____ pipe gives the plumber an easier method of joining cast iron in areas where space is limited.
   a. Gray iron
   b. Ductile
   c. No-hub
   d. Hub-and-spigot

2. _____ pipe and fittings are recommended for structures that are subject to vibration, settling, or excessive corrosion.
   a. Standard
   b. Extra-heavy
   c. Service-weight
   d. Ductile

3. Tees are typically dimensioned with two numbers, as in 4 × 3. The first number indicates _____.
   a. the length of the straight-through
   b. the length of the branch
   c. the diameter of the straight-through opening
   d. the diameter of the branch opening

4. Using a(n) _____ allows for smoother flow of wastes through the piping system.
   a. increaser
   b. trap
   c. sweep
   d. ⅛ bend

5. A _____ prevents sewer gas from escaping into the building.
   a. wye
   b. tee
   c. bend
   d. trap

6. A(n) _____ connects branches to the main stack at 45-degree angles.
   a. wye
   b. bend
   c. sweep
   d. inlet

7. To change to a smaller-size pipe in a run, use a _____ fitting.
   a. double cross
   b. closet flange
   c. reducer
   d. long bend

8. A _____ connects the water closet to the drainage pipe system.
   a. wye
   b. long sweep
   c. closet bend
   d. closet flange

9. A(n) _____ may connect a larger pipe to the top of the stack to prevent frost buildup from closing the opening in the vent stack.
   a. inlet
   b. increaser
   c. ¼ bend
   d. flange

10. No-hub and hub-and-spigot cast-iron pipe are made in _____ and _____ foot lengths.
   a. 5, 10
   b. 10, 15
   c. 6, 12
   d. 8, 12

4.0.0 ◆ HANGERS AND SUPPORTS

Hangers are used for horizontal or vertical support of pipes and piping. The principal purpose of hangers and brackets is to keep the piping in alignment and prevent it from bending, swaying, or distorting. Plumbing codes require pipe to be securely fastened horizontally and vertically. You

Plumbing Codes and Cast-Iron Pipe

Plumbing codes are designed to protect the health and safety of the general public. Codes vary from location to location, but all are based on sound principles of sanitation and safety.

Plumbing codes regulate every aspect of cast-iron pipe and fitting selection and installation. It is critical that you become familiar with the state, city, or local plumbing codes that apply to the area where you work. Some fittings may be allowed in one area that are strictly forbidden in another area. For example, a running trap (see *Figure 15*) may be illegal in your area but not in other areas. Installing a running trap could be a costly mistake.

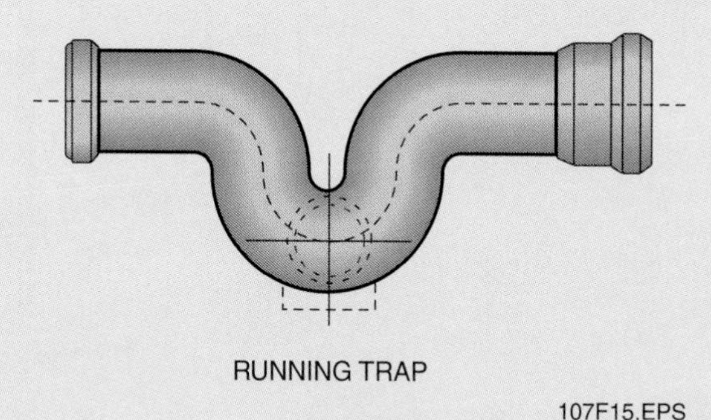

Figure 15 ◆ Running trap.

In selecting supports and fasteners, refer to the local code and building specifications. You as the plumber must select and use the support and fastener that are best suited for the job. You are also responsible for installing hangers or fasteners at the correct intervals along the length of the pipe run. The code will specify what needs to be done. If there are unusual circumstances, the project specifications may direct you to make changes from generally accepted plumbing practices.

Never work without knowing the applicable codes and detailed project specifications!

need to be aware of the code requirements and follow them. Check the blueprints and specifications for any unusual pipe support requirements.

4.1.0 Types of Hangers and Supports

The hangers and supports you use will depend on the size of the pipe that needs to be supported and the type of material to which the hanger or support is attached. A variety of hangers are available for horizontal runs (see *Figure 16*) and vertical runs (see *Figure 17*). When several pipe runs are close together and parallel, a pipe trapeze (see *Figure 18*) or brackets made from prepunched channel (see *Figure 19*) can be used.

4.2.0 Uses

When you are selecting the types of hangers or supports to use, you must determine whether you are working horizontally or vertically. All hangers

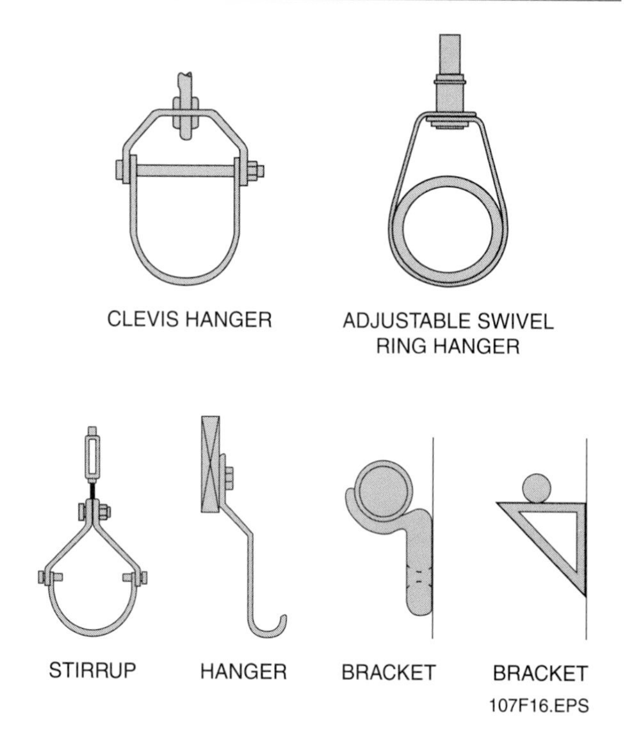

Figure 16 ◆ Horizontal pipe hangers.

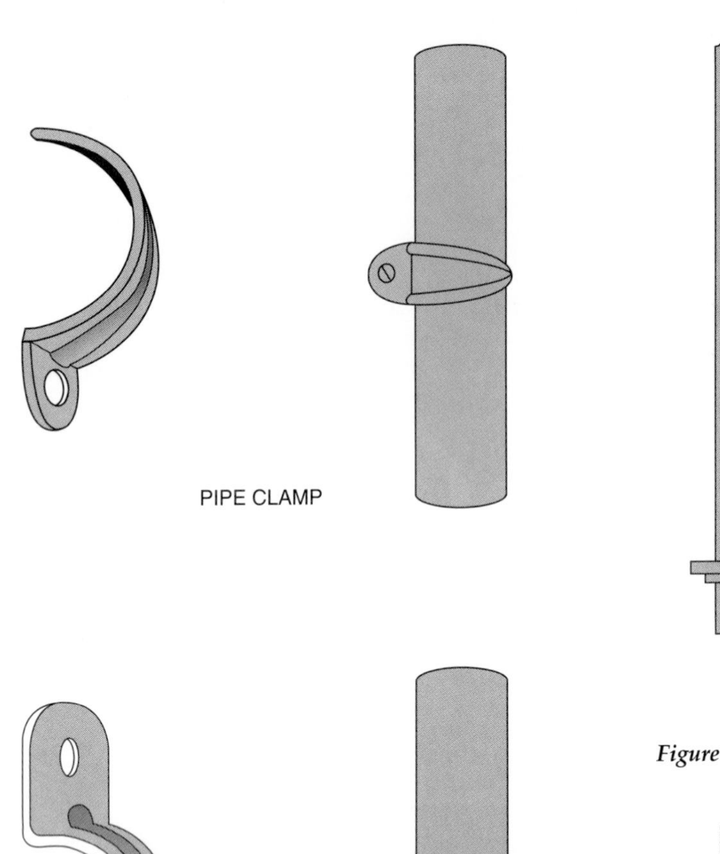

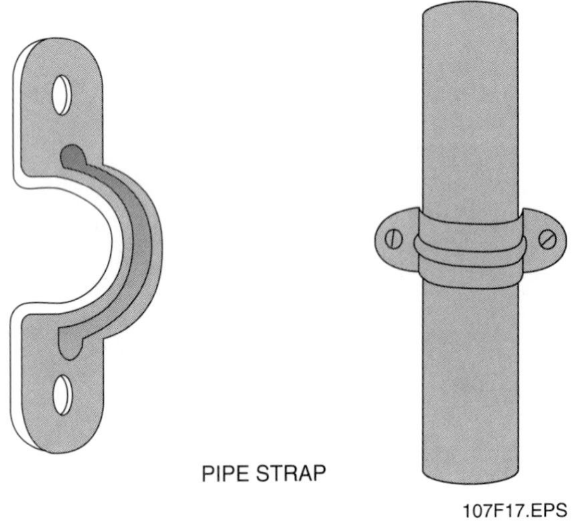

Figure 17 ◆ Vertical pipe supports.

Figure 18 ◆ Pipe trapeze.

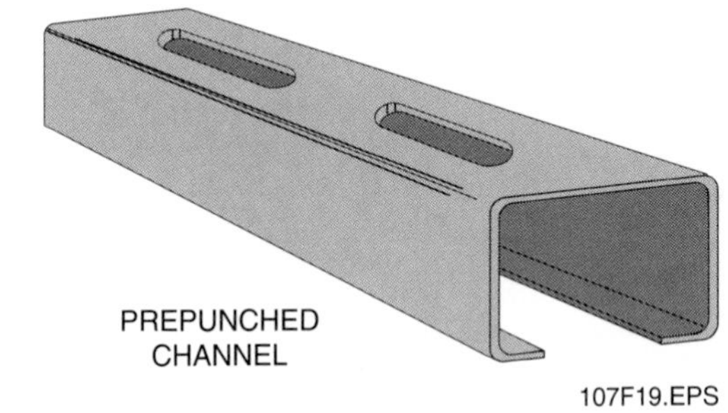

Figure 19 ◆ Channel.

and supports keep pipe in alignment, but horizontal pipes must also be kept from moving back and forth or swaying. Selection for vertical pipe runs must also take into account the vertical load, or total weight, of the pipe.

4.2.1 Horizontal Pipe Runs

Most authorities and plumbing codes specify that pipe must have hangers to support it at each joint. Be sure to check the governing code.

Hangers are used to keep the pipe in alignment and to prevent it from swaying. Hangers need to be as close to the hub as possible.

When cast-iron pipe is suspended with more than 18 inches between nonrigid hangers, you must keep the pipe from swaying by using **sway braces** (see *Figures 20* and *21*).

Horizontal hangers can be attached to wooden structures with screws, lag screws, or large nails. For fastening pipe to I-beams, bar joists, and other metal structures, you can use beam clamps or C-clamps (see *Figure 22*).

Many types of horizontal support hangers may be used, depending on the conditions and depending on what the contractor, architect, or engineer specifies. No matter what kind of support you use on a horizontal run of pipe, be sure to maintain a proper **slope** of ⅛ to ¼ inch per foot. (Slope will be explained in more detail later in this module.)

4.2.2 Vertical Pipe Runs

Vertical hangers are used to support vertical runs of pipe and to maintain pipe alignment. A friction

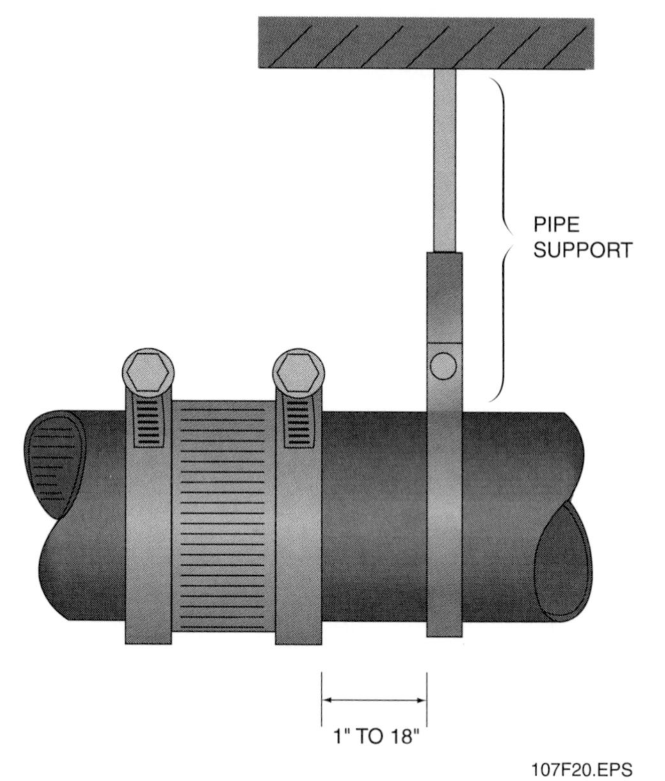

Figure 20 ◆ Hanger distance.

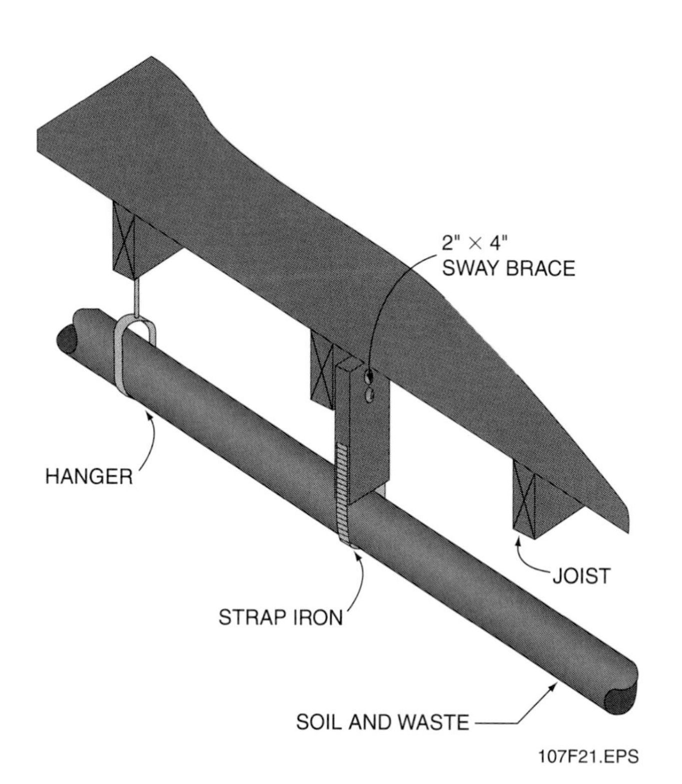

Figure 21 ◆ Sway brace.

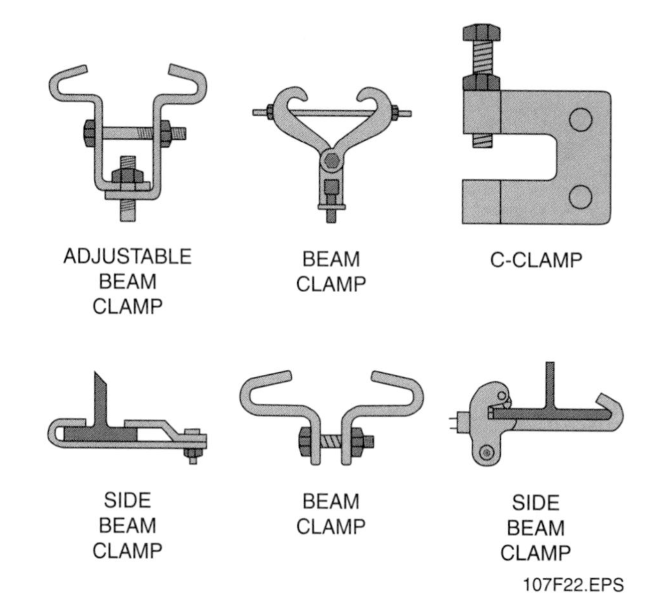

Figure 22 ◆ Beam clamps.

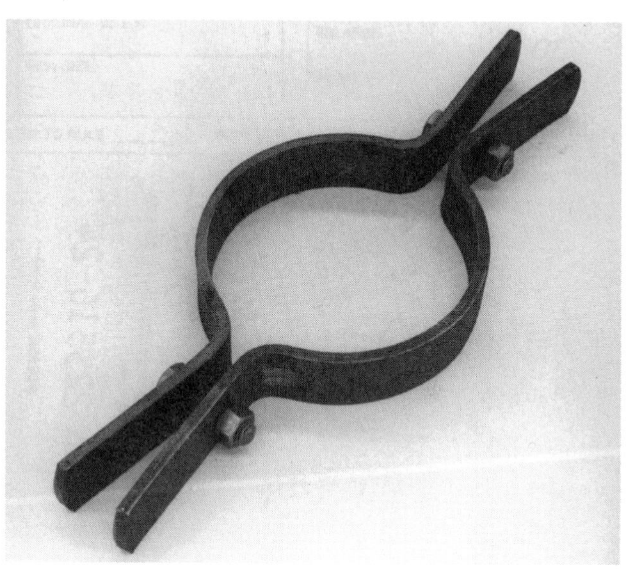

Figure 23 ◆ Friction clamp.

clamp is used primarily to support the vertical load of the pipe (see *Figure 23*).

4.3.0 Installation

Fastening devices must be used to attach pipe hangers or supports to wood, masonry, concrete, and steel. Because of the weight of cast-iron pipe, the holding power of different fasteners must be considered, and the correct fastener for each job must be selected.

4.3.1 Wooden Structures

Nails, screws, and bolts are used to fasten pipe in wooden structures. Nails can be used for vertical and horizontal pipe runs only if vibration and strength are not factors. Screws and bolts have more holding power than nails.

4.3.2 Masonry and Concrete Structures

You may use various expansion anchors or threaded masonry fasteners to fasten support hangers to masonry and concrete. Use a **rotary hammer drill** with a special masonry bit to drill into the masonry (see *Figure 24*). After drilling the hole, insert a screw anchor (see *Figure 25*), a plastic expansion anchor (see *Figure 26*), or a threaded masonry anchor (also called a Tapcon; see *Figure 27*) in the hole. Then fasten the screw into the insert, which expands the insert to hold the screw tightly in the masonry. Toggle bolts can be used to attach pipe supports to hollow masonry units such as concrete block (see *Figure 28*).

107F24.TIF

Figure 24 ◆ Rotary hammer drill.

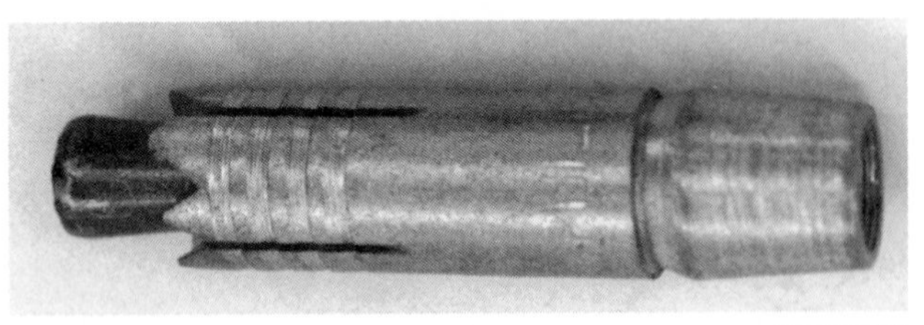

107F25.TIF

Figure 25 ◆ Screw anchor.

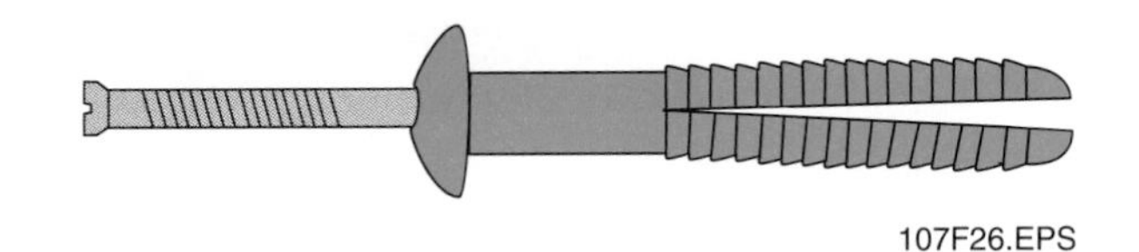

107F26.EPS

Figure 26 ◆ Plastic expansion anchor.

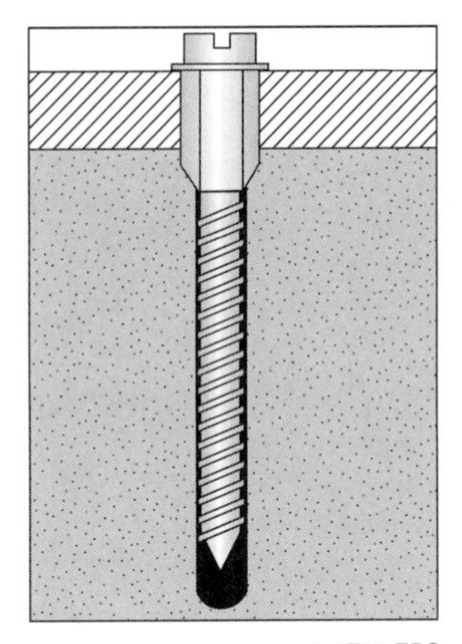

107F27.EPS

Figure 27 ◆ Threaded masonry anchor.

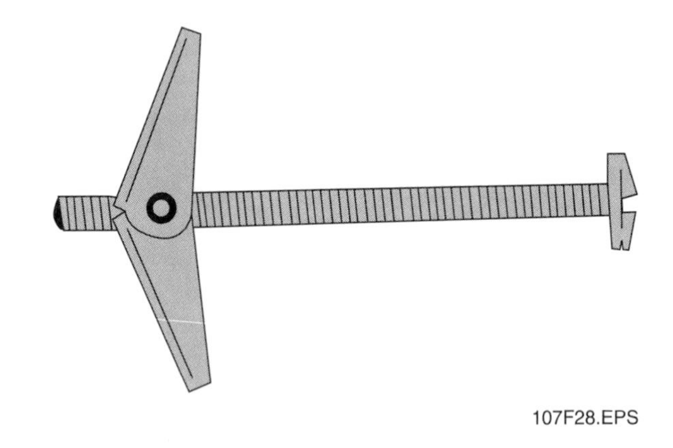

107F28.EPS

Figure 28 ◆ Toggle bolt.

ON THE LEVEL

Powder-Actuated Tools

A special tool that can be used to anchor pipe supports to concrete, masonry, and steel is the **powder-actuated fastening tool** (see *Figure 29*). This tool shoots a fastener into the structure by exploding a blank .22-caliber shell. Care must be taken to use the correct size of shell and nail. Too heavy a shell could drive a nail completely through the steel beam. No one is permitted to use a powder-actuated fastening tool without a certificate issued by the manufacturer of the tool. Manufacturers' representatives visit shops and job sites to conduct training in their equipment and to certify people in the use of the equipment.

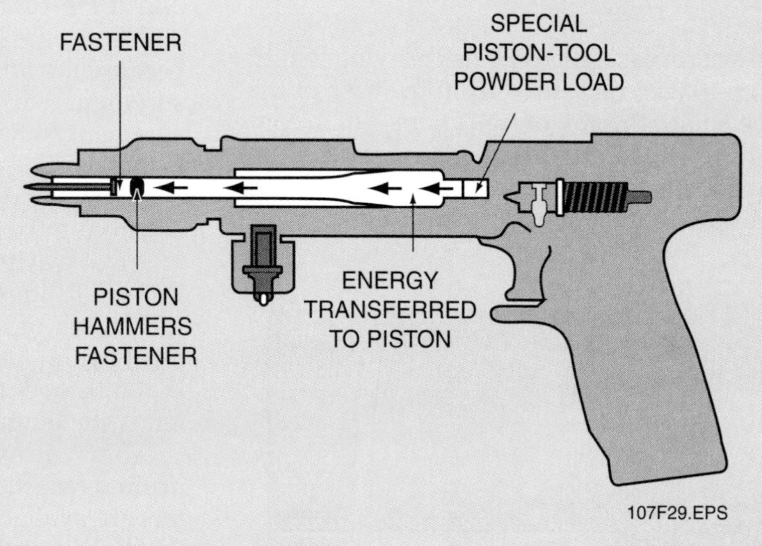

Figure 29 ◆ Powder-actuated fastening tool.

Review Questions

Section 4.0.0

1. When several pipe runs are close together and parallel, a(n) _______ can be used to support the pipe.
   a. bracket
   b. anchor
   c. sway brace
   d. pipe trapeze

2. A _______ clamp is used primarily to support the vertical load of the pipe.
   a. friction
   b. clevis
   c. support
   d. C

3. Most authorities and plumbing codes require hangers to support horizontal pipe at each ______ .
   a. wye
   b. tee
   c. joint
   d. bend

4. _______ are used to keep the pipe in alignment and to prevent it from swaying.
   a. Anchors
   b. Friction clamps
   c. Toggle bolts
   d. Hangers

5. No matter what method of support is used on a horizontal run of pipe, be sure to maintain the proper ________.
   a. slope
   b. joint
   c. anchor
   d. code

5.0.0 ◆ MEASURING, CUTTING, JOINING, AND ASSEMBLING

This section describes procedures for measuring, cutting, joining, and assembling cast-iron pipe using **lead and oakum joints.**

5.1.0 Measuring and Cutting

You must be sure to measure and cut pipe carefully. Inaccurate measuring and improper or sloppy cutting cost a business money. If a pipe is cut too short, it is useless. If a pipe is cut improperly, it cannot be used to make a secure joint. Measuring and cutting hub-and-spigot pipe is more involved than working with no-hub pipe.

5.1.1 Measuring and Cutting Hub-and-Spigot Pipe

Hub-and-spigot cast-iron pipe has a laying length of 5 feet or 10 feet, measured from the base of the hub to the end of the spigot (see *Figure 30*). In connecting long runs containing many sections of pipe, the **insertion length** must be added to the length of pipe to get a correct measurement. The insertion length is the length of the spigot end that is inserted into the hub.

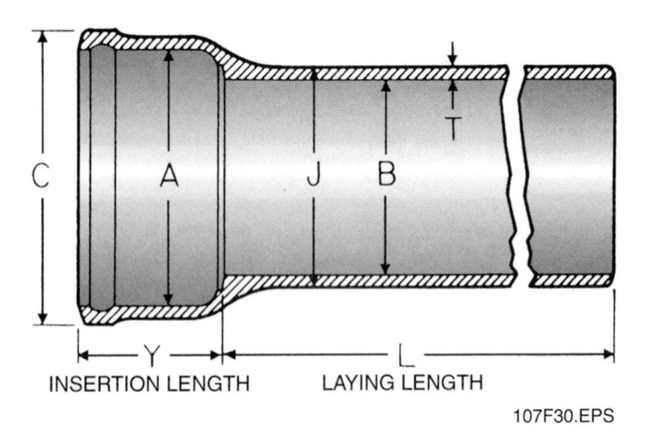

Figure 30 ◆ Dimensions.

Insertion lengths for the most commonly used pipe sizes are 2½ inches for 2-inch pipe; 2¼ inches for 3-inch pipe; and 3 inches for 4-, 5-, or 6-inch pipe.

You must know how to locate the center of a fitting to be able to measure pipe correctly. The center of a fitting is the point where the lines of flow through the fitting intersect (see *Figure 31*).

In a drainage system, if a ¼ bend is to be connected to a ⅛ bend fitting with a length of soil pipe, and if the center-to-center measurement must be 36 inches, the length of pipe is less than 36 inches because the fittings take up part of the overall measurement (see *Figure 32*). To make a proper measurement, place the fittings together as shown in *Figure 33*. Then measure the distance between the center marks on the fittings and subtract that number from the overall measurement (36 inches).

A fast way to figure pipe length is to use a ruler. Put the fittings close together. Then place the 36-inch mark of the ruler on the center mark of the right fitting (see *Figure 34*). The mark on the ruler that falls over the center mark of the fitting on the left is the length to cut the pipe.

After you measure the pipe, you cut it to the desired length with a soil pipe cutter. Soil pipe cutters are available in a variety of models (see *Figure 35*). Each cutter can be used to cut a variety of pipe sizes. The soil pipe cutter uses a length of chain, each link of which contains a cutting wheel. When the chain is tightened around the pipe, the cutters press into the metal and break the pipe cleanly.

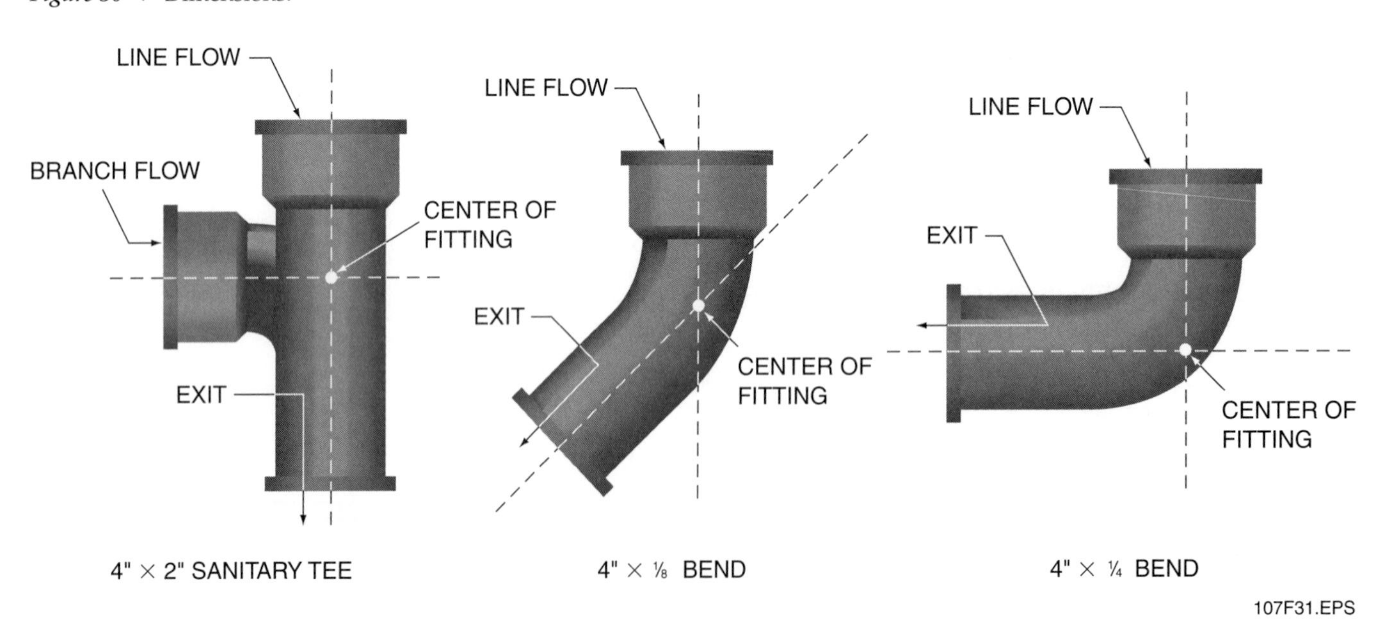

Figure 31 ◆ Finding the center of a fitting.

5.1.2 Measuring and Cutting No-Hub Pipe

Measuring no-hub cast-iron pipe is rather simple. Using a folding rule or tape measure, measure the exact distance between the fitting connection and the pipe, or from pipe to pipe, and subtract the allowance for the ridge on the inside of the neoprene gasket (see *Figure 36*). No-hub pipe should be used without cutting it, if possible.

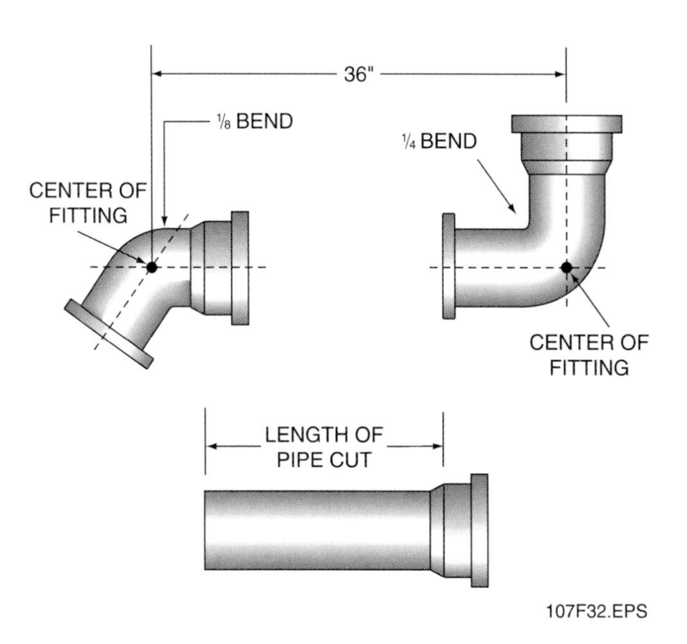

Figure 32 ◆ Figuring the cutting length.

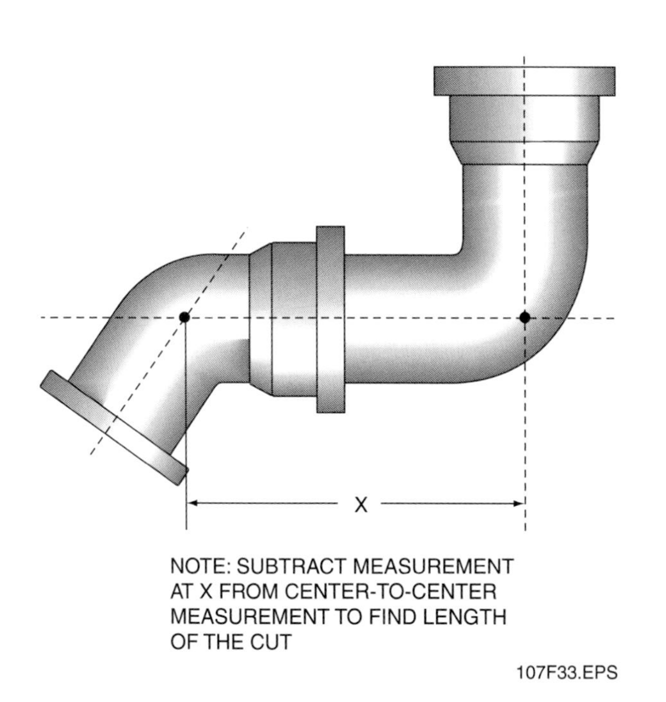

Figure 33 ◆ Placing fittings together.

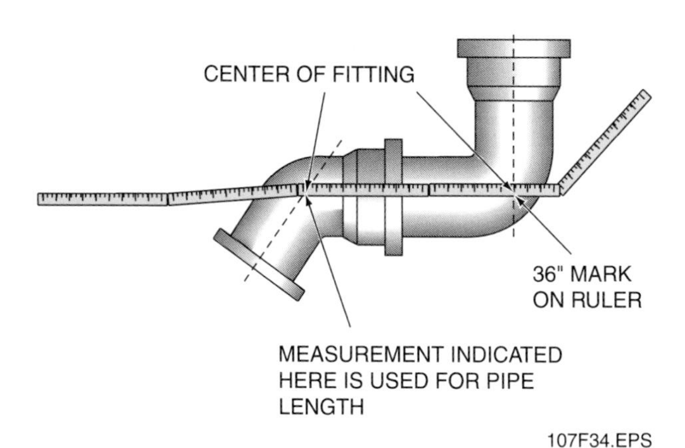

Figure 34 ◆ Using a folding rule.

Figure 35 ◆ Soil pipe (snap) cutter.

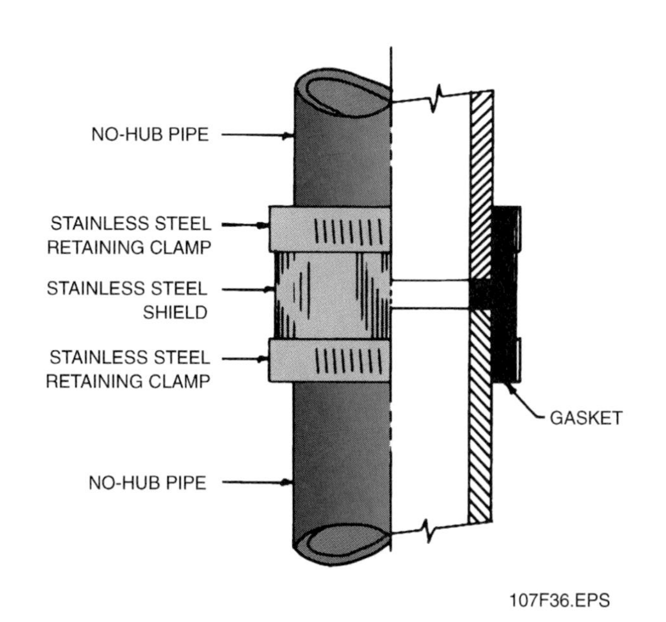

Figure 36 ◆ Gasket allowance.

Plan the layout of the DWV system in advance. Each **coupling** restricts the flow of wastes and adds to the cost of the job. A well-planned job can eliminate unnecessary connections. A typical no-hub vertical modular system is shown in *Figure 37.*

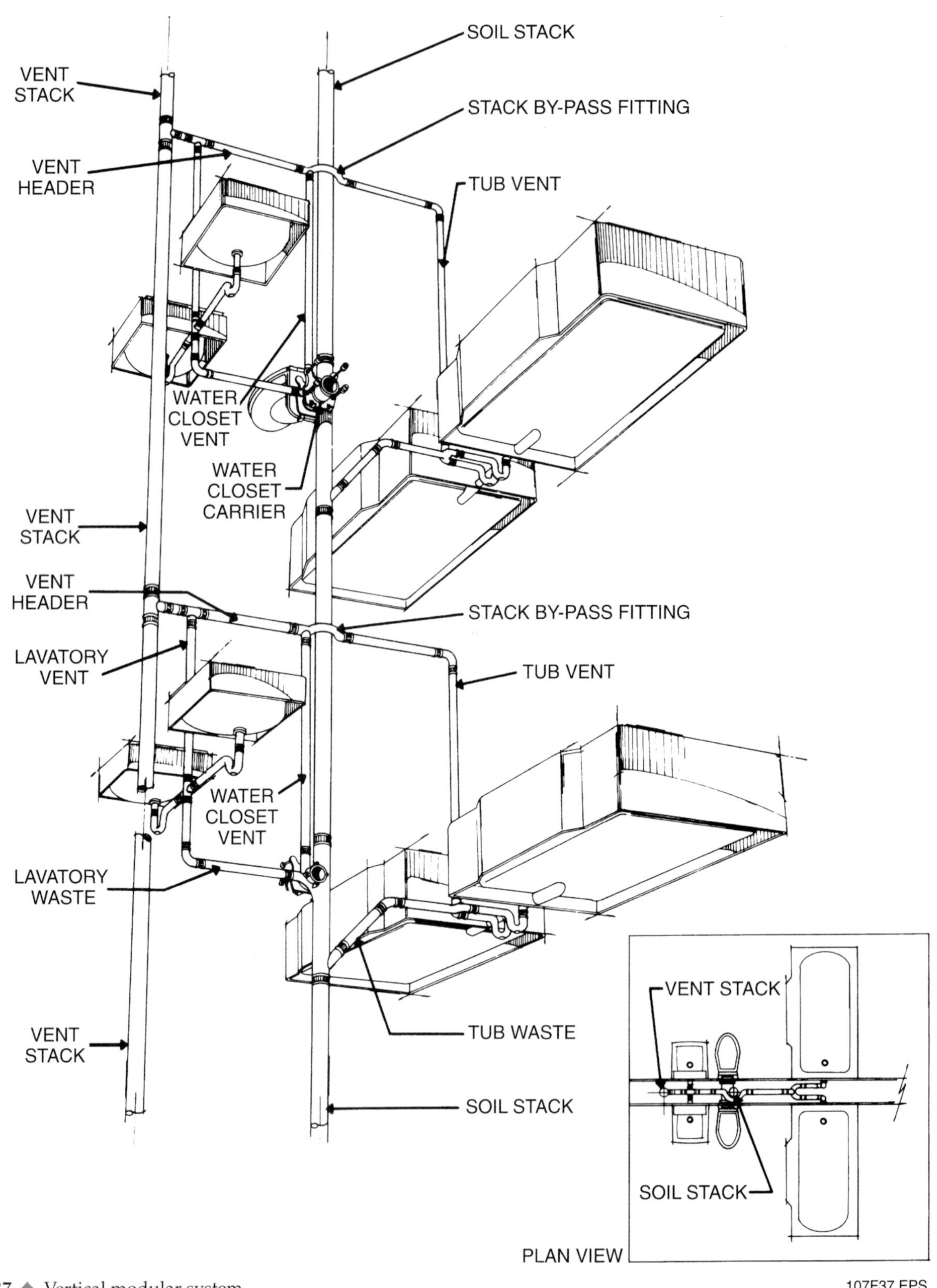

Figure 37 ◆ Vertical modular system.

5.2.0 Joining

This section describes the techniques and procedures used to join hub-and-spigot pipe and no-hub pipe. Each type of pipe is discussed separately.

5.2.1 Joining Hub-and-Spigot Pipe

Hub-and-spigot pipe can be joined by either of two methods: the compression joint (see *Figure 38*) or the lead and oakum joint (see *Figure 39*).

A compression joint is a fast method of joining cast-iron pipe and fittings. This method requires the use of pipe and fittings that do not have a bead on the spigot end.

For a compression joint, the plumber inserts a neoprene gasket into the hub end of the pipe. The spigot end (without a bead) is then pushed or drawn into the hub by a chain puller, lead hammer, or pushing bar. This causes the neoprene gasket to compress and fill in any air space between the two sections of pipe. The result is a leakproof, rotproof, and pressureproof joint that can absorb vibrations and can be deflected or bent up to 5 degrees without leaking or failing.

To make a compression joint, follow these steps:

Step 1 Clean the neoprene gasket and pipe ends.

Step 2 Insert the gasket in the hub.

Step 3 Apply rubber lubricant to the inside of the gasket.

Step 4 Force the spigot end of the pipe into the gasket.

Lead and oakum joints are leakproof, rotproof, strong, and flexible. Plumbers have long recognized the waterproofing qualities of **oakum fiber.** Molten lead, when poured over oakum in the hub end of the pipe, completely seals and locks the joint. After the lead has cooled a minute or two, it can be caulked into the joint. *Table 1* lists the amounts of materials needed per joint for various sizes of pipe.

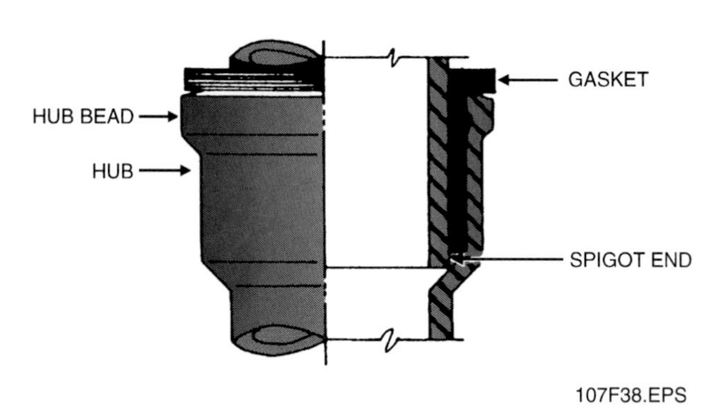

Figure 38 ◆ Compression joint.

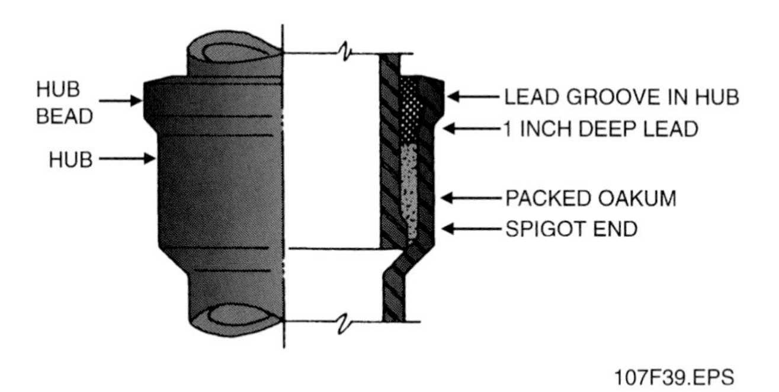

Figure 39 ◆ Lead and oakum joint.

Table 1 Material Equivalents

Pipe size (in.)	Oakum of Equivalent (16-in. sections)	Lead (lbs.)
2	3	1½
3	4½	2¼
4	5	3
5	6½	3¾
6	7½	4½
8	9½	6
10	12	7½

ON THE LEVEL

Lead and Oakum Joints

Lead and oakum joints require special tools. These tools are discussed below.

The **running rope,** or **joint runner,** is made of fiber. It is clamped around the pipe next to the hub using a **spring clamp** (see *Figure 40*). The joint runner keeps the hot, molten lead in the hub during the pouring operation. The pipe joint runner is only used to make upside-down and horizontal joints (see *Figures 41* and *42*).

The propane-melting furnace is used to melt lead for making soil pipe joints (see *Figure 43*). It burns liquefied petroleum (LP) gas. Safety is of the utmost importance in operating a propane furnace. The manufacturer's operating instructions and maintenance procedures must be followed explicitly.

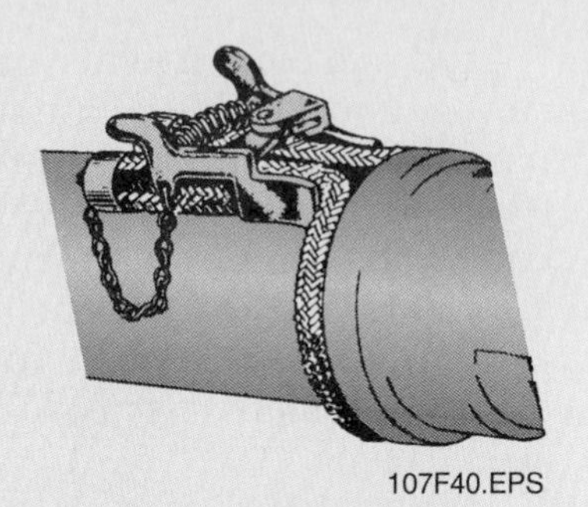

Figure 40 ◆ Running rope and spring clamp.

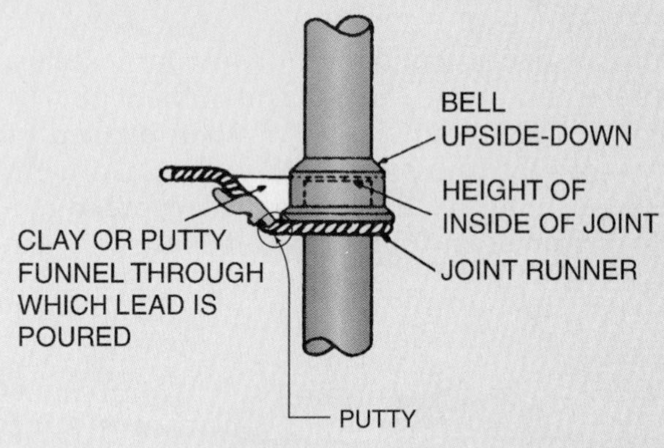

Figure 41 ◆ Upside-down joint.

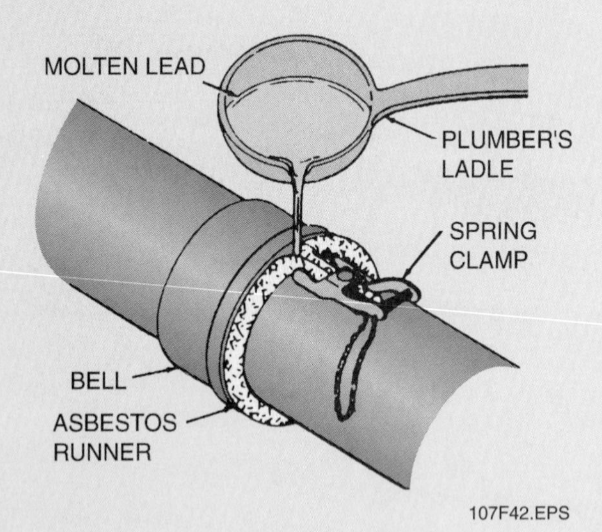

Figure 42 ◆ Horizontal joint.

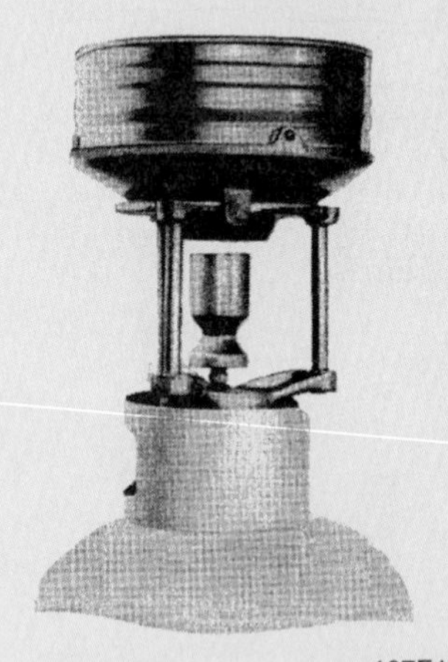

Figure 43 ◆ Propane-melting furnace.

The **pot** is placed on the head of the furnace. It is the container in which the lead is melted. A **ladle** is used to dip the molten lead from the pot and pour it into the joint (see *Figure 44*).

Yarning irons are used to pack the oakum fiber by hand into the joint (see *Figure 45*); a hammer is not used.

Packing irons, which are broader and thicker than yarning irons, are used to pack oakum with a hammer (see *Figure 46*).

Caulking irons are used to drive the lead firmly into the joint (see *Figure 47*). This is necessary because lead shrinks when it cools. Therefore, to produce a tight seal, the lead must be caulked to fill the joint.

Different shapes and sizes of caulking irons are necessary for different sizes of pipe and varying locations of joints. For example, the **ceiling caulking iron** is the only caulking iron that can be used to caulk a joint next to the ceiling (see *Figure 48*).

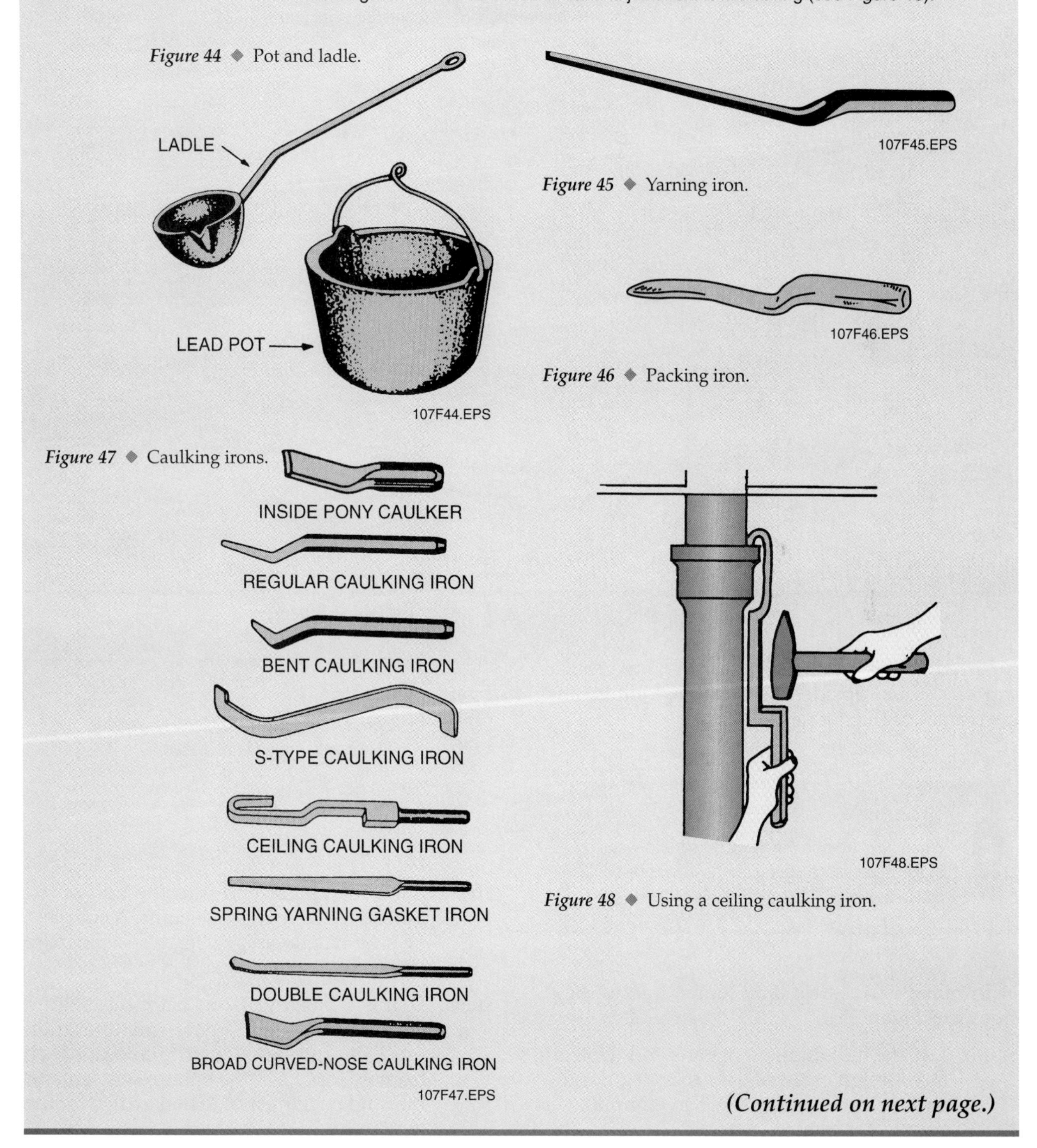

Figure 44 ◆ Pot and ladle.

Figure 45 ◆ Yarning iron.

Figure 46 ◆ Packing iron.

Figure 47 ◆ Caulking irons.

Figure 48 ◆ Using a ceiling caulking iron.

(Continued on next page.)

There are "inside" and "outside" caulking irons. **Outside caulking irons,** which angle to the right, are used to shape the lead to the inside of the hub. **Inside caulking irons,** which angle to the left from the bottom up, are used to shape the lead near the spigot. See *Figures 49* and *50.*

Caulking irons must have a good caulking surface. The surface must be inspected often and reshaped on a bench grinder when needed to maintain the correct angle. This ensures high-quality, reliable joints. Cleaning the irons and protecting them with a thin film of oil makes them easier to use and also makes them last longer.

The **pickout iron** has a diamond-shaped flat point (see *Figure 51*). It is used to remove the lead and oakum from a joint.

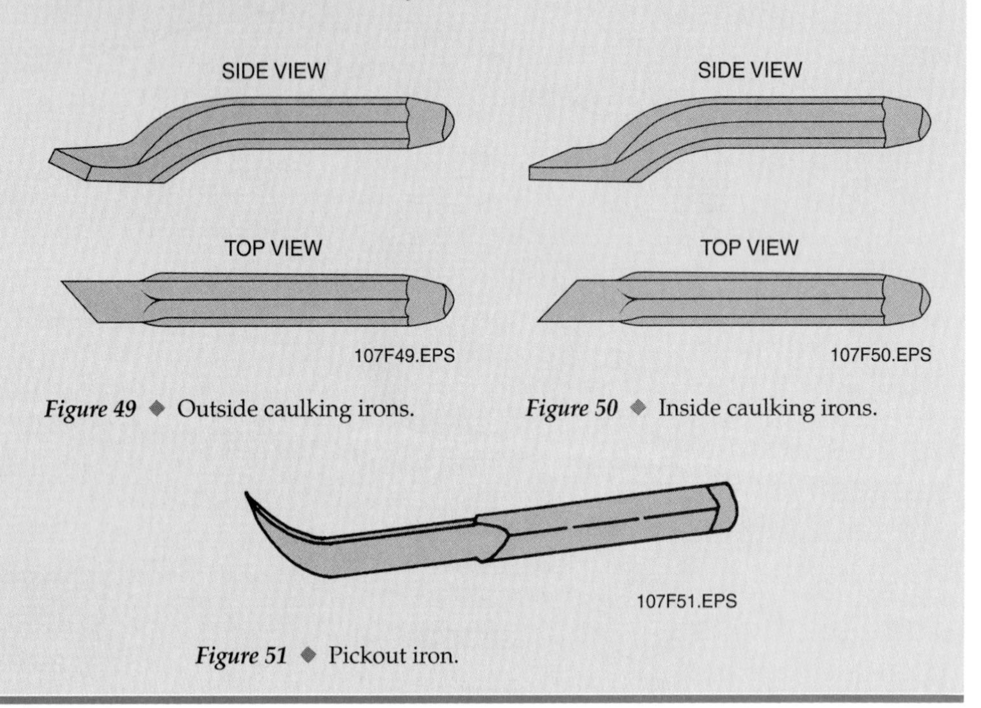

Figure 49 ◆ Outside caulking irons.

Figure 50 ◆ Inside caulking irons.

Figure 51 ◆ Pickout iron.

Listed below are the steps for making vertical caulked joints, upside-down caulked joints, and horizontal caulked joints.

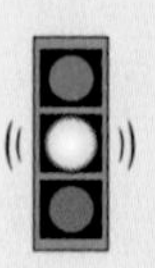

CAUTION

When you are pouring lead, be sure to wear safety glasses, gloves, high-top shoes or boots, and long pants.

To make vertical caulked joints, follow these steps (see *Figure 52*):

Step 1 Dry the hub-and-spigot ends and wipe off any foreign materials. If necessary, dry the ends with a heating torch to eliminate all traces of moisture.

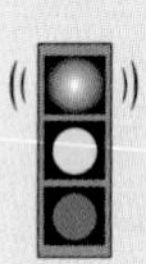

WARNING!

Moisture can cause molten lead to explode out of a joint. Serious injuries can result.

Step 2 Slide the spigot end into the hub of the other pipe and align the joint. A cut piece of pipe has no spigot bead, so take extra care to center the cut end in the hub.

Step 3 Using a yarning iron, pack the oakum around the pipe. Repeat this operation until the hub is packed to about 1 inch from its top. Pack the oakum with a hammer and packing iron to make a bed for the molten lead.

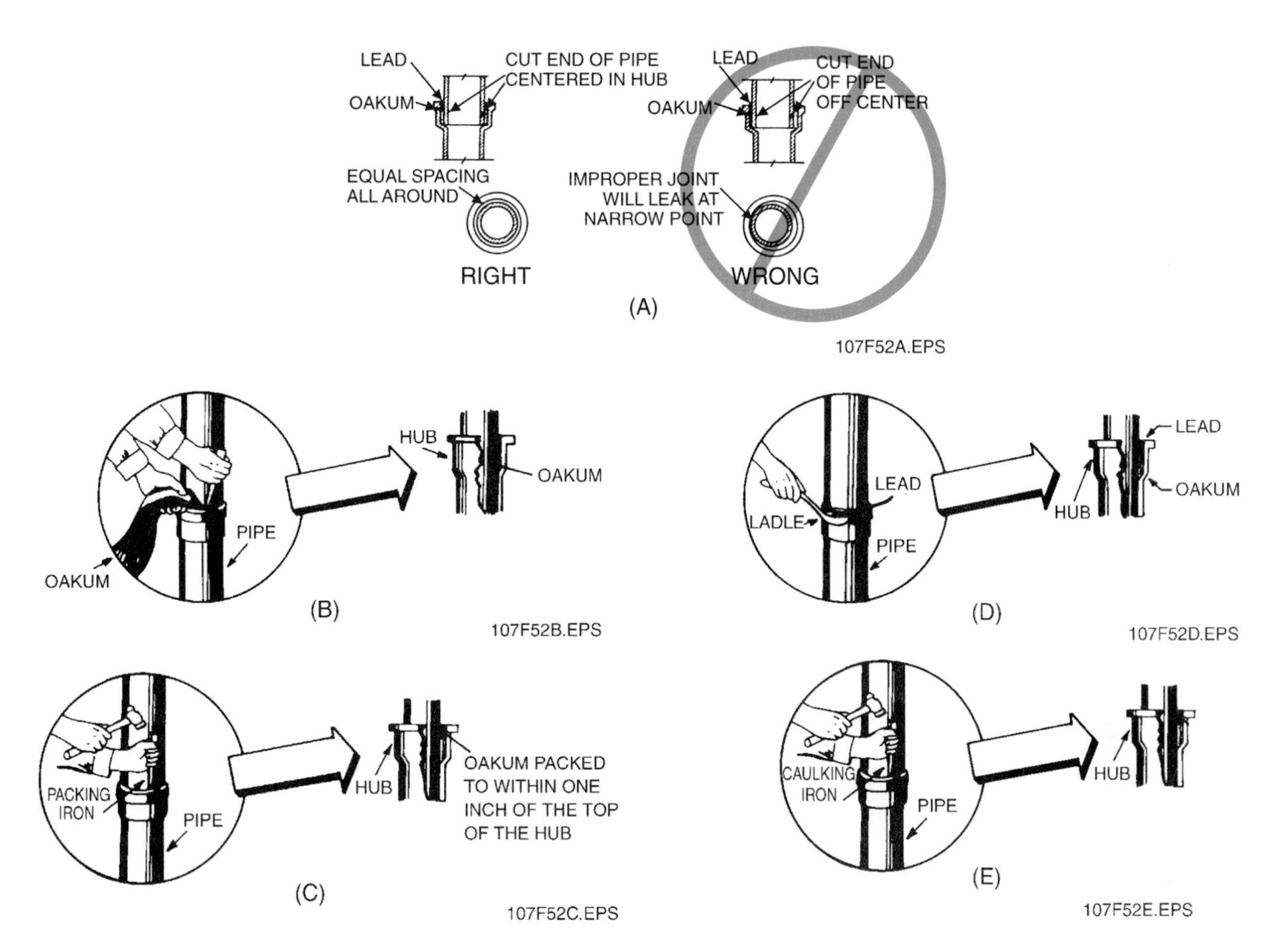

Figure 52 ◆ Making vertical caulked joints.

Step 4 Using the plumber's ladle, carefully pour the molten lead into the joint. Dip enough lead to fill the joint in one pouring. Allow a minute or two for the molten lead to harden and to change in color from royal blue to a dull gray. Usually, one pound of lead is melted for each inch of pipe size.

Step 5 Caulk the joint using the outside caulking iron first and then the inside caulking iron. The first four blows should be struck 90 degrees apart around the joint to set the pipe. Drive the lead down on the oakum and into contact with the spigot surface on one edge and with the inner surface of the hub on the other. Use firm but light hammer blows.

Caulking the lead too tightly may create pressures high enough to crack the pipe. If this occurs, the broken section must be replaced.

To make upside-down caulked joints, follow these steps (see *Figure 53*):

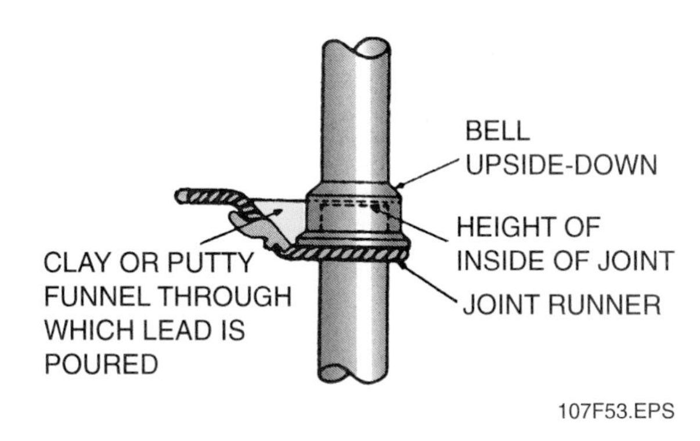

Figure 53 ◆ Making an upside-down caulked joint.

Step 1 Prepare the ends and pack the oakum as in an upright vertical joint.

Step 2 Clamp the joint runner around the pipe. Raise the ends and the clamp of the joint runner. Build a funnel in the raised end. Construct the funnel out of fire clay, putty, or plaster. The funnel must be as high as or higher than the inside height of the lead portion of the joint.

Step 3 Pour the lead into the joint, using the funnel. While the lead is still hot, remove the runner, trim off the surplus with a cold chisel, and caulk the lead to the outside and inside walls of the joint.

To make horizontal caulked joints, follow these steps (see *Figure 54*):

Step 1 Prepare the ends of the pipe and pack the joint with oakum as in vertical joints.

Step 2 Clamp the joint runner in place around the pipe and fill the joint with molten lead.

Step 3 After the lead hardens, remove the runner and trim off the surplus.

Step 4 Caulk the joint just as you would a vertical joint, but use the inside iron first.

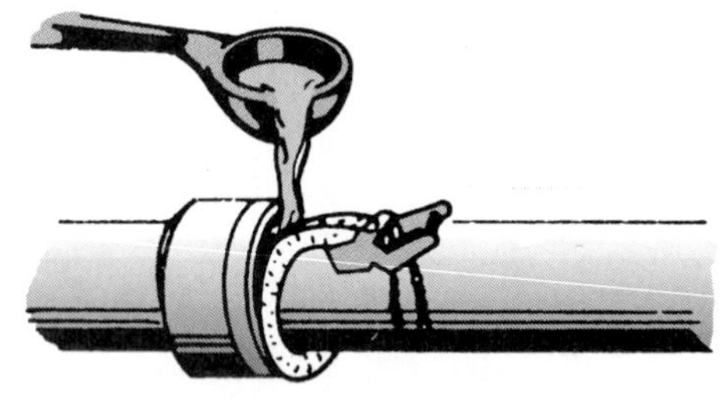

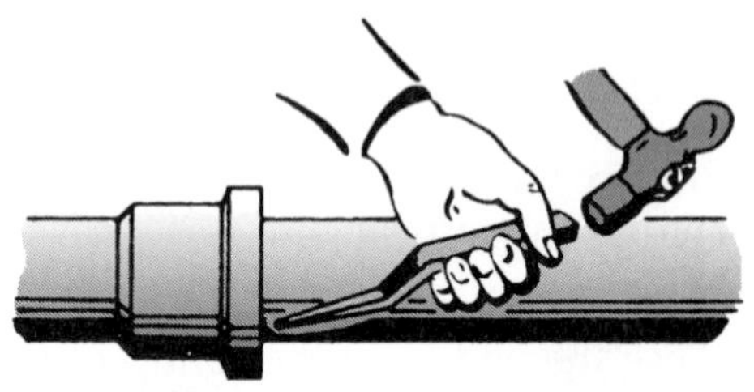

Figure 54 ◆ Making horizontal caulked joints.

5.2.2 Joining No-Hub Pipe

Couplings are used to connect no-hub pipe (see *Figure 55*). The couplings have three components: a gasket, a stainless steel shield, and clamps. Gaskets and clamps are manufactured to conform to the regulations of the Cast Iron Soil Pipe Institute. The gaskets are made of neoprene and must be labeled as neoprene and must have the Institute's trademark and patent number (3, 233, 922).

The gasket is flexible, with a ridge on the inside diameter to control the distance the gasket can be slipped onto the end of no-hub pipe.

The shield and clamps are made of stainless steel. They also should be labeled with "all stainless" or with recognized abbreviations and with the institute's trademark and patent number (3, 233, 922). Notice that the neoprene gasket is surrounded by the corrugated stainless steel shield, with two clamps around the gasket and shield. Heavy-duty couplings (see *Figure 55*) have four clamps and are available with either a 5/16-inch or 3/8-inch nut.

No-hub pipe joining requires two special tools: a soil pipe cutter and a **torque wrench.** The soil

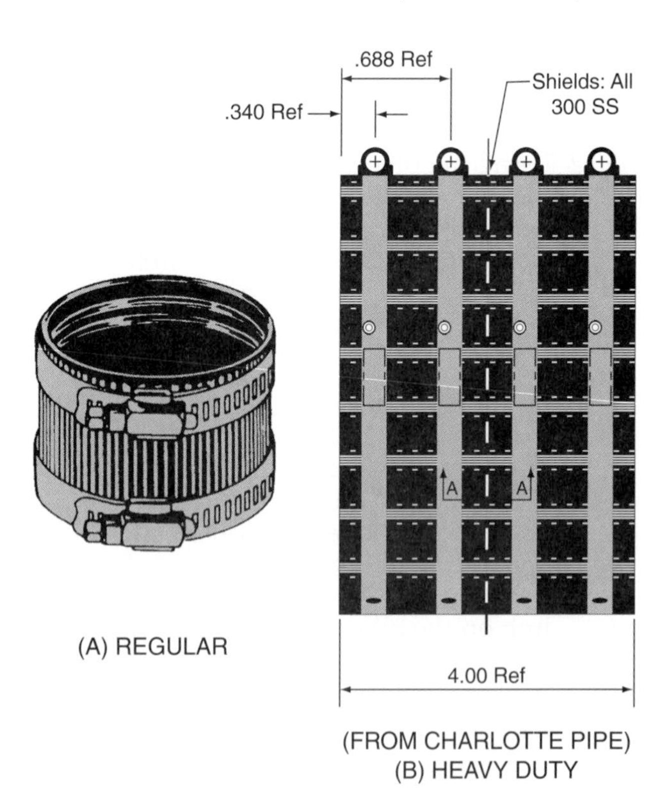

Figure 55 ◆ Couplings and heavy-duty couplings.

pipe cutter is the same tool used to cut hub-and-spigot pipe (see *Figure 35*).

The torque wrench, which is used to tighten the coupling clamps, has a 5/16-inch nut driver (see *Figure 56*). Each clamp needs to be tightened to the correct **torque**, or approximately 60 inch-pounds of force. Torque is the amount of holding pressure the clamp is exerting on the stainless steel shield. Use a torque wrench that is designed to disengage at 60 inch-pounds; you will hear a clicking sound when the wrench disengages.

When you are tightening the coupling, be sure that the torque is applied evenly. Switch the wrench from one clamp to the other. Tightening one clamp to 60 inch-pounds before you begin to tighten the other clamp could cause the gasket to become misaligned, and the coupling might leak.

An electric drill attachment is available that tightens both clamps at once, guaranteeing that the clamping pressure is evenly distributed between the two clamps (see *Figure 57*).

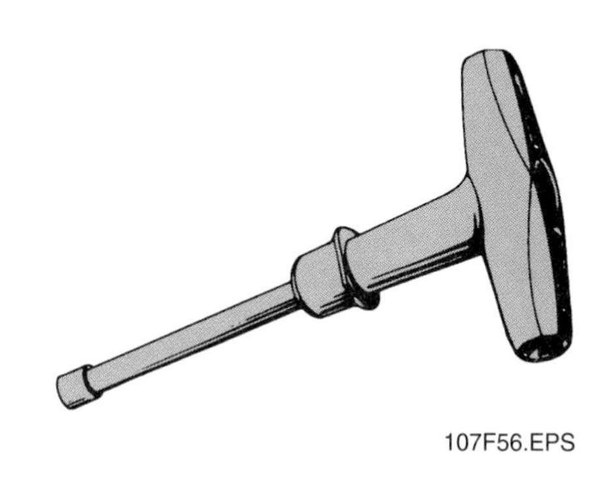

107F56.EPS

Figure 56 ◆ Torque wrench.

107F57.TIF

Figure 57 ◆ Electric drill attachment.

5.3.0 Assembling

Once you have measured accurately and cut cleanly, the pipe and fittings are ready to assemble. For both types of cast-iron pipe (hub-and-spigot and no-hub), it is critical to the success of the system to keep the pipe aligned. Without proper alignment, waste may not flow through the system properly and leaks may develop.

5.3.1 Assembling Hub-and-Spigot Pipe and Fittings

Assembling hub-and-spigot pipe and fittings requires:

- Proper alignment and slope
- Watertight joints
- Proper anchoring of the pipe and fittings

Proper alignment means that the centerline in every straight run of pipe is perfectly straight and that pipe and fittings meet at the correct angle. Pipe alignment is necessary to guarantee tight joints.

Horizontal runs of pipe must be installed to provide a slope (fall) that allows the waste to drain away. DWV piping systems depend on gravity to flow waste out of the piping system. Without the correct slope, gravity is not effective. Slope, as referred to here, is the downward slant of the pipe. Branch pipes generally require a slope of 1/8 inch to 1/4 inch per foot on all horizontal runs. At this slope, the waste in the branch moves fast enough to make sure that self-cleaning takes place in the pipe.

If the slope is more than 1/4 inch per foot, liquid and solid wastes may separate. When separation occurs, the heavier solid waste is deposited in the bottom of the pipe instead of flowing away with the liquid. When the slope is less than 1/8 inch per foot, the waste flow slows, and the branch tends to become blocked.

You will learn more about determining the slope or grade required for DWV pipes in the module *Introduction to DWV Systems.*

When the pipe is perfectly aligned, it must be secured to the building structure. This will ensure that the pipe remains aligned.

5.3.2 Assembling No-Hub Pipe and Fittings

Always clean the area of the pipe and fitting that the gasket is to cover. Dirt, mud, sand, or any other material between the gasket and the pipe can cause a leaky connection.

Pipe alignment is necessary to guarantee tight joints. The slope must also be considered, as described earlier.

As with hub-and-spigot pipe, when the pipe is perfectly aligned, it must be secured to the building structure. This will ensure that the pipe remains aligned.

When you have proper alignment and the coupling is in place, tighten the clamps evenly by alternating from one clamp to the other until each clamp has about 60 inch-pounds of torque. A special torque wrench attachment for an electric drill, as discussed earlier, is recommended because it tightens both clamps at the same time (see *Figure 57*).

Did You Know?

Cast-iron piping installations can be tested by a water test (hydrostatic test), air pressure test, smoke test, or peppermint test. Check the plumbing code for your area to determine which test should be applied. If possible, test only one floor at a time.

Codes vary on testing requirements. Some require a 10-foot head of water, and others require a 5-foot head. First, plug all openings except for one vertical pipe. Use one vertical pipe in the system and extend it at least 5 feet higher than the horizontal drain piping. If code requires a 10-foot head, extend the pipe at least 10 feet. Fill this pipe with water. This checks that the joints in the system do not leak. An inspector may require you to remove cleanout plugs or caps at different points along the system piping to make sure that the water has reached everywhere in the drainage system.

Air testing is similar to water testing, except that the pipes are filled with compressed air. A pressure gauge at the test plug will tell you whether any pressure has been lost. Some inspectors may inject smoke into the system to locate leaks. Others may use a peppermint-spiked smoke. Peppermint is so strong that any leaks are easily located by the odor.

Inspectors are especially careful to check for cross connections, defective or inferior materials, or poor craftsmanship. The inspection must be done before the pipes are covered.

Review Questions

Section 5.0.0

1. The _____ of a fitting is the point at which the lines of flow through the fitting intersect.
   a. bend
   b. sweep
   c. center
   d. joint

2. In connecting long runs containing many sections of hub-and-spigot pipe, you must add the _____ to the length of pipe to get a correct measurement.
   a. telescoping length
   b. hub length
   c. lay length
   d. insertion length

3. The _____ has a length of chain, each link of which contains a cutting wheel.
   a. rotary hammer
   b. soil pipe cutter
   c. soil pipe wrench
   d. torque wrench

4. A(n) _____ provides a fast method of joining hub-and-spigot cast-iron pipe and fittings.
   a. compression joint
   b. caulked joint
   c. lead and oakum joint
   d. insertion joint

5. The _____ keeps the hot, molten lead in the hub during the pouring operation.
   a. yarning iron
   b. joint runner
   c. oakum fiber
   d. lead wrench

6. Usually you should melt _____ pound(s) of lead for each inch of pipe size to make a caulked joint.
   a. three
   b. one-half
   c. two
   d. one

7. The _____ used to join no-hub pipes have three components: a gasket, a stainless steel shield, and clamps.
   a. compression joints
   b. couplings
   c. neoprene joints
   d. unions

8. If the no-hub plumbing materials are not _____ correctly, they might not have been manufactured according to industry standards.
   a. joined
   b. cut
   c. labeled
   d. fitted

9. Horizontal runs of pipe must be installed with _____ to allow waste to drain away.
   a. a slope
   b. supports
   c. joints
   d. a run

10. The _____ is used to tighten the coupling clamps when you are joining no-hub cast-iron pipe.
   a. snap wrench
   b. clamp wrench
   c. pipe wrench
   d. torque wrench

Summary

Cast iron refers to iron products that are formed using a mold. Two common types of iron are used to make cast-iron pipes and fittings: gray cast iron and ductile cast iron. Either type can be used to form hub-and-spigot and no-hub cast-iron pipes in lengths of 5 or 10 feet. The pipes are made in both service weight and extra-heavy weight, depending on the application.

The hub-and-spigot pipe has a bell shape or enlargement on one end. The spigot end of the next pipe in the run fits into the bell. The pipes are joined using either a compression joint or a lead and oakum joint to produce a watertight seal. No-hub cast-iron pipe has no enlargement on either end and is joined using a coupling consisting of a neoprene gasket, stainless steel shield, and clamps. The two sections of no-hub pipe are joined by inserting two ends of pipe into the coupling and then tightening the clamps securely to form a watertight seal.

Pipe may be joined using fittings. The fittings may change the direction of the pipe run, attach a vent or branch line to the main line, change the size of the pipe in a run, or connect a fixture to the drainage piping.

Bends are fittings that change the direction of the drainage or vent piping. Branches such as a wye and a tee are used to join two or more pipes to the main pipe. Wyes connect branches to the main stack at 45-degree angles. Tees connect pipe at right angles or 90-degree angles. When you are changing the size of a pipe in a run, you would use an increaser to connect to a larger pipe and a reducer to connect to a smaller pipe. Traps are installed between the drainage line and fixtures or drains. A closet bend is used to connect the water closet to the main drainage pipe.

Installed cast-iron pipe must be supported to prevent it from bending, sagging, or otherwise losing its alignment. Hangers or supports such as brackets, clamps, and pipe trapezes secure the pipe in either a horizontal or vertical position. Supports are attached to the building frame using anchors. The type of anchor used depends on the building frame material. Supports can be fastened to wooden, concrete, and masonry frames.

Planning ahead saves time and money when you are installing cast-iron pipe. Accurate measurements reduce the chances that you will waste pipe by making the wrong cut. Several techniques are used to measure pipe correctly before the pipe is cut and joined. Taking into account the depth of a fitting makes the difference in having a pipe too long or too short for the application. Always measure and mark before you cut.

As with all aspects of plumbing, the selection and installation of cast-iron pipe is strictly regulated by building and plumbing codes. Codes vary widely between locations. It is critical to check local codes that affect your work. Incorrect selection or installation will cause the job to fail inspection by plumbing code officials.

PROFILE IN SUCCESS

John Tuck

Field Foreman
Eastern Mechanical Contractors
Atlanta, GA

John is a native of Atlanta, Georgia. He grew up and continues to live and work there. The company for which he works, Eastern Mechanical, specializes in plumbing, heating/cooling process piping, utility piping (gas and fire mains, storm water pipes), gas piping, and refrigeration. While in high school, John, with an eye on his future, attended vocational training classes to get a better grip on construction skills. He started as a helper and today supervises 35 workers doing everything in plumbing installation from welding to pipe fitting.

How did you become interested in the construction industry?

I realized that high school was not going to prepare me to make a decent living. I felt that I needed to develop better skills than those available through the curriculum. I attended vocational education classes for two years while still in high school. The classes covered carpentry, masonry, plumbing, and wiring. They were enough to spark my interest in the construction industry, and I discovered that I had a knack for plumbing. I found I was very good at reading prints and putting together fittings. I had studied drafting in high school (both mechanical and architectural), so that helped me to read drawings. I continued with training programs after I started working to improve my skills further. I attended an apprenticeship program through a national plumbing association.

It has taken a long time, but success does come to those who keep working at it. In construction, everyone starts at the bottom. It's a long, hard crawl, but if you stick to it, you can make a great living. When I started, I worked in residential service. Soon, I decided I wanted to work on bigger jobs. I started doing commercial and eventually light industrial projects that went from the ground up. It was very satisfying to do a job from the very beginning and see it through to completion.

What do you think it takes to be a success in your trade?

Success takes one thing: attitude. You have to want to do what you are doing. I don't care if it is bagging groceries just so long as you like what you are doing. You have to be interested. You have to commit your heart to the job. Lots of people are content to scrape by, just learning enough to get a particular job finished. People who are driven want to develop skills and abilities. But I am not a believer in the school of hard knocks as the only way to learn. It is better to get into training programs where you learn from experienced craftspeople. It is sharing who they are, where they have been, and what they have learned that makes the learning experience a success. I try to learn from each job I do. I want to improve for the next job.

What are some of the things you do in your role of field foreman?

I am in charge of making sure the job is installed correctly. This means that the company president comes to me with a set of plans and specs and says, "Go to it." I coordinate the construction sequence. I do a take-off or estimate for materials for each of the systems (underground, aboveground, waste, and water). This take-off is let out for bids. Once the bids are back, I evaluate the contractors and make my selections. After that I order materials and make sure everything gets where it is needed when it is needed. Another part of my job is assigning people to work crews and giving them jobs to do. I move from area to area to answer questions, inspect work, and correct mistakes. I try to lead by example. I stay busy and always do my best work.

What do you like most about your job?

I like the freedom. Every job is different. Each job involves different site conditions, people, contractors, systems, piping requirements, and even weather. The variety is great. I am not trapped inside an office with a boss looking over my cubicle, or stuck in a plant tightening bolts on an assembly line. Every day has new problems to solve. This allows me to be creative. I can change to fit the situation, to improve the workflow. If I see a better way to get a job done, I can make it happen. It's a great feeling.

What would you say to someone entering the trades today?

Don't become discouraged. It's a long road. You can't see where you are going at the beginning because you are behind a hill. It takes perseverance and dedication to move up that hill. But, once you get to the top, there is no end to the possibilities. The potential for income is fantastic. The prospect for having a satisfying and fulfilling career is there. The trick is to keep at it, no matter what. Keep driving yourself to get better. It pays off.

GLOSSARY

Trade Terms Introduced in This Module

Backwater valve: A specially designed cast-iron check valve used in DWV systems.

Bend: Fitting that changes the direction of piping.

Caulking iron: Tool used to drive the lead firmly into a lead and oakum joint when working with hub-and-spigot cast-iron pipe and fittings.

Ceiling caulking iron: Specially shaped tool used to drive the lead firmly into a lead and oakum joint when working with hub-and-spigot cast-iron pipe and fittings next to a ceiling.

Check valve: Valve used to prevent backward flow of fluid in a piping system.

Closet bend: Fitting that connects the water closet to the main drainage piping.

Closet flange: Fitting used to connect the water closet to the closet bend.

Compression joint: Joint formed using a neoprene gasket inserted into the bell end of a hub-and-spigot pipe. The pressure applied between the two joined pipes forces the gasket to compress and fill in any air gaps in the joint.

Coupling: Fitting used to connect two lengths of no-hub cast-iron pipe.

Double sanitary tee: A tee fitting that changes the direction of flow from horizontal to vertical and that has openings for two branch lines. Also called a sanitary cross.

Double wye: Fitting that changes direction at 45 degrees. It connects horizontal waste pipes and branch drains to the building main. It has openings for two branch lines.

Ductile iron: A type of cast iron in which magnesium is added to the molten gray iron to reform the graphite flakes into nodular shapes and reduce brittleness.

Heel inlet: A bend that has an inlet to connect a smaller pipe to the main line. The inlet is located at the base of the curve or heel of the bend.

Hub-and-spigot cast-iron pipe: Pipe that has a bell or enlargement at one end of the pipe where the spigot (smooth end) of the next pipe slides in to form a joint. Also known as bell-and-spigot pipe.

Increaser: A fitting used to increase the size of a straight-through line of pipe. It is often used for the vent stack before it goes through the roof to reduce the chance of frost clogging the vent opening in very cold climates.

Insertion length: Length of a pipe that fits into the fitting or other joint in assembling a pipe run. The measurement must be figured into the calculation before cutting a piece of pipe for joining.

Inside caulking iron: Tool used to shape the lead near the spigot in a lead and oakum joint.

Joint runner: See *running rope.*

Ladle: Spoon-like tool used to dip the molten lead from the pot and pour it into the joint.

Lead and oakum joint: Joint for hub-and-spigot cast-iron pipe that uses molten lead poured over a rope made of oakum to form a watertight seal between the two pieces of pipe.

Long bend: Bend fitting with one leg longer than the other.

No-hub cast-iron pipe: A pipe with no enlargement or bell on either end.

Oakum fiber: A caulking material of old hemp rope or jute fibers that have been treated with tar or a tar derivative.

Outside caulking iron: Tool used to shape the lead to the inside of the hub or bell of a hub-and-spigot pipe.

Pickout iron: Tool with a diamond-shaped flat point used to remove the lead and oakum from a joint.

Pot: Vessel used to hold the molten lead.

Powder-actuated fastening tool: Tool that uses a .22-caliber shell to drive nails or fasteners into the framework of a structure. May only be used by certified workers.

Reducer: A fitting used to reduce the size of a pipe. The reducing end may be one or two pipe sizes smaller than the other end. It may be used to decrease the flow within the drainage system.

Rotary hammer drill: Drill with a pounding action that lets you drill into concrete, brick, or tile. It rotates and hammers at the same time and drills much faster than regular drills.

Running rope: Fiber rope that is clamped around the pipe using a spring clamp to keep the hot, molten lead in the hub when making a lead and oakum joint.

Sanitary cross: See *double sanitary tee.*

Sanitary tee: A tee fitting for pipe that has a slight curve in the 90-degree transition to channel the flow of wastewater or sewage from a branch line to the main line.

Service weight (SV): A lightweight kind of cast-iron pipe. Service weight refers to the kind of pipe, not its actual weight, which can vary.

Slope: Measurement of the fall of a length of pipe from level. The change in level is expressed in degrees of an angle.

Soil pipe cutter: Heavy-duty tool used for cutting cast-iron pipe.

Spring clamp: Tool used to clamp the running rope around the pipe next to the hub when making lead and oakum joints.

Sway brace: A support that keeps lengths of cast-iron pipe from swaying when they are suspended.

Sweep: Fittings with a greater radius of the curve than a bend. Used for a smooth change of direction.

Torque: The turning or twisting force applied to an object—like a nut, bolt, or screw—using a socket wrench or screwdriver to bring the object to an approved rate of tightness. Torque is measured in inch-pounds (in.-lb.) and foot-pounds (ft.-lb.). A torque wrench may cover torque ranges from 40 in.-lb. to 2,000 ft.-lb.

Torque wrench: Type of wrench with a gauge or other means to indicate the amount of rotating force applied to a fastener, such as a nut, as it is turned.

Trap: A device or fitting designed to provide a liquid seal that will prevent the back-passage of air but will allow the passage of sewage or wastewater.

Wye: Fitting used to make a 45-degree direction change in horizontal drain and waste pipes when connecting to the building main drain.

Yarning iron: Tool used to pack the oakum fiber into the joint when making lead and oakum joints to connect hub-and-spigot cast-iron pipe.

ANSWER KEY

Answers to Review Questions

Sections 2.0.0–3.0.0

1. c
2. b
3. c
4. c
5. d
6. a
7. c
8. c
9. b
10. a

Section 4.0.0

1. d
2. a
3. c
4. d
5. a

Section 5.0.0

1. c
2. d
3. b
4. a
5. b
6. d
7. b
8. c
9. a
10. d

NCCER CRAFT TRAINING USER UPDATES

The NCCER makes every effort to keep these textbooks up-to-date and free of technical errors. We appreciate your help in this process. If you have an idea for improving this textbook, or if you find an error, a typographical mistake, or an inaccuracy in the NCCER's Craft Training textbooks, please write us, using this form or a photocopy. Be sure to include the exact module number, page number, a detailed description, and the correction, if applicable. Your input will be brought to the attention of the Technical Review Committee. Thank you for your assistance.

Instructors – If you found that additional materials were necessary in order to teach this module effectively, please let us know so that we may include them in the Equipment/Materials list in the Instructor's Guide.

Write: Curriculum Revision and Development Department
National Center for Construction Education and Research
P.O. Box 141104, Gainesville, FL 32614-1104

Fax: 352-334-0932

E-mail: curriculum@nccer.org

Craft ______ Module Name ______

Copyright Date ______ Module Number ______ Page Number(s) ______

Description ______

______

______

______

(Optional) Correction ______

______

______

(Optional) Your Name and Address ______

______

______

Module 02108-00

Carbon Steel Pipe and Fittings

Course Map

This course map shows all of the modules in the first level of the Plumbing curriculum. The suggested training order begins at the bottom and proceeds up. Skill levels increase as you advance on the course map. The local Training Program Sponsor may adjust the training order.

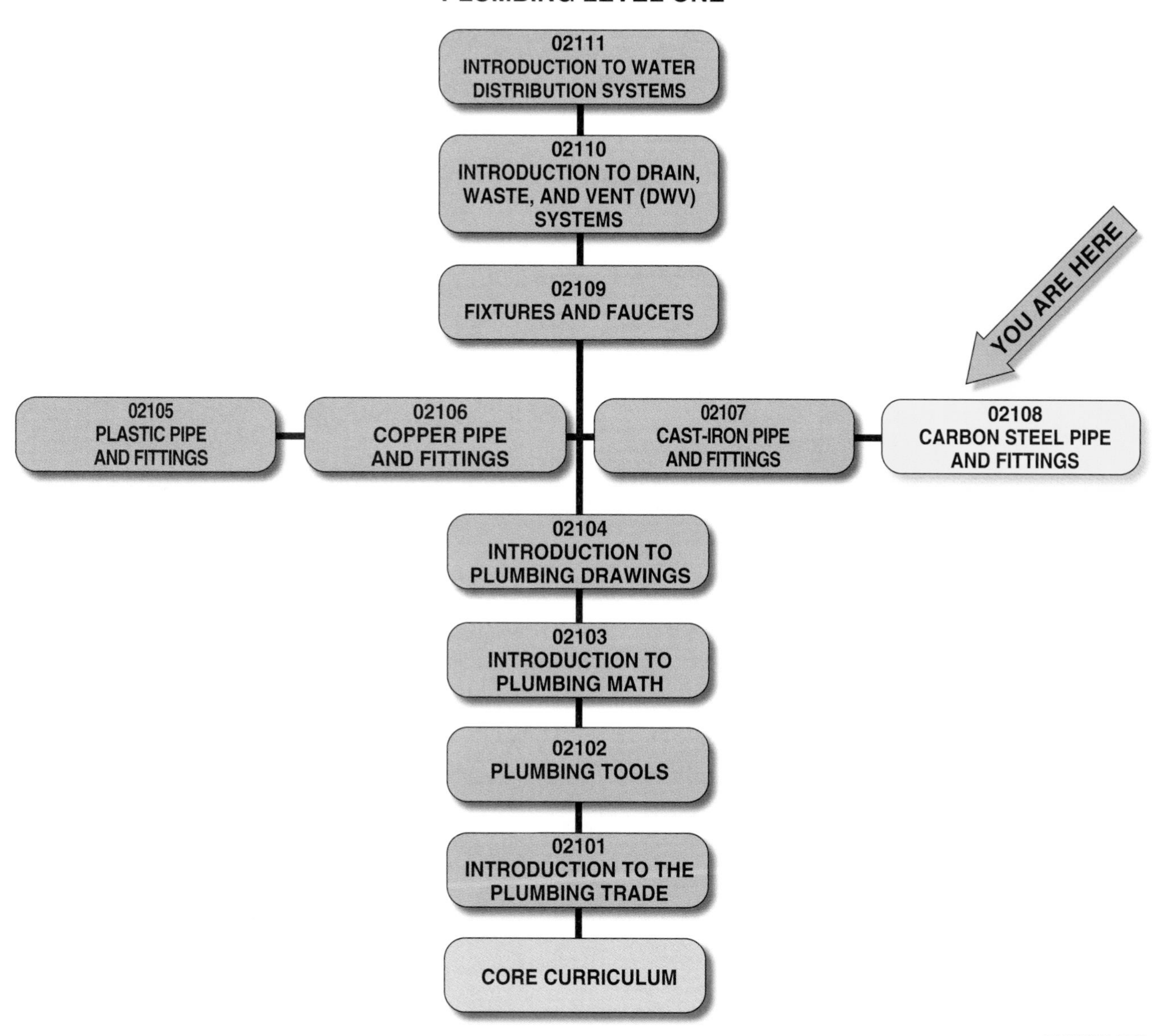

108CMAP.EPS

Copyright © 2000 National Center for Construction Education and Research, Gainesville, FL 32614-1104. All rights reserved. No part of this work may be reproduced in any form or by any means, including photocopying, without written permission of the publisher.

MODULE 02108 CONTENTS

Figures

Tables

MODULE 02108

Carbon Steel Pipe and Fittings

Objectives

When you finish this module, you will be able to:

1. Identify the common types of materials and schedules used with carbon steel piping.
2. Identify common types of fittings and valves used with carbon steel piping.
3. Identify the various techniques used in hanging and supporting carbon steel piping.
4. Demonstrate the ability to properly measure, cut, and join carbon steel piping.
5. Identify the hazards and safety precautions associated with carbon steel piping.

Prerequisites

Before you begin this module, it is recommended that you successfully complete task modules: Core Curriculum; Plumbing Level One, Modules 02101 through 02107.

Materials Required for Trainees

- *Carbon Steel Pipe and Fitting Dimensions* book
- Steel pipe in different sizes
- Steel pipe fittings
- Steel pipe cutter
- Threading die and stock
- Cutting oil
- Tape measure
- Marking tool
- Two pipe wrenches of appropriate size
- Appropriate personal protective equipment

1.0.0 ◆ INTRODUCTION

The ability to work with steel pipe is a basic plumbing skill that incorporates many other skills. Running steel pipe efficiently and properly requires knowledge of math, the ability to use both hand and power tools safely, and attention to detail. This module emphasizes the basic skills required to measure, cut, ream, thread, and assemble carbon steel pipe. This module also introduces you to grooved piping.

Steel pipe has many uses in the field. Some of these uses are hot and cold water distribution, steam and hot water heating systems, gas and air piping systems, and drainage and vent systems.

2.0.0 ◆ MATERIALS

The two common types of carbon steel pipe are **black iron pipe** and **galvanized pipe.**

Black iron pipe, which gets its color from the carbon in the steel, is most often used as gas or air-pressure pipe. You use black iron pipe where corrosion will not affect its uncoated surfaces.

Galvanized pipe is steel pipe that has been dipped in molten zinc. This protects the pipe surfaces from abrasive, corrosive materials and gives the pipe a dull, grayish color. You use galvanized pipe most often when plumbing specifications require steel pipe.

Steel

Iron ore is the base material for steel alloy. Depending on the types of elements added to the steel alloy, there are three basic kinds of steel: carbon steel, low-alloy steel, and high-alloy (stainless) steel. The term "carbon steel" is often used to avoid confusion with stainless steel. Carbon steel has up to 1.65 percent manganese or small quantities of silicon, aluminum, copper, and other elements. Manganese helps prevent brittleness and improves strength, hardness, and flexibility.

Carbon steel is produced by one of the following four processes: open-hearth, electric furnace, basic-oxygen, or Bessemer.

The open-hearth process has been the main method of producing steel for many years. An open-hearth furnace is a huge, shallow tank lined with firebrick. A fuel (gas, oil, powdered coal) is mixed with hot air in burners and blasted across the tank. Molten iron, steel scrap (recycled steel), and limestone are melted together in the tank. After the removal of impurities collected by the limestone, the steel is ready.

The electric furnace process uses tremendous charges of electricity to produce steel. The furnace is loaded with scrap iron. An electrical arc is produced using electrodes. The resulting heat (6,000°F, or 3,312°C) melts the scrap iron and produces the steel alloy.

The basic-oxygen process is the method used most often today. Molten iron from a blast furnace (a giant oven lined with firebrick) is poured into the basic-oxygen furnace. Pure oxygen is blown onto the metal through a tube called an oxygen lance. The oxygen combines with the impurities and is removed from the furnace, leaving the steel.

The Bessemer process, developed in the 1850s, produced steel by blowing air through molten iron at high pressure. The process used oxygen to draw off impurities, but it was not efficient enough to remove all the unwanted elements. This process is no longer used in the United States.

After the liquid steel has been created in the furnaces, it must be further processed and cast into solid form before manufacturers can use it.

Did You Know?

The concept of grooved piping dates back to World War I, when it was necessary for armies in the field to join pipe quickly and efficiently. Since then, grooved piping has undergone continuous development. Grooved pipe is now a primary means of joining pipe. Steel pipe, cast-iron pipe, and ductile iron pipe may have grooved joints.

2.1.0 Schedules and Sizes

Steel pipe is measured by the **nominal size,** which is based on the approximate inside diameter (ID) of the pipe, or by the outside diameter (OD). The cross sections in *Figure 1* show the inside and outside diameters for 1-inch nominal-size steel pipe.

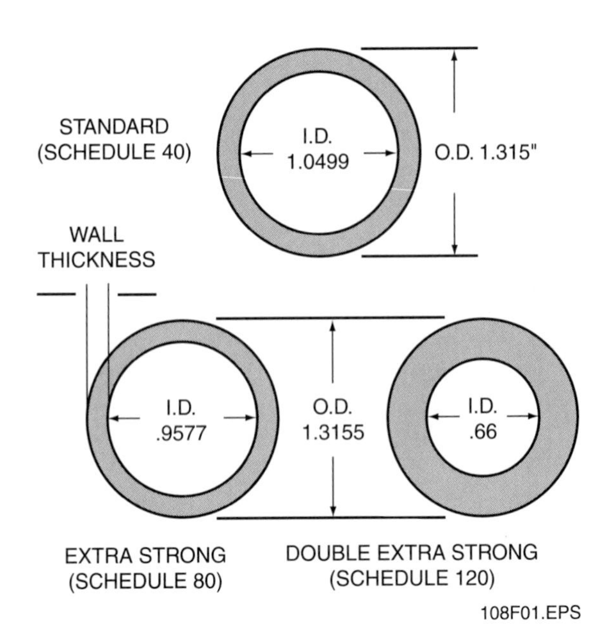

Figure 1 ◆ Inside and outside diameter.

Nominal is the word normally used in the plumbing trade to describe an approximation based on the inside diameter.

Steel pipe is manufactured in the following nominal sizes: ⅛ inch, ¼ inch, ½ inch, ¾ inch, 1 inch, 1¼ inches, 1½ inches, 2 inches, 2½ inches, 3 inches, 4 inches, 5 inches, 6 inches, 8 inches, 10 inches, and 12 inches. Pipe larger than 12 inches is measured by its outside diameter.

Steel pipe for threading is manufactured in three weights or strengths: **standard weight, extra strong,** and **double extra strong.** Extra strong is referred to by some manufacturers as extra heavy, and double extra strong is referred to as double extra heavy.

Because all weights for a particular size of pipe have the same outside diameter, the same threading dies will fit all three weights of each size of pipe.

The wall thickness of the pipe and the inside diameter differ with each weight. The thicker the wall, the smaller the inside diameter, but the more pressure the pipe will withstand. Standard weight will be adequate in most plumbing situations. However, the two stronger weights must be used when pressures require it. *Table 1* gives some dimensions of Schedule 40 (standard) and Schedule 80 (extra strong) steel pipe.

Steel pipe is manufactured in lengths of 21 feet, threaded or unthreaded. Larger sizes (6 inches and above) come in random lengths of 18 to 22 feet.

2.1.1 Labeling

For ease of handling, pipe is shipped in bundles. The number of lengths per bundle is determined by the size of the pipe. A tag is attached to each bundle, labeling the number of feet within the bundle. Each length of steel pipe that is at least 1½ inches must be labeled with the name or brand of the manufacturer, its length, and ASTM A 120, which identifies that it meets American Society for Testing and Materials standards. Carbon steel pipe 1½ inches and smaller must have that information on the tag attached to a bundle.

2.1.2 Threads

The standard thread used on pipe is V-shaped, with an angle of 60 degrees, very slightly rounded at the top (see *Figure 2*). The taper is 1/32 inch per inch of length. Tapered pipe threads produce a pressure-tight connection. When they are tight, they also produce a mechanically rigid piping system.

Table 1 Dimensions of Schedule 40 and Schedule 80 Steel Pipe

	Dimensions of Schedule 40 (Standard Weight) Steel and Wrought Iron Pipe			Dimensions of Schedule 80 (Extra Strong) Steel and Wrought Iron Pipe		
Nominal Diameter, Inches	**Actual Inside Diameter, Inches**	**Actual Outside Diameter, Inches**	**Weight per Foot Pounds**	**Actual Inside Diameter, Inches**	**Actual Outside Diameter, Inches**	**Weight per Foot Pounds**
⅛	0.269	0.405	0.25	0.215	0.405	0.31
¼	0.364	0.540	0.43	0.302	0.540	0.54
⅜	0.493	0.675	0.57	0.423	0.675	0.74
½	0.622	0.840	0.86	0.546	0.840	1.09
¾	0.824	1.050	1.14	0.742	1.050	1.47
1	1.049	1.315	1.68	0.957	1.315	2.17
1¼	1.380	1.660	2.28	1.278	1.660	3.00
1½	1.610	1.900	2.72	1.500	1.900	3.63
2	2.067	2.375	3.66	1.939	2.375	5.02
2½	2.469	2.875	5.80	2.323	2.875	7.66
3	3.068	3.500	7.58	2.900	3.500	10.25
3½	3.548	4.000	9.11	3.364	4.000	12.50
4	4.026	4.500	10.80	3.82	4.500	14.98
5	5.047	5.583	14.70	4.81	5.563	20.78
6	6.065	6.625	19.00	6.76	6.625	28.57

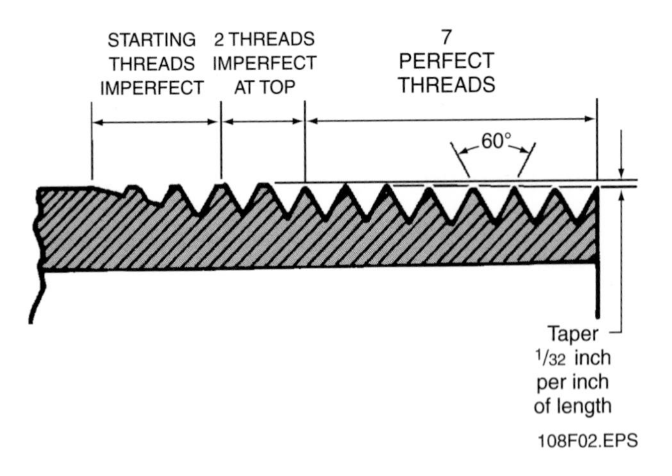

Figure 2 ◆ American standard thread.

Notice in *Figure 2* that the first seven threads are perfect threads. They are sharp at the top and bottom. The next two threads are imperfect at the top and perfect at the bottom. The rest of the threads are not perfect at the bottom or the top. The imperfect threads are those threads that were not completely cut. They have no sealing power. If the perfect threads are marred or broken, they will lose their sealing power.

Did You Know?

Pipe manufacturers adopted the American Standard Pipe and Pipe Thread Dimensions in 1886. This decision standardized the thread size on all manufactured pipe products.

The number of threads per inch required for each pipe size is given in *Table 2*. Notice the number of usable threads required.

Review Questions

Section 2.0.0

1. The type of carbon steel pipe most often used as gas or air-pressure pipe, where corrosion will not affect its uncoated surfaces, is called _____ pipe.
   a. stainless steel
   b. black iron
   c. steel alloy
   d. galvanized

Table 2 American Standard Pipe and Pipe Thread Dimensions

Nominal Pipe Size	Threads Per Inch	Number of Usable Threads*	Hand-Tight Engagement** (Inches)	Thread Makeup Wrench and Hand Engagement** (Inches)	Total Length of External Thread (Inches)
1/8	27	7	3/16	1/4	3/8
1/4	18	7	1/4	3/8	9/16
3/8	18	7	1/4	3/8	5/8
1/2	14	7	5/16	1/2	3/4
3/4	14	8	5/16	9/16	13/16
1	11 1/2	8	3/8	11/16	1
1 1/4	11 1/2	8	7/16	11/16	1
1 1/2	11 1/2	8	7/16	3/4	1
2	11 1/2	9	7/16	3/4	1 1/16
2 1/2	8	9	11/16	1 1/8	1 9/16
3	8	10	3/4	1 3/16	1 5/8
3 1/2	8	10	13/16	1 1/4	1 11/16
4	8	10	13/16	1 5/16	1 3/4
5	8	11	15/16	1 3/8	1 13/16
6	8	12	15/16	1 1/2	1 15/16
8	8	14	1 1/16	1 11/16	2 1/8
10	8	15	1 3/16	1 15/16	2 3/8
12	8	17	1 3/8	2 1/8	2 9/16

*Rounded to the nearest thread.

**Rounded to the nearest 1/16 inch.

2. Steel pipe that has been dipped in molten zinc to protect the pipe surfaces from abrasive and corrosive materials is called _____ pipe.
   a. carbon steel
   b. steel alloy
   c. galvanized
   d. black iron
3. Schedule 40 steel pipe is _____ weight.
   a. extra heavy
   b. double extra heavy
   c. standard
   d. service
4. Schedule 80 steel pipe is _____ weight.
   a. double extra heavy
   b. extra strong
   c. standard
   d. heavy
5. All weights for a particular size pipe have the same _____ .
   a. outside diameter
   b. inside diameter
   c. thread diameter
   d. nominal size

3.0.0 ◆ COMMON FITTINGS AND VALVES

Plumbing codes require that different types of fittings be used in water supply and distribution systems (**pressure fittings**) and drain, waste, and vent (DWV) systems (**drainage fittings**). These fittings are used to change the direction of flow in either system. Valves regulate the flow of water in a plumbing system.

3.1.0 Fittings and Valves for Threaded Pipe

Pipe fittings for steel pipe are generally made of malleable iron or cast iron. Malleable iron has a slight temper produced by prolonged **annealing** of ordinary cast iron. This process makes the iron tough. Also, it can be bent or pounded to some extent without breaking.

Iron pipe fittings are manufactured in two types: ordinary or pressure fittings and recessed or drainage fittings (see *Figure 3*). Drainage fittings have a drainage pattern design and recessed threads that should be used only on soil and waste piping systems.

Drainage fittings differ from ordinary fittings in several ways. The insides of drainage fittings are smooth and are shaped for easy flow. Their shoulders are recessed so that when piping is screwed in, it fills the recessed area and forms an unbroken internal contour. Drainage fittings are also designed so that horizontal lines entering them will have a slope of ¼ inch per foot. Ideally, a pipe screwed into a drainage fitting completely fills the recessed cavity. However, this condition is rarely achieved. Illustration 2 of *Figure 3* shows the condition of most drainage joints.

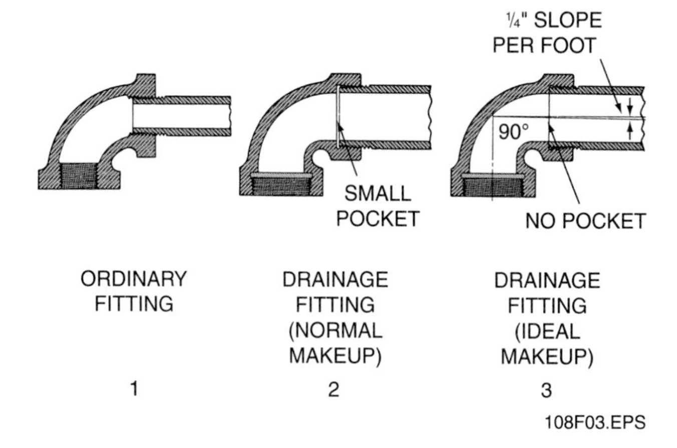

Figure 3 ◆ Steel pipe fittings.

More common pressure fittings are used for hot and cold water distribution systems, steam and hot water heating systems, and gas and air piping systems. In any of these systems, slight obstructions on the inside surface of the pipe and fitting joint do not affect the flow of materials. These fittings are discussed below.

3.1.1 Tees

A tee is a fitting used to make a branch that is 90 degrees to the main pipe. You can buy tees in a number of sizes and patterns. If all three outlets are the same size, the fitting is called a regular tee, shown in *Figure 4*. If outlet sizes vary, the fitting is called a reducing tee, shown in *Figure 5*. Tees are specified by giving the straight-through (run) dimensions first, then the side-opening dimensions. For example, a tee with a run outlet of 2 inches, a run outlet of 1 inch, and a branch outlet of ¾ inches is known as a *2 × 1 × ¾ tee* (always state the large run size first, the small run size next, and the branch size last). Tees are also available with male threads on a run or branch outlet.

3.1.2 Elbows

Elbows, often called ells, are used to change the direction of the pipe. The most common ells,

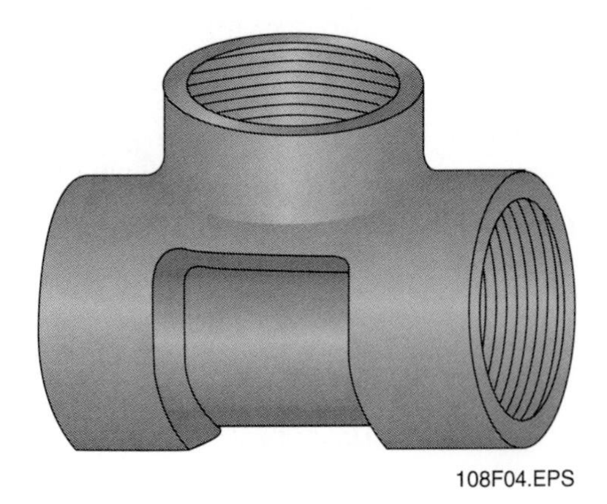

Figure 4 ◆ Regular tee.

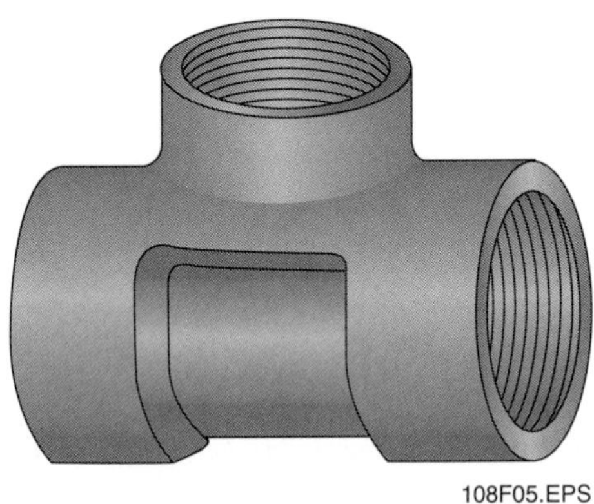

Figure 5 ◆ Reducing tee.

shown in *Figure 6*, are the 90-degree ell; the 45-degree ell; the street ell, which has a male thread on one end; and the reducing ell, which has outlets of different sizes. Ells are also available to make 11¼-degree, 22½-degree, and 60-degree bends.

3.1.3 Unions

Unions make it possible to disassemble a threaded piping system. After disconnecting the union, you can unscrew the length of pipe on either end of the union. There are various types of steel pipe unions. The two most common are the ground joint and the flange.

Two pipes are connected by a **ground joint union** (see *Figure 7*) by screwing the thread and shoulder pieces of the union onto the pipes. The collar then draws both the shoulder and thread parts together. This union creates a gastight or watertight joint.

With a **flange union** (see *Figure 8*), the flanges screw to the two pipes to be joined. Then the flanges are pulled together by nuts and bolts. A

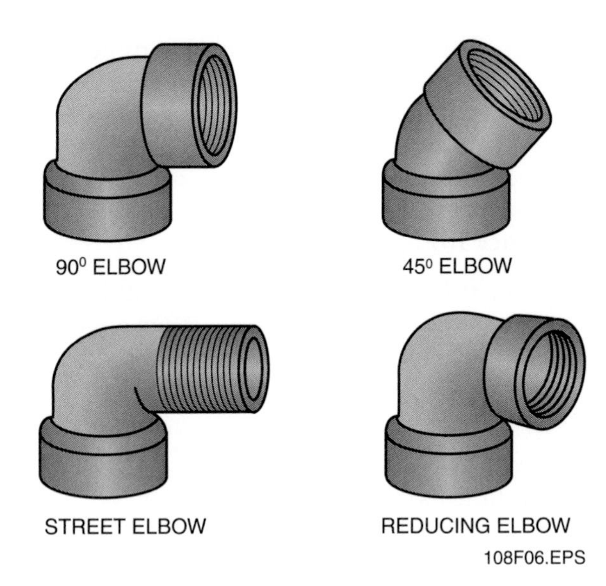

Figure 6 ◆ Elbows.

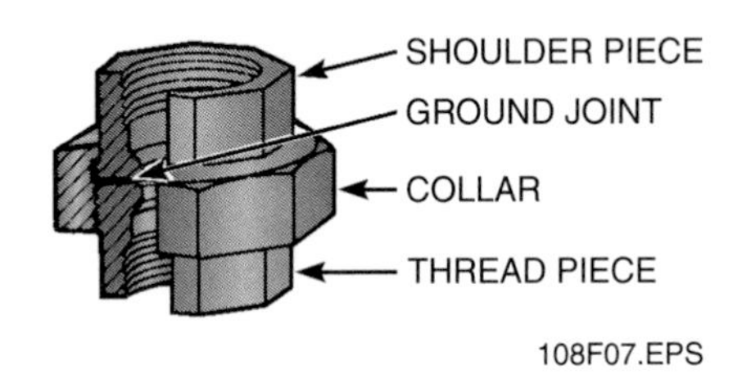

Figure 7 ◆ Ground joint union.

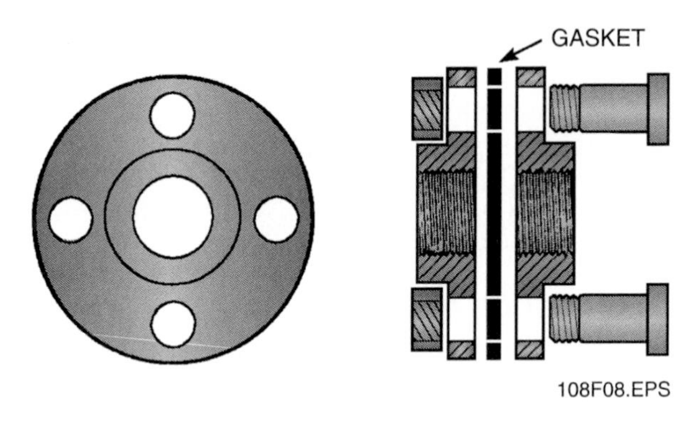

Figure 8 ◆ Flange union.

gasket between the flanges makes this connection gastight or watertight.

3.1.4 Couplings and Other Common Fittings

Couplings (see *Figure 9)* are short fittings with female threads in both openings. They are used to connect two lengths of pipe in making straight runs. Couplings cannot be used in place of unions because they cannot be disassembled.

Other common fittings include **nipples, crosses, plugs, caps,** and **bushings.**

Nipples (see *Figure 10*) are pieces of pipe, 12 inches or less in length, that are threaded on both ends and are used to make extensions from a fitting or to join two fittings. Nipples are manufactured in many sizes, beginning with the close or all-thread nipple.

Crosses (see *Figure 11)* are four-way distribution devices.

Plugs are male-threaded fittings used to close openings in other fittings and fit on the female end of a pipe or nipple. There are a variety of heads (square, slotted, and hexagon) found on plugs, as shown in *Figure 12.*

Caps (see *Figure 13*) are fittings with a female thread that are used for the same purpose as plugs, except that the cap fits on the male end of a pipe or nipple.

Bushings (see *Figure 14*) are fittings with a male thread on the outside and a female thread on the inside. They are usually used to connect the male end of a pipe to a fitting of a larger size. The ordinary bushing has a hexagon nut at the female end.

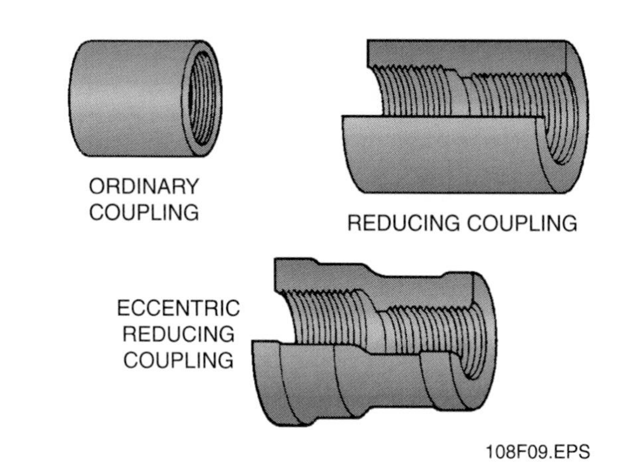

Figure 9 ◆ Couplings.

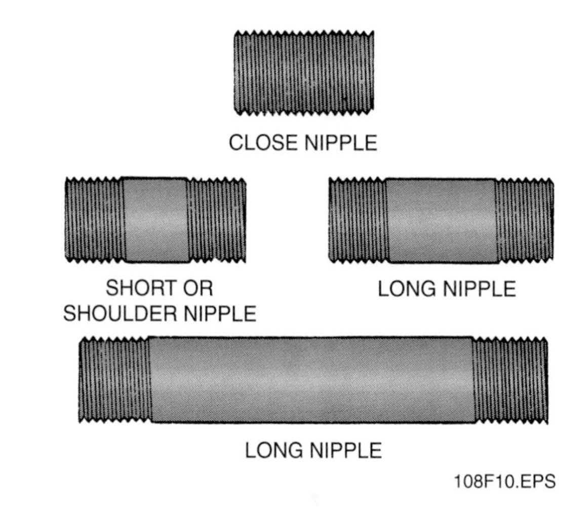

Figure 10 ◆ Nipples.

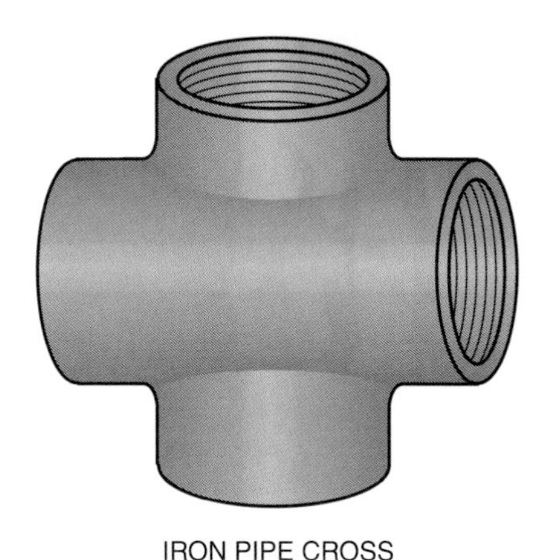

Figure 11 ◆ Cross.

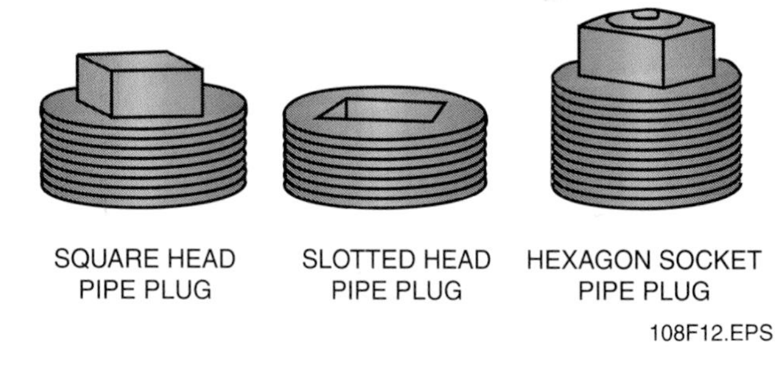

Figure 12 ◆ Plugs.

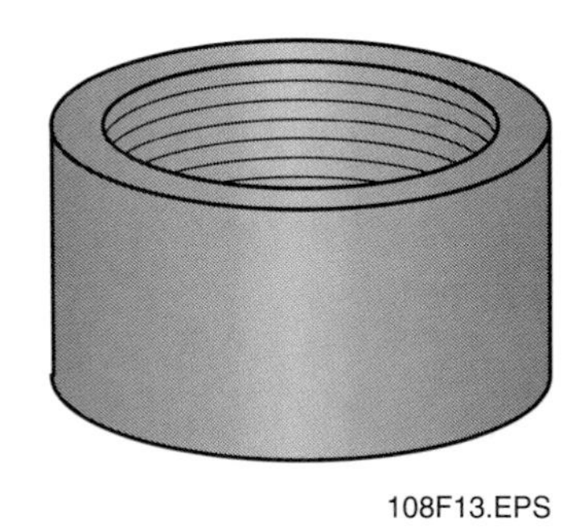

Figure 13 ◆ Cap.

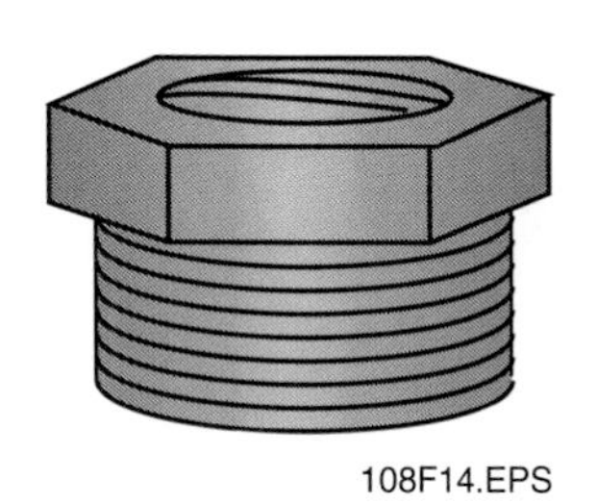

Figure 14 ◆ Bushing.

3.2.0 Fittings and Valves for Grooved Pipe

A grooved pipe joint consists of the following components:

- Specially grooved pipe ends cut by special machines to exact manufacturer's standards
- A synthetic rubber gasket
- Lubricant
- A coupling consisting of two housings

A wide variety of fittings and valves are available with grooved ends. Special gaskets are also available for various service conditions.

It is important to consult the manufacturer's design specifications before using a grooved pipe system.

3.3.0 Threaded Valves

All common valves are manufactured for threaded pipe. The most commonly used are the **gate valve, globe valve, ball valve,** and **plug valve** (or gas service valve). *Figure 15* shows these types of valves.

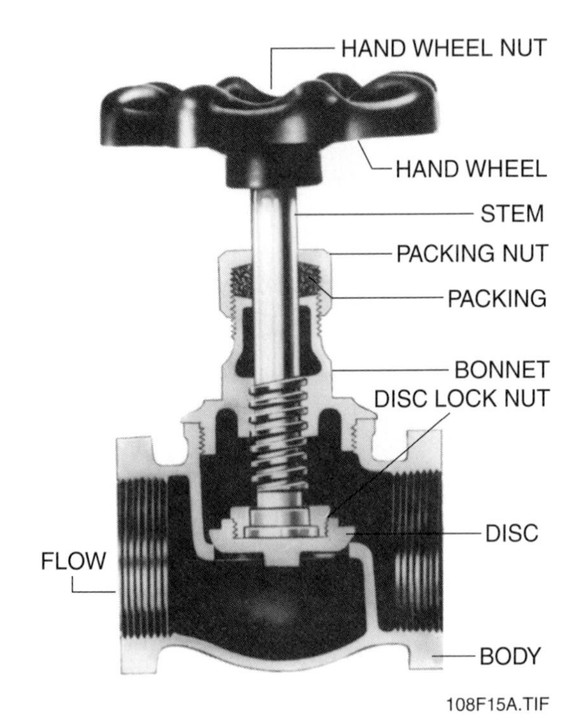

108F15A.TIF

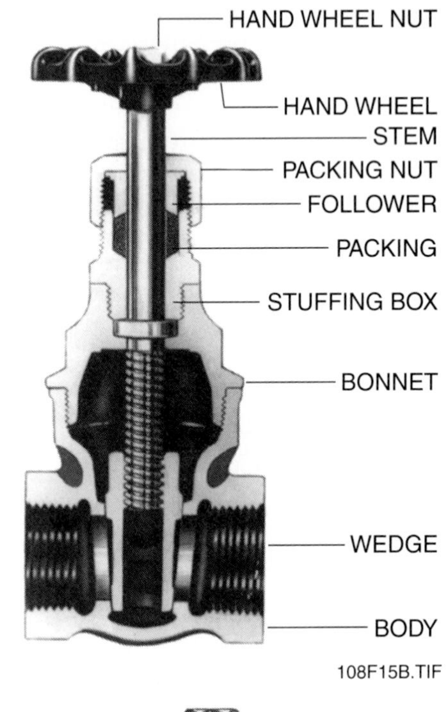

108F15B.TIF

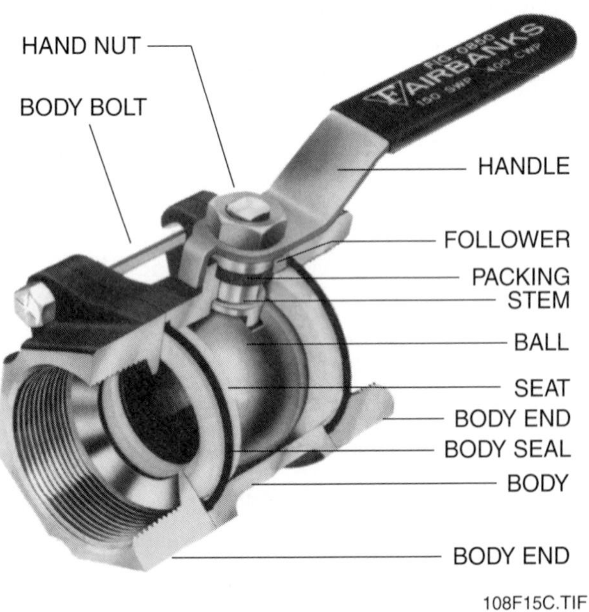

108F15C.TIF

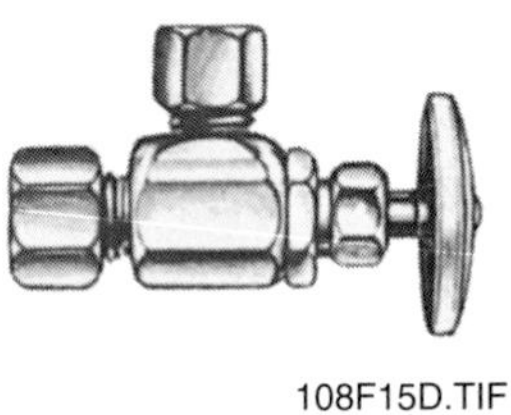

108F15D.TIF

Figure 15 ◆ Types of valves.
(A) Globe valve.
(B) Gate valve.
(C) Ball valve.
(D) Plug valve.

Gate valves control flow by moving a gate at a right angle to the flow. They are best suited for main supply lines and pump lines. They can be used in lines for steam, water, gas, oil, and air.

Globe valves are recommended for general service on steam, water, gas, and oil, where frequent operation and close control of flow are required. To close the valve, turn the handle clockwise until the valve stem firmly seats the washer or disc between the valve stem and the valve seat. This stops the flow of gas or liquid.

Ball valves are used to control the flow of gases and liquids. They are also used where quick cutoffs for line maintenance are required. The ball part of the valve rotates into the open or closed position.

Plug valves (or gas service valves) are similar to ball valves in how they work, but plug valves use a tapered plug with a hole or slot instead of the ball. This valve also can be turned on or off with a quarter turn. Plug valves are most often used for natural gas service and stops.

Review Questions

Section 3.0.0

1. A fitting that has female thread in both openings and is used to connect two lengths of straight pipe is a _____ .

a. union
b. plug
c. coupling
d. bushing

2. A female-threaded fitting used to close off the male end of a pipe or a nipple is a _____ .

a. plug
b. cross
c. bushing
d. cap

3. A ____ is a four-way distribution fitting.

a. nipple
b. plug
c. cross
d. cap

4. A _____ fitting is smooth on the inside and shaped for easy flow.

a. drainage
b. bushing
c. pressure
d. compression

5. _____ valves are recommended for general service on steam and water.

a. Gate
b. Globe
c. Ball
d. Plug

4.0.0 ◆ HANGERS AND SUPPORTS

The plumber must install and support all pipe so the pipe and joints remain leakproof. Improper support can cause joints to sag, and the added stress increases the possibility that the pipe will leak, break, or crack. The plumber needs to know how to attach pipe to wood, masonry (including tile and concrete), and steel surfaces.

4.1.0 Types and Uses of Hangers and Supports

Different types of supports and hangers are designed for holding and supporting pipe in either a horizontal or vertical position. One or more hangers manufactured for the purpose should be used to support carbon steel pipe.

The basic components of hangers are pipe attachments, connectors, and structural attachments.

4.1.1 Pipe Attachments

Pipe attachments touch or connect directly to the pipe. Rings, clamps, and clevises are among the attachments that can support piping from the ceiling horizontally (see *Figure 16*).

Pipe hooks, U hooks, J hooks, tube straps, tin straps, perforated band irons, half clamps, suspension clamps, hold down clips, and pipe clamps can support piping on wood-frame construction. *Figure 17* shows some of these supports.

Vertical pipe should be supported at each floor level using riser clamps (see *Figure 18*).

Other pipe attachments include universal pipe clamps (see *Figure 19*) and standard 1⅝-inch or 1½-inch channels (see *Figure 20*). The clamps come in standard finishes of mild steel and electrogalvanized.

4.1.2 Connectors

Connectors link the pipe attachment to the structural attachment. The two main groups of connectors are (1) rods and bolts and (2) other rod attachments.

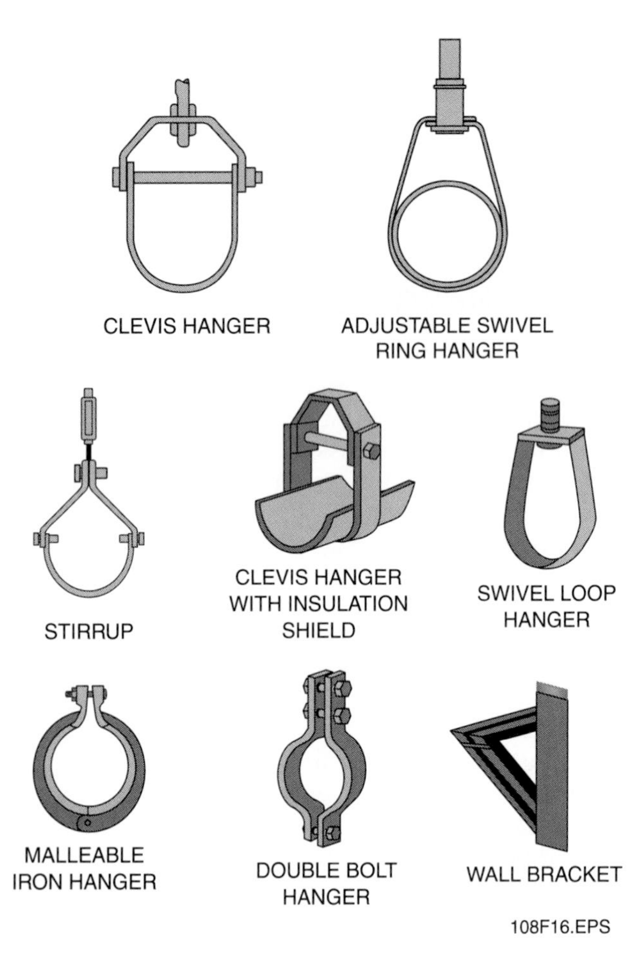

Figure 16 ◆ Hangers used for horizontal support.

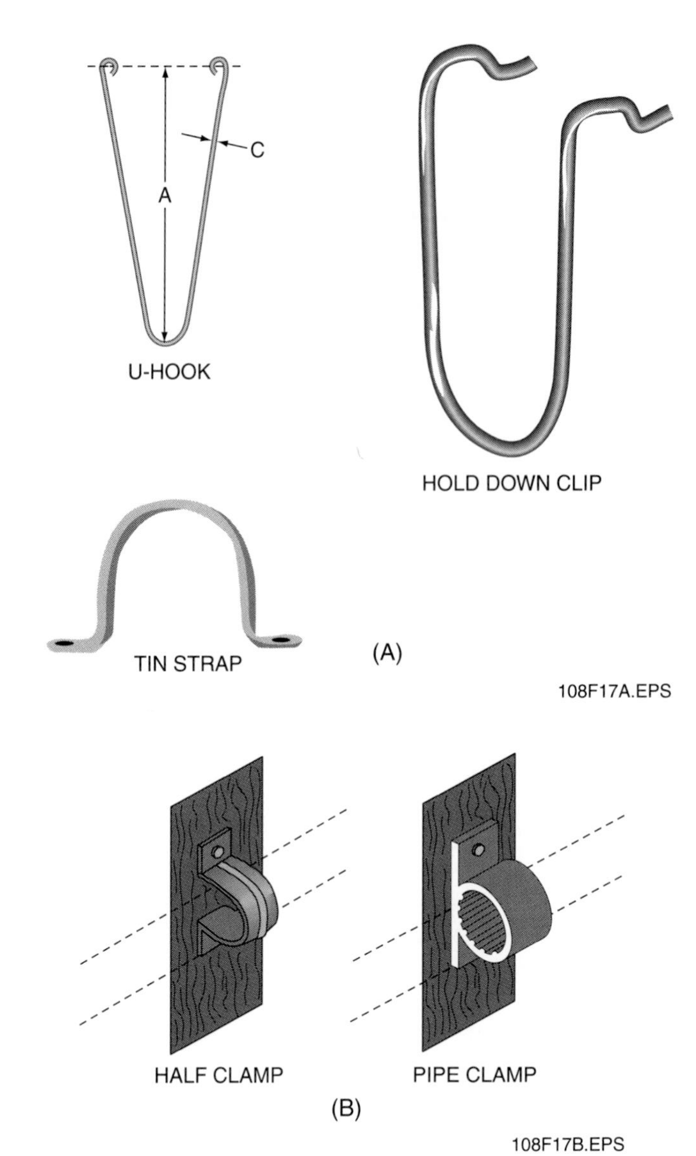

Figure 17 ◆ Pipe supports for wooden frames.

Rod attachments, which are available in several sizes, include eye sockets, extension pieces, rod couplings, hanger adjusters, turnbuckles, clevises, and eye rods (see *Figure 21*). Eye sockets are used with hanger rods. You will use extension pieces when you are attaching hanger rods to beam clamps and other types of building attachments.

Rod couplings and reducing rod couplings support pipe lines where it is possible to connect to an existing stud.

4.1.3 Structural Attachments

Structural attachments are the anchors and anchoring devices that hold the pipe hanger assembly securely to the building structure. These include powder-actuated anchors, concrete inserts, beam clamps, C-clamps, beam attachments, various brackets, ceiling flanges and plates, plate washers, and lug plates.

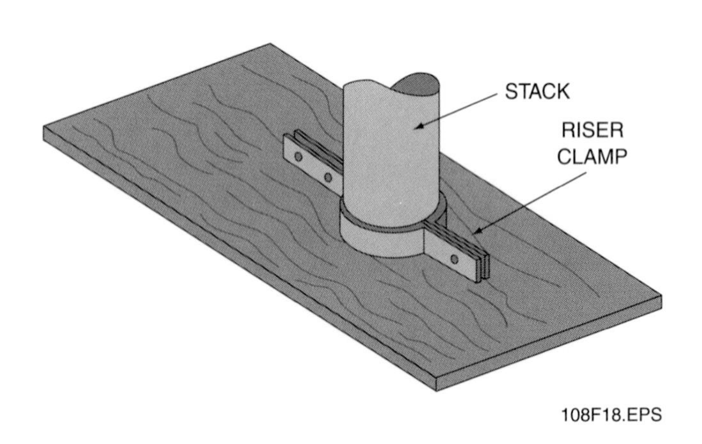

Figure 18 ◆ Vertical pipe supports.

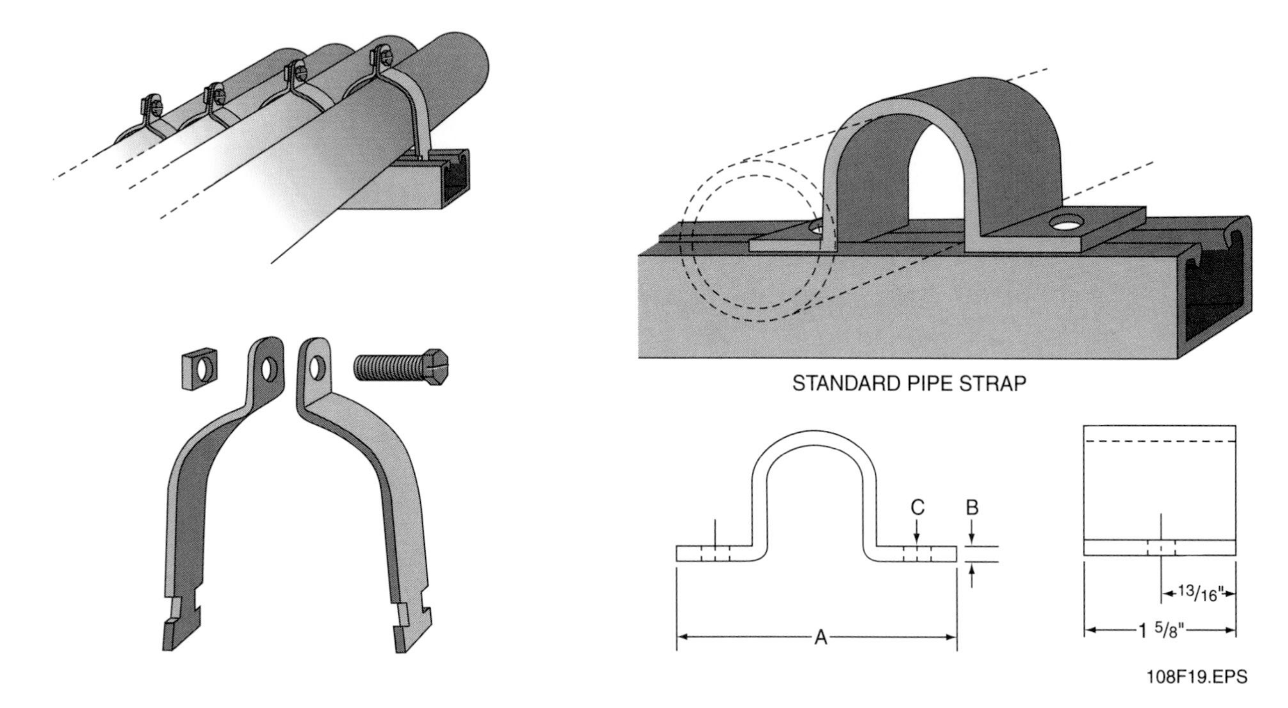

Figure 19 ◆ Universal pipe clamps.

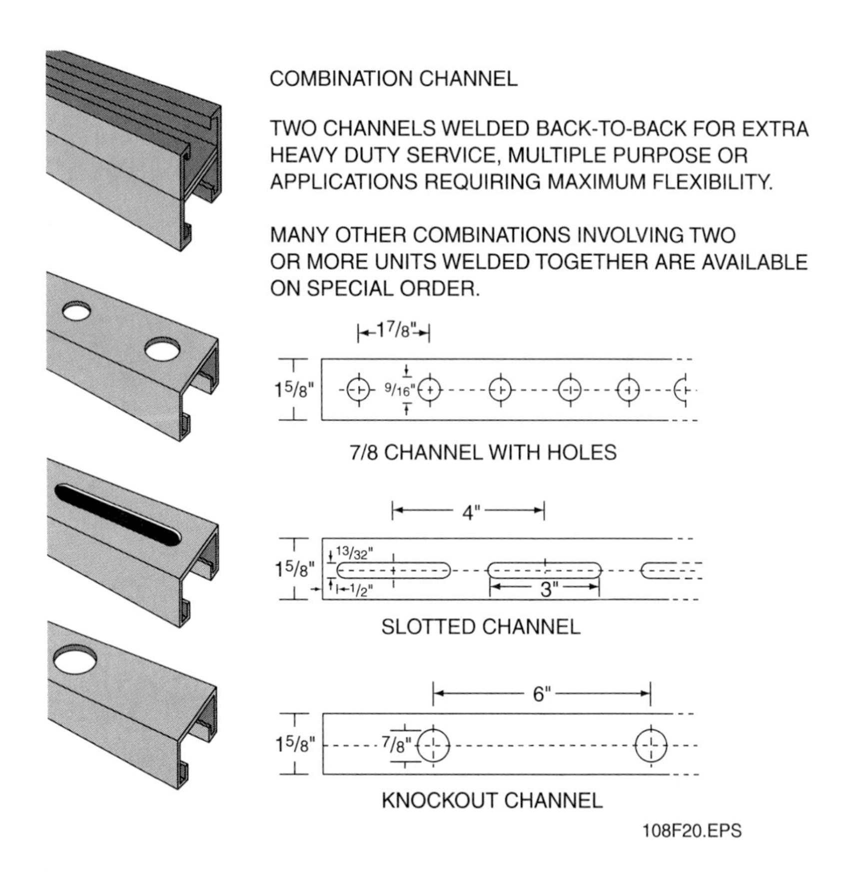

Figure 20 ◆ Pipe channels.

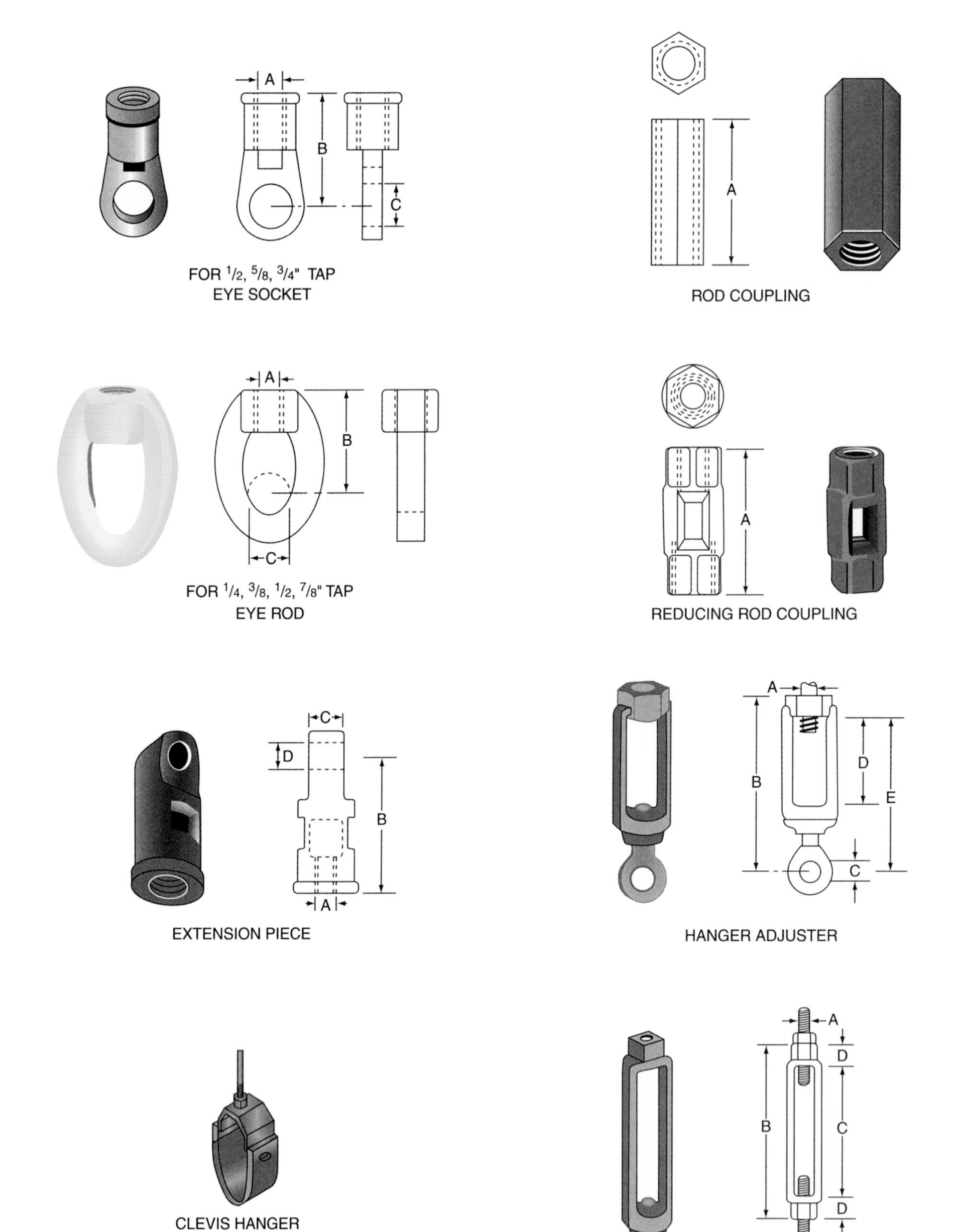

Figure 21 ◆ Connectors.

Wedge anchors, sleeve anchors, and stud anchors (see *Figure 22*) can be used to fasten the piping system to concrete or brick. Non-drilling anchors and self-drilling anchors also can be used in concrete. Do not use these anchors in new concrete that has not had enough time to cure.

In steel-framed buildings, clamp-like devices are often used to attach hangers for smaller pipes. These include beam clamps, I-beam clamps, and steel C-clamps.

In wood-framed buildings, nails and screws typically are used to attach pipe hangers and supports to the structure. The riser clamp, pipe clamp, or pipe strap is placed over the pipe and secured to the structural frame for vertical pipe runs. Plumbers' tape is nailed directly onto the floor joists.

4.1.4 Installation

Vertical piping should be supported at sufficient intervals to keep the pipe in alignment. How you support vertical piping depends on the size of pipe and whether it needs to be supported at or between the floor lines. It also depends on local code requirements (see *On the Level: Local Code Requirements*).

Vertical pipe can be supported at each floor line with riser clamps (see *Figure 23*). Vertical runs between floor levels can be supported with either an extension ring hanger and wall plate (see *Figure 24*) or a one-hole strap (see *Figure 25*).

Figure 23 ◆ Riser clamps.

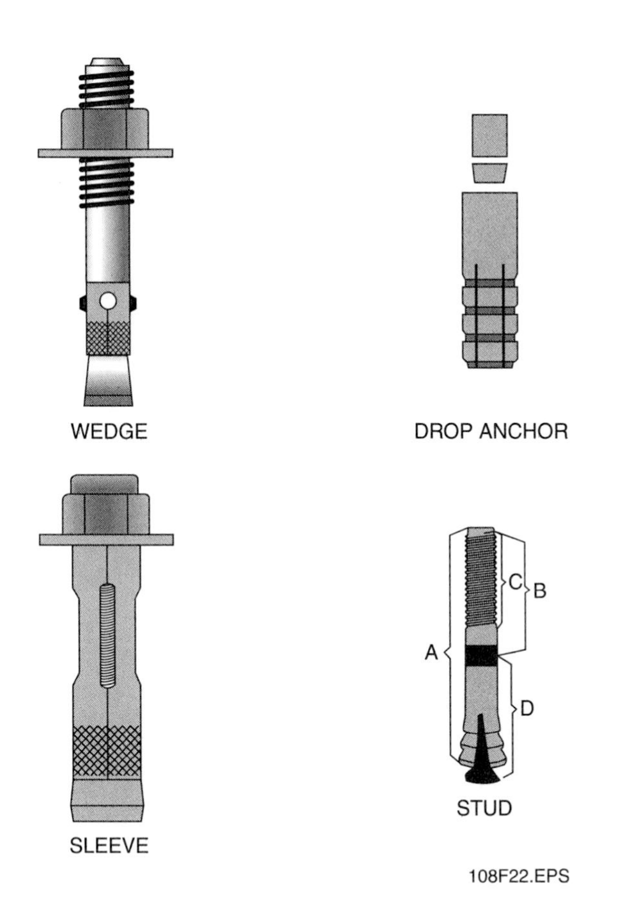

Figure 22 ◆ Anchors for concrete or brick.

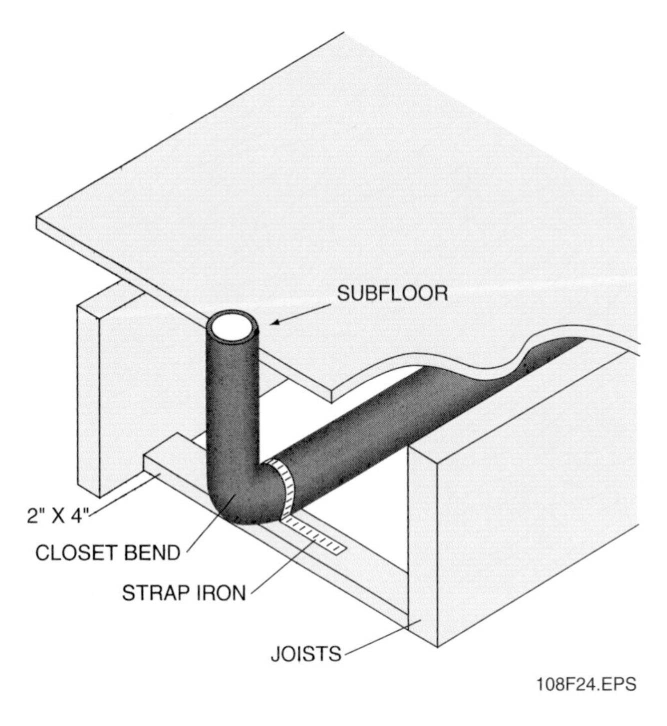

Figure 24 ◆ Extension ring hanger and wall plate.

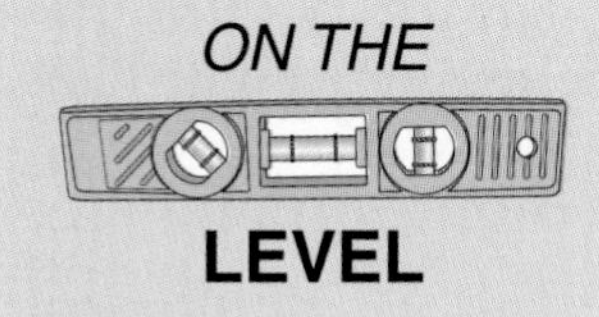

Local Code Requirements

Consult the local code and manufacturer's recommendations for spacing intervals when installing pipe hangers and supports. Codes vary widely from one area to another. Often you can find spacing requirements in the specifications section of the building plans. Failing to support pipe adequately will get a rejection from the plumbing inspector. Poor workmanship or misunderstandings will earn a rejection as well.

You must also check the codes to know which type of fasteners are allowed in your area. Again, you will often find this information in the specifications. Double-check both sources. If they don't agree, ask your supervisor.

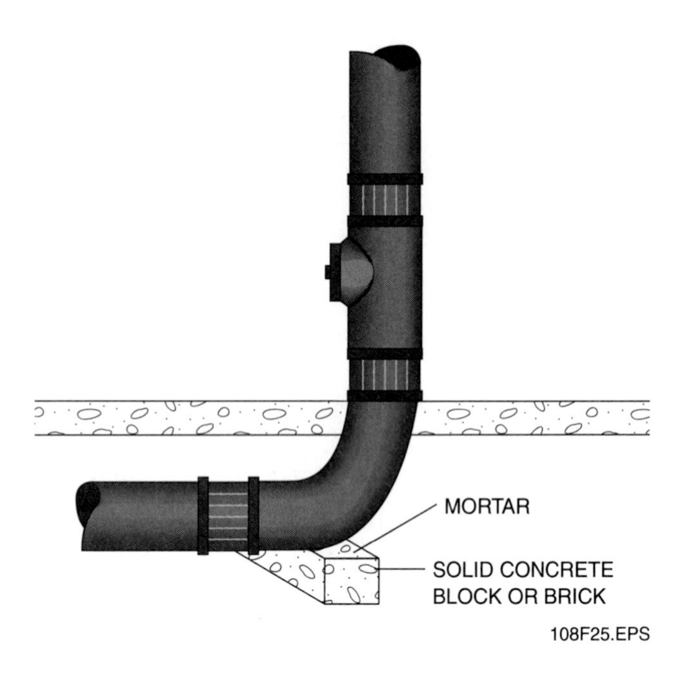

Figure 25 ◆ One-hole strap.

Review Questions

Section 4.0.0

1. The basic components of hangers are _____ .
   a. pipe attachments, connectors, and structural attachments
   b. pipe hooks, U hooks, and J hooks
   c. connectors, couplings, and anchors
   d. clamps, pipe attachments, and pipe hooks
2. Carbon steel pipe should be supported with _____ .
   a. pipe clamps and eye rods
   b. rods and bolts
   c. one or more hangers manufactured for the purpose
   d. a sleeve anchor
3. Vertical pipe should be supported at each floor level with a _____ clamp.
   a. half
   b. suspension
   c. pipe
   d. riser
4. _____ are used to fasten the piping system to concrete.
   a. Clamps
   b. Anchors
   c. Risers
   d. Connectors
5. Proper _____ are required for pipes not to sag, bend, or move out of alignment.
   a. anchors
   b. attachments
   c. supports
   d. slopes

5.0.0 ◆ MEASURING, CUTTING, JOINING, AND ASSEMBLING STEEL PIPE

This section describes the major steps in working with steel pipe. These steps are measuring, cutting, reaming, threading, joining, and assembling.

5.1.0 Working with Threaded Pipe

When you are working with threaded pipe, you must know how to measure a length of pipe for proper installation. You must take into account the length of pipe that gets screwed into the fitting (thread engagement) or your pipe will be cut too short. You must also calculate the size of the fitting.

5.1.1 Measuring

Threaded pipe can be measured by a variety of methods:

- **End-to-end:** Measure the full length of the pipe, including both threaded ends.
- **End-to-center:** Use for pipe that has a fitting screwed on one end only; pipe length is equal to the measurement *minus* the end-to-center dimension of the fitting *plus* the length of thread engagement.
- **Center-to-center:** Use with a length of pipe that has fittings screwed onto both ends; pipe length is equal to the measurement *minus* the sum of the end-to-center dimensions of the fittings *plus* twice the length of the thread engagement.
- **End-to-face:** Use for pipe that has a fitting screwed on one end only; pipe length is equal to the measurement *plus* the length of the thread engagement.
- **Face-to-face:** Use for same situation as center-to-center measurement; pipe length is equal to the measurement *plus* twice the length of the thread engagement.
- **End-to-back:** Use for pipe that has a fitting screwed on one end only; pipe length is equal to the measurement *plus* the length of the screwed-on fitting, *plus* the length of the thread engagement.
- Center-to-face: Use with a length of pipe that has fittings screwed onto both ends; pipe length is equal to the measurement from the center of one of the screwed-on fittings to the face of the opposite fitting, *plus* twice the length of the thread engagement.
- Face-to-back: Use with a length of pipe that has fittings screwed onto both ends; pipe length is equal to the measurement, *plus* the distance from the face to the back of one screwed-on fitting, *plus* twice the length of the thread engagement.

Figure 26 shows the different measuring techniques. It is important to allow for fitting dimensions (end-to-end or end-to-center) and for length of thread engagement (see *Table 3*).

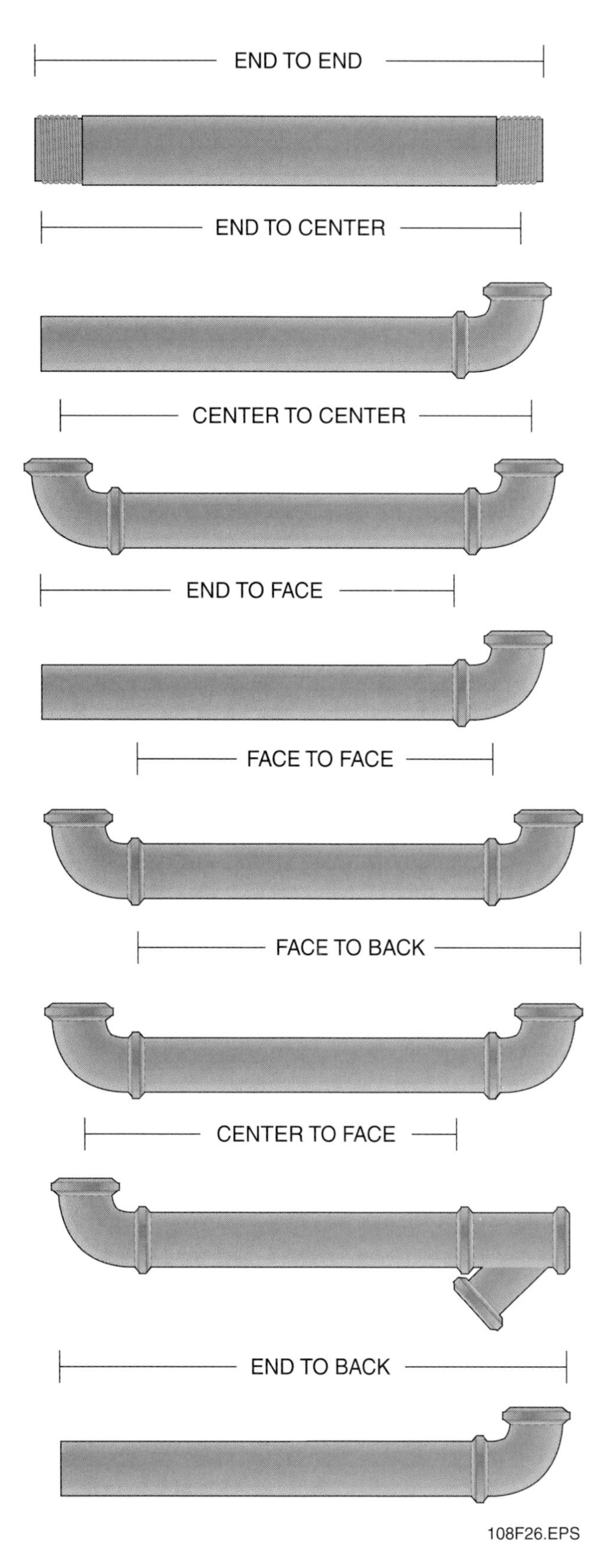

Figure 26 ◆ Measuring methods.

Table 3 Thread Engagement

Size of Pipe (Inches)	Outside Diameter (Inches)	Number of Threads per Inch	Total Length of Threads (Inches)	Effective Length (Inches) (Approx.)	Thread Engagement (Inches) (Approx.)
1/4	0.54	18	5/8	3/8	3/8
3/8	0.675	18	5/8	7/16	3/8
1/2	0.840	14	13/16	9/16	1/2
3/4	1.05	14	13/16	9/16	1/2
1	1.315	11 1/2	1	11/16	9/16
1 1/4	1.660	11 1/2	1	11/16	5/8
1 1/2	1.9	11 1/2	1	3/4	5/8
2	2.375	11 1/2	1 1/16	3/4	11/16
2 1/2	2.875	8	1 9/16	1 1/8	15/16
3	3.5	8	1 5/8	1 3/16	1
4	4.5	8	1 3/4	1 5/16	1 1/16
5	5.56	8	1 13/16	1 3/8	1 3/16
6	6.625	8	1 15/16	1 1/2	1 1/4

5.1.2 Cutting

Pipe cutters (see *Figure 27*) may have from one to four cutting wheels. A single cutting wheel requires enough area for the cutter to be rotated all the way around the pipe. The more cutting wheels a cutter has, the less rotation it requires to cut the pipe. In addition to the cutting wheels, pipe cutters have an adjusting screw and at least two guiding wheels.

To operate the pipe cutter, revolve it around the pipe and tighten the cutting wheel ¼ revolution with each turn. Avoid overtightening the cutting wheel. Overtightening could damage the wheel or cause it to break.

If the tube or pipe looks mashed after it has been cut, replace the cutting wheel or wheels. Lubricating oil needs to be applied periodically to all movable parts on the pipe cutter for it to operate smoothly.

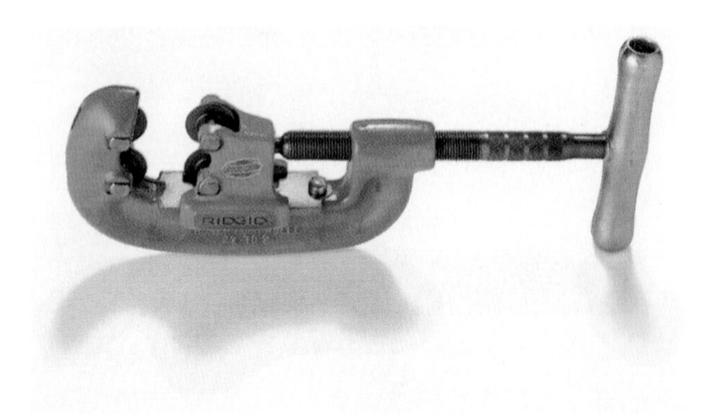

108F27.TIF

Figure 27 ◆ Pipe cutter.

5.1.3 Reaming

After a piece of pipe is cut, a burr will be left on the inside of the pipe. A **reamer** is used to remove the burr. Reamers may be straight or spiral, as shown in *Figure 28*.

If the burr is not reamed off, it will collect deposits and slow the flow of liquid in the pipe. Because reamers are tapered, one reamer can deburr many sizes of pipe. Reaming is always required for pipe installed in gas, air, steam heating, or fire suppression systems.

5.1.4 Threading

There are two types of pipe threaders: hand threaders and power threaders. Hand threaders

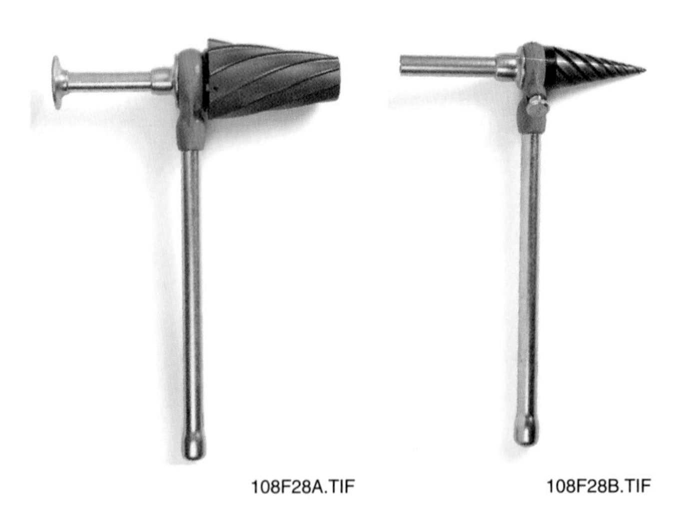

108F28A.TIF 108F28B.TIF

Figure 28 ◆ Reamers.

are made up of two parts: the die and the stock, as shown in *Figure 29*. Dies are used to cut the threads. The stock is the device that holds the die. Although it is usually used manually, a hand threader could be used with the power drive or power vise shown in *Figure 30*, which threads automatically.

A pipe-threading machine (see *Figure 31*) is a power threader that is used when large quantities of pipe require threading. This machine rotates, threads, cuts, and reams pipe. The pipe is mounted through the machine chuck, and then the chuck is tightened.

A number of tools can be used to hold pipe while you are threading it. The standard **yoke vise**, shown in *Figure 32*, is probably the most common vise used by plumbers. Its jaws hold pipe firmly and prevent it from turning. Depending on its size, a yoke vise can handle pipe ⅛ inch to 3½ inches in diameter. The teeth may leave marks, so you would use this vise only on pipe that will not show.

The **chain vise**, shown in *Figure 33*, is used in the same way as the standard yoke vise, but the chain vise can hold much larger pieces of pipe than the standard yoke vise. The chain must be

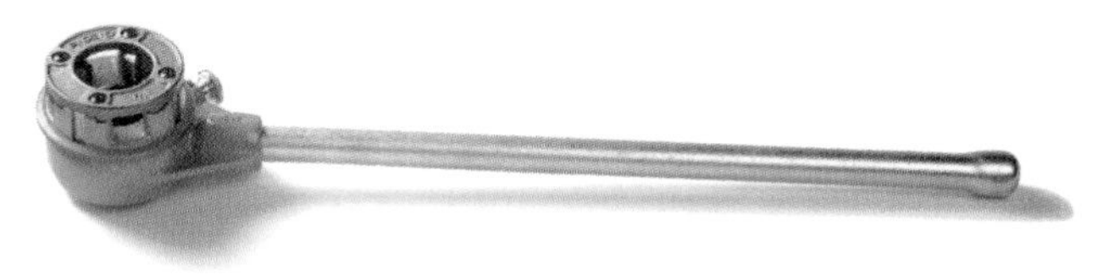

108F29.TIF

Figure 29 ◆ Pipe die and stock.

108F30.TIF

Figure 30 ◆ Power drive.

108F32.TIF

Figure 32 ◆ Yoke vise.

108F31.TIF

Figure 31 ◆ Pipe-threading machine.

108F33.TIF

Figure 33 ◆ Chain vise.

kept oiled or it will become stiff. A stiff chain will make the chain vise operate poorly.

Pipe wrenches are used to grip and turn round stock. They have teeth that are set at an angle. This angle allows the teeth of the wrench to grip and permits the wrench to turn in only one direction.

There are three common types of pipe wrenches:

- The **straight pipe wrench** (see *Figure 34*)
- The **offset pipe wrench** (see *Figure 35*), used in confined spaces
- The **compound leverage pipe wrench** (see *Figure 36*), used when extra strength is needed to turn pipe assemblies (as the hook jaw turns the pipe one way, an offset chain trunnion assembly applies pressure in the opposite direction)

Pipe tongs, shown in *Figure 37,* are typically used on large pipe. A variety of chain wrenches are available for large pipes as well (see *Figure 38*).

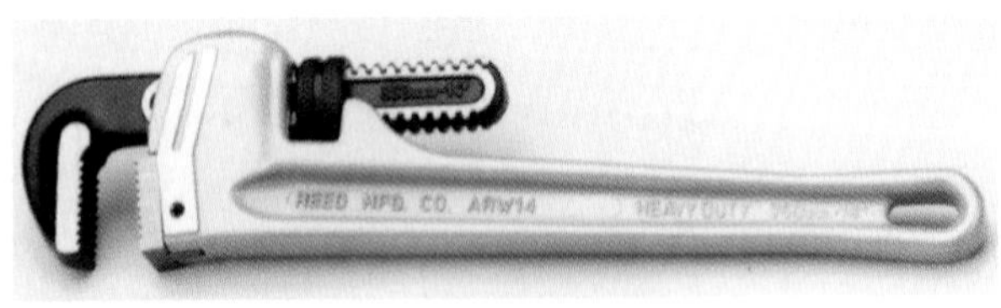

108F34.EPS

Figure 34 ◆ Straight pipe wrench.

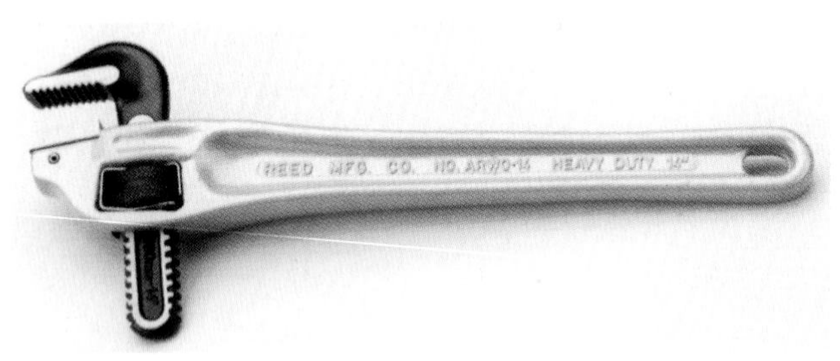

108F35.EPS

Figure 35 ◆ Offset pipe wrench.

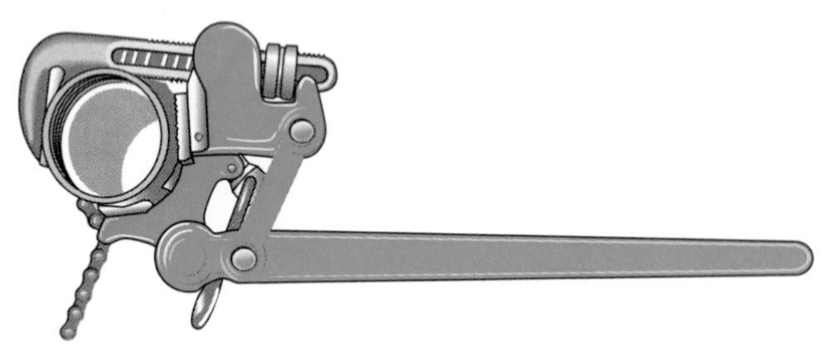

108F36.EPS

Figure 36 ◆ Compound leverage pipe wrench.

The chain in both pipe tongs and a chain wrench must be oiled often to prevent it from becoming stiff or rusty.

Strap wrenches, shown in *Figure 39,* are used to hold chrome-plated or other types of finished pipe. The strap wrench does not leave jaw marks or scratches on the pipe. Resin applied to the strap adds to the wrench's holding power and reduces slippage.

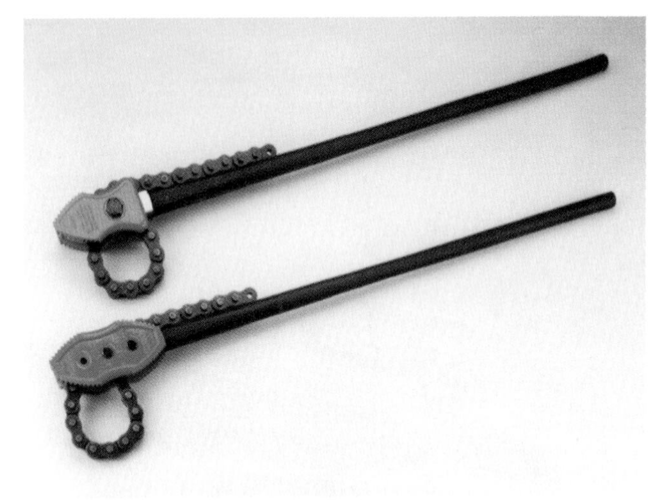

108F37.EPS

Figure 37 ◆ Pipe tongs.

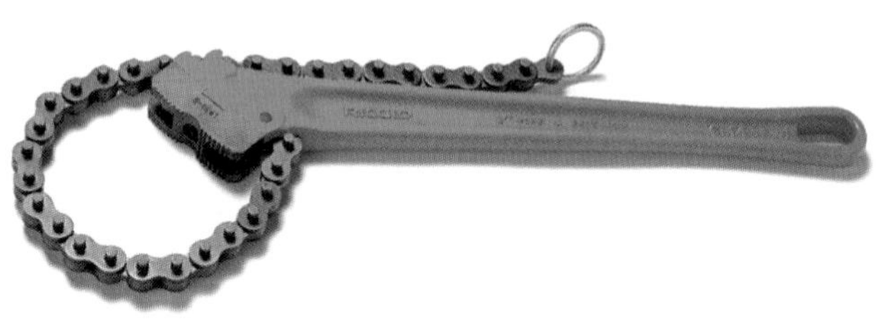

108F38.TIF

Figure 38 ◆ Chain wrench.

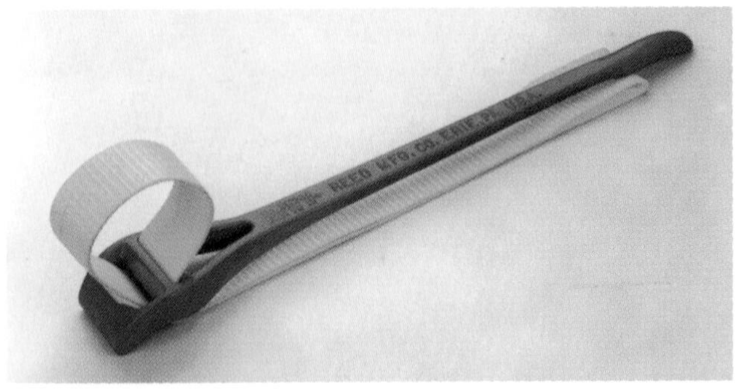

108F39.EPS

Figure 39 ◆ Strap wrench.

To cut threads using a hand die and stock, follow these steps (see *Figure 40*):

Step 1 Select the correct size of die for the pipe being threaded.

Step 2 Inspect the die to make sure the cutters are free of nicks and wear.

Step 3 Lock the pipe securely in a vise.

Step 4 Slide the die over the end of the pipe, guide-end first.

Step 5 Push the die against the pipe with the heel of one hand. Take three or four short, slow, clockwise turns. Be careful to keep the die pressed firmly against the pipe. When enough thread is cut to keep the die firmly against the pipe, apply some thread-cutting oil. This oil prevents the pipe from overheating from friction, and it lubricates the die. Oil the threading die every two or three downward strokes.

Step 6 Back off a quarter turn after each full turn forward to clear out the metal chips. Continue until the pipe projects one or two threads from the die end of the stock. Having too few threads is as bad as having too many threads.

Step 7 Remove the die by rotating it counter-clockwise.

Step 8 Wipe off excess oil and any chips.

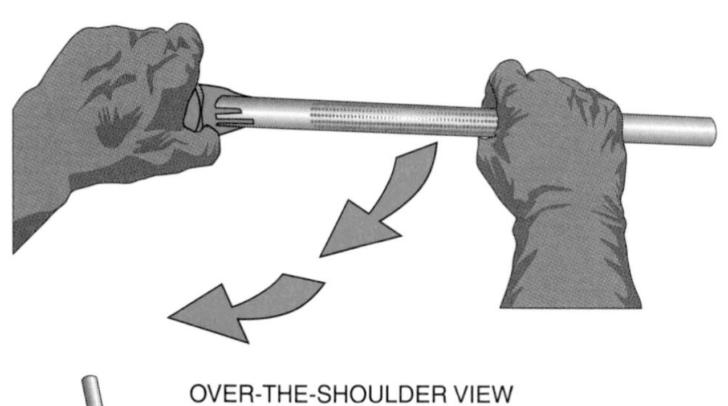

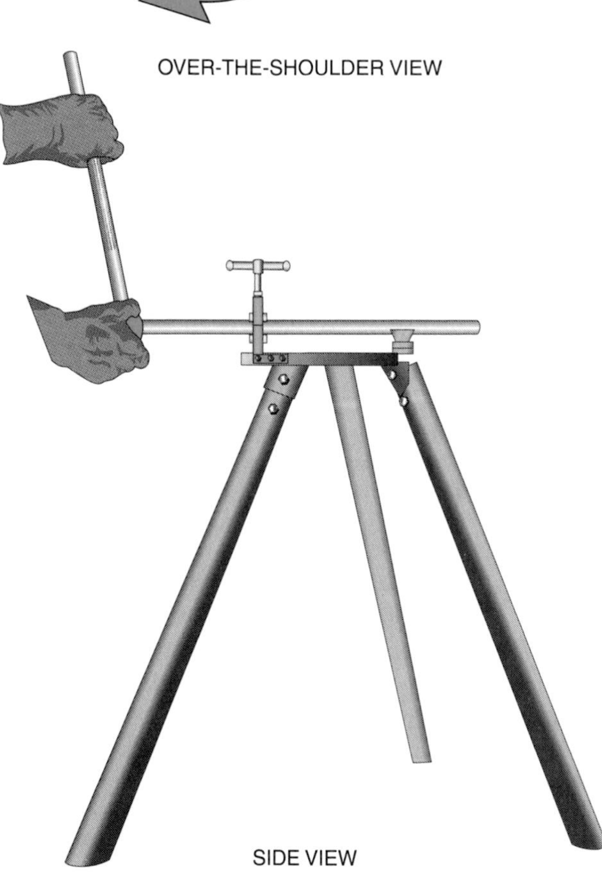

Figure 40 ◆ Cutting threads using a hand die and stock.

CAUTION

The chips made when you are threading pipe are sharp and can cut you. Use a rag, not your bare hand, to wipe off excess oil.

To cut threads using a **power-threading machine**, follow these steps (see *Figure 41*):

Step 1 Select and install the correct size of die and inspect it for nicks.

Step 2 Mount the pipe into the machine chuck. Long pipe must have additional support.

Step 3 Check the pipe and die alignment.

Step 4 Cut threads until two threads appear at the other end of the die. Stop the threading action. Apply cutting oil during the threading operation.

Step 5 Back off the die until it is clear of the pipe.

Step 6 Remove the pipe from the machine chuck. Be careful not to mar the threads.

Step 7 Wipe the pipe clean of oil and metal chips.

Each threading machine is slightly different. Become familiar with the manufacturer's operating procedures before you attempt to operate the machine.

5.1.5 Assembling

Plumbers use **pipe joint compound**, sometimes called pipe dope, to seal a joint and provide lubrication for assembly. **Teflon tape** made specifically as a pipe joint sealer may also be used.

ON THE LEVEL

Pipe Dope

Pipe joint compound is available in a wide variety of materials. The most common pipe joint compound is used for domestic water lines. Joint compound is designed to resist interaction with substances moving through the system. Depending on the type of gas or liquid moving through the pipe system, manufacturers recommend different kinds of joint compound. Always read factory recommendations before using the product.

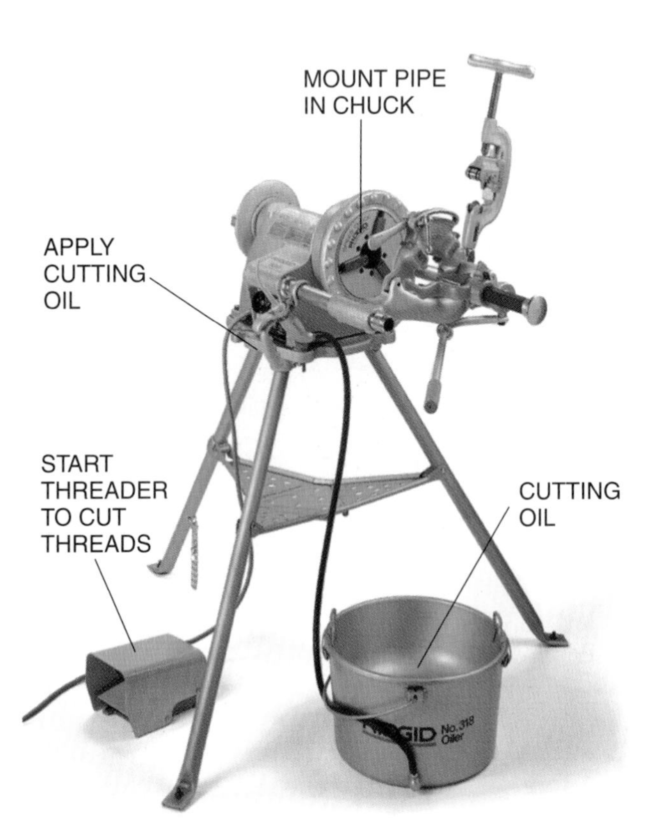

108F41.TIF

Figure 41 ◆ Cutting threads using a power-threading machine.

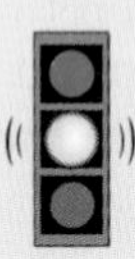

CAUTION

Become thoroughly familiar with the maintenance and safety instructions for any power machine. A poorly maintained machine can become a safety hazard. Dirt and metal chips can get stuck in the cutting dies and produce inadequate threads. Clogged components may also send debris flying. These flying objects can blind you or a co-worker.

To assemble threaded steel pipe, follow these steps (see *Figure 42*):

Step 1 Apply pipe joint compound or Teflon tape only to the male threads of the pipe or fit-

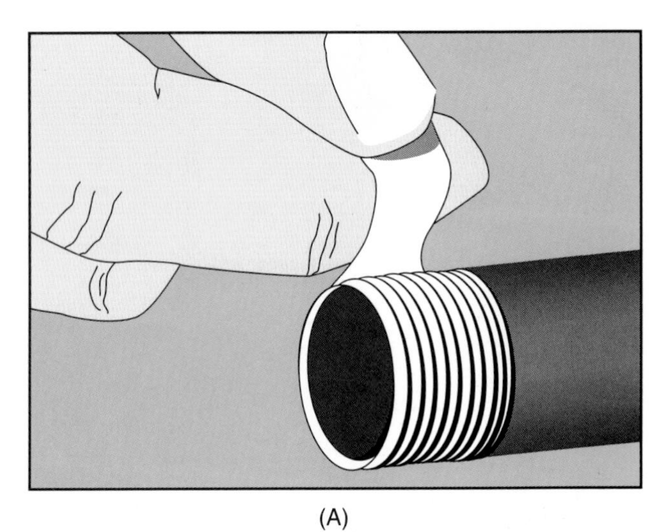

(A)

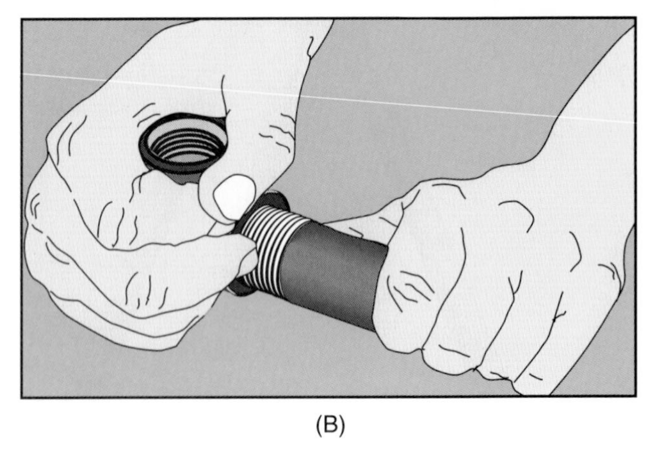

(B)

108F42.EPS

Figure 42 ◆ Assembling threaded steel pipe. (A) Apply Teflon tape to male threads in a clockwise direction. (B) Start the fitting into the threaded pipe by hand.

tings before you assemble a pipe connection. Do not apply the compound to the female threads of either a pipe or a fitting. The twisting motion used to join the pipes will ball up the pipe dope or tape if it is applied to the female threads. The debris will get stuck inside the pipe system.

Step 2 If you are using Teflon tape, apply the tape in a clockwise direction, the same direction as the fitting turns. Be sure to check the local gas codes to see if the use of tape is permitted.

Step 3 Start the fitting into the threaded pipe by hand. Turn the fitting clockwise. Finish tightening the fitting with a pipe wrench.

5.2.0 Working with Grooved Pipe

The same procedures can be used to cut grooved pipe as you used for threaded carbon steel pipe. The steps for joining are the following (see *Figure 43*):

Step 1 Check to make sure that the gasket is suitable for the intended use. Some manufacturers color-code their gaskets. Apply a

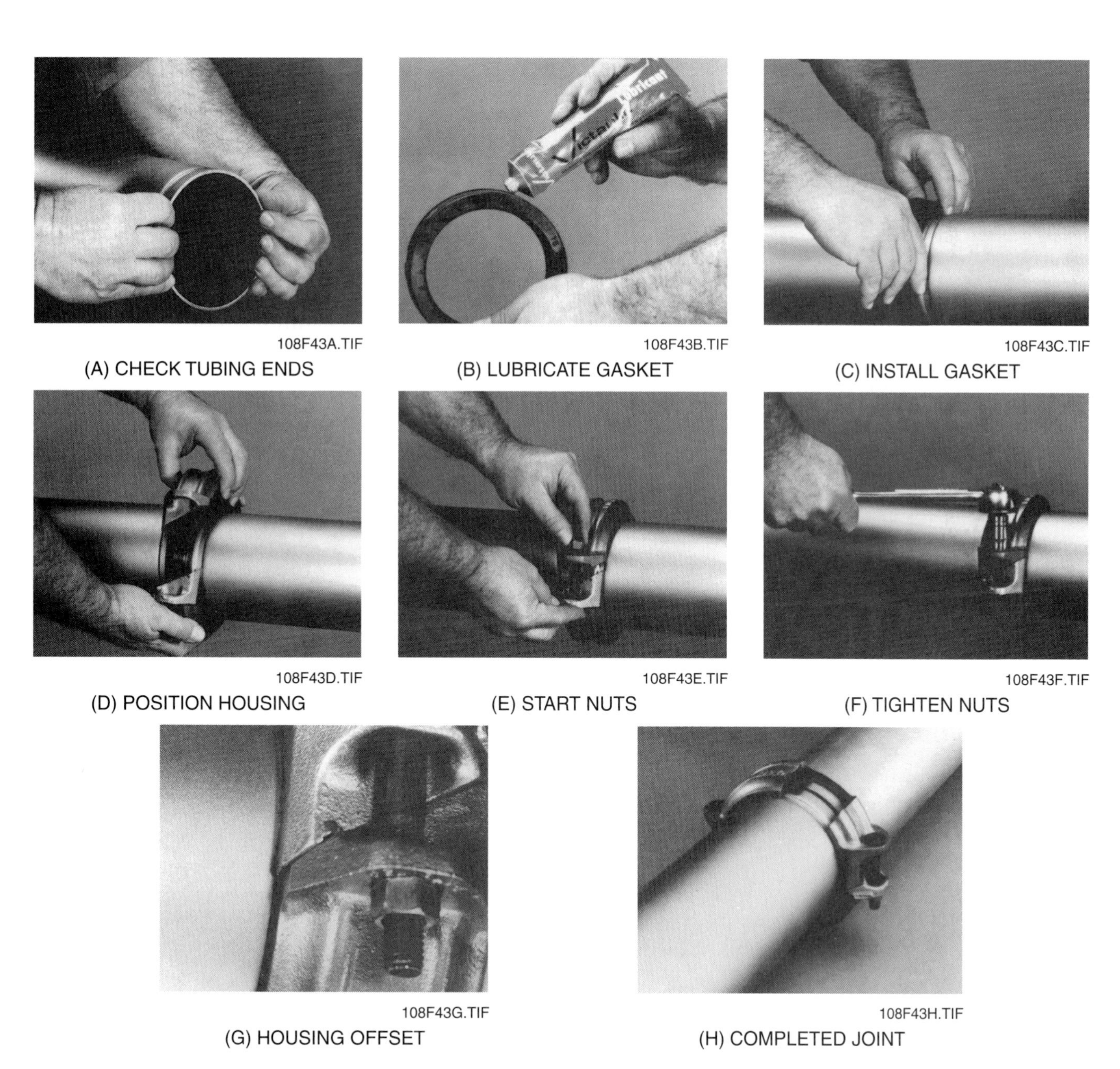

Figure 43 ◆ Joining grooved pipe.

thin coat of lubricant to the gasket lips and the outside of the gasket.

Step 2 Check the pipe ends. To make sure you will have a leakproof seal, the ends must be free from indentations, projections, or roll marks.

Step 3 Install the gasket over the pipe end. Be sure the gasket lip does not hang over the pipe end.

Step 4 Align and bring the two pipe ends together. Slide the gasket into position, and center it between the grooves on each pipe. Be sure that no part of the gasket extends into the groove on either pipe.

Step 5 Assemble the housing segments loosely, leaving one nut and bolt off to allow the housing to swing over the joint.

Step 6 Install the housing, swinging it over the gasket and into position into the grooves on both pipes.

Step 7 Insert the remaining bolt and nut. Be sure that the bolt track head engages into the recess in the housing.

Step 8 Tighten the nuts alternately and equally to maintain metal-to-metal contact at the angle bolt pads.

Review Questions

Section 5.0.0

1. When measuring pipe, you must take the _____ into account or your pipe will be cut too short.
   a. burr
   b. weight of pipe
   c. type of pipe
   d. thread engagement

2. If a tube or pipe looks mashed after it has been cut, you should _____ the cutting wheel or wheels.
   a. sharpen
   b. replace
   c. polish
   d. adjust

3. Use a _____ to clean the burr out of the inside of a pipe.
   a. pipe cutter
   b. threader
   c. reamer
   d. yoke

4. A(n) _____ has two parts: a die to cut threads and a stock to hold the die.
   a. hand threader
   b. pipe cutter
   c. offset wrench
   d. chain vise

5. Never put pipe joint compound or Teflon tape on the _____ of a fitting.
   a. threaded end
   b. female threads
   c. male threads
   d. inside

Summary

Carbon steel pipe is used in many modern plumbing applications. Carbon steel is a result of the process of combining iron with carbon to produce a steel alloy. Depending on the types of elements added to the steel alloy, there are three basic types of steel: carbon steel, low-alloy steel, and high-alloy (stainless) steel.

Black iron pipe is used most often for gas and air-pressure applications. Galvanized steel pipe is used most often for plumbing hot and cold water distribution, steam and hot water heating systems, and gas and air piping systems. The zinc coating protects the galvanized steel pipe from abrasive or corrosive materials.

Steel pipe is categorized by both its size (inside or outside diameter) and its strength. Schedule 40 is a standard weight pipe. Schedule 80 is an extra strong weight. Schedule 120 is a double extra strong weight.

Threads used on all carbon steel pipe are standardized. This means thread sizes on all manufactured pipe products are the same.

Iron pipe fittings are made for both pressure and drainage pipe systems. Pressure pipe fittings cannot be used in drainage systems because they do not have smooth enough inside surfaces. Drainage fittings are specifically designed for smooth flow.

Fittings are used to join lengths of pipe, change direction of runs of pipe, cap off open pipe, and add branches to main pipe lines. Valves are often attached to the ends of pipe to provide access to fluid flowing through the system.

All pipe that is installed vertically or horizontally requires adequate support to keep it from sagging, bending, and losing proper alignment. Codes and building specifications tell you which hangers, fasteners, and connectors are appropriate for different applications. Always check the code before you install plumbing.

Threaded pipe can be measured in several ways, depending on whether there are fittings on one or two ends. Incorrect measuring will mean your pipe is cut too long or too short for the installation. A variety of tools are available for working with and cutting carbon steel pipe. Familiarity with tool use and maintenance will ensure that your workmanship is of the best quality. A plumber is only as good as his or her tools.

As with all aspects of plumbing, the selection and installation of carbon steel pipe is strictly regulated by building and plumbing codes. Codes vary widely between locations, so you must always check local codes that affect your work. Incorrect selection or installation will cause the job to fail inspection by plumbing code officials.

PROFILE IN SUCCESS

Will Seilhamer

Plumbing
President, J & W Mechanical Service, Inc.
Hagerstown, MD

Born and raised in Hagerstown, Maryland, Will entered the Navy in 1962 as a machinist's mate on a destroyer with a 300-man crew. Although he wanted to train in air-conditioning and refrigeration systems, Will was assigned to the engine room, where he learned a variety of technical plumbing and pipefitting skills. He started his own business in 1984 and continues with it today.

How did you become interested in the construction industry?
When I went into the Navy, I wanted to learn a trade. As a machinist's mate, I could have worked in the refrigeration part of the ship or in the engine room. I ended up in the engine room, where I learned how to work with pressure and steam. We kept the turbines that drove the ship in working order. We did boiler repair, worked on pump operations, and maintained the ship's steam systems. We worked in shifts of four hours on and eight hours off. It was hard work, but it was a great opportunity. I traveled to 37 countries and went around the world once. I also spent a year in Vietnam.

After my discharge in 1967, I went back to Hagerstown where I applied to a mechanical company and got a job as a plumber. I enrolled in the Associated Builders and Contractors (ABC) apprenticeship-training program in 1968 and got my journeyman's license in 1972.

The company expanded to specialize in boiler systems and in high-pressure steam systems and repairs. Because of my Navy experience, they switched me from plumbing to the new division. At that time, an older man—my mentor—was in charge. Shortly afterward, he had a heart attack and retired. Since I had the most experience, I was put in charge of the division. I stayed with the company for 13 years.

In 1984, I started my own contracting business and have been doing that for the last 16 years. My former company had been involved with ABC, so I joined the Cumberland Valley chapter. I was most interested in the apprenticeship-training program. I became more involved with ABC by serving on committees and review groups. I served as the chairman of the local apprenticeship program for approximately 8 years. In 1996, I was elected president of the Cumberland Valley chapter. Currently, I participate on several committees and chair the Construction Education Trust Committee.

What do you think it takes to be a success in your trade?
The most important characteristic you need is the ability to stick with it through four years of apprenticeship training. Some people entering the trades don't want to return to the classroom, even though it is only one or two nights a week. You have to be willing to work through that. It takes a strong work ethic. You also need a good employer who supports your efforts and is committed to training.

Theory, hands-on skills, and good math skills are important in the plumbing trade. You need all three to become a qualified craftsman.

What are some of the things you do in your role of company president?
I wear a half-dozen hats as a company president. We are a small company with about 28 employees. I help run the office operations, do the estimates for new jobs, manage field coordination and scheduling, and supervise some jobs. My position requires that I oversee several projects.

I am also active in the ABC chapter. I give presentations to sophomores and juniors in area high schools. Some of my employees accompany me to speak on what it is like to work in the trades.

I am firmly committed to the success of the apprenticeship-training program. I enjoy the professional activities associated with membership. We get together and network at functions with our contractors, competitors, and potential customers. It is a great activity.

What do you like most about your job?
I love what I am doing and I love the plumbing trade. I like building things and seeing the finished product. Working in the construction industry gives me a great sense of accomplishment and the opportunity to work with a lot of different contractors. It is like being on a sports team—everyone pulls together to get the project finished. It makes me proud to be part of that.

What would you say to someone entering the trades today?
The economy is going along really well these days. For that reason, the trades seem to be at the bottom of people's career lists. I would say that people don't realize that the trades can provide a great career and a great living. You can earn excellent wages, achieve advancement, and provide a good future for your family.

You have to spend a lot of hours and work hard to be successful. You need drive and willpower. Most of all, you have to stick with it. Keep learning new technology; don't get behind in the advancements the industry is making. And always strive to be better.

GLOSSARY

Trade Terms Introduced in This Module

Annealing: A process of tempering steel through heating and slow cooling to toughen and reduce brittleness.

Ball valve: A valve used to control the flow of gases and liquids. It is installed in piping systems where quick shutoffs for in-line maintenance may be necessary.

Black iron pipe: Steel pipe with a black color used for gas and air-pressure pipe. Should not be used where corrosion will affect its uncoated surface.

Bushing: Fitting used to connect the male end of a pipe to a fitting of a larger size. Has a male fitting on the outside and a female fitting on the inside.

Cap: Fitting with female threads, used to close off the male end of pipe or nipple.

Center-to-center: Measurement from the center of one fitting to the center of another fitting on the opposite end of a length of pipe.

Chain vise: Tool used to cut pipe by holding the pipe in a vise while cutting with a chain that has cutting wheels in each link.

Compound leverage pipe wrench: Tool with a hook-like jaw that turns the pipe one way while an offset chain trunnion assembly applies pressure in the opposite direction.

Coupling: Short fitting with female threads in each end, used to connect two lengths of pipe when making straight runs.

Cross: Fitting that connects four lengths of pipe in a cross-like shape in a distribution system.

Double extra strong (double extra heavy): An industry standard measurement of the weight or strength of steel pipe. Also referred to as Schedule 120.

Drainage fitting (recessed fitting): Fitting designed with smooth inside surfaces and shaped for easy flow to form an unbroken internal contour for use on drainage systems.

Elbow (ell): Fitting used to change the direction of the pipe. The most common elbows are 90-degree, 45-degree, street ell, and reducing ell.

End-to-back: Measurement from one end of a pipe with a fitting screwed on one end plus the length of the thread engagement at the other end.

End-to-center: Measurement from the center of a fitting to the end of the threads on the opposite end of a length of pipe.

End-to-end: Measurement from one end of a pipe to the other end, including both threaded ends.

End-to-face: Measurement from the end of a fitting to the end of the threads on the opposite end of a length of pipe.

Extra strong (extra heavy): An industry standard measurement of the weight or strength of steel pipe. Also referred to as Schedule 80.

Face-to-face: Measurement from the end of a fitting attached to a length of pipe to the end of a fitting on the opposite end.

Flange union: Fitting that connects two pipes. A flange is screwed onto the end of each pipe that needs to be joined and then bolted together with nuts and bolts. A gasket in between flanges makes it a leakproof seal.

Galvanized pipe: Pipe coated with zinc to provide a protective coating that resists corrosion or oxidation (rust).

Gate valve: A valve in which the flow is controlled by moving a "gate" or disc that slides in machined grooves at right angles to the flow.

Globe valve: A valve in which the flow is controlled by moving a circular disc against a metal seat that surrounds the flow opening.

Ground joint union: Fitting that joins two pieces of pipe by screwing the thread and shoulder pieces of the fitting onto the pipe. The collar piece is then tightened to join the sections of pipe into a watertight joint.

Nipple: Short length of pipe with threads at both ends, used to make extensions from a fitting or to join two fittings.

Nominal size: Approximate measurement in inches of the inside diameter (ID) of steel pipe up to 12 inches. (Pipe larger than 12 inches is measured by its outside diameter [OD].) The designation used to specify the size of pipe is not necessarily equal to the exact inside diameter.

Offset pipe wrench: Type of pipe wrench used in confined spaces where a regular pipe wrench won't fit.

Pipe cutter: Tool with one to four cutting wheels that can be rotated around a pipe to cut it.

Pipe joint compound (pipe dope): Sealing material used when joining lengths of pipe.

Pipe tongs: Wrench-like tool with a length of chain, used to tighten large pipe.

Plug: Male-threaded fitting, used to close openings in other fittings.

Plug valve: A valve used to control the flow of gases and liquids. Similar to a ball valve, but uses a tapered plug with a hole or slot instead of the ball.

Power-threading machine: Machine used when large quantities of pipe require threading. The machine rotates, threads, cuts, and reams pipe.

Pressure fitting (ordinary fitting): Any type of fitting used for steam, gas, or water supply piping. Not designed for use in drainage piping.

Reamer: Tool used to smooth and remove burrs from the inside of the cut end of a piece of pipe.

Standard weight: An industry standard measurement of the weight or strength of steel pipe. Also referred to as Schedule 40.

Straight pipe wrench: Common type of wrench, used in all plumbing applications.

Strap wrench: Wrench used to hold chrome-plated or other types of finished pipe so that there are no jaw marks or scratches on the pipe.

Teflon tape: Sealing tape used to wrap the threads of a pipe before joining to ensure a watertight, secure fit.

Yoke vise: Tool with jaws that firmly hold a piece of pipe and prevent it from turning.

ANSWER KEY

Answers to Review Questions

Section 2.0.0

1. b
2. c
3. c
4. b
5. a

Section 3.0.0

1. c
2. d
3. c
4. a
5. b

Section 4.0.0

1. a
2. c
3. d
4. b
5. c

Section 5.0.0

1. d
2. b
3. c
4. a
5. b

NCCER CRAFT TRAINING USER UPDATES

The NCCER makes every effort to keep these textbooks up-to-date and free of technical errors. We appreciate your help in this process. If you have an idea for improving this textbook, or if you find an error, a typographical mistake, or an inaccuracy in the NCCER's Craft Training textbooks, please write us, using this form or a photocopy. Be sure to include the exact module number, page number, a detailed description, and the correction, if applicable. Your input will be brought to the attention of the Technical Review Committee. Thank you for your assistance.

Instructors – If you found that additional materials were necessary in order to teach this module effectively, please let us know so that we may include them in the Equipment/Materials list in the Instructor's Guide.

Write: Curriculum Revision and Development Department
National Center for Construction Education and Research
P.O. Box 141104, Gainesville, FL 32614-1104

Fax: 352-334-0932

E-mail: curriculum@nccer.org

Craft ______ Module Name ______

Copyright Date ______ Module Number ______ Page Number(s) ______

Description ______

______

______

______

(Optional) Correction ______

______

______

(Optional) Your Name and Address ______

______

______

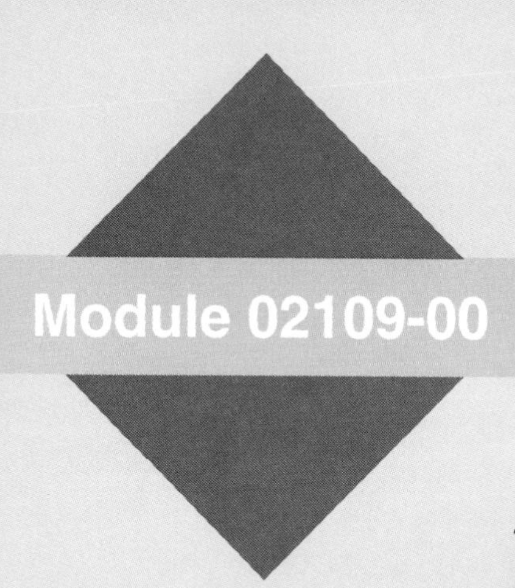

Module 02109-00

Fixtures and Faucets

Course Map

This course map shows all of the modules in the first level of the Plumbing Curriculum. The suggested training order begins at the bottom and proceeds up. Skill levels increase as you advance on the course map. The local Training Program Sponsor may adjust the training order.

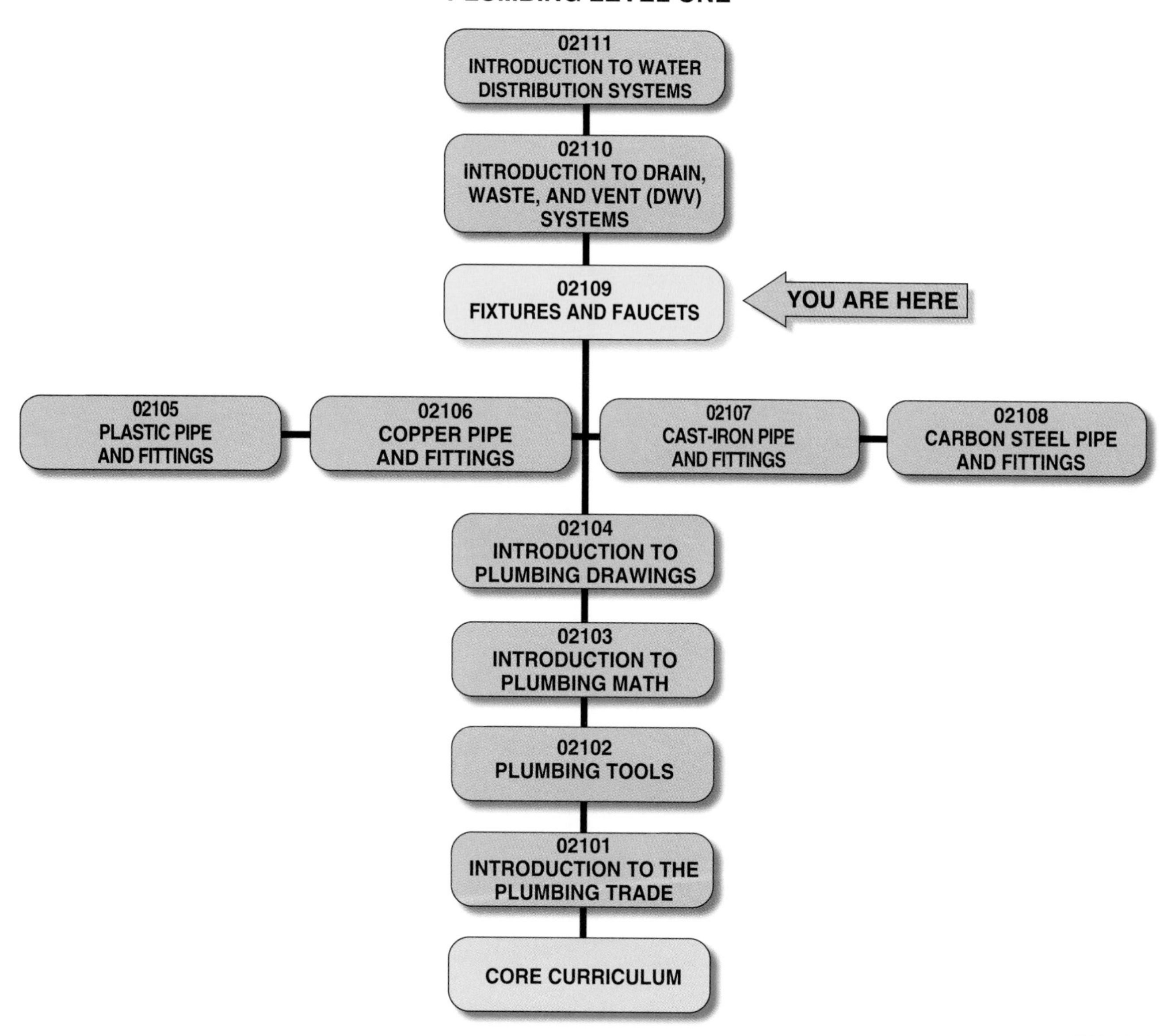

109CMAP.EPS

Copyright © 2000 National Center for Construction Education and Research, Gainesville, FL 32614-1104. All rights reserved. No part of this work may be reproduced in any form or by any means, including photocopying, without written permission of the publisher.

MODULE 02109 CONTENTS

Figures

MODULE 02109

Fixtures and Faucets

Objectives

When you finish this module, you will be able to:

1. Identify the basic types of materials used in the manufacture of plumbing fixtures.
2. Discuss common types of sinks, lavatories, and faucets.
3. Discuss common types of bathtubs, bath-shower modules, shower stalls, and shower baths.
4. Discuss common types of toilets, urinals, and bidets.
5. Discuss common types of drinking fountains and water coolers.
6. Discuss common types of garbage disposals and domestic dishwashers.

Prerequisites

Before you begin this module, it is recommended that you successfully complete task modules: Core Curriculum; Plumbing Level One, Modules 02101 through 02108.

Materials Required for Trainees

1. This module
2. Appropriate personal protective equipment
3. Sharpened pencil and paper

1.0.0 ◆ INTRODUCTION

A plumber comes in contact with many different types of fixtures and faucets. A fixture, in a plumbing installation, is a device that receives water from a water supply line. A faucet is a fixture that is used for drawing water from a pipe.

As a plumber, you need to be familiar with the different types of fixtures and faucets that are available on the market for use in plumbing installations. Plumbing fixtures include lavatories and sinks, bathtubs, shower stalls and shower baths, toilets and water closets, urinals, bidets, garbage disposals, domestic dishwashers, laundry trays, drinking fountains and water coolers, and service sinks and mop basins. Faucets include utility faucets, bathroom faucets, and kitchen faucets.

Various types of fixtures are used in all residential, commercial, and industrial buildings. For this reason, plumbers need to be concerned with more than just the techniques used for installing these fixtures. They need to be aware of the construction and working principles of the various fixtures so that they can make proper water supply and waste connections and choose the right fixtures and faucets for each job. For instance, commercial fixtures are designed to withstand more use than residential fixtures.

This module is divided into two parts. The first part covers the materials commonly used to make fixtures, the various kinds of fixtures available, how each fixture is used, and the basic principles of operating each fixture. This information will help you understand how to choose the right fixture for installation. The second part of this module covers

Water Conservation

The most practical approach to water shortages is to conserve water. This means using—and wasting—less. The development of new water sources is very expensive. A variety of devices have been developed to conserve both water and energy in the face of increasing demand worldwide. On a close-to-home level, 1.6 gallon water conservation water closets have replaced the old water closets that typically used 3 to 5 gallons to flush the bowl.

Title 18 CFR 803, the Conservation of Power and Water Resources Act (revised in April 1999), defines the steps that state and local governments must take to reduce the strain on water resources.

the kinds of faucets that are available and how each is used, so that you can choose the proper faucet for each installation.

2.0.0 ◆ MATERIALS USED TO MAKE FIXTURES

Modern plumbing fixtures are made from durable, corrosion-resistant, nonabsorbent materials that have smooth, easily cleaned surfaces. In some manufacturing of plumbing fixtures, these materials are used alone (in **stainless steel** kitchen sinks, for instance). In other manufacturing, these materials are used in combinations (**fiberglass**-reinforced plastic shower stalls, for example). The basic materials used in manufacturing plumbing fixtures include **china (vitrified porcelain)**, steel coated with **porcelain enamel**, **cast iron** coated with porcelain enamel, stainless steel, acrylic plastic, and fiberglass reinforced with plastic.

2.1.0 Porcelain

Molded *vitrified porcelain,* also called *china* or *vitreous clay,* is used in the more expensive fixtures. Porcelain is made from a mixture of fine clay (called kaolin), quartz, feldspar, and silica. The materials are fused, or vitrified, into a glass-like state. The complete process of fusion, or vitrification, requires a temperature of 2600°F (1426°C). The process produces a sanitary surface that is strong, nonporous, and easily cleaned when used alone or when used as a coating for steel or cast iron.

2.2.0 Porcelain Enamel

Porcelain is also used in its liquid, or enamel, form in combination with other materials, such as steel or cast iron. Manufacturers use a bonding process that fuses the two materials together. This creates a more durable and functional material for making fixtures.

The result of this process is often called porcelain enamel, but the terms *glass lining* and *vitreous enamel* are sometimes used instead.

2.3.0 Cast Iron

Some plumbing fixtures—such as bathtubs, sinks, and lavatories—are made from **gray iron,** also known as cast iron, a metal that is ideally suited for plumbing fixtures because of its low cost and its ability to be cast in a wide variety of shapes.

2.4.0 Sheet Steel

Lavatories, sinks, and bathtubs can also be made from **sheet steel.** In a low-cost process, these fixtures are formed by stamping the sheet steel into the shape of the fixture. The surfaces of these plumbing fixtures are then fused with a porcelain enamel finish, which is glossy, decorative, protective, and sanitary. These fixtures are usually less expensive, and also less durable (they chip easily), than those made of cast iron.

2.5.0 Stainless Steel

Stainless steel is another metal commonly used to make plumbing fixtures, such as kitchen sinks. Although it is difficult to form, it is an excellent material for plumbing fixtures. It doesn't need additional coating to produce a sanitary surface because the stainless steel itself has a durable, silver-like finish. Unlike enameled surfaces, stainless steel will not chip.

Today's Materials

Many of the bathtubs, shower stalls, and hot tubs installed today are made to look like natural materials such as stone and granite, but they are actually made from acrylics (plastics).

Dupont's Corian® is an acrylic-based premium solid surface material that is easy to maintain, resists damage, and can be shaped into countertops, work surfaces, sinks and bowls, shower and tub surrounds, and wainscoting and windowsills.

2.6.0 Plastics

Plastics are used more and more often now for plumbing fixtures because of their versatility, light weight, and relatively low cost, although the finish is not as durable as enameled cast iron. The plastics most commonly used today are acrylic plastic and plastic-reinforced fiberglass. Plastic-reinforced fiberglass is used to make one-piece showers and bathtubs, which include both the fixture and the adjoining walls. The strength of the product comes from the glass fibers. Fiberglass tubs and shower stalls are lightweight and easy to install. A plastic gel coating installed during manufacturing gives them a shiny, enamel-like surface. Marbleized colored acrylic plastic sheets are made into decorative lavatories that look like marble. These fixtures are attractive, and they are better than marble fixtures because they do not absorb water.

3.0.0 ◆ BASIC TYPES OF FIXTURES

This section describes each of the basic types of fixtures, including lavatories and sinks, bathtubs, shower stalls, toilets and water closets, urinals, bidets, garbage disposals, domestic dishwashers, laundry trays, drinking fountains and water coolers, and service sinks and mop basins, as well as some appliances.

3.1.0 Lavatories and Sinks

Lavatories are plumbing fixtures designed for installation in bathrooms and other locations primarily for washing your hands and face. They have the most variety of all the plumbing fixtures in color, size, and shape. Installing lavatories requires a **waste pipe** and a small-diameter vent pipe.

Sinks are shallow, flat-bottomed plumbing fixtures that are designed for use in food preparation and dishwashing. Kitchen sinks come in single, double, and triple compartments, usually installed in a countertop (see *Figure 1*). They are

109F01.EPS

Figure 1 ◆ Double-compartment sink.

Americans with Disabilities Act (ADA)

The ADA has very explicit requirements to improve the accessibility of all facilities to people with disabilities. The technical requirements contained in the *ADA Accessibility Guidelines for Buildings and Facilities* apply to all public accommodations such as restaurants, movie theaters, stores, and medical facilities.

The requirements are based on space allowances and reach ranges for people with various kinds of disabilities. The requirements affect everything from sinks and toilets to water coolers and grab bars.

manufactured from a variety of materials and are available in different shapes, sizes, colors, and styles.

Lavatories and sinks fall into four major styles according to methods of installation (see *Figure 2*). These include:

- Ledge-type or wall-hung
- Self-rimming
- Built-in rim
- Undercounter

A wall-hung lavatory, or lavatory carrier, is a lavatory that is fastened to the wall by a support bracket. The bracket is made from stamped steel or cast iron and is supplied by the manufacturer. This type of lavatory is normally installed with the **flood-level rim** 31 inches above the floor. The flood rim is the top edge of the fixture. If water fills higher than the flood rim, it will spill over onto the countertop and floor.

Several different styles of wall-hung lavatories are available (see *Figure 3*). These include the following:

- Corner
- Raised-back
- Ledge or shelf-back
- Wheelchair-accessible

Figure 2 ◆ Major styles of lavatories or sinks. (A) Ledge-type or wall-hung. (B) Self-rimming. (C) Built-in rim. (D) Undercounter.

Figure 3 ◆ Wall-hung lavatories. (A) Corner. (B) Raised-back. (C) Ledge or shelf-back. (D) Wheelchair-accessible.

Wheelchair-accessible lavatories are installed with the flood-level rim 34 inches above the floor and the drain fittings offset or moved back toward the wall rather than dropping directly down from the fixture.This puts the lavatory at an accessible height and also permits the wheelchair to slide easily under the lavatory. The offset fittings allow room for the user's legs under the lavatory. These lavatories are fitted with special faucets that have long blade handles that extend toward the user and do not require grasping and turning to turn on the water (see *Figure 4*). This is particularly necessary for users who have restricted mobility or who may suffer from arthritis.

The self-rimming lavatory or sink is set directly on the countertop. This sink or lavatory is manufactured with the mounting rim as part of the overall structure. It requires no external mounting frame to secure it to the countertop. Kitchen and bar sinks are examples of self-rimming designs.

The built-in rim lavatory has a metal rimming frame that is set in a stainless steel rim. This rim is used to support the bowl in the countertop.

The undercounter lavatory has no rim. It is mounted below an acrylic plastic or marble countertop. Brackets placed underneath the counter support this style.

The vanity-top lavatory (see *Figure 5*) is an undercounter-style lavatory that has a one-piece washbasin and countertop that is set on top of a vanity cabinet. Vanity-tops are manufactured from marbleized acrylic plastic and are available in a variety of colors and sizes.

Pedestal lavatories (see *Figure 6*) are decorative lavatories that sit on pedestals rather than on countertops or vanities. They may be mounted to the floor or to the wall. They must be firmly mounted according to the manufacturer's specifications.

3.2.0 Bathtubs

Bathtubs (see *Figure 7*) are plumbing fixtures designed as receptacles or containers for water. They are shaped to fit the human body and are used for bathing.

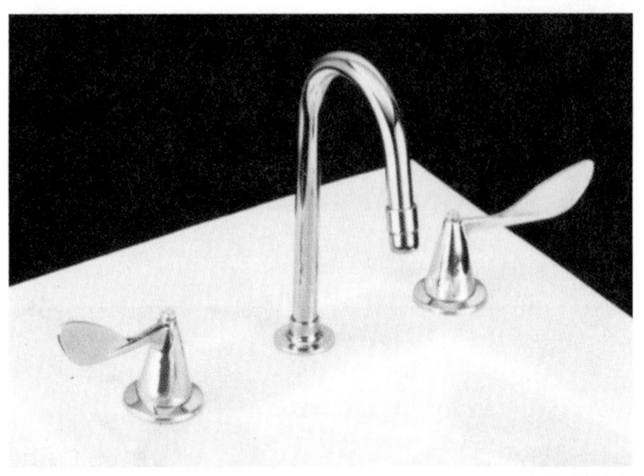

109F04.EPS

Figure 4 ◆ Accessible faucet.

109F05.EPS

Figure 5 ◆ Vanity-top lavatory.

109F06.EPS

Figure 6 ◆ Pedestal lavatory.

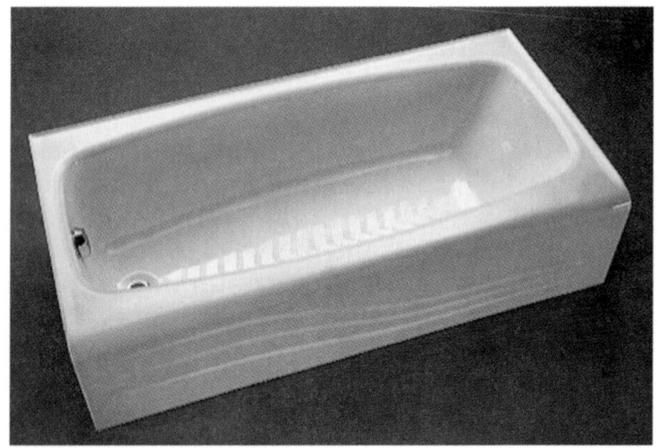

109F07.EPS

Figure 7 ◆ Bathtub.

Bathtubs are identified as being either **right-hand bathtubs** or **left-hand bathtubs,** depending on where the waste opening is. A right-hand tub has the drain on the right end of the tub as you face its finished side or apron. A left-hand tub has the drain on the left end of the tub.

To provide proper sanitation and to reduce maintenance, the exterior surface of the tub must be durable enough to withstand frequent cleanings, and the tub should have a slope of ⅛ inch per foot to the drain outlet.

Bathtubs are available in several sizes and shapes, but the standard bathtub is usually 5 feet long. Some tubs are manufactured with a special nonslip or slip-resistant safety feature to help reduce the danger of falls. Specialty bathtubs (such as whirlpools) are available with jets installed around the tub to provide high-pressure outlets for water therapy massage (see *Figure 8*).

Bathtubs are available in several styles, including built-in bathtubs and bath-shower modules (see *Figure 9*). The built-in bathtub is permanently built into the walls and floor of the bathroom. Bath-shower modules are usually made in one piece, although some models are available with the walls and tub as separate units. Because of their size, bath-shower modules are usually installed only in new homes.

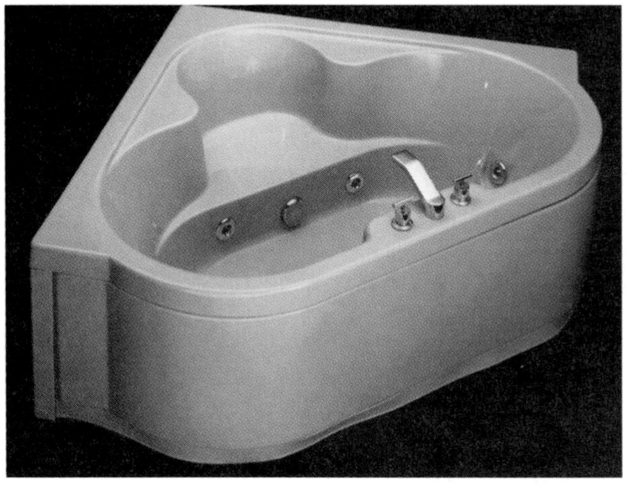

109F08A.EPS

(A)

LOCATE HYDROJET 15" DOWN BECAUSE OF THE RECESSED AREA SHOWN

AIR BLOWER

INSTALL AIR BLOWER A MINIMUM OF 12" ABOVE WATER LEVEL

ELECTRIC HEATER

FILTER

RETURN LINE

2-SPEED PUMP

SUCTION LINE

(B)

109F08B.EPS

Figure 8 ◆ Whirlpool bathtub. (A) Photo. (B) Schematic.

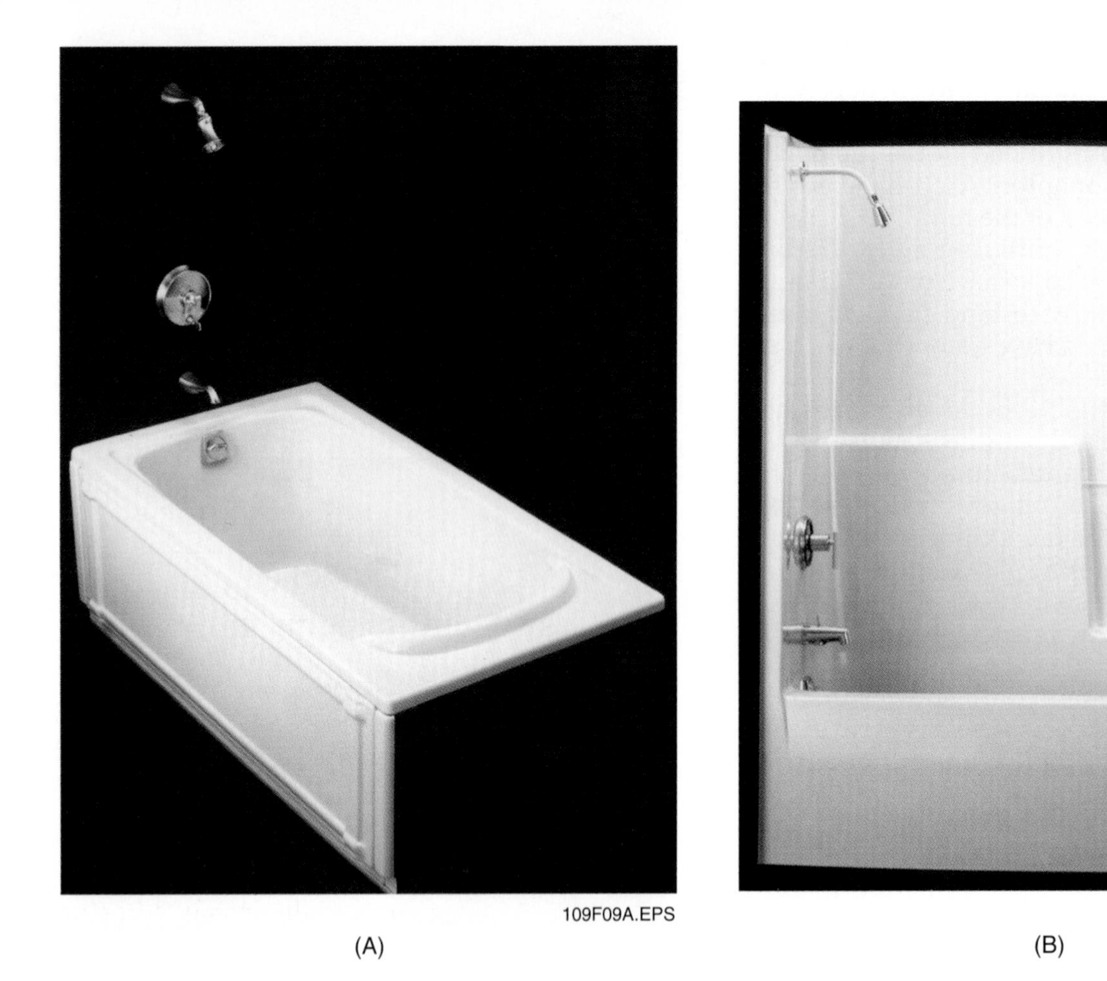

Figure 9 ◆ Bathtubs. (A) Built-in bathtub. (B) Bath-shower module.

Bath-shower modules are manufactured from plastic-reinforced fiberglass. The walls are usually high enough to eliminate the need for conventional wall material such as ceramic tile. Because these modules have no joints, cleaning the tub after installation is easy.

3.3.0 Shower Stalls

Shower stalls or shower baths are available in many sizes, shapes, and colors. They are stalls or baths in which water is showered from above onto the user's body. Shower stalls with seats are available. Residential shower baths consist of one shower head installed in a small enclosed space. Showers in industrial buildings, schools, gymnasiums, and similar facilities usually have a series of shower heads installed in a large shower room.

The walls of shower stalls are made of waterproof materials such as gel-coated fiberglass, enameled steel, or glazed tile. They can be either freestanding or built-in (see *Figure 10*).

Built-in showers are of two types. One consists of a premade shower enclosure. The other has walls constructed of glazed ceramic tile. A safety pan is required under the installation. Check your local plumbing code requirements for size, installation of shower stalls and floors (base), and materials. Not all product types and sizes are permitted in all areas.

Shower enclosures are made either as one-piece enclosures or with the walls and base as separate units. The walls and base of the shower enclosure are assembled on the job site. Bases for shower stalls are usually made of **terrazzo,** fiberglass, cast stone, or enameled steel. The dimensions of these bases are usually 36 inches by 36 inches or 36 inches by 48 inches with a slope of ¼ inch per foot to a center drain.

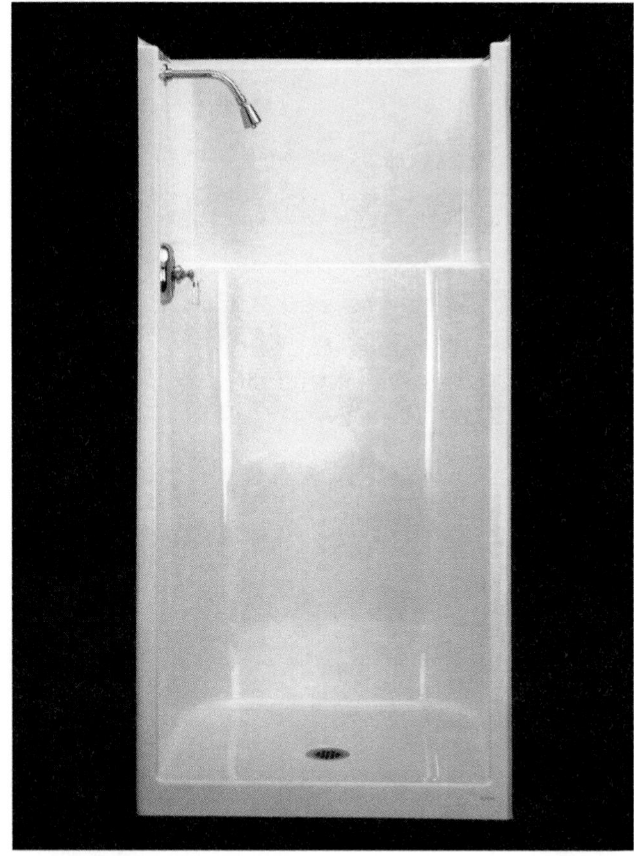
109F10.EPS

Figure 10 ◆ Built-in shower.

Review Questions

Sections 2.0.0–3.3.0

1. A common nonabsorbent and sanitary finish for plumbing fixtures is _____.
   a. stainless porcelain
   b. porcelain enamel
   c. metal rimming
   d. fiberglass enamel

2. A _____ is required under a shower installation.
   a. flood-level rim
   b. safety pan
   c. waste pipe
   d. shower enclosure

3. For proper drainage, a bathtub must have a slope of _____ per foot to the drain outlet.
   a. ¼ inch
   b. ⅔ inch
   c. ⅛ inch
   d. ¾ inch

4. Corner lavatories, raised-back lavatories, and ledge lavatories are examples of _____ lavatories.
   a. wall-hung
   b. accessible
   c. self-rimming
   d. undercounter

5. A _____ lavatory is made with the mounting rim as part of the overall structure.
   a. built-in
   b. vanity-top
   c. pedestal
   d. self-rimming

3.4.0 Toilets and Water Closets

Water closets or toilets are water-flushed plumbing fixtures designed to carry away solid organic waste.

Water closets come in both round-front and elongated shapes. Elongated water closets are used in public facilities. Both round-front and elongated shapes are used in residences. If you are installing a water closet for use by a person with a disability, the rim height should be 17½ inches.

Water closets are usually manufactured from vitreous china. They are usually installed one of two ways (see *Figure 11*):

- As a floor mount, where the water closet is secured directly to the floor and connected to the drainage system pipe by a **flange**
- Wall-hung, where the water closet is suspended from the wall on supporting chair carriers

Other types of water closets are still found, such as the old-style wall-mounted chain-pull tank and the corner tank (American Standard). Commercially available toilets today are all 1.6-gallon water-conserving tanks. But many of the older style toilets are installed in commercial and residential buildings, so you will need to be familiar with them.

3.4.1 Basic Operating Principles

The water closet is designed to carry away waste. When you flush a toilet, the water moves through passages in the bowl under pressure or gravity. The force of the water cleans the bowl. The water then moves the solid waste through the trap and into the drain. A siphoning action removes the solid waste from the bowl. This siphoning action

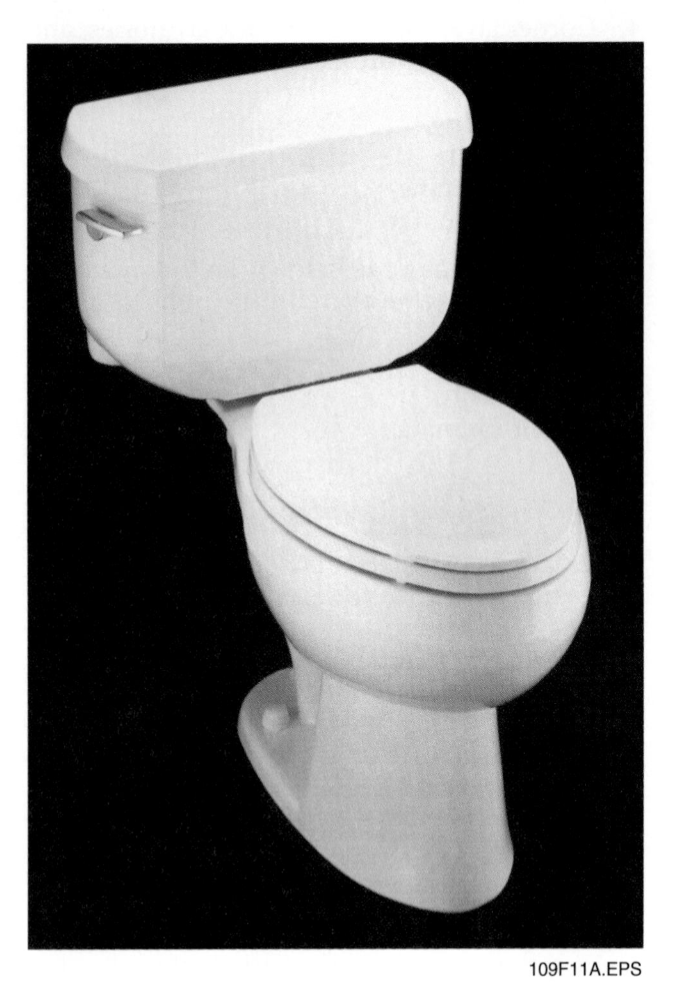

109F11A.EPS

(A)

109F11B.EPS

(B)

Figure 11 ◆ Water closet installations. (A) Floor mount. (B) Wall-hung.

Older-Style Toilets

The Conservation of Power and Water Resources Act has greatly influenced the design of water closets. Newer models have been developed that cut water use from 3 to 5 gallons per flush to 1.6 gallons. These newer models are a combination of siphon-jet and washdown styles. Some models use a pressure tank that greatly increases the efficiency of the lower-water flushing action.

Older-style toilets use older operating principles. You are not permitted to sell or install these preconservation-style water closets because they use too much water. But you will still find these older models in many residential and commercial settings.

Washdown Water Closet

The washdown water closet (see *Figure 12*) is inexpensive and simple in construction. It is the least efficient and the noisiest of all water closets. The waste passageway is located at the front of the bowl, which causes the front portion of the closet to protrude over the passageway. It is more susceptible to staining and contamination because the flat surface at the front of the bowl area is above the water level. For these reasons, many municipal codes no longer allow installation of washdown water closets.

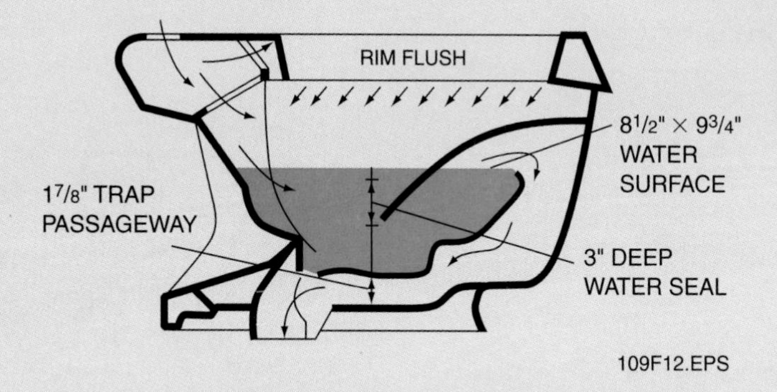

Figure 12 ◆ Washdown water closet.

Siphon Water Closets

Siphon-type water closets use a jet of water to speed up the siphon action, along with a downstream leg that is longer than the upstream leg. This creates a partial vacuum, which helps pull the waste from the bowl.

Siphon water closets come in three main categories or designs—the reverse-trap water closet (see *Figure 13*), the siphon-jet water closet (see *Figure 14*), and the siphon-action water closet (see *Figure 15*). A fourth category, the blowout water closet (see *Figure 16*), is used only in commercial plumbing installations.

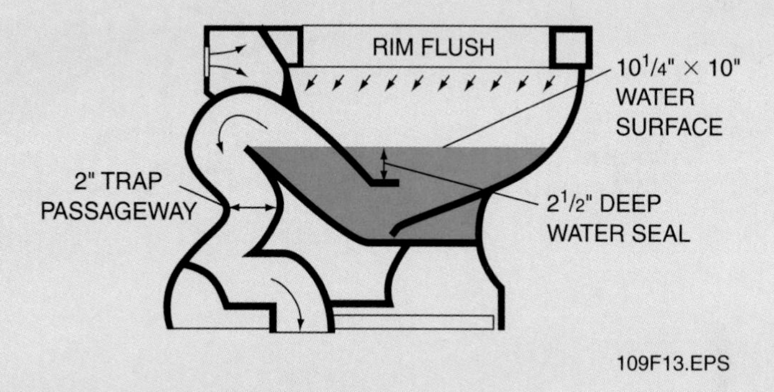

Figure 13 ◆ Reverse-trap water closet.

The reverse-trap water closet looks and flushes like the siphon-action water closet. However, the water surface area, passageway, and water seal of the reverse-trap are smaller than those of the siphon-jet or siphon-action water closets. These features allow the reverse-trap water closet to operate on less water than the other two designs. The reverse-trap water closet has a waste passageway that measures 2 inches in diameter, a water surface area of 10¼ inches by 10 inches, and a 2½-inch-deep water seal.

The siphon-jet water closet has a larger passageway and more water surface area than the reverse-trap closet. This reduces the tendency for waste articles to become stuck and clog the passageway. The siphon-jet water closet does not require a head of water in the bowl to flush out solids. Instead, the stream of water that the closet delivers spins to the outlet of the trap. This type of water closet has a waste passageway that measures 2⅝ inches in diameter, a water area of 12 inches by 10½ inches, and a 3-inch-deep water seal.

The siphon-action water closet is usually manufactured with a one-piece closet that combines the closet and flush tank in one unit, which makes it more compact than other types of closets. It has a water surface area that measures 12 inches by 11½ inches, and it flushes quietly. It has a trap passageway diameter of 2 inches and a 3-inch-deep water seal.

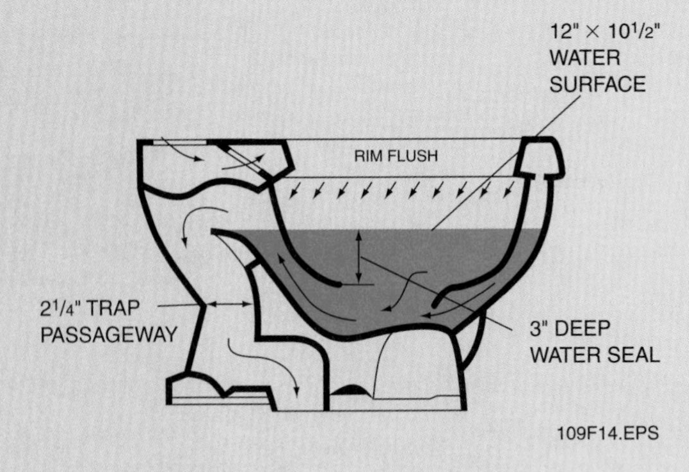

Figure 14 ◆ Siphon-jet water closet.

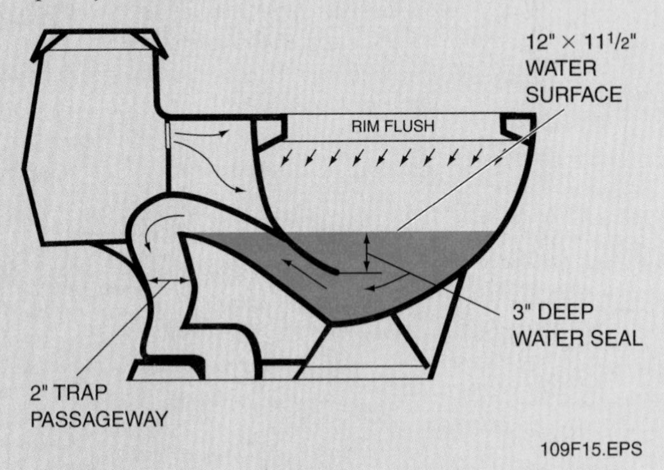

Figure 15 ◆ Siphon-action water closet.

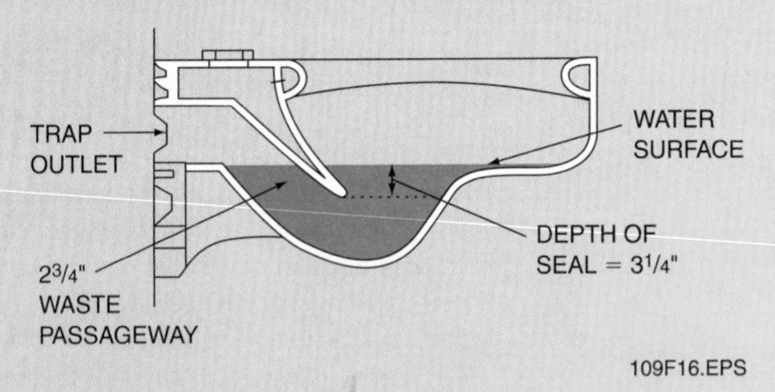

Figure 16 ◆ Blowout water closet.

In the blowout water closet, which is usually a wall-hung bowl, the flushing action depends entirely on a large jet of water being directed to the inlet of the trap passageway. The force of the jet of water draws the contents of the bowl into the inlet trap passageway. The jet then forces or blows the waste into the outlet passageway to flush the bowl. The blowout water closet's main advantage is its ability to flush large amounts of toilet paper without becoming clogged.

The blowout water closet has a waste passageway that measures 2¾ inches in diameter, a water surface area of 11¼ inches by 14½ inches, and a 3¼-inch-deep water seal.

occurs when the water moves through the downward leg of the trap, creating a drop in the atmospheric pressure at the trap outlet. This partial vacuum, along with the head of water from the bowl, sweeps the wastes out.

The siphoning action continues to draw water out of the bowl even after it slows to a trickle. As the water flow through the trap lessens, air is allowed to rush into the trap. This causes a break in the vacuum and equalizes the pressure on both sides of the trap. The **flush valve** or flush tank delivers enough additional water to reseal the trap.

Although this is basically how all water closets operate, each type produces the flushing action in a slightly different way. Several older-style types are described in the *On the Level* feature. Each water closet has its own special uses and advantages.

3.5.0 Urinals

Urinals are water-flushed plumbing fixtures designed to receive urine directly. They are commonly installed in public restrooms for men. Because the turnover in users is so rapid, and because the bulk of a water tank is not practical in these installations, urinals are usually fitted with flush valves (see *Figure 17*).

Urinals are manufactured from vitreous china or stainless steel. Like water closets, urinals are available with a variety of flushing actions. In urinals, these include washouts and siphon jets.

The heights of urinals for people with disabilities are determined by local codes.

3.5.1 Basic Operating Principles

A urinal must have an adequate flushing device. The flushing device must be able to completely remove the urine from the fixture after it is used, to prevent the spread of disease, fouling of the urinal, and offensive urine odor. Urinals may be flushed with a hand valve, a flushometer valve (see *Figure 17*), or a flush tank. Electric eye (which uses an infrared sensor) and automatic flush devices are also available (see *Figure 18*).

Two main types of urinals are washout urinals and siphon-jet urinals.

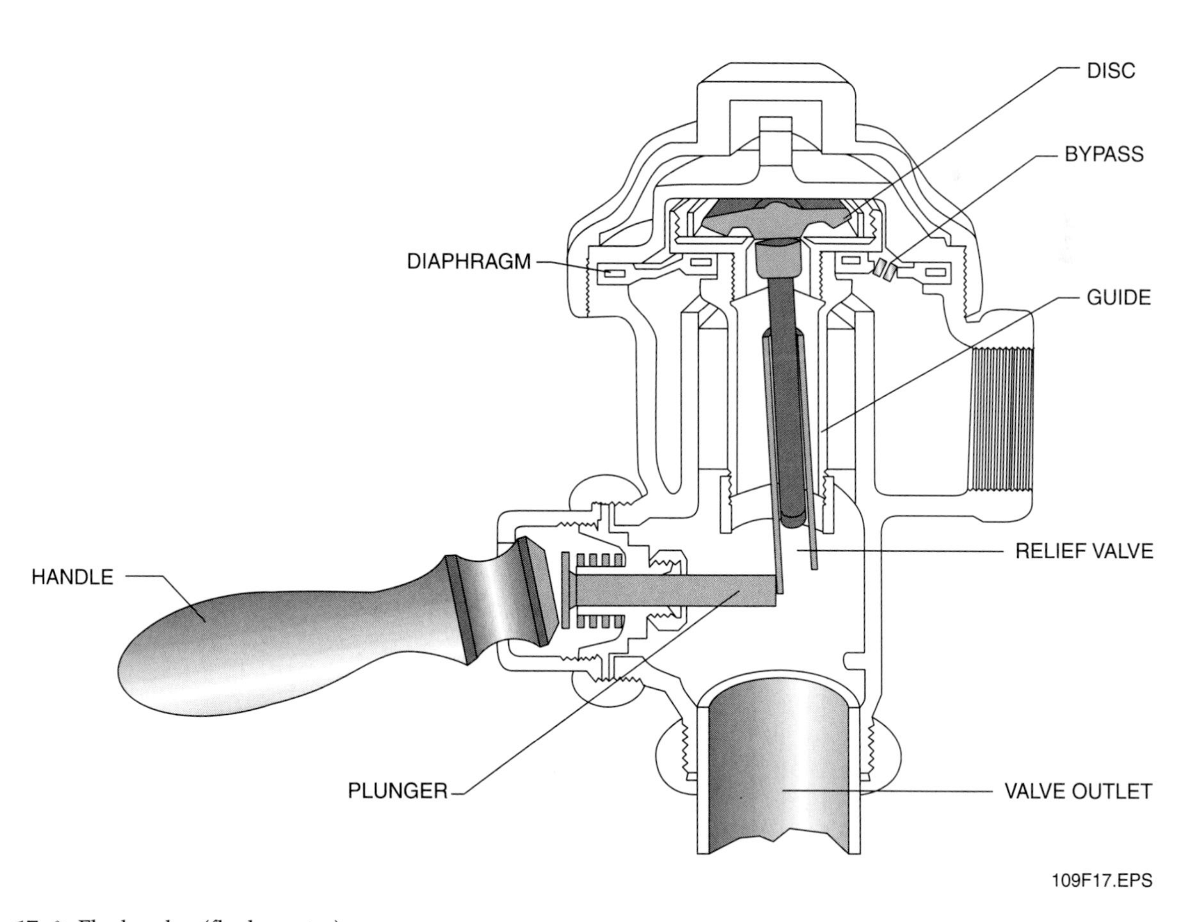

Figure 17 ◆ Flush valve (flushometer).

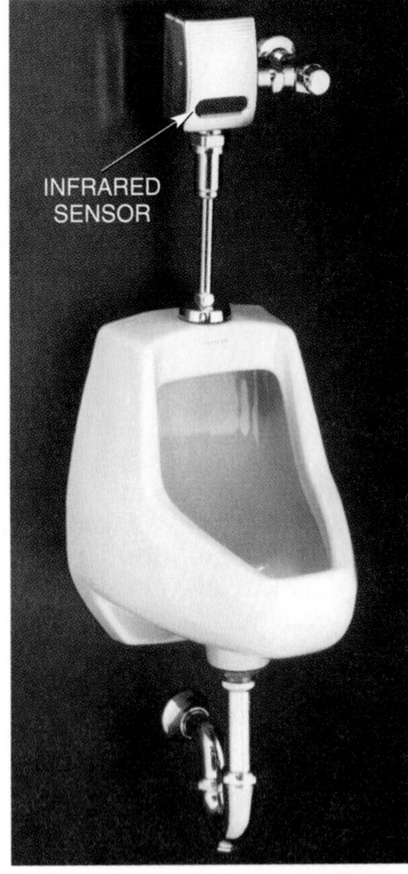

Figure 18 ◆ Automatic flush device on siphon-jet urinal.

Figure 19 ◆ Washout urinal.

The washout urinal washes waste out of the trap rather than flushing it out. The washing water enters the top of the urinal, flows out through openings in the top rim of the urinal into the fixture, spreads across the back of the urinal, and flows by gravity out through the urinal trap at the base and into the sanitary drainage system.

Washout urinals are characterized by the restricted opening over the trap inlet at the base of the urinal (see *Figure 19*). The purpose of these small openings over the trap is to prevent the trap from becoming plugged by debris in the urinal. These restricted openings may be of two designs—either a beehive or small openings cast in the china. Washout urinals are available in two styles—wall-hung washout urinals with a built-in outlet and bottom-outlet wall-hung washout urinals.

Wall-hung washout urinals have built-in 2-inch P-traps. The restrictive trap openings are cast into the china at the bottom of the urinal. Wall-hung washout urinals require a 2-inch waste pipe and a small-diameter **vent pipe.**

Another type of wall-hung washout urinal has a bottom outlet. It has a beehive strainer over a drain opening that is attached to a separate exposed 1½-inch trap. The bottom-outlet wall-hung washout urinal requires a small-diameter vent pipe. Some codes prohibit its use because it has an invisible trap seal.

Siphon-jet urinals use a flushing action that is similar to the one used in siphon-jet water closets. This type of urinal has a large opening over the trap inlet located in the bottom of the urinal.

Siphon-jet urinals are available in two styles—wall-hung siphon-jet urinals (see *Figure 18*) and women's urinals (see *Figure 20*). *Figure 18* also shows an infrared sensor, which detects when a person is standing in front of the urinal and when the person has stepped away, then automatically triggers the flushing action.

Siphon-jet urinals flush 1 gallon or less per use. This feature is a result of the Conservation of Power and Water Resources Act. You will find infrared urinals in commercial settings, such as airports and movie theaters, where user volume typically is very high.

Wall-hung siphon-jet urinals usually have a 2-inch built-in trap. They require a small-diameter vent pipe.

109F20.TIF

Figure 20 ◆ Women's urinal.

Did You Know?

The women's urinal is designed so that it can be straddled for use, so it does not have a seat like a typical water closet. Women's urinals are available in either floor-set or wall-hung versions. The floor-set style is recessed slightly below the finished floor grade to provide for drainage. The wall-hung urinal is supported by a concealed metal carrier or other permitted backing so that no strain is transmitted to the pipe connection. Both styles have built-in traps.

Most of these urinals were installed in the 1960s, and they are sometimes still found in older public buildings. It's unlikely you'll install a women's urinal, but you may encounter some that need maintenance.

3.6.0 Bidets

Bidets (pronounced bih-*days* and shown in *Figure 21*) are companion fixtures to water closets. They are used for personal hygiene in bathing the external genitals and posterior parts of the body. Because of their height and style, they are particularly useful for older persons or persons recovering from an illness who have difficulty using a shower or tub.

109F21.TIF

Figure 21 ◆ Bidet.

The bidet has long been a popular plumbing fixture in European and South American bathrooms, and it is becoming increasingly popular in modern bathrooms in the United States. It is fitted with cold and hot water faucets so that the user can regulate water temperature and rate of flow. The user sits astride the bidet facing the faucets. The bidet is also fitted with a pop-up stopper that allows the user to let the water accumulate in the bowl. The bidet is rinsed through a rim flushing action similar to that used in a water closet and has a 1½-inch trap.

Review Questions

Sections 3.4.0–3.6.0

1. A floor-mount water closet is secured directly to the floor and connected to the drainage system pipe by a _____.
   a. trap
   b. flange
   c. flush valve
   d. chair carrier

2. A siphon-jet urinal is designed to flush with _____ gallon(s) per flush.
   a. 2
   b. 1.6
   c. 0.5
   d. 1

3. A newer model water closet is a combination of a washdown and a _____ style toilet and is designed to reduce water consumption per flush.
   a. wall-hung
   b. siphon
   c. siphon-jet
   d. vacuum

4. Because a water tank is not practical, a _____ is usually installed to clear a urinal.
   a. flush tank
   b. flush valve
   c. flush vacuum
   d. flush closet

5. A _____ water closet is often used in commercial plumbing installations because of its ability to flush large amounts of toilet paper without becoming clogged.
   a. washdown
   b. siphon-action
   c. blowout
   d. siphon-jet

Figure 22 ◆ Garbage disposal.

3.7.0 Garbage Disposals

A garbage disposal or food waste disposal (see *Figure 22*) is an electric device used to grind food waste into a pulp that is then discharged into the drainage system. It is designed to be used with running water supplied by the kitchen sink faucet. Garbage disposals are mounted under the kitchen sink (or under one compartment of a two-compartment sink) in place of the basket strainer assembly that would otherwise lead to the drainage fixture. Disposals may be connected so that the waste is discharged into a separate P-trap or to a continuous waste fitting. A fitting is used to bring the wastewater from a dishwasher through the garbage disposal, as required by code.

When installing a garbage disposal in a double sink, the plumber must use a directional tee with an internal baffle that joins the disposal waste and the waste from the other sink compartment (see *Figure 23*). The tee and baffle prevent the disposal waste from backing up into the other sink compartment.

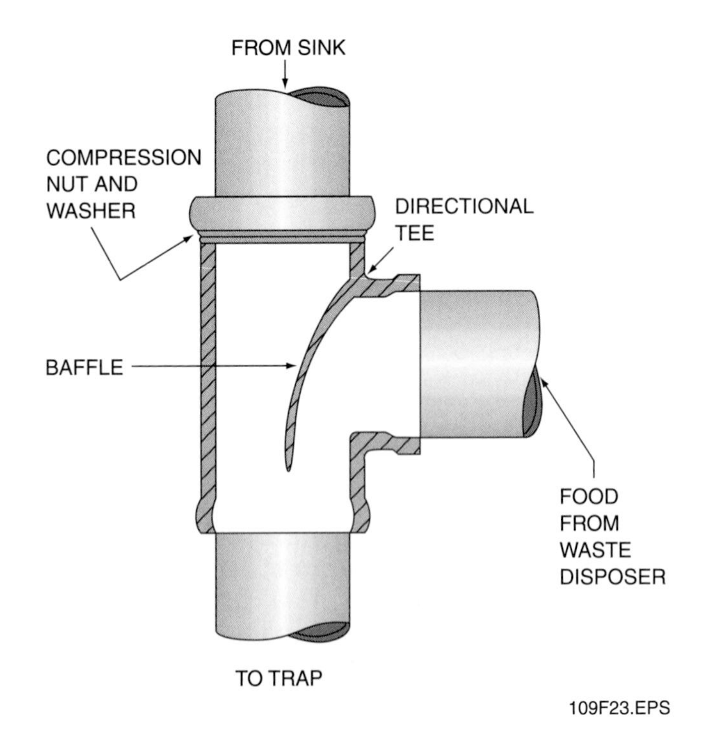

Figure 23 ◆ Garbage disposal directional tee.

It is important for the plumber to remind the homeowner that a large quantity of water should be run through the garbage disposal when it is in use. This will help wash the food waste pulp out of the drainage pipe and keep the drainpipe from getting clogged.

3.8.0 Domestic Dishwashers

A domestic dishwasher (see *Figure 24*) is an electric appliance used in the home for washing dishes. The most common sizes of dishwashers are 18-inch and 24-inch units. Most dishwashers installed in homes use a pump-out waste disposal system. This system can be installed with a rubber hose or copper tubing connecting the dishwasher waste unit to a dishwasher tailpiece under the kitchen sink basket strainer. If the sink is equipped with a garbage disposal unit, the dishwasher can be connected into the dishwasher drain connection on the garbage disposal. In either type of connection, some codes require a vacuum breaker or air gap fitting in the dishwasher discharge piping (see *Figure 25*) so that waste will not back up into the dishwasher.

3.9.0 Laundry Trays

Laundry trays or tubs are usually installed in the laundry rooms of homes. They are fitted with cold water, hot water, and drain connections. They are used for washing clothes and other household items and for receiving waste from automatic clothes washers. They can also be used to store water from automatic clothes washers that are equipped with a water reuse cycle.

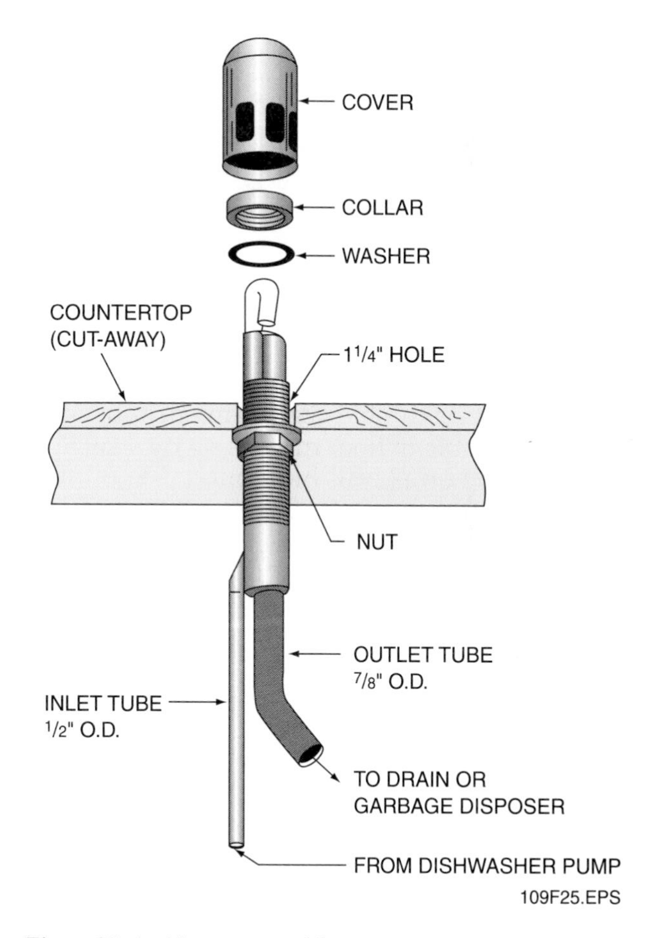

Figure 25 ◆ Air gap assembly.

Figure 24 ◆ Domestic dishwasher.

ON THE LEVEL

More Than Meets the Eye

Plumbers are also called on to install other fixtures and appliances. Typically, a plumber is required to install the washer box for hooking up the washing machine (see *Figure 26*). Plumbers rough-in the hot and cold water and a drain. Another option for draining a washing machine is a laundry tray.

It is also fairly common to have a plumber install refrigerator icemaker connections. Plumbers may also help install private swimming pool water supply and disposal systems.

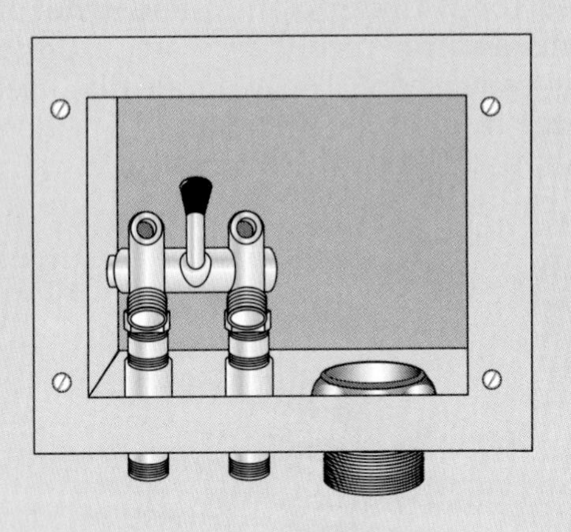

109F26.EPS

Figure 26 ◆ Washer boxes.

Laundry trays are available in either single-compartment or double-compartment tubs, and they are available in floor models set on steel legs (see *Figure 27*) or as wall-hung models mounted on a bracket that attaches to the wall. Both models come supplied with the proper mounting hardware.

3.10.0 Service Sinks and Mop Basins

Service sinks (see *Figure 28*) and mop basins (see *Figure 29*) are plumbing fixtures that are installed in commercial buildings for building maintenance personnel to use. They are installed in janitors' closets and building maintenance areas.

A service sink, or slop sink, has a deep basin that is large enough to hold a scrub pail. These sinks are used to fill and empty scrub pails, rinse mops, and dispose of cleaning water. The sinks are usually made of enameled cast iron, and they come supplied with a standard P-trap.

Mop basins, or mop receptors, are service sinks that are set on the floor of a janitors' closet. They are somewhat more convenient to use than service sinks because the mop pail does not have to be lifted in and out of the fixture. Mop basins can be made of enameled cast iron, terrazzo, fiberglass, or ceramic tile.

109F27.EPS

Figure 27 ◆ Laundry tray—floor type.

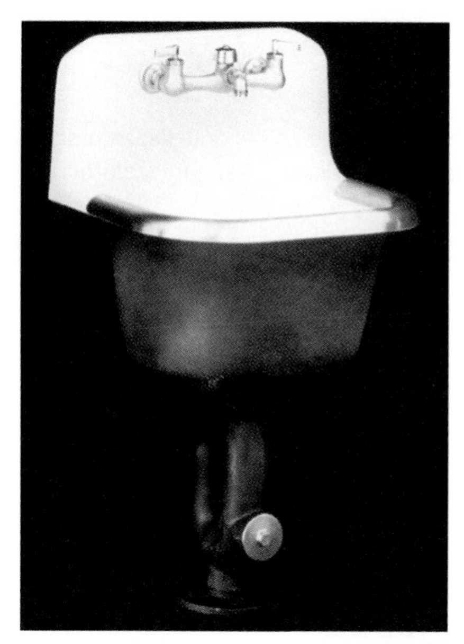

109F28.TIF

Figure 28 ◆ Service sink.

109F29.TIF

Figure 29 ◆ Mop basin.

3.11.0 Drinking Fountains and Water Coolers

Drinking fountains are plumbing fixtures that deliver a stream or jet of water through a nozzle (or bubbler) without cooling it (see *Figure 30*). They provide a convenient source of sanitary drinking water. Drinking fountains are available in a variety of designs.

A drinking fountain that is combined with a water-cooling unit is called a water cooler (see *Figure 31*). These units require an electrical hookup to

109F30.TIF

Figure 30 ◆ Accessible drinking fountain.

109F31.TIF

Figure 31 ◆ Water cooler.

ON THE LEVEL

Local Plumbing Codes

Local plumbing codes govern construction and installation of drinking fountains and water coolers. Typical plumbing codes require the following sanitary features:

- The bowl of the fountain must be constructed of a nonabsorbent material.
- A mouth protector must be provided over the mouthpiece.
- The drinking water must be delivered at an angle so that no water can fall back onto the nozzle.
- The nozzle must be located above the flood-level rim of the fountain.
- The water supply to the fountain must be entirely separated from the waste.

power the water-cooling system. The drinking water is cooled to the desired temperature, usually around 50 degrees F, before it is delivered to the nozzle.

Both drinking fountains and water coolers are available in a variety of styles, but the wall-hung fixture is the most common style. They are manufactured from vitreous china, enameled cast iron, or stainless steel.

Review Questions

Sections 3.7.0–3.11.0

1. The nozzle on a drinking fountain must be angled so that water cannot _____.
   a. spill over the flood rim
   b. fall back onto the nozzle
   c. splash the drinker's face
   d. pool above the waste drain

2. To prevent disposal waste from backing up into the other compartment in a double sink, you must install a _____.
   a. vacuum breaker
   b. directional tee with an internal baffle
   c. strainer with a directional tee
   d. waste valve

3. It is important to install a(n) _____ in the dishwasher discharge piping.
   a. vacuum breaker
   b. internal baffle
   c. directional tee
   d. discharge pipe

4. A _____ typically has a deep basin that is large enough for a scrub pail.
   a. utility sink
   b. mop basin
   c. laundry tray
   d. service sink

5. _____ can be used to store water from washers that are equipped with a water reuse cycle.
   a. Slop sinks
   b. Service sinks
   c. Mop basins
   d. Laundry trays

4.0.0 ◆ FAUCETS

A plumber comes in contact with many kinds of faucets. A faucet is any fixture used for drawing water from a pipe.

Because faucets are used in all residential, commercial, and industrial buildings, it is important for a plumber to understand what kinds of faucets are available, how they work, and how and where they are used. Familiarity with the different sizes, shapes, and styles of faucets is essential so that you can choose the right kind for each installation.

Faucets fall into three basic classifications:

- Utility faucets
- Bathroom faucets
- Kitchen faucets

In all three classifications there are two types of faucets—compression faucets and noncompression faucets, which are covered in more detail later in this module.

4.1.0 Utility Faucets

Utility faucets are found in boiler rooms, in laundry rooms, and on outside walls. They are usually manufactured from extra-grade brass, white metallic alloy, or **thermoplastic** materials. Most utility faucets are not plated because appearance is not important. Common utility faucets include:

- sediment and boiler drains (see *Figure 32*)
- laundry faucets used for laundry tubs, utility sinks, service sinks, and mop sinks

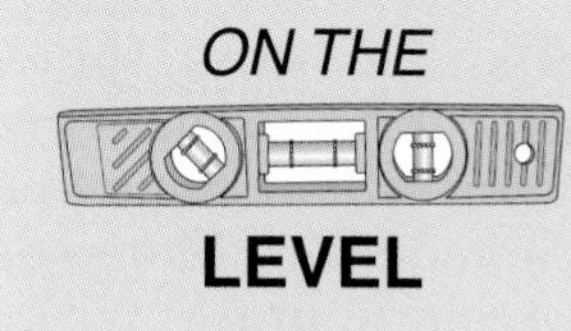

Regulating Water Flow

Compression faucets control the flow of water by compressing (pressing) a washer between the **valve stem** and the **valve seat.** The stem has an enlarged threaded portion. This threaded portion fits threads into the body so that the stem moves up or down as the handle is turned.

A noncompression or washerless faucet regulates the flow of water by obstructing the water. When the faucet is open, the openings in two disks are matched, allowing the water to flow through. When the faucet is closed, the disks rotate and cut off the flow of water.

The water outlet end, or **bibb,** for a utility faucet is made with either a plain (threadless) bibb or a hose (threaded) bibb as shown in *Figure 33.* The bibb has a downward angle that directs the water toward the ground. Most common utility compression faucets have female iron pipe threads on the end that is attached to the water pipe during installation.

The plain bibb compression faucet has a flange designed to fit flush against siding or wall covering. It has hidden female pipe fittings on the water inlet end.

HC-409 SINGLE MALE DRAIN BOILER SIZE 1/2" 3/4"

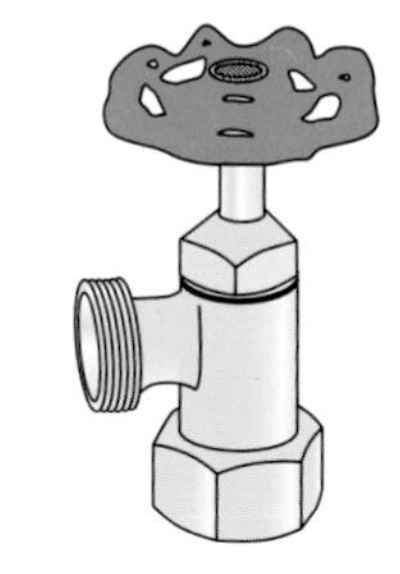

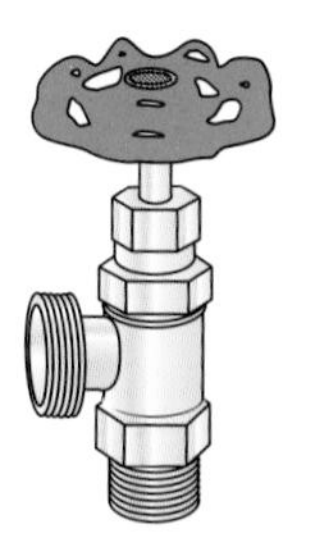

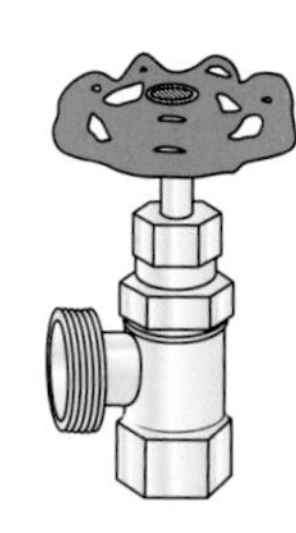

Figure 32 ◆ Boiler drains.

Sill cocks are sometimes called hose bibbs. In commercial applications, they're often specified as wall hydrants (see *Figure 34*).The weatherproof lawn faucet, or freezeproof compression faucet (see *Figure 35*), is used to provide water access out-

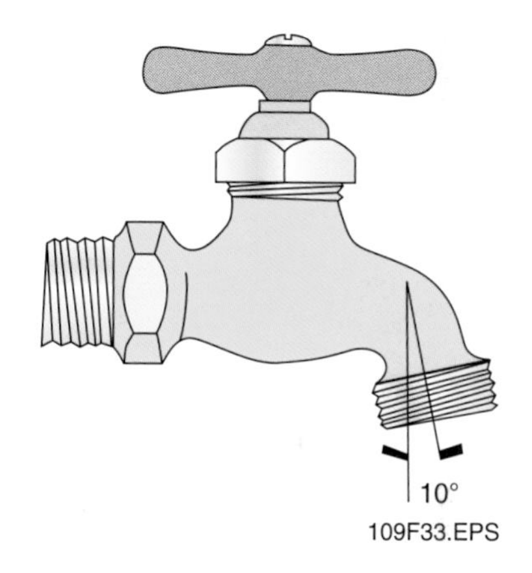

Figure 33 ◆ Threaded hose bibb.

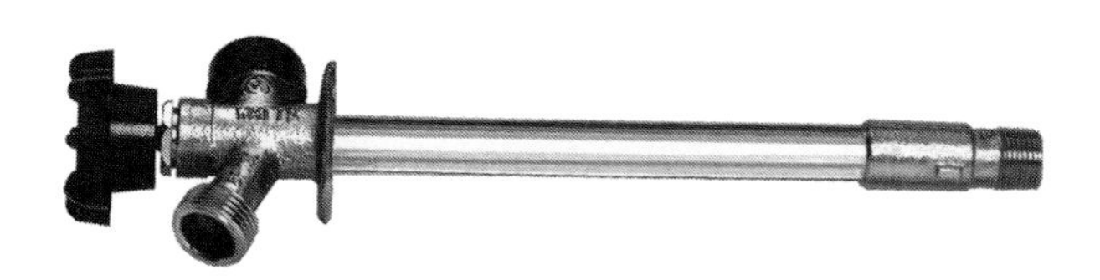

Figure 34 ◆ Wall hydrant.

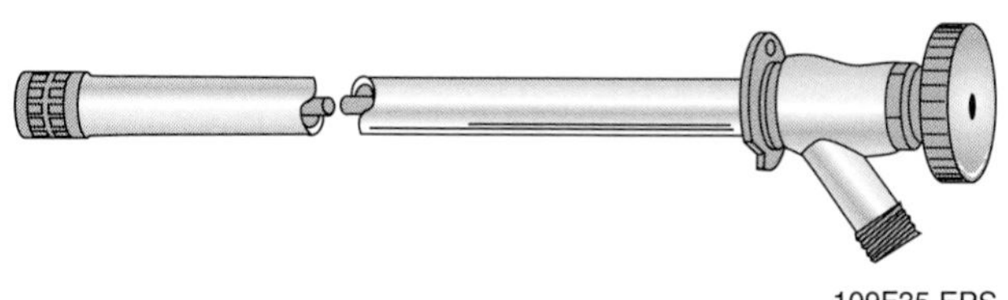

Figure 35 ◆ Weatherproof lawn compression faucet.

side a building. These faucets attach to the water pipe with male threads. The faucets are run through the outside wall of the structure, leaving only the spigot, or faucet head, exposed to the weather. When the water is shut off, it drains from the spigot through the extension tube to the valve seat, which is inside the building.

It is important to install the lawn or freezeproof faucet properly so that it does not freeze from constant exposure to the weather. The valve seat should be well within the protection of the structure, installed according to the manufacturer's specifications.

Compression faucets are suitable for moderate and heavy pressure. This type of faucet is usually used for water outlets. Compression faucets control the flow of water by compressing a **seat washer** between the valve stem and the valve seat, as shown in *Figure 36*. The compression occurs when the valve stem is turned down into corresponding threads in the body of the faucet.

4.2.0 Bathroom and Kitchen Faucets

Faucets and accessories for kitchen sinks, showers, and bathroom lavatories and fixtures are made of many different materials and come in a variety of decorator shapes and sizes. Materials used for making faucets include brass, white metallic alloys, and different types of plastics, such as nylons and acrylics.

The distance between the hot and cold water inlets will vary depending on where the openings are in the sink or lavatory (see *Figure 37*). This distance, usually measured in even numbers, ranges from 4 inches to 12 inches. The most common bathroom faucets have 4-inch or 8-inch centers. The most common center used for a kitchen faucet is 8 inches.

These faucets can be classified in two main categories: compression and noncompression. These two types will be covered in more detail in a later section.

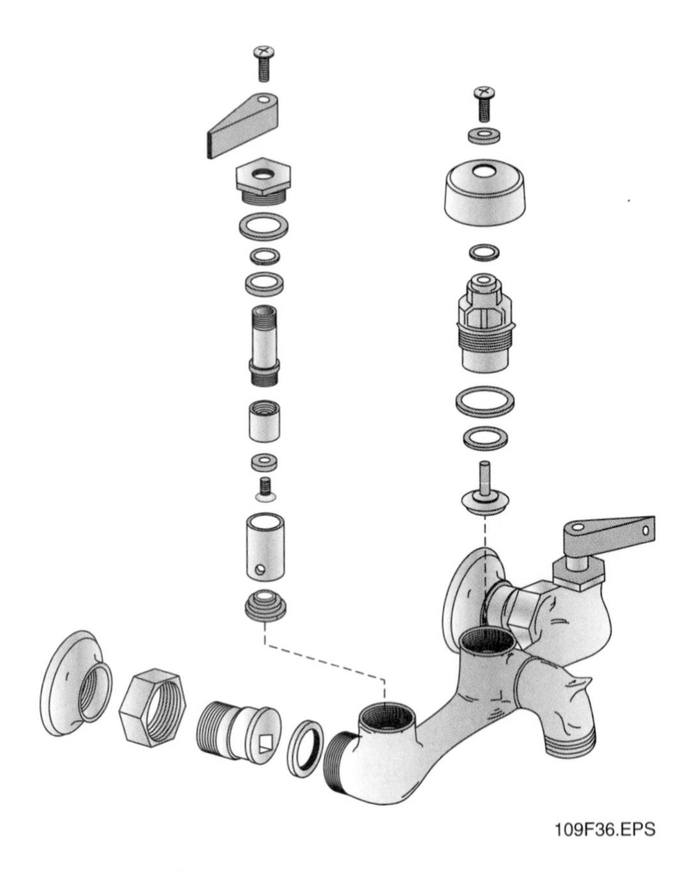

Figure 36 ◆ Compression faucet.

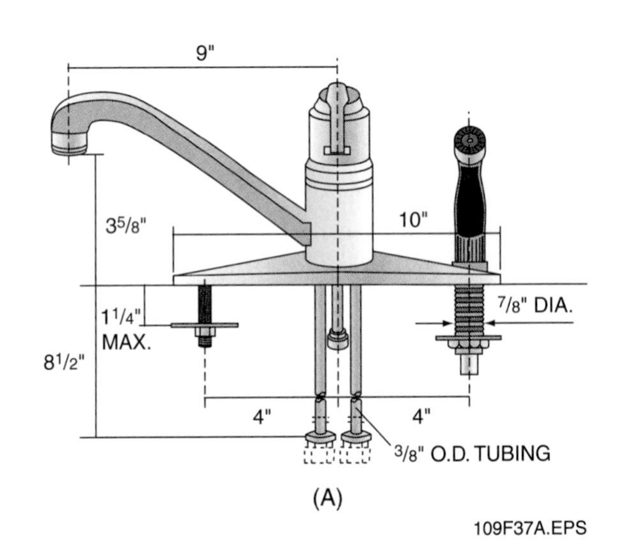

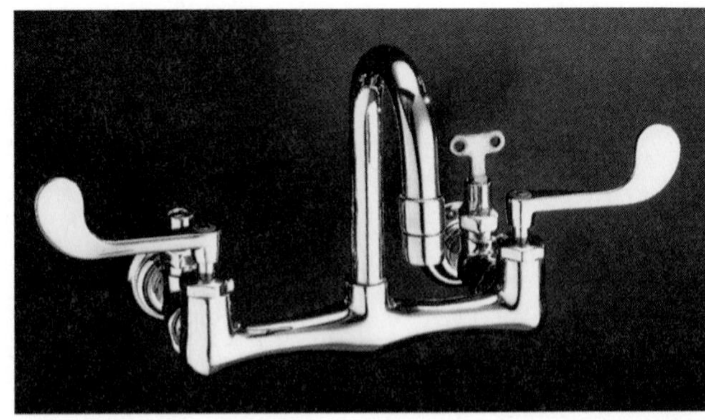

Figure 37 ◆ Common center for kitchen faucets.

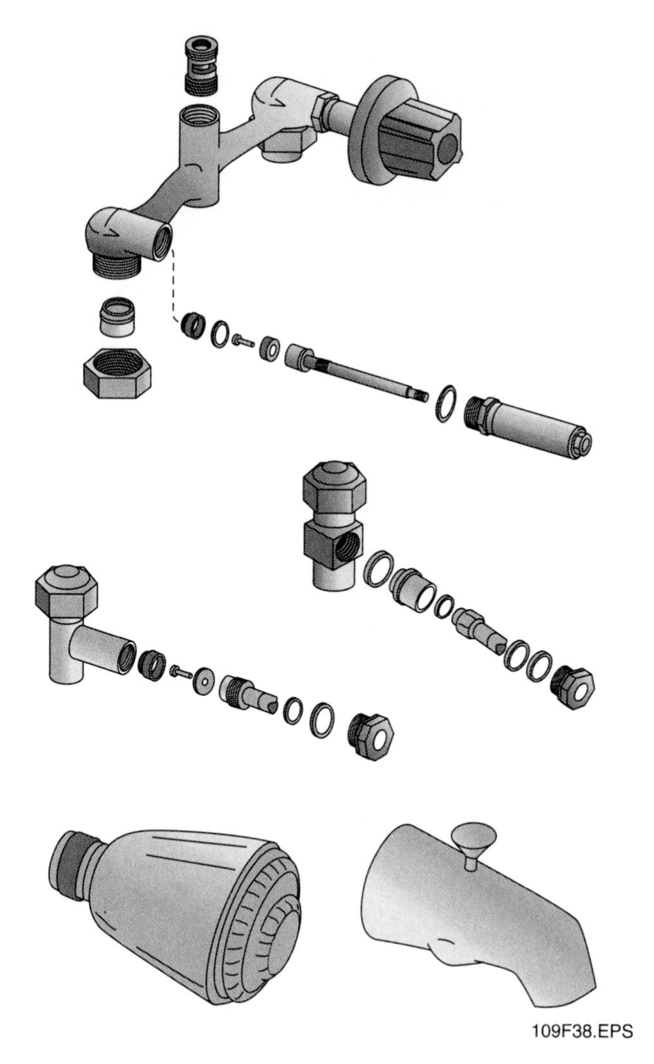

Figure 38 ◆ Combination shower and bath fitting.

Figure 39 ◆ Single-control pressure mixing valve.

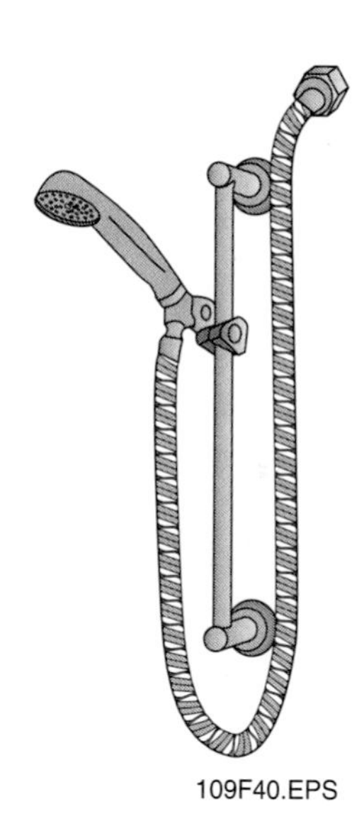

Figure 40 ◆ Combination fitting with flexible water line.

4.2.1 Combination Shower and Bath Fittings

A combination shower and bath fitting holds both faucet and shower head, as shown in *Figure 38*. A **diverter**, located in the spigot or spout, sends the water to the shower head. When the diverter is lifted, it blocks the flow of water to the spout. This forces the water up into the shower head. Water pressure and temperature are controlled by dual hot and cold valves. Some units use a single-control pressure mixing valve to adjust the water's flow-rate and temperature (see *Figure 39*).

Another type of combination shower uses a flexible water line. With a flexible water line, the shower head can easily be raised, lowered, or turned in any direction. The head can either be handheld or placed on a wall-mounted bracket (see *Figure 40*).

4.3.0 Compression and Noncompression Faucets

When compression faucets are used for kitchen sinks or bathroom lavatories, they usually combine two compression valves and a mixer, which mixes the hot and cold water and delivers it through a mixing spout common to both valves (see *Figures 41 and 42*). The drain valve is controlled with a plunger located between supply valves, as shown in *Figure 43*. Some older dwellings may have sinks and lavatories that use individual compression valves and faucets for the hot and cold water.

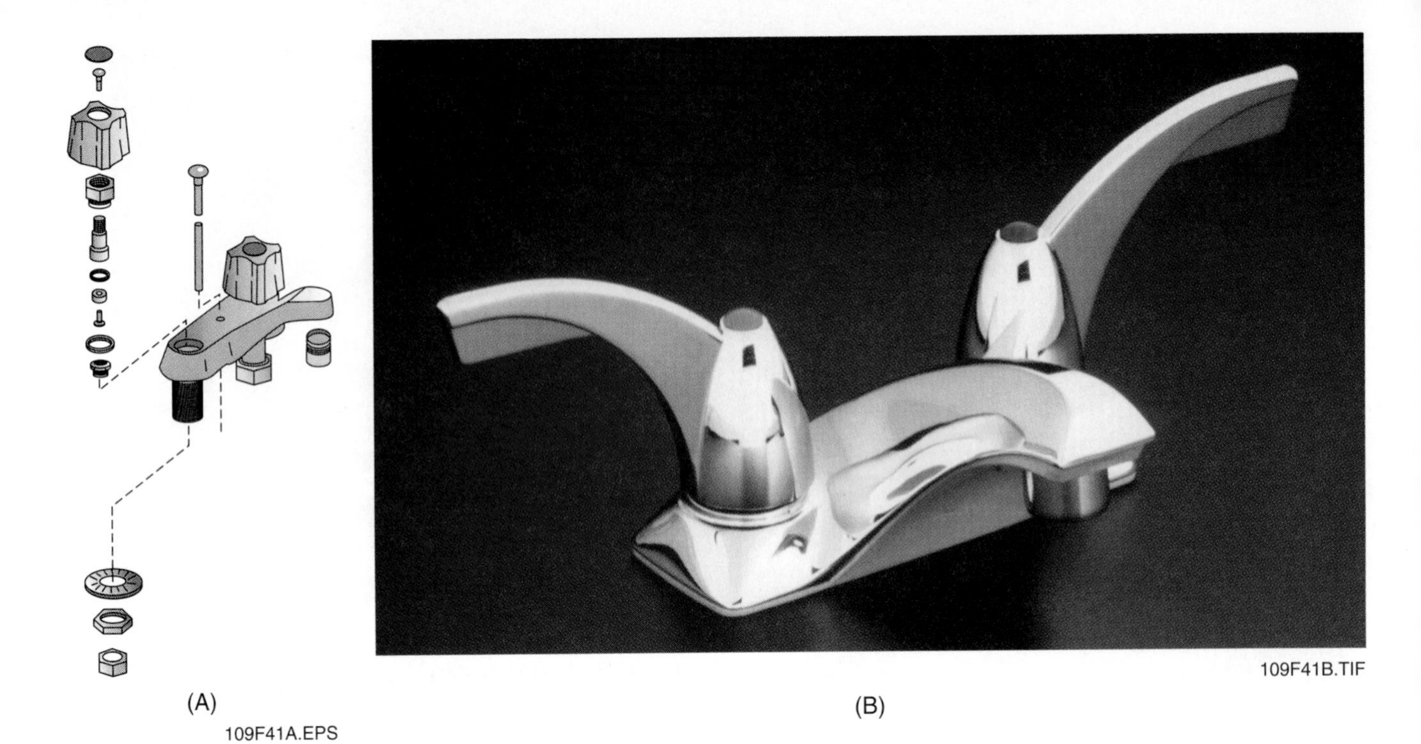

Figure 41 ◆ Compression faucets for bathroom.

These faucets are installed in pairs—one for hot water, one for cold water.

In some cases, to conserve water, it may be more practical to install a self-closing compression faucet, as shown in *Figure 44*. The self-closing

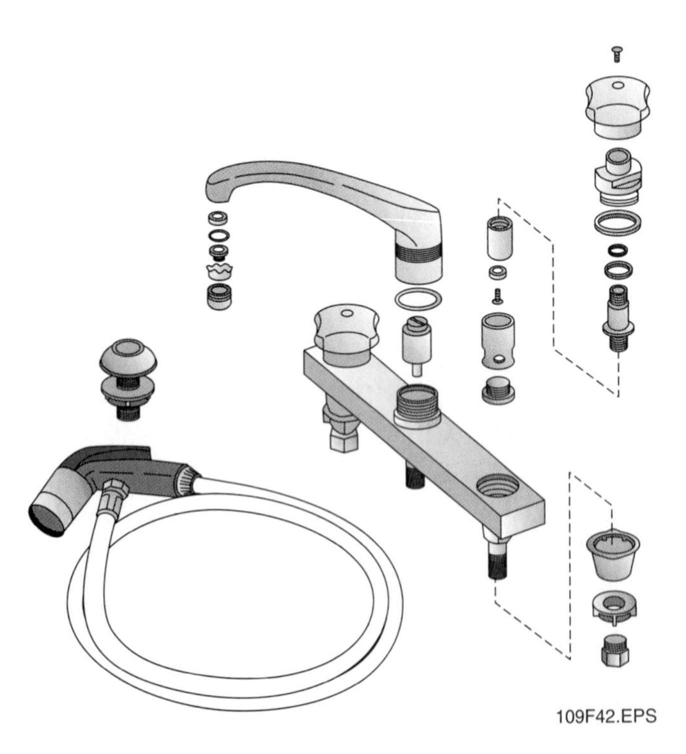

Figure 42 ◆ Compression kitchen faucet with sprayer hose.

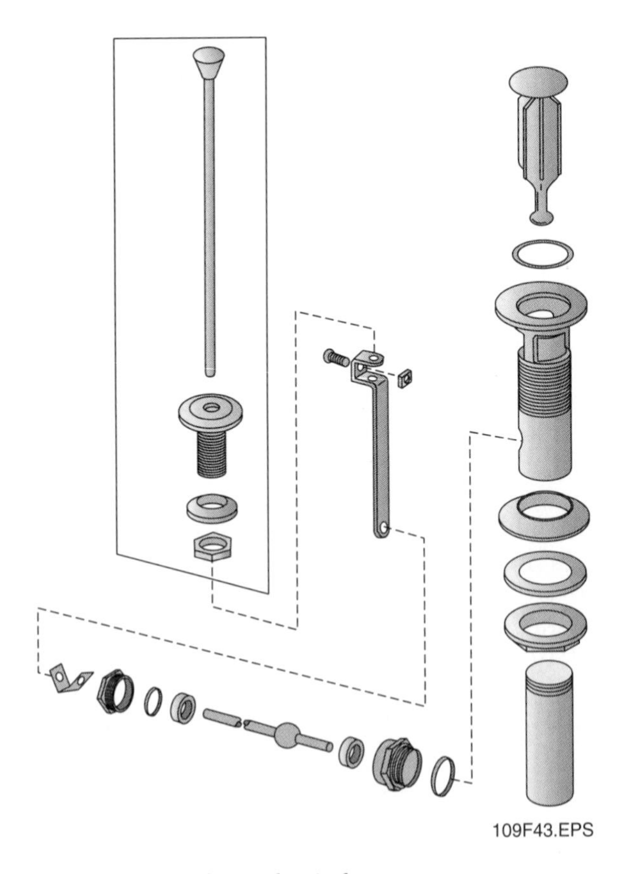

Figure 43 ◆ Part of a mechanical pop-up.

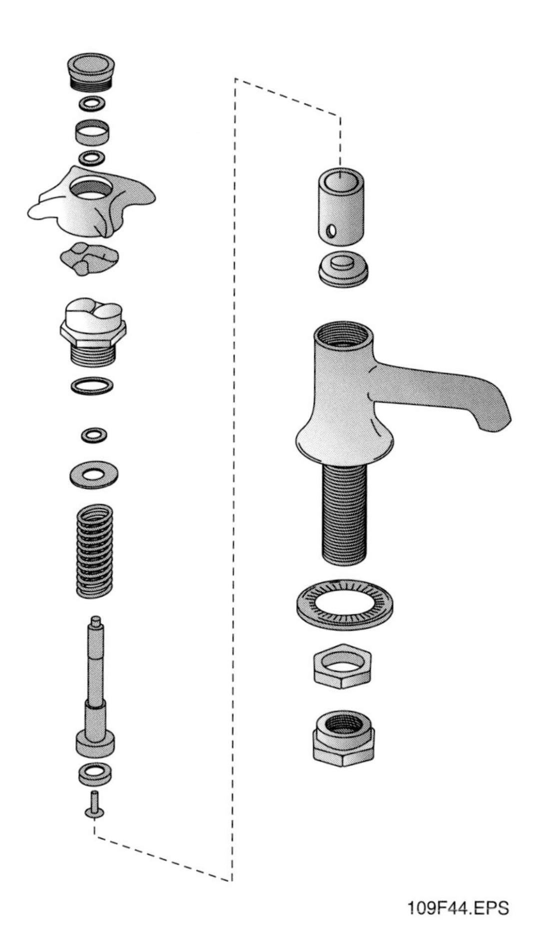

109F44.EPS

Figure 44 ◆ Self-closing compression faucet.

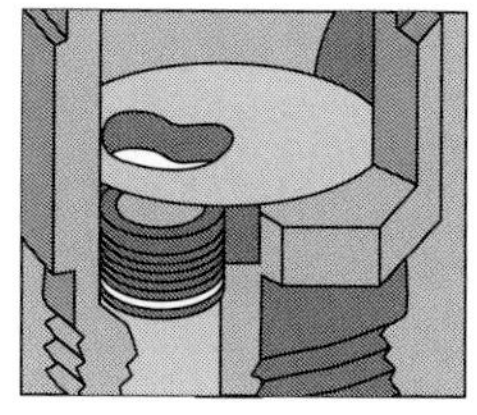

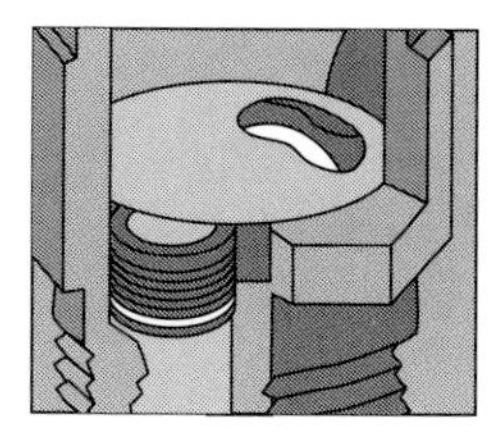

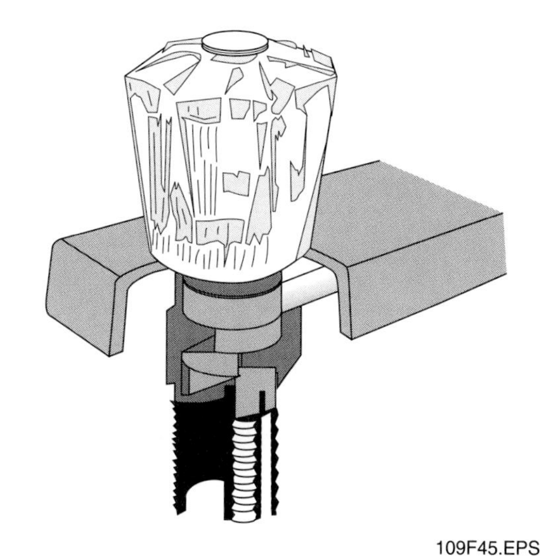

109F45.EPS

Figure 45 ◆ Washerless faucet water flow control.

faucet ensures more economical use of water. It automatically closes the valve as soon as the handle is released. A self-closing faucet may present some problems over time, though. It has a tendency to receive too little water flow, which can cause lime and other minerals to build up in the system. As a result, the faucet begins to malfunction.

Noncompression (port-control) faucets do not rely on the compression of a washer to control the flow of water. Water flow is controlled by rotating balls, rotating cylinders, or rotating discs. Because there is no repeated compressing of a washer, these faucets usually require less maintenance than compression faucets.

Washerless, single-control, and pushbutton styles are among the most common noncompression faucets available. Washerless faucets control water flow by matching openings in two separate discs located in the valve. While one disc remains stationary, the other disc rotates with the band control or lever, as shown in *Figure 45*. Although these faucets are called "washerless," an **O-ring** may be used to prevent water leakage from around the valve assembly.

Single-control faucets (see *Figure 46*) control both the hot and cold water with one hand control. These faucets have become more and more popular because of their convenience. The most popular designs include the rotating ball faucet and the rotating cylinder faucet.

The rotating ball faucet uses a rotating ball in place of a compression washer. The ball is made of metal or plastic and contains several different sizes of holes or openings. As the user rotates the single lever control, these openings align with the hot and cold water ports.

Water mixture and flow is controlled by the position of the lever. Moving the lever to the right causes cold water to flow. Moving it to the left causes hot water to flow. Pushing the lever back increases the rate of flow.

The rotating cylinder faucet is sometimes called a cartridge faucet. This type of faucet controls the water temperature and the rate of water flow by using a balancing piston. The balancing piston adjusts its position when pressure of either hot or cold water changes (see *Figure 47*). As the user makes changes in the water temperature and

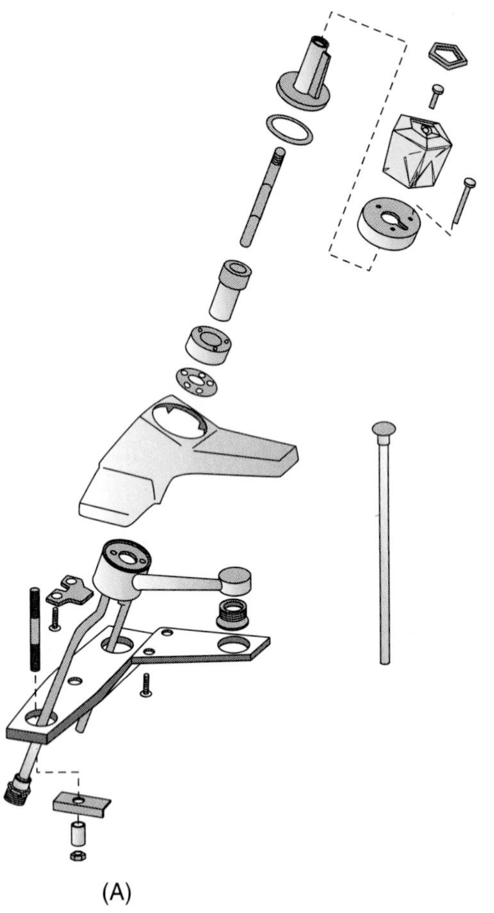

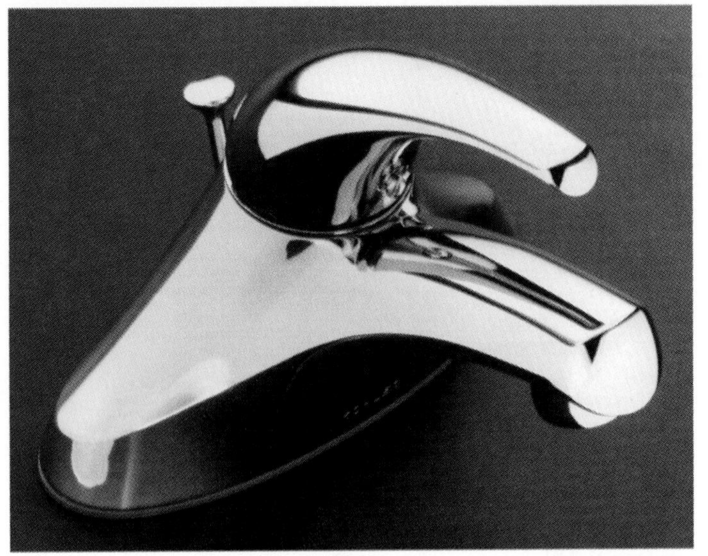

109F46B.TIF

(B)

Figure 46 ◆ Single-control faucets.

pressure by either turning or pulling the single handle, the balancing piston moves to the right (cold) or to the left (hot). This ensures that the flow of water remains constant.

Another type of single-control faucet uses a ceramic disc to control the flow of water. The ceramic disc is sealed inside the cartridge. Because ceramic is very hard and is not likely to become eroded, it gives long-wearing and trouble-free service.

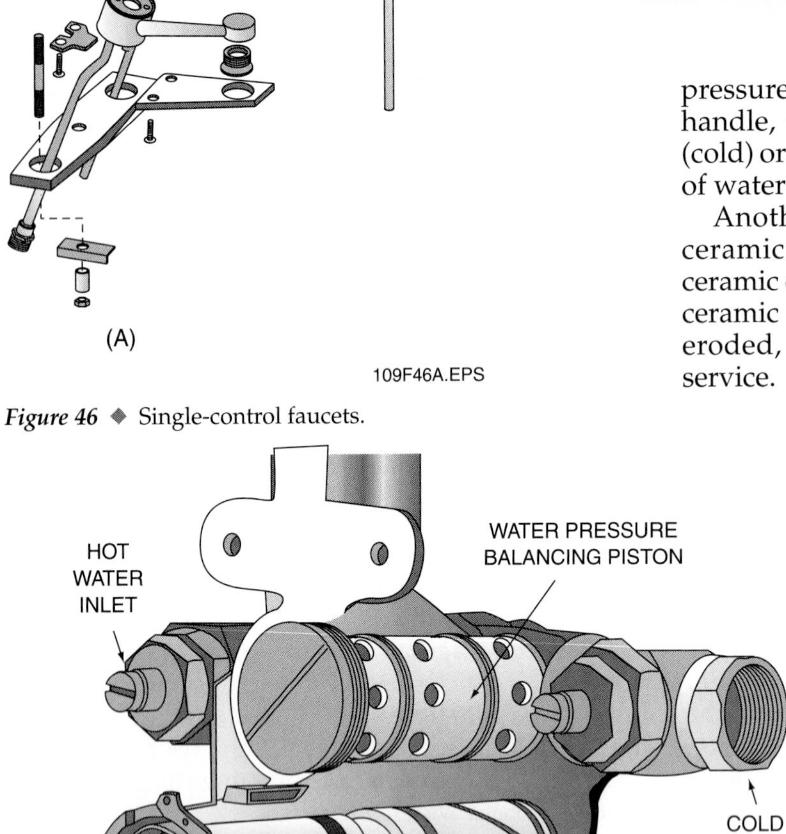

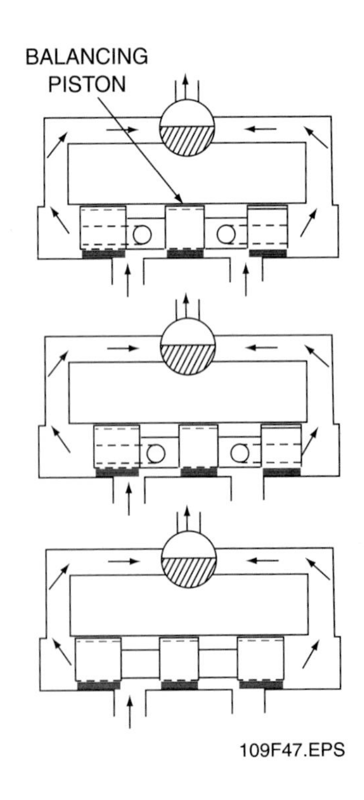

109F47.EPS

Figure 47 ◆ Cartridge faucet.

Review Questions

Section 4.0.0

1. _____ faucets control the flow of water by pressing a seat washer between the valve stem and the valve seat.
   a. Valve
   b. Rotating ball
   c. Cylinder
   d. Compression

2. A rotating ball faucet uses a rotating ball in place of a _____.
   a. valve
   b. compression washer
   c. disc
   d. piston

3. Washerless faucets control water flow by matching openings in two separate discs located in the _____.
   a. cylinder
   b. piston
   c. diverter
   d. valve

4. _____ faucets have a tendency to receive too little water flow, which can cause lime and other minerals to build up in the system.
   a. Plain bibb
   b. Compression
   c. Self-closing
   d. Weatherproof

5. Utility faucets are usually manufactured from _____, white metallic alloy, or thermoplastic materials.
   a. gray iron
   b. sheet steel
   c. copper
   d. extra-grade brass

Summary

Fixtures in plumbing installations are devices that receive water from a water supply line. Fixtures are used in every kind of building—residential, commercial, and industrial. Therefore, it is essential for you to understand the basics of these fixtures—how they are manufactured, how and where they are used, why certain ones are chosen for particular uses, how they operate, and how they connect with water supplies and waste outlets. In this module you have learned these basic facts about fixtures.

You also have learned about faucets, which are a vital element of all water systems. Faucets should be chosen and located according to the needs of the system. It is important to be able to identify various types of faucets. Plumbers must have a basic understanding of the parts and how they work. Plumbers also need to be able to install the faucet properly and repair it. Repairs and installation will be dealt with in the module explaining installation of valves and faucets.

PROFILE IN SUCCESS

Ed Cooper

Plumbing
Supervisor, TD Industries
Dallas, TX

Ed was born in west Texas. In that part of the country, most people worked in the oil patch or in ranching. The construction industry was slow. Ed tried both oil and ranching, but found he kept coming back to plumbing. His father was a plumber and a tough boss. Ed realized he really wanted to succeed in a craft that demonstrates personal skill and a dedication to hard work.

How did you become interested in the construction industry?

After trying my hand at ranching and working in the oil fields, I decided to follow the craft that my father practiced. I didn't have any formal training like the apprenticeship programs available today. In fact, when I started, the primary technology for joining cast iron pipe was still the lead and oakum process. After I left the service, I decided to try my own plumbing company. There wasn't much construction going on in west Texas, so I moved to Dallas. My business combined remodeling, new construction, and service jobs. For the five years I was in business for myself, I enjoyed the remodeling projects the most. Then I had an opportunity to join TD Industries and broaden my experience into commercial, industrial, and process piping. I had not been exposed to this type of plumbing and wanted to learn more.

Unfortunately, I injured my back and was not able to do any more of the heavy lifting required in many plumbing jobs. At that point, I had to refocus my career goals. This opened up all sorts of possibilities for me. I continued to learn about new technologies and materials. I also began to develop my supervisory and management abilities.

What do you think it takes to be a success in your trade?

You have to like what you are doing. In plumbing, like many of the crafts, it is your dedication to hard work that enables you to succeed. If you are just trying to make a living, you could work flipping hamburgers. If you are determined to work with your hands, meet challenges, and develop a craft, then by all means, pursue a plumbing career. There is a lot more to plumbing than installing pipe. You have to be able to work with others, deal successfully with customers, learn new techniques, and keep up with technological developments. Craftwork is hard work. You have to *want* to work at it.

What are some of the things you do in your role of supervisor?

Mostly I manage the work of groups of plumbers. This means that I have to schedule jobs, assign work, monitor progress, and ensure a quality product. I do very little plumbing in my job. As a supervisor, I get to work with different people. I am essentially a coach and a teacher to my crews. I am also involved with association education and training and teach classes for the annual continuing education requirement for plumbing licenses.

Management skills include helping others to set and meet goals for both professional development and personal achievement. It is my job to make sure the workers understand the importance of goal setting and what opportunities arise from successfully meeting your goals.

What do you like most about your job?

I like coaching instead of being a boss. Teaching keeps me sharp. For years, plumbing was viewed as a stagnant field. For the last 15–20 years, however, technology has surged ahead. There are new methods and materials to learn; more are being developed every day. It is the personal satisfaction of passing on knowledge that really makes my job great. In the old days, the plumbers on the job wouldn't teach the new people stuff until they thought it was the right time. Today, I can help young people advance as fast as their hard work and dedication take them.

What would you say to someone entering the trades today?
First, I would say, visit a job site. Get an idea of what is involved in plumbing. There is a lot of hard work involved in being a successful plumber. You have to be realistic about the type of work and the conditions. You can't just do it for the money. You have to like to work outside, in a physical environment.

You also have to learn continually. After you get your license, you must keep it up to date by studying and training more each year. If you put in the work and effort, you can make a good living in plumbing.

GLOSSARY

Trade Terms Introduced in This Module

Bibb (or bib): A faucet with an outlet at a downward angle, generally used for hose connections on the outside of a building.

Cast iron: An iron alloy cast in sand molds and machined to make many building products, such as pipe, pipe fittings, fixtures, ornamentation, and fencing.

China: See *vitrified porcelain.*

Diverter: A device that redirects the flow of water, as in a combination shower and bath fitting. When the diverter is engaged, water flows through the shower head instead of through the spigot.

Faucet: A fixture that is used to draw water from a pipe.

Fiberglass: Spun filaments of glass woven into yarn or roving (twisted strands) and textile materials such as cloth or mats. Fiberglass cloth saturated with a plastic (vinyl resin or epoxy) can be pressed into molds to make products such as bathtubs and shower enclosures.

Fixture: A device that receives water from a water supply line.

Flange: A rim on one end of a length of pipe that provides a connection point to another length of pipe or piece of equipment (such as a water closet). Bolts or studs with nuts are used to hold two flanges together, with some type of gasket between them. The flange is soldered or solvent welded to the drainpipe stub to connect the fixture to the drain, waste, and vent system.

Flood-level rim: The rim of a plumbing fixture from which overflow will spill, such as the top edge of a sink.

Flush valve (flushometer): A device activated by direct water pressure that discharges a certain amount of water to fixtures for flushing.

Gray iron: A type of iron with a controlled carbon content used mostly for water main piping and for sanitary waste piping within a structure. Also called *cast iron.*

Laundry trays: Fixed tubs, usually installed in the laundry rooms of homes; they are used for washing clothes and for receiving wastewater from automatic clothes washers.

Left-hand bathtub: Has the drain on the left end of the tub as you face its length.

O-ring: A circular gasket that is placed in grooves to provide maximum sealing effectiveness.

Porcelain enamel: Liquid used to coat materials such as steel and cast iron to create a vitrified porcelain (china) finish.

Right-hand bathtub: Has the drain on the right end of the tub as you face its length.

Seat washer: A disc that fits at the end of the valve stem and is used to regulate the flow of water through a faucet.

Sheet steel: Steel that has been rolled to a given thickness in a sheet-like form. Can be used to mold fixtures.

Stainless steel: A corrosion-resistant steel alloy containing nickel and/or chrome, often in combination with small amounts of other elements.

Terrazzo: A type of floor surface made by embedding small pieces of marble or other hard stone in a mortar base. When hardened, the surface is ground and polished smooth, although it may be left slightly rough to create a nonslip surface.

Thermoplastic: A plastic that becomes soft and pliable when heated and hard and rigid when cooled again. This type of plastic is often used for utility faucets.

Valve seat: Opening in a faucet through which the water flows. When the handle is turned to open the faucet, the valve stem raises the seat washer up and clears the opening for water to flow.

Valve stem: Part of a faucet that rises and lowers when the handle is turned; it opens or closes the faucet outlet.

Vent pipe: Small-diameter pipe used to permit the escape of air in a DWV system.

Vitrified porcelain: Mixture of fine clay, quartz, feldspar, and silica that is heated to 2600°F (1426°C) to create a nonporous, glass-like finish. Also known as *china.*

Waste pipe (fixture branch): The pipe from the trap of a plumbing fixture to the vent pipe, carrying nonbody wastes to the building drain.

ANSWER KEY

Answers to Review Questions

Sections 2.0.0–3.3.0

1. b
2. b
3. c
4. a
5. d

Sections 3.4.0–3.6.0

1. b
2. d
3. c
4. b
5. c

Sections 3.7.0–3.11.0

1. b
2. b
3. a
4. d
5. d

Section 4.0.0

1. d
2. b
3. d
4. c
5. d

NCCER CRAFT TRAINING USER UPDATES

The NCCER makes every effort to keep these textbooks up-to-date and free of technical errors. We appreciate your help in this process. If you have an idea for improving this textbook, or if you find an error, a typographical mistake, or an inaccuracy in the NCCER's Craft Training textbooks, please write us, using this form or a photocopy. Be sure to include the exact module number, page number, a detailed description, and the correction, if applicable. Your input will be brought to the attention of the Technical Review Committee. Thank you for your assistance.

Instructors – If you found that additional materials were necessary in order to teach this module effectively, please let us know so that we may include them in the Equipment/Materials list in the Instructor's Guide.

Write: Curriculum Revision and Development Department
National Center for Construction Education and Research
P.O. Box 141104, Gainesville, FL 32614-1104

Fax: 352-334-0932

E-mail: curriculum@nccer.org

Craft ________ Module Name ________

Copyright Date ________ Module Number ________ Page Number(s) ________

Description

(Optional) Correction

(Optional) Your Name and Address

Module 02110-00

Introduction to Drain, Waste, and Vent (DWV) Systems

Course Map

This course map shows all of the modules in the first level of the Plumbing Curriculum. The suggested training order begins at the bottom and proceeds up. Skill levels increase as you advance on the course map. The local Training Program Sponsor may adjust the training order.

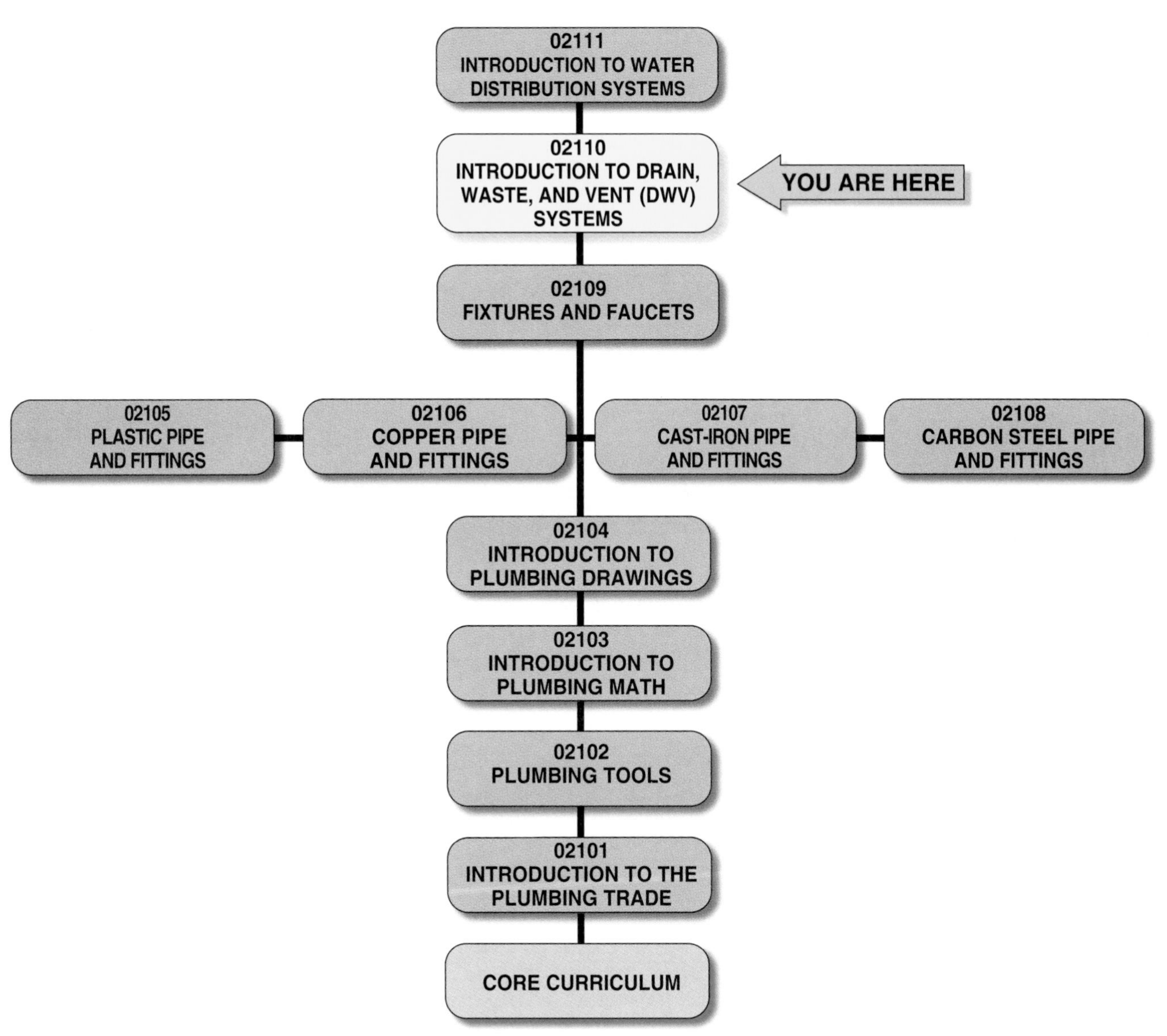

110CMAP.EPS

Copyright © 2000 National Center for Construction Education and Research, Gainesville, FL 32614-1104. All rights reserved. No part of this work may be reproduced in any form or by any means, including photocopying, without written permission of the publisher.

MODULE 02110 CONTENTS

Figures

Table

MODULE 02110

Introduction to Drain, Waste, and Vent (DWV) Systems

Objectives

When you finish this module, you will be able to:

1. Explain how waste moves from a fixture through the drain system to the environment.
2. Identify the major components of a drainage system and describe their functions.
3. Identify types and parts of a trap, explain the importance of traps, and how traps lose their seals.
4. Identify the various types of DWV fittings and describe their applications.

Prerequisites

Before you begin this module, it is recommended that you successfully complete task modules: Core Curriculum; Plumbing Level One, Modules 02101 through 02109.

Materials Required for Trainees

1. This module
2. Appropriate personal protective equipment
3. Sharpened pencil and paper

1.0.0 ◆ INTRODUCTION

When doing any plumbing work, the plumber should have a broad understanding of drainage systems. This module describes the flow of waste products from the building, to the treatment facilities, and back into the ecosystem (our streams, rivers, and lakes). This cycle begins at the fixture drains that connect to the drain, waste, and vent (DWV) piping system. Waste, either liquid or in solution with solids, enters the DWV system from the fixture drains and flows into the building's pipe system. The pipe system is designed to remove this waste safely from the building's interior. You will be introduced to the factors that influence DWV system design and how different types of drains, fittings, vents, and pipe are used to move waste out of a building. You will also learn installation requirements that prevent malfunctions in the system. Plumbers also install storm drainage systems that carry rainwater from roofs and open areas to the storm sewers. The design and installation of storm drainage systems will be covered in another module.

Sanitary drainage systems can be divided into three parts: (1) the pipes inside the building, usually referred to as the *DWV system;* (2) the drain pipe buried outside the building, which is called the *building sewer;* and (3) the public sewer, which carries the building wastes to the treatment plant and eventually back to the ecosystem. See *Figure 1* for an overview of a typical community sewer system.

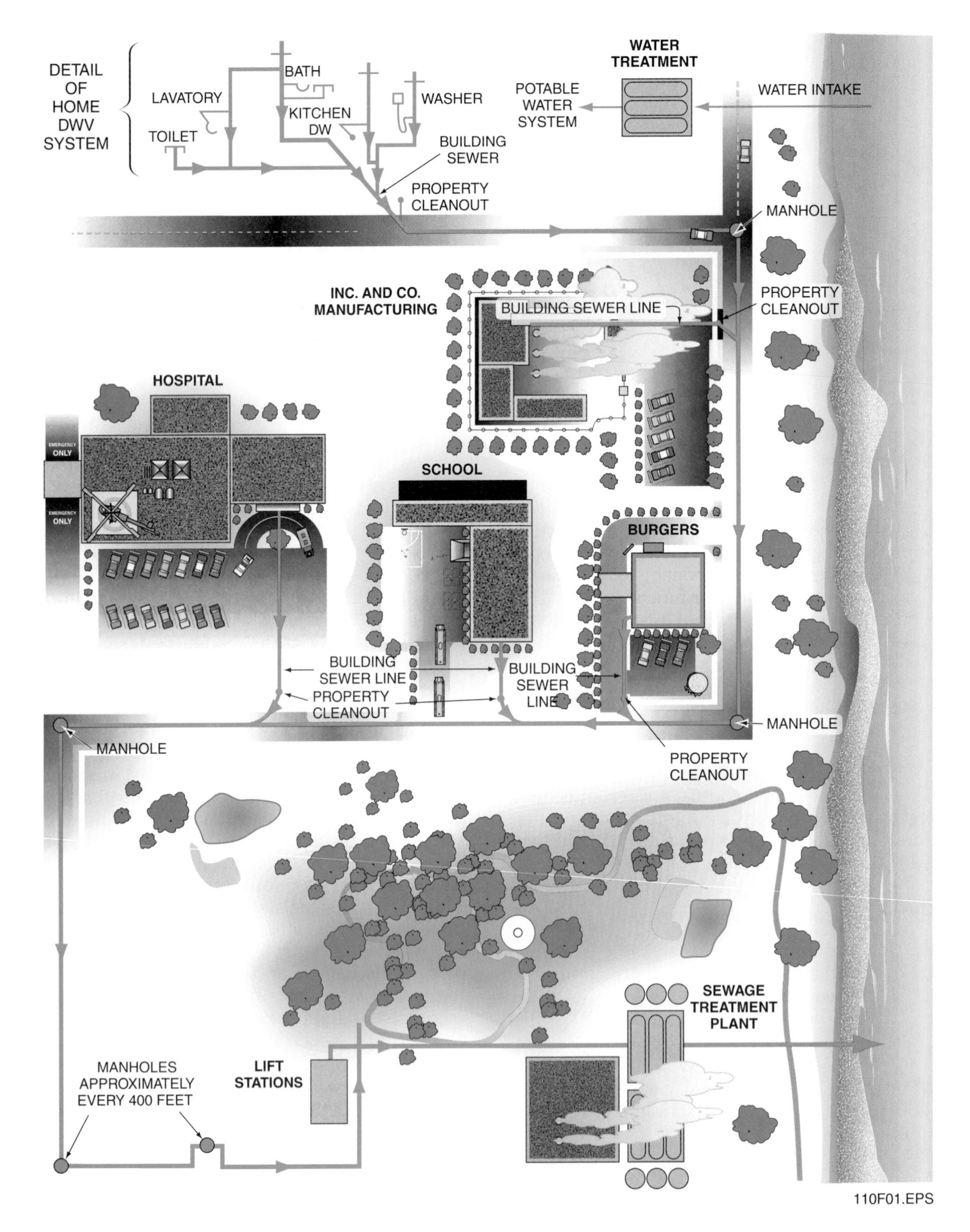

Figure 1 ◆ Overview of a typical community sewer system.

2.0.0 ◆ DWV SYSTEMS

Plumbers design, install, and maintain DWV systems inside buildings and the building sewers buried outside on the property. Usually the municipality is responsible for installing and maintaining the public sewers, lift stations, and treatment plants.

The DWV system inside a building is a network of pipes designed to remove the wastes from plumbing fixtures and drains in a safe, reliable, and efficient way. There are many names for each of the pipes and fittings in this network. Following is a list of the major components that are illustrated in *Figure 2*.

- Building drain
- Soil stack
- Stack vent
- Individual vents
- Fixture branches
- Fixture drain or trap arm
- Traps
- Bends or elbows
- Tees
- Wyes
- Couplings, reducers, and adapters
- Cleanouts

3.0.0 ◆ FIXTURE DRAINS

Fixture drains connect the fixture to the building's DWV piping system. Many fixtures have drains that strain the wastewater before it enters the drainage piping. Examples of fixture drains include a basket strainer for a kitchen sink, PO (pop-up) plugs for lavatories, and other strainers for bidets and showers (see *Figure 3*).

Did You Know?

PO originally stood for plug opening. This term comes from the early type of lavatory fixtures that used a stopper to seal off the drain. Now lavatory drains with pop-up assemblies are more common than the older chain and stopper types.

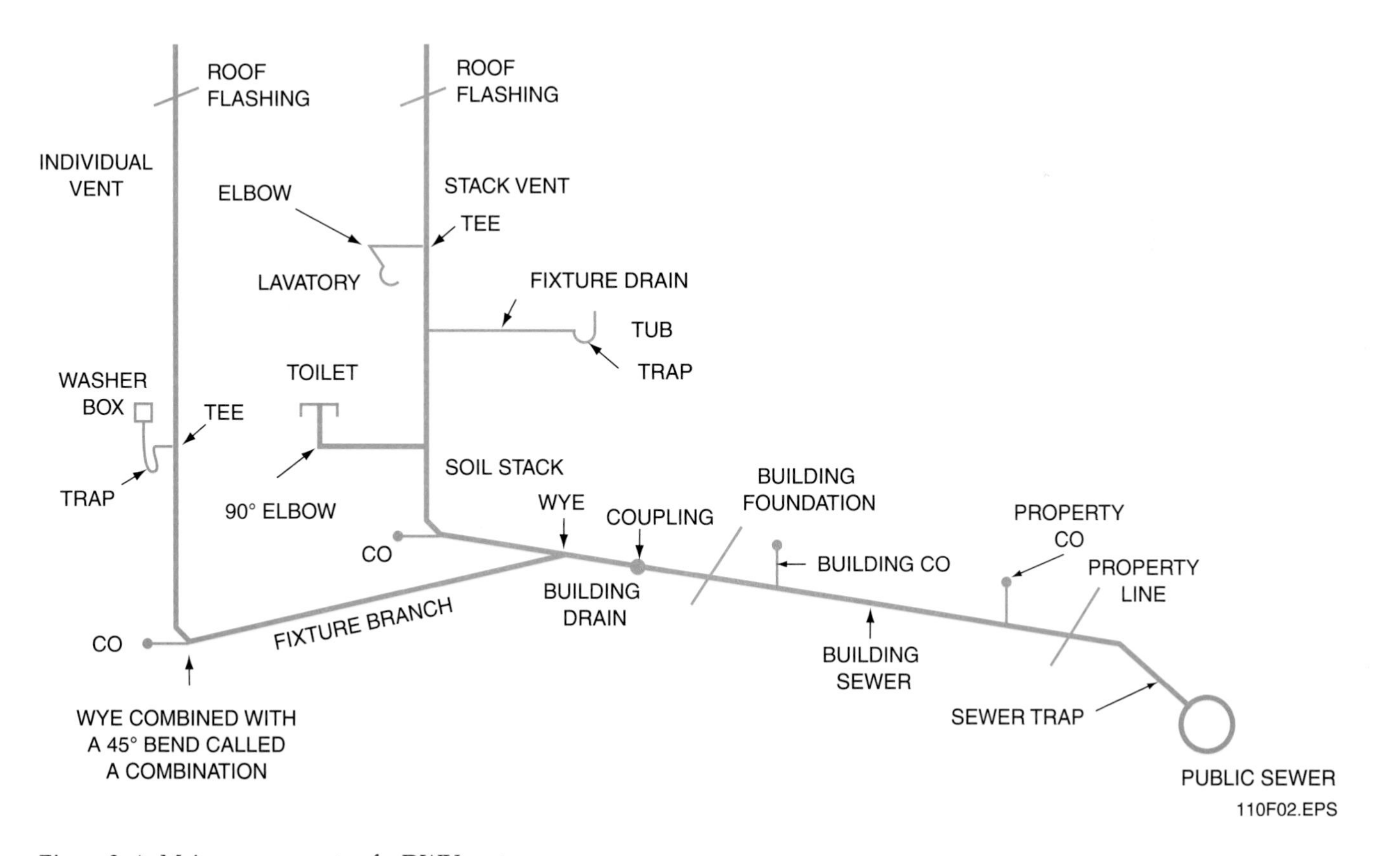

Figure 2 ◆ Major components of a DWV system.

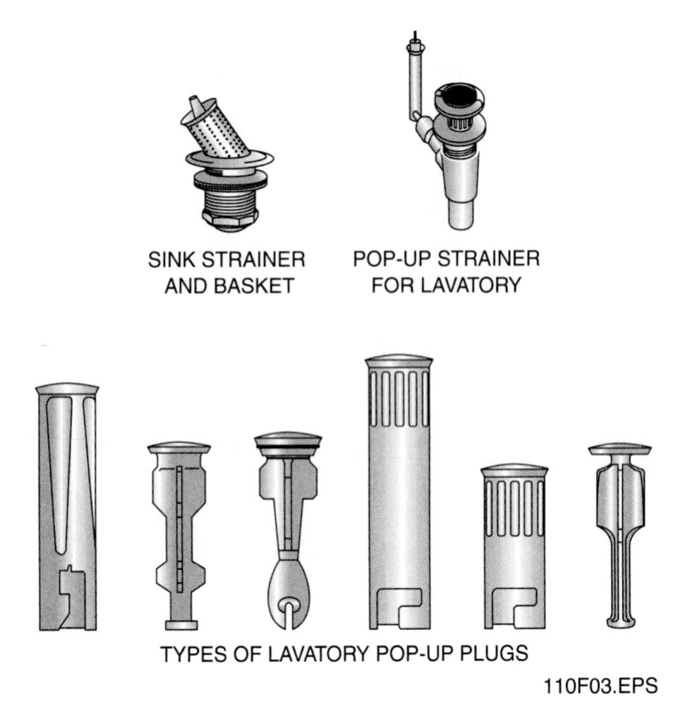

Figure 3 ◆ Common fixture drains.

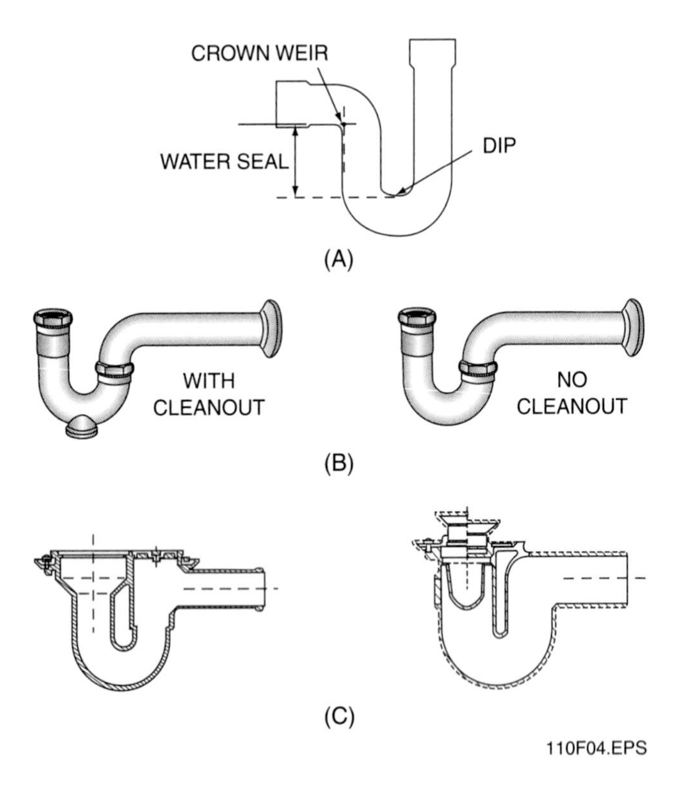

Figure 4 ◆ Common types of P-traps. (A) One-piece trap. (B) Two-piece traps. (C) P-trap type floor drains.

4.0.0 ◆ TRAPS

Traps and vents protect the safety of our homes and buildings. The water seal in a trap protects us from airborne pathogens (germs), foul odors, and potentially explosive sewer gases.

Traps are important components of the DWV piping system. A trap is a fitting or device that provides a liquid seal of 2 inches to 4 inches; this seal prevents sewer gases from leaking back into the building but should not affect the flow of sewage or wastewater through the drain. As will be explained later in this module, trap seals are protected by vents.

Modern codes require that every plumbing fixture must have a trap that protects the fixture and user from the sanitary drain system. In most cases, each fixture must have its own trap. Some fixtures, such as water closets and many urinals, have integral or built-in traps.

Fixture traps are the heart of any DWV system. For a fixture trap to function properly, it must empty completely, be self-cleaning, have a smooth interior waterway, and be accessible for cleanout. The depth of the seal and the amount of water normally held in the trap are important factors in trap design. The fixture trap, which creates a water seal, requires a vent system to protect it from **siphonage** and **back pressure.**

4.1.0 Common Types of Traps

The **P-trap** is the most commonly used trap. P-traps can be one-piece or two-piece with a union nut. P-traps can also have a cleanout. Several styles of P-trap are shown in *Figure 4.* P-traps designed to be attached directly to fixtures are usually 1¼ inches and 1½ inches for sinks and lavatories; 1½ inches to 2 inches for tubs and showers; and 2 inches to 3 inches for floor drains. P-traps can be made of brass, brass with chrome plating, plastic, copper, cast iron, malleable iron, and glass. Floor drains may be designed as P-traps.

ON THE LEVEL

Don't Get Trapped by an Old Trap

Drum traps used to be installed for bathtubs and lavatories. Most experienced plumbers have seen drum traps (see *Figure 5*) in older buildings. This metal cylinder was installed in the fixture waste line, with the top of the trap sticking up through the finished floor for easy access. The drum trap is now prohibited by most plumbing codes. Use of a drum trap must be approved by the local authority before installation. It may only be used with fixtures specifically designed for drum traps.

Another kind of trap that can no longer be used is the **S-trap** (see *Figure 6*). Improvements in plumbing design have replaced this trap with more efficient and effective P-traps. The long downstream leg of the S-trap tends to promote siphonage (explained later in this module). The plumber could make the trap more effective by increasing the depth of the seal, but this led to other problems: with greater depth there was a greater chance that solids would stay in the trap. Fungus growth is also a problem in traps that are too deep.

(**Nonsiphon traps** are available with deeper trap seals for cases where the plumbing system is subjected to abnormal changes in pressure or to a lot of evaporation.)

Two types of S-traps were used: the full S-trap and the ¾ S-trap. Many older homes still have S-traps. Plumbing supply stores stock these traps for service but not for new construction installation.

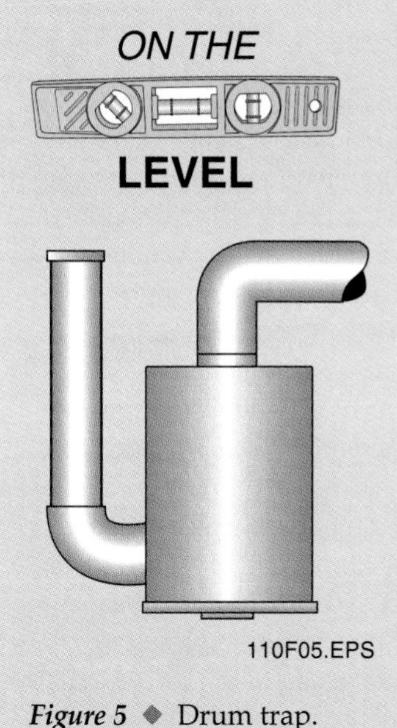

Figure 5 ◆ Drum trap.

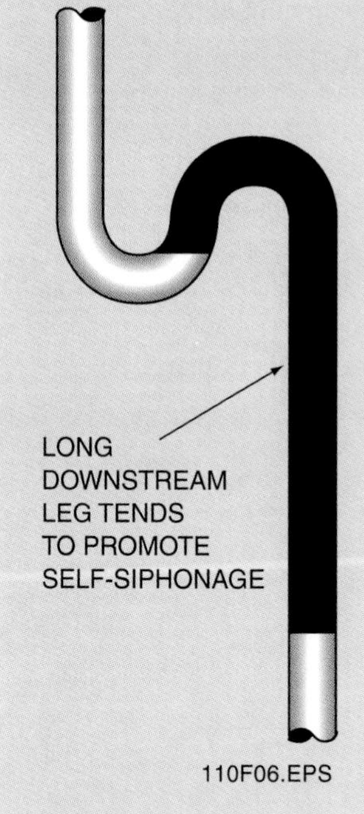

Figure 6 ◆ S-trap.

Double Trapping

Water closets (toilets) have integral traps as part of their design. Traps built into water closets (see *Figure 7*) should never be attached to another trap. Double trapping can cause blockage in the DWV piping.

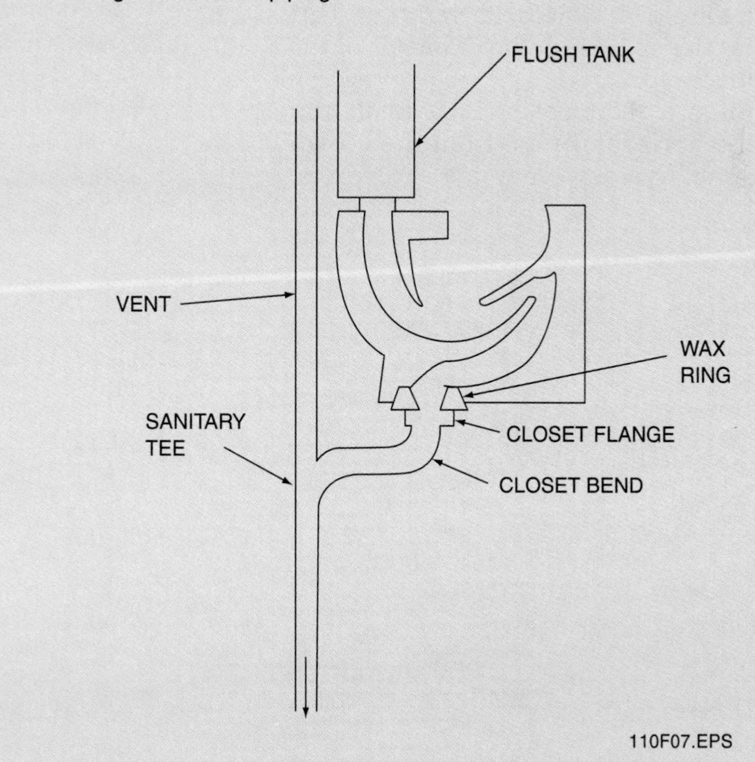

Figure 7 ◆ Integral or built-in trap.

4.2.0 Parts of Traps

The basic parts of a trap are shown in *Figure 8*. The parts include the following:

- Inlet—where water enters from the fixture
- Top dip—the inside curve of the pipe under the inlet
- Bottom dip—the bottom of the lowest curve of the pipe beneath the inlet
- **Crown weir** (sometimes called the **trap weir**)—the point in the curve of the trap that is directly below the crown (highest point); this is the point at which the water level will normally remain when the fixture is not discharging through the trap
- Fixture drain (also called the *trap arm*)—the point where wastewater leaves the trap and goes into the drainage piping

4.3.0 Trap Installation Requirements

Even though local codes govern specific installation requirements, such as dimensions and location of traps, there are some typical trap installation requirements.

Generally, the vertical distance from the fixture outlet to the crown weir may not exceed 24 inches. The second critical dimension is the horizontal distance from the crown weir to the trap vent. This distance varies depending on the diameter of the trap (see *Figure 9* and *Table 1*).

The third important dimension is the total drop (or **fall**) in the horizontal pipe from the crown weir to the vent. This drop may not exceed one

Did You Know?

In the earliest home plumbing fixtures (in the 1850s), the main safeguards against odors and sewer gases were handmade traps that the plumber installed in the drains of individual fixtures. These traps often lost their water seals because of siphonage and back pressure (described in this section) and became ineffective. All efforts to prevent seal loss failed, because the principle of venting fixture drains (covered later in this module) was not known at the time.

In the early 1900s, the problems with fixture traps led health officials to require a secondary safeguard: the installation of building traps on each sanitary or combined building sewer. Without this additional safeguard, rats were able to travel freely from one building to another. Building traps became the second line of defense against rats in the sewer systems.

This requirement was a big advance at the time. But since the development of modern collection, drainage, and venting systems, most model codes don't require building traps. In fact, many codes actually *prohibit* building traps. The only exceptions are in areas where sewer gases are extremely corrosive, or where the sewer gases contain high explosive gas content, creating a risk of explosions in the public sewer system that might, for instance, blow off manhole covers and cause considerable damage.

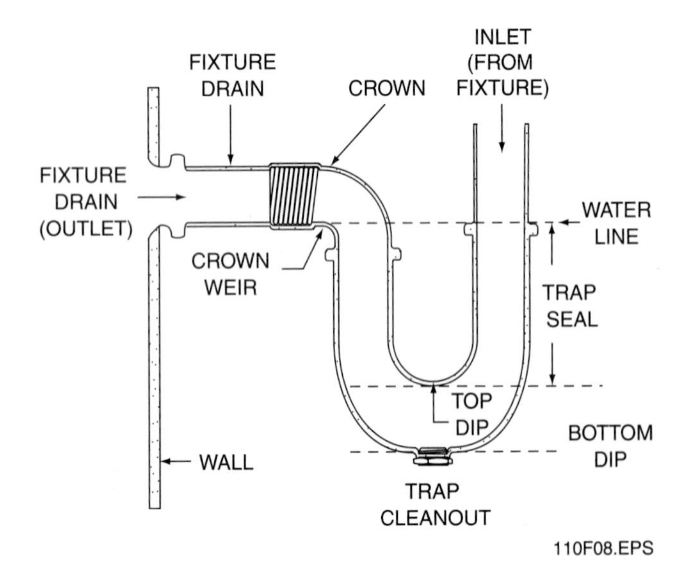

Figure 8 ◆ Parts of a trap.

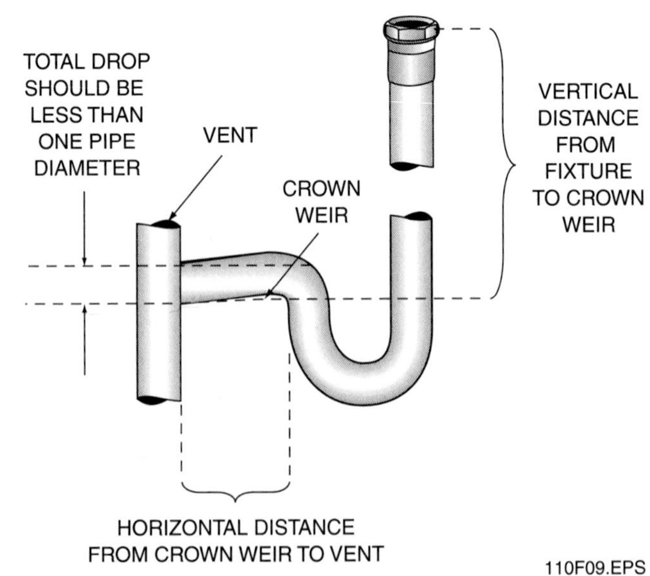

Figure 9 ◆ Critical dimensions of a trap vent.

Plumbing Codes

Plumbing codes protect the safety, health, and welfare of all U.S. citizens. Although there are many variations on what codes require, all codes are based on principles of sanitation and safety.

There is no single national plumbing code whose requirements are adopted by all states and localities, but model codes are developed and revised on a regular basis. States and other jurisdictions can use these model codes as a basis for developing their own plumbing codes. DWV piping requirements are also defined by counties and municipalities and are enforced by their inspectors.

You must always check the codes in the area where you are working. Model codes, for example, vary widely on how far vents have to be from traps and on what size fixture drains must be. Consult your local code before you start a job. Some variations in model codes are shown in *Table 1*.

Table 1 Horizontal Distance of Fixture Trap from Vent at Slope of ¼ Inch per Foot

Size of Fixture Drain	Size of Trap	Distance from Trap		
		Standard Plumbing Code (1994) International Plumbing Code (1997)	Uniform Plumbing Code (1997)	National Standard Plumbing Code (1996)
1¼"	1¼"	3'6"	2'6"	3'6"
1½"	1½"	5'0"	3'6"	5'0"
2"	2"	6'0"	5'0"	8'0"
3"	3"	10'0"	6'0"	10'0"
4" and larger*	4"	12'0"	10'0"	12'0"

*Size of Fixture Drain specification is 4" only (not 4" and larger) for Standard Plumbing Code (1994) and International Plumbing Code 1997.

Almost all codes require you to install cleanouts to provide access to all parts of the drainage system so that you can remove obstructions. Cleanouts range from removable plugs in horizontal drainage piping to manhole covers in the building sewer. Codes vary widely on the specifics, so you must check your local code's requirements.

Under most codes, cleanout openings cannot be used to install new fixtures, unless there is another cleanout of equal accessibility and capacity, and you have written approval of the plumbing inspector. On pipes that are 4 inches or less in diameter, many codes specify that cleanouts must be the same size as the pipe they serve. For larger pipes, the size of the cleanout must be at least 4 inches in diameter.

The various plumbing codes require that **cleanout fittings** be installed at specified locations within the DWV piping system and that the fittings be accessible.

Most codes have requirements for cleanout locations in horizontal runs of pipe. In horizontal drain lines 4 inches in diameter or less, cleanouts must be installed no more than 50 feet apart. For lines larger than that, cleanouts cannot be placed more than 100 feet apart.

Some codes require that cleanouts also be provided at or near the foot of each waste or soil stack and near the junction of the building drain and the building sewer.

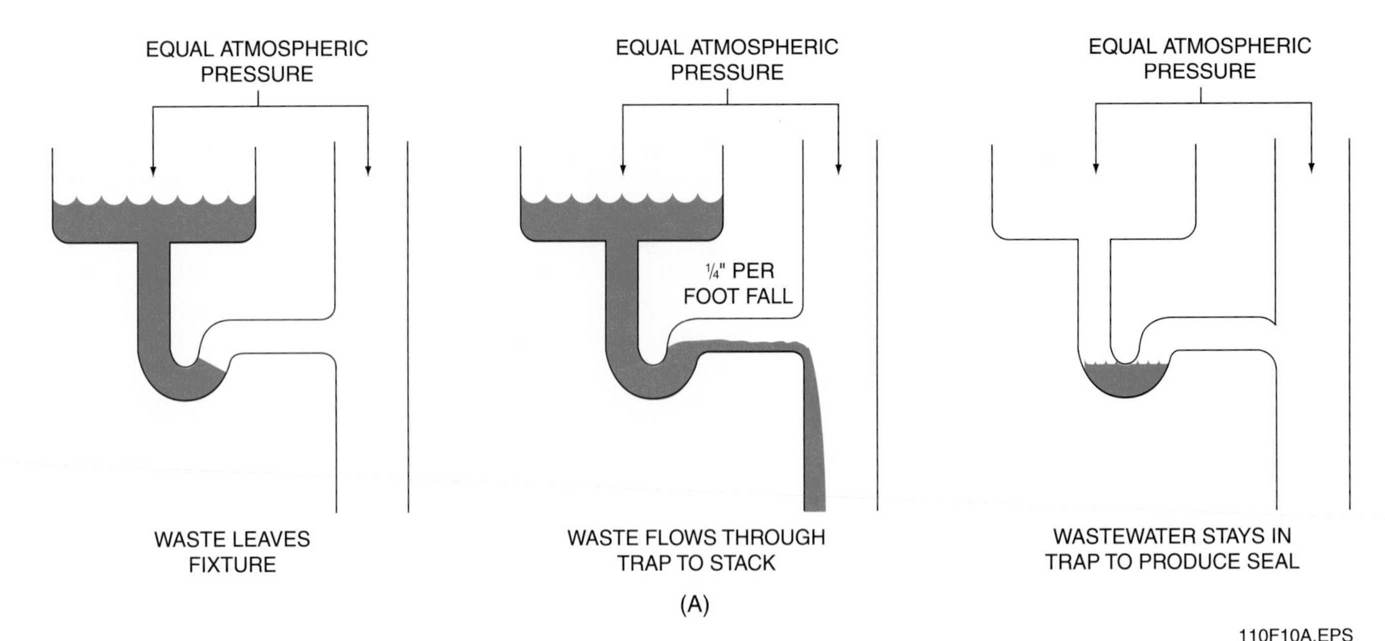

pipe diameter. Usually, this fall is ¼ inch per foot. If the horizontal leg is installed with greater drop, the trap is likely to siphon.

Many P-traps are manufactured in two pieces. The J-bend piece joins with the fixture tailpiece and is secured with a gasket and compression nut called a slip-joint washer and a slip-joint nut. The outlet end, called the *trap arm,* joins the horizontal drain, which connects to the vent.

Generally, traps should be installed for each plumbing fixture that does not have a built-in trap. Sometimes, codes will permit one trap to serve more than one fixture. For example, three lavatories that are 30 inches or less apart may be connected to one trap. A similar situation exists when sinks containing two or three compartments are installed.

4.4.0 Why a Trap Loses Its Seal

A thorough understanding of how traps function helps the plumber understand how important it is to install traps correctly. Also, the plumber can use this knowledge to determine what causes a trap to malfunction.

First, consider what occurs when a trap functions properly. Waste from the fixture flows into and through the trap, as shown in *Figure 10.* The trap is refilled with the last of the wastewater to leave the fixture. This water provides the necessary liquid seal. For the trap to function this way, the pressure on both sides of the trap must remain nearly equal. Water tends to flow in a level line. This level line is called the **hydraulic gradient.** The trap weir must always be installed lower than the top of where the fixture drain enters the vent line.

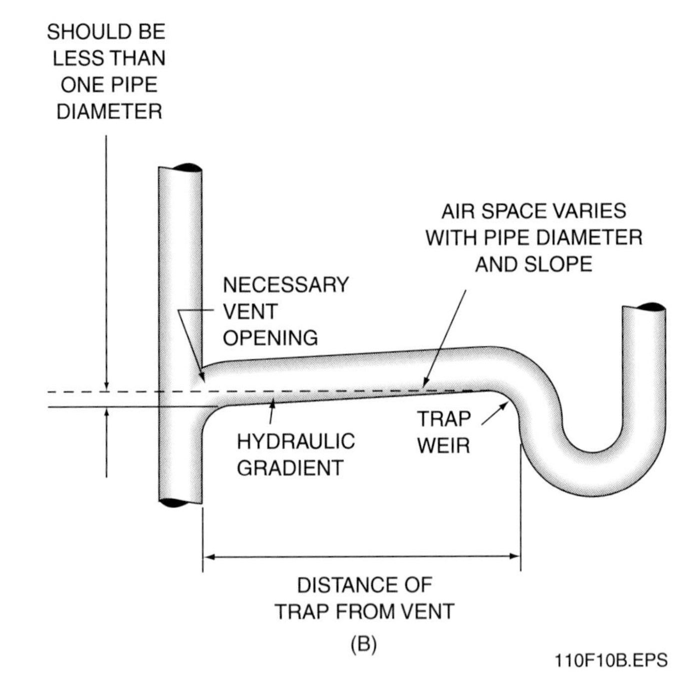

Figure 10 ◆ (A) How a trap works. (B) Hydraulic gradient.

A trap may lose its seal in a number of ways—through siphonage, back pressure, evaporation, **capillary attraction,** wind effect, or trap failure. A plumber can prevent both siphonage and back pressure by properly designing and installing the DWV system.

4.4.1 Siphonage

If the trap is not properly vented, the trap is likely to siphon. Siphonage occurs when there is a difference between the pressure inside and outside the DWV piping. This pressure difference pushes the water that is normally held in the trap into the DWV piping system. Generally, siphonage occurs when the DWV piping is improperly vented or the vent is blocked (see *Figure 11*). As the waste leaves the trap, an area of reduced pressure is created in the drainage piping. Because of the difference in pressure, the water is forced from the trap. This destroys the trap seal. Siphonage happens when there is too much fall and the crown weir is higher than the top of where the fixture drain enters the vent.

4.4.2 Back Pressure

Back pressure can cause a trap seal to break. Back pressure is pressure inside the DWV piping that is greater than atmospheric pressure (see *Figure 12*). If enough wastewater from a fixture enters the stack so that a slug of water forms a moving plug, the air in the stack below the plug is compressed. This excess pressure tries to escape through the trap in fixture B as shown in the figure. To prevent normal back pressure from destroying the trap seal, the stack must be properly sized, and the trap must be properly sized and protected by a vent. Also, the seal must be deep enough; generally a trap seal of 2 to 4 inches is required.

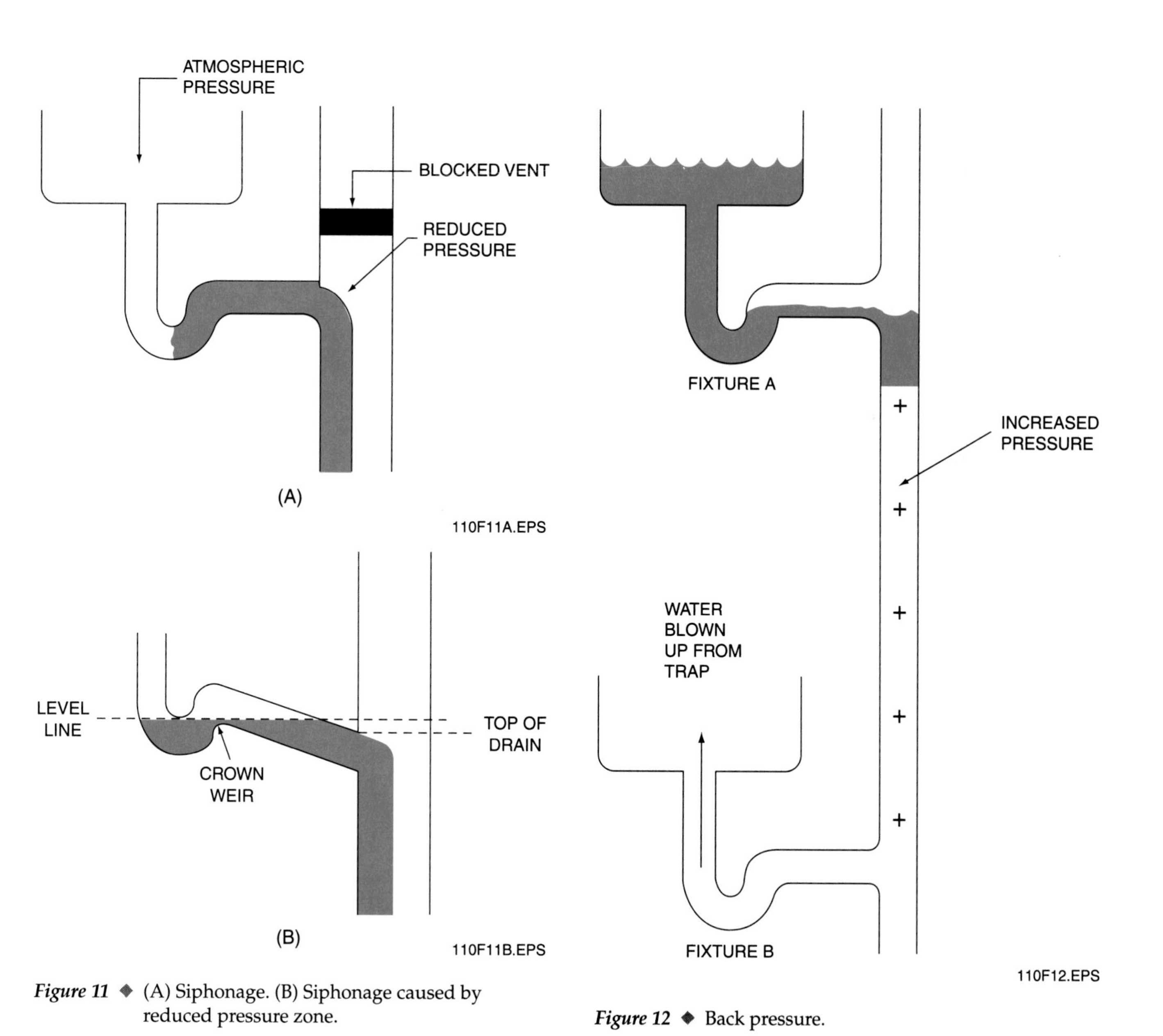

Figure 11 ◆ (A) Siphonage. (B) Siphonage caused by reduced pressure zone.

Figure 12 ◆ Back pressure.

4.4.3 Evaporation

A trap may also lose its seal as a result of evaporation. This is most likely to happen in traps that are seldom used. The water evaporates, causing the seal to break. If the DWV piping is properly designed, evaporation will only become a problem during long periods of nonuse. Unused floor drains are often the cause when sewer gas enters a structure. If you anticipate this condition, you can install extra-deep traps. Many codes require trap primers where evaporation of the trap seal is likely. Trap primers are connected to a regularly used water line and emit water through a small tube into the trap so that it can keep its seal.

4.4.4 Capillary Attraction

Capillary attraction may cause a trap seal to be broken if a porous material such as string or paper is caught in the trap (see *Figure 13*). The porous material acts as a wick and draws the water out of the trap by capillary action. Cleaning the trap will solve this problem.

4.4.5 Wind Effect

Wind effect is one of the least likely ways a trap will lose its seal. In situations where there are strong upward or downward air currents, the pressure or suction of the moving air may cause the water in the trap to rise or fall. If it rises enough to spill over into the waste pipe, less water remains in the trap and the seal is weakened. This could lead to trap seal failure because lower than normal back pressures could break the seal.

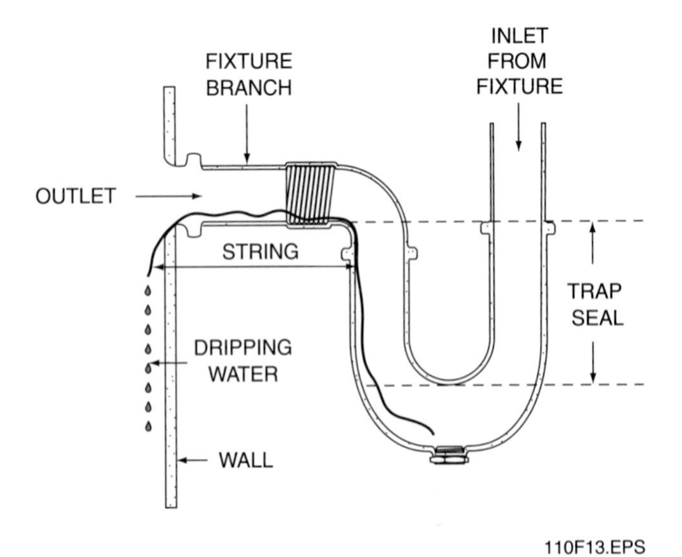

Figure 13 ◆ Capillary attraction.

4.4.6 Trap Failure

A more common source of waste and sewer gas leaking into a building is a leak in the trap. This can be caused by a crack in the trap, by worn washers, or by a broken nut, solder joint, or glue joint.

Review Questions

Sections 2.0.0–4.4.0

1. Because double-trapping promotes blockage in a DWV pipe, _____ should never be attached to another trap.
   a. drum traps
   b. P-traps
   c. water closets
   d. fixtures

2. _____ occurs when there is reduced pressure inside and outside the DWV piping.
   a. Backflow
   b. Evaporation
   c. Capillary action
   d. Siphonage

3. A properly sized stack will prevent _____ from breaking the trap seal.
   a. wind effect
   b. back pressure
   c. backflow
   d. capillary attraction

4. The _____ is the point at which the water level will normally remain when the fixture is not discharging through the trap.
   a. crown weir
   b. inlet
   c. bottom dip
   d. fixture

5. Generally, a trap seal depth of _____ inches is required.
   a. ½ to 1
   b. 2 to 4
   c. 4 to 6
   d. 6 to 8

5.0.0 ◆ VENTS

Every trap requires a vent of some type. Vent pipes are critical for plumbing fixtures to function correctly as part of the sanitary drainage system. Venting prevents back pressure or siphonage from breaking the water trap seals serving the fixtures. All the vent pipes of a building create the vent system and are connected to the drain pipes. The system may include one or more pipes. Vents are installed to provide a free flow of air and to maintain equalized pressure throughout the drainage system. There are many types of vents, one of which is shown in *Figure 14*. Different types of vents will be discussed later in your training.

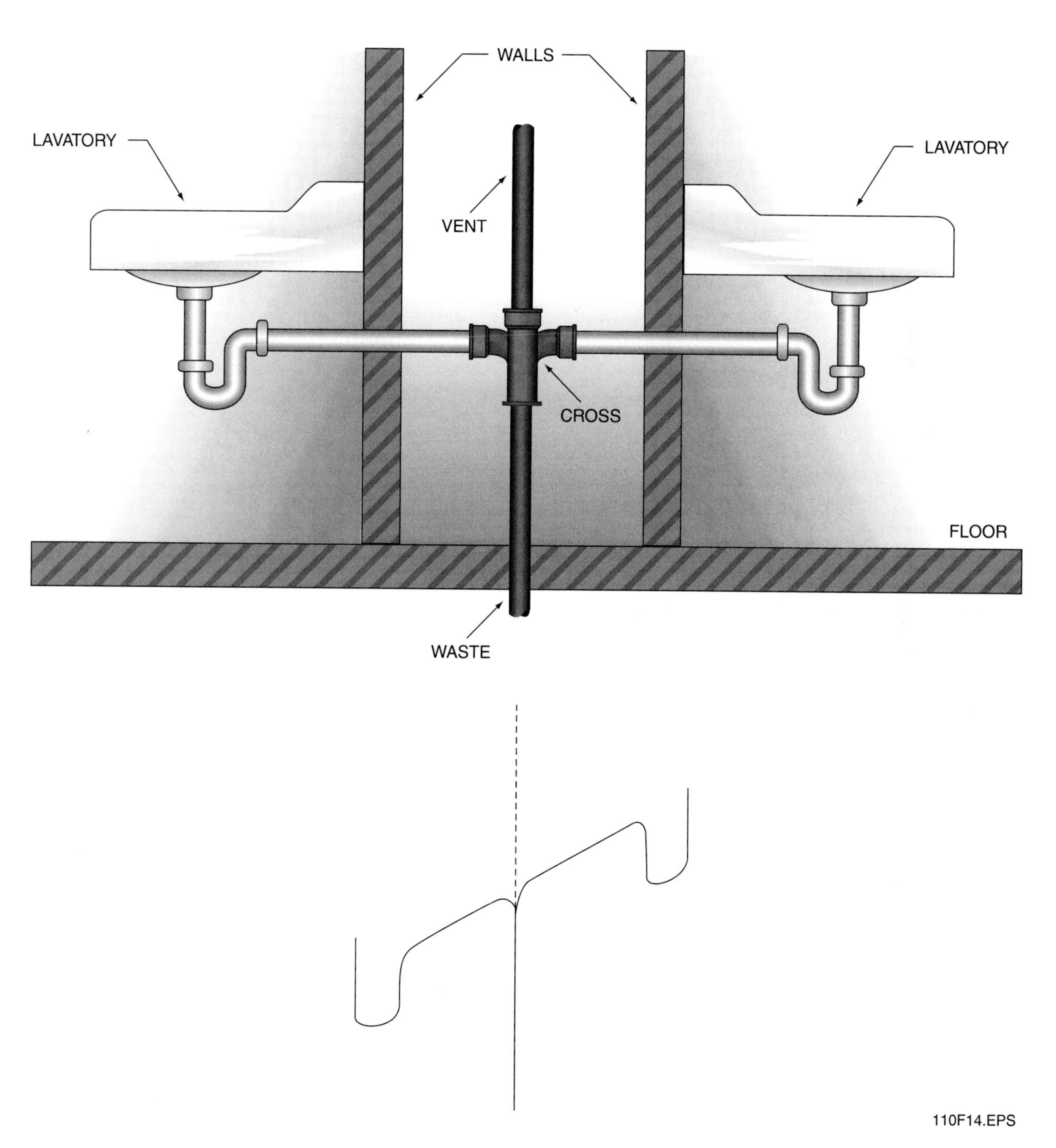

Figure 14 ◆ Common vent.

5.1.0 Distance from Trap to Vent

If a vent is placed too far from the trap, the vent opening will fall below the crown weir because of the fall of the waste pipe. This situation could cause the trap to self-siphon and lose its seal. To prevent this, most codes publish tables based on pipe fall and pipe diameter. These tables give the maximum distances allowed between the pipe and the vent. Check your local code for details.

6.0.0 ◆ SIZING DRAINS AND VENTS

Drainage systems fall into two major categories: storm water drains and building drains. Storm water drains collect storm water from roofs and pavement. This water is either held for on-site disposal or is disposed of at a certain rate into a storm sewer system. The sizing of storm drainage systems is based on expected rainfall in a given geographic area. Building drainage systems must be appropriately sized based on the expected water use in the building. Plumbers must understand the mechanics of fluid flow in pipes to properly size pipes, drains, and vents.

Vents are used in plumbing systems to balance pressure in the piping network. This balance of pressure is necessary to prevent the fixture traps from losing their seals. For the vents to function properly, plumbers must size the vents correctly according to how many fixtures are connected, the number of drainage fixture units (how much water discharges into the drain per minute), and the length of the vent pipe. If the plumber installs inadequately sized vents, the plumbing system will not work properly.

You will learn how to size drainage and venting systems as you advance through the levels of the plumbing curriculum.

Review Questions

Sections 5.0.0–6.0.0

1. _____ balance the pressure within the drainage system to maintain the water seals in fixture traps.
   a. Fixture drains
   b. Vents
   c. P-traps
   d. Cleanouts

2. Determining the correct distance from the trap to the vent prevents the vent opening from falling below the _____.
   a. crown weir
   b. fixture drain
   c. seal
   d. pipe diameter

3. _____ are used by states and jurisdictions as a basis for developing their own plumbing codes.
   a. National plumbing codes
   b. Uniform plumbing codes
   c. Standard plumbing codes
   d. Model codes

4. Tables based on fall and pipe diameter give the maximum _____ allowed between the crown weir and the vent.
   a. sizes
   b. distances
   c. fittings
   d. drains

5. If the total drop between the crown weir and the fixture vent exceeds one pipe diameter, the trap is likely to _____.
   a. evaporate
   b. crack
   c. siphon
   d. pressurize

7.0.0 ◆ FITTINGS AND THEIR APPLICATIONS

Fittings are devices used to connect pipe. Those used in DWV systems are called **drainage fittings.** This section provides information about the various types of DWV fittings and their uses within the DWV piping system.

7.1.0 DWV Fitting Materials

DWV fittings are made from many different materials, including copper, brass, lead, steel, cast iron, clay, glass, and various types of plastic. Cast iron and plastic are the most commonly used materials (see *Figure 15*). Some codes may prohibit the use of certain materials for certain applications, so be sure to check local code requirements before installing a DWV piping system. Not all fittings are available in all materials.

Although the fitting material may vary, fittings of the same design have the same names. For example, a plastic **sanitary tee** and a cast-iron sanitary tee are basically identical, even though they are made from different materials.

Common fittings for copper pipe for DWV purposes include 90-degree ells, 45-degree ells, 22½-degree ells, male adapters, tees, cleanout tees, reducing tees, and reducers. Common DWV copper valves are butterfly and gauge.

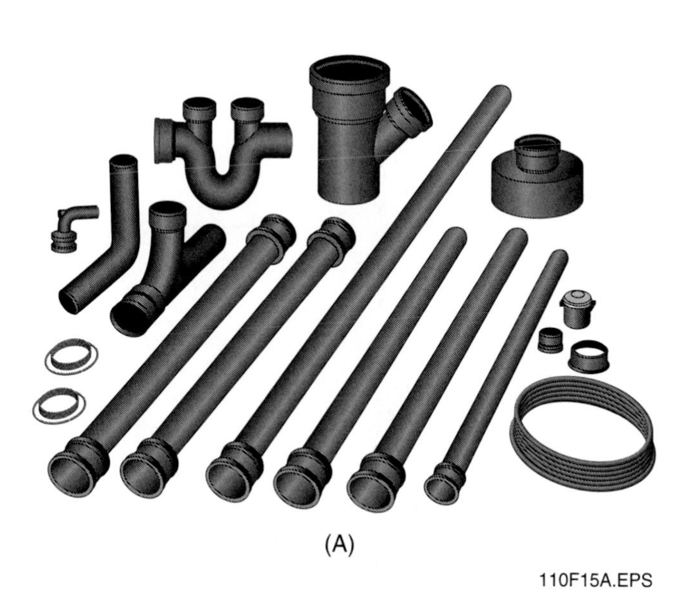

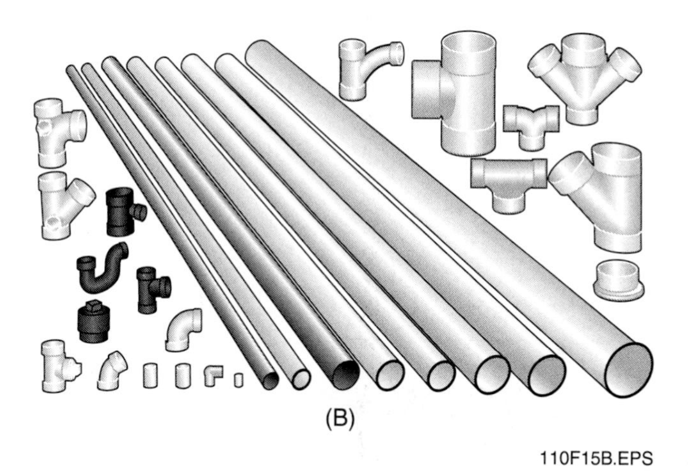

Figure 15 ◆ (A) Cast-iron pipe and fittings and (B) PVC pipe and fittings.

7.2.0 General DWV Fitting Requirements

Most codes state that drainage fittings may not slow or block the flow of materials in the pipe. Because of this, sanitary drainage fittings are made with a sweeping design to allow the smooth flow of material in the system, as shown in *Figure 16.*

Another code requirement is that the direction of hub-type fittings should not go against the flow of the system—that is, the wastes should flow from the bell to the spigot end of the pipe.

7.3.0 DWV Fittings

Sanitary fittings are used to connect DWV branches to the main DWV system. The branch inlets of these fittings may be reducing (going from a larger pipe to a smaller pipe). If so, they can be joined to the system without reducers.

7.3.1 Bends

The term bend often is used in reference to cast iron fittings. With other types of fittings, the term *elbow* is more common.

Bends are used to change the direction of a **run** of pipe. They are available in 1/16, 1/8, 1/6, 1/5, and

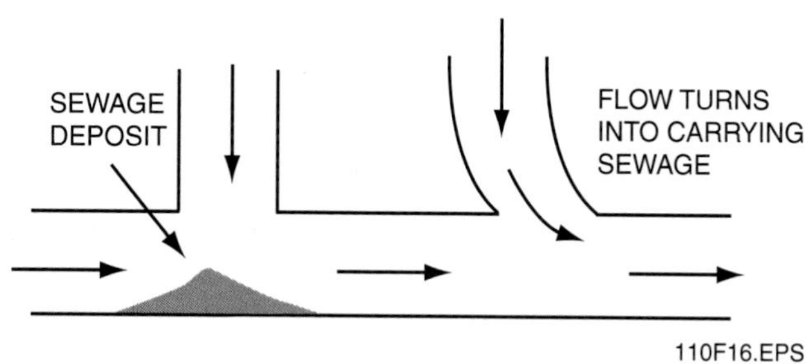

Figure 16 ◆ Sweeping design.

¼ bends, as shown in *Figure 17*. The ⅕ bend is available only in cast-iron pipe.

The plumbing design tells you which bends are used and where they are placed in the system.

Bends are expressed as fractions of a complete circle. A circle contains 360 degrees. To determine the number of degrees a given bend turns, multiply the fraction by 360 degrees. Thus, a ¼ bend turns 90 degrees (¼ × 360 degrees = 0.25 × 360 degrees = 90 degrees).

Bends are available with **heel inlets** and **side inlets** (see *Figure 18*) to allow smaller lines to be connected to the bend. Be sure to check local code requirements. It is important to note that a high or low pattern side inlet bend cannot be used as a vent if the inlet is horizontal.

Bends with side inlets are available with single and double side inlets. To determine whether the inlets are right or left inlets, place the spigot of the bend down and look through the hub (or bell) end (in the same direction water would be flowing down the drain), as shown in *Figure 19*. Left inlets will be on the left side, right inlets on the right side.

Quarter bends are available in three patterns, as shown in *Figure 20*. The basic fitting is simply called a ¼ bend.

Short sweep ¼ bends and **long sweep ¼ bends** may be used at the base of DWV stacks. They are required by the various codes because the longer radius of their turn greatly decreases flow resistance and back pressure. For tall stacks, the long sweep ¼ bend generally is required. Always check the local codes to determine which bend must be used in your area.

Double ¼ bends (also called a twin ell), shown in *Figure 21*, are used to collect and combine the flow from two opposite runs into a single run of

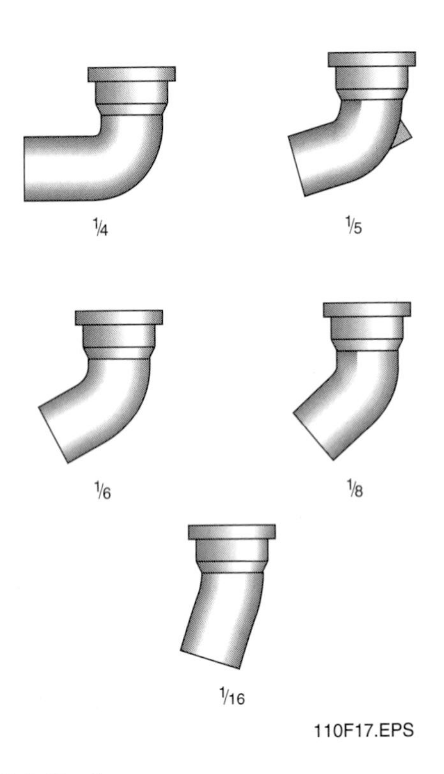

Figure 17 ◆ Bends.

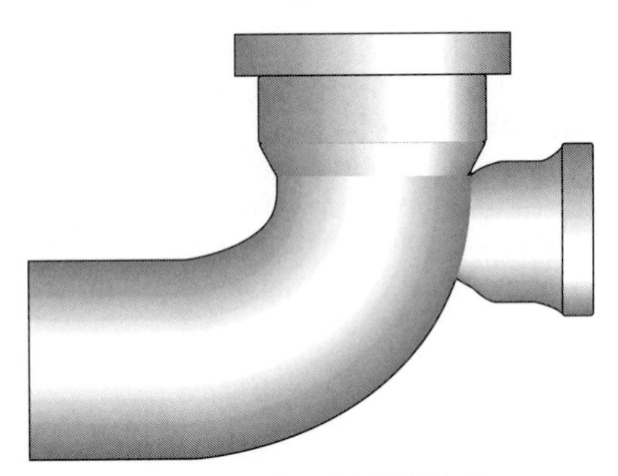

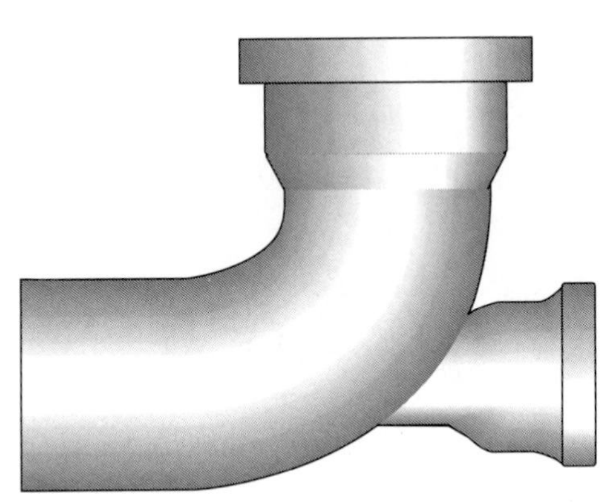

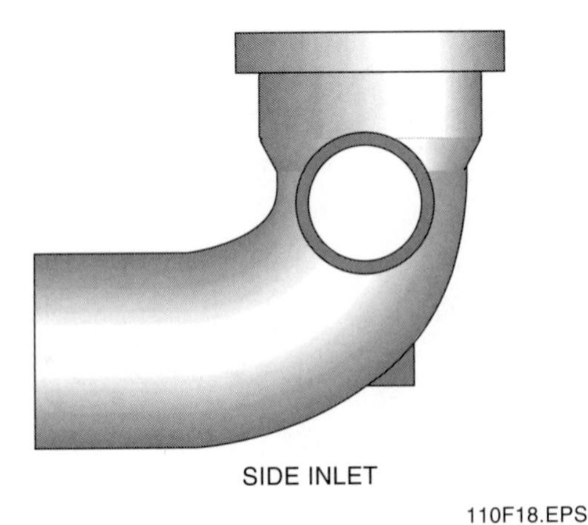

Figure 18 ◆ Bends with heel and side inlets.

pipe. Note the direction of flow in *Figure 21.* Be sure to check the local code requirements.

Codes allow use of **vent ells** (see *Figure 22*) only in vent lines. If they were placed in drainage or waste lines, their sharp turn radius design would severely restrict the flow of materials. They are available only in plastic fittings.

7.3.2 Adapters

DWV fittings are most often used with pipe made from the same material. However, adapters (see *Figure 23*) can be used to join pipe of different materials. Adapters may be used to join copper tubing to galvanized iron pipe, for instance, or plastic to cast-iron pipe.

7.3.3 Cleanouts

Fittings are also available for cleanouts. Cleanout fittings have internal threads on the branch fitting to accept a threaded cleanout plug. A cleanout adapter (see *Figure 24*) may also be installed in one end or branch of a fitting to provide a cleanout access. In both cases, a cleanout plug is provided to permit access to the DWV piping system and to prevent leakage.

Sanitary fittings are available in single and double patterns. The double pattern is used to connect two branch lines entering the system from opposite directions. This allows for central placement of horizontal runs and vertical runs.

Sanitary branch fittings consist of tees, wyes, and combinations of the two.

7.3.4 Tees

Sanitary tees, shown in *Figure 25,* are used for branches that run from horizontal to vertical.

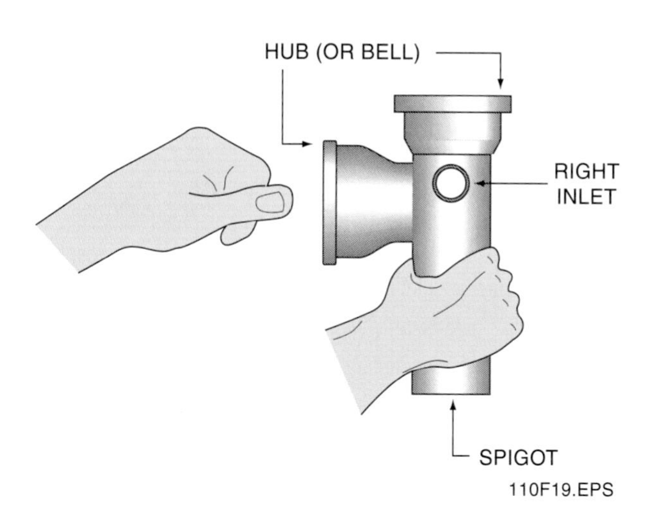

Figure 19 ◆ Determining hand of inlet.

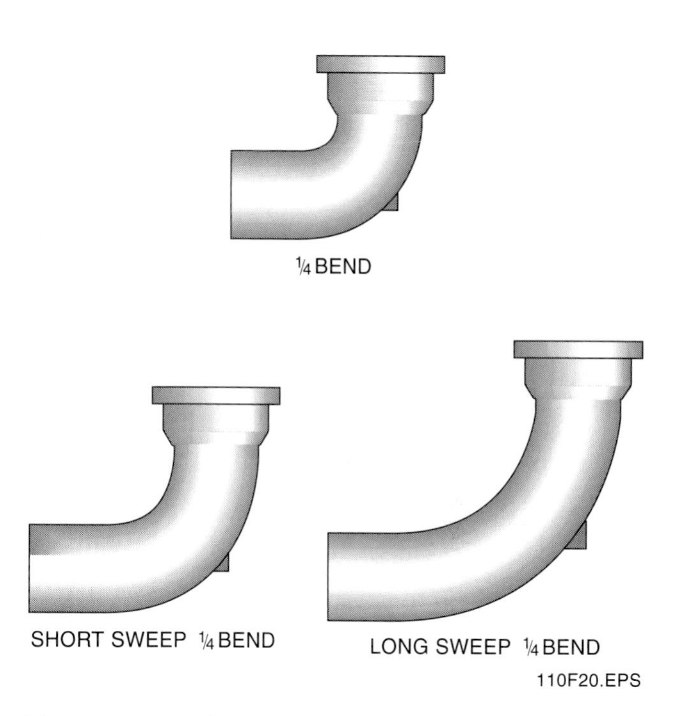

Figure 20 ◆ Available quarter bends.

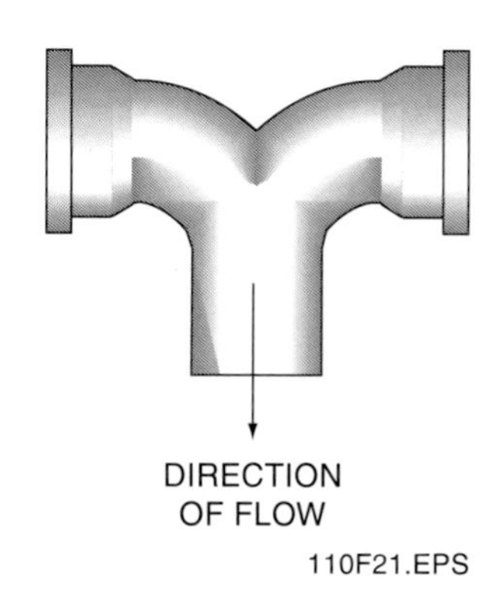

Figure 21 ◆ Double quarter bend.

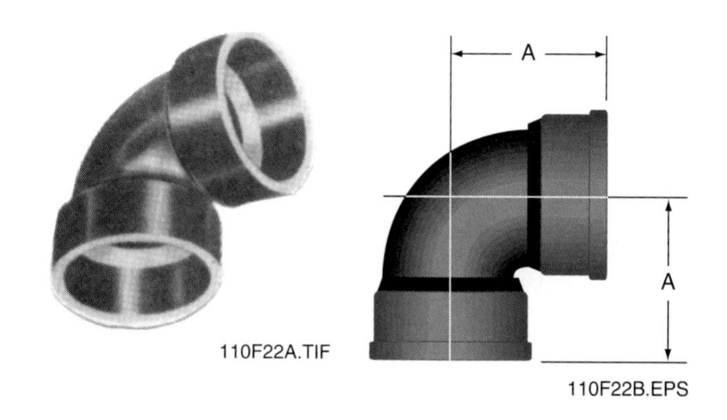

Figure 22 ◆ Vent ells.

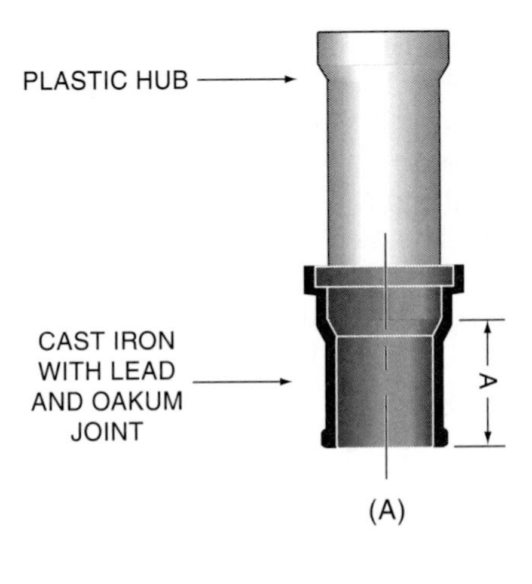

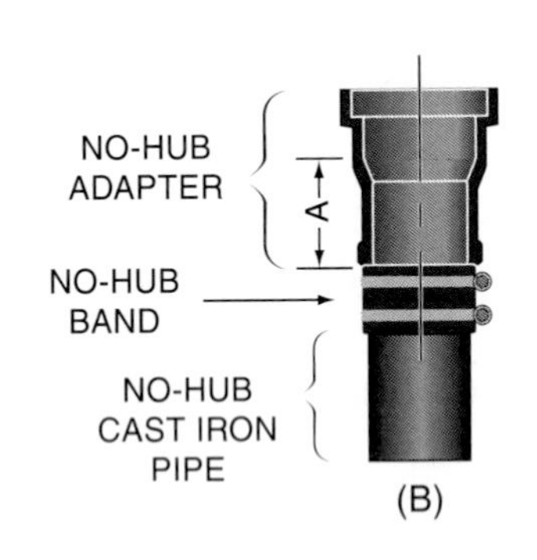

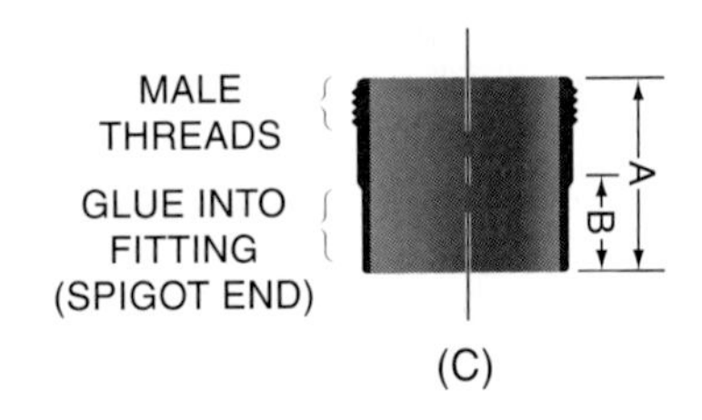

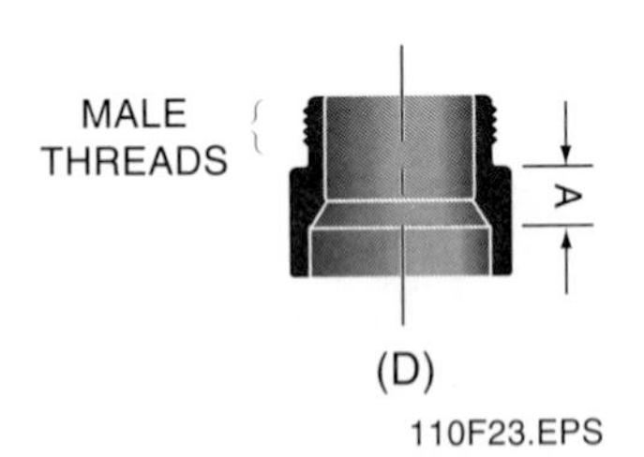

Figure 23 ◆ Adapters. (A) Hub adapter—cast iron. (B) No-hub adapter. (C) Male adapter with spigot. (D) Male adapter with hub.

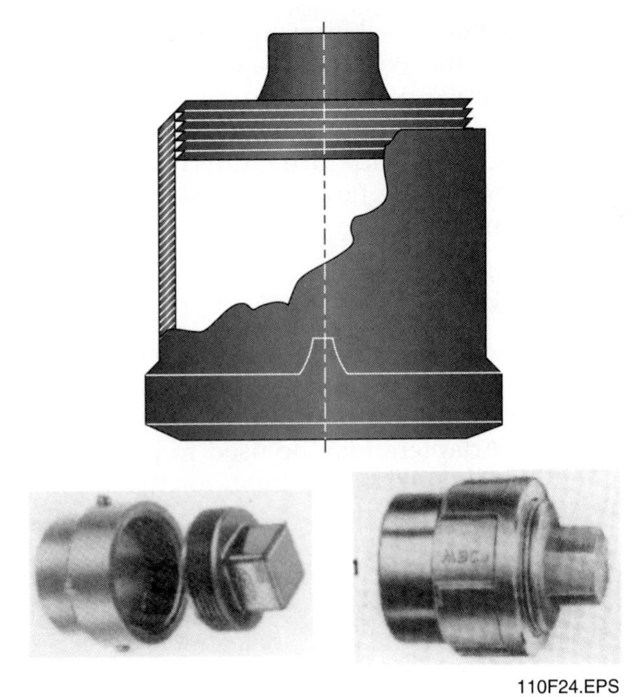

Figure 24 ◆ Cleanout adapter.

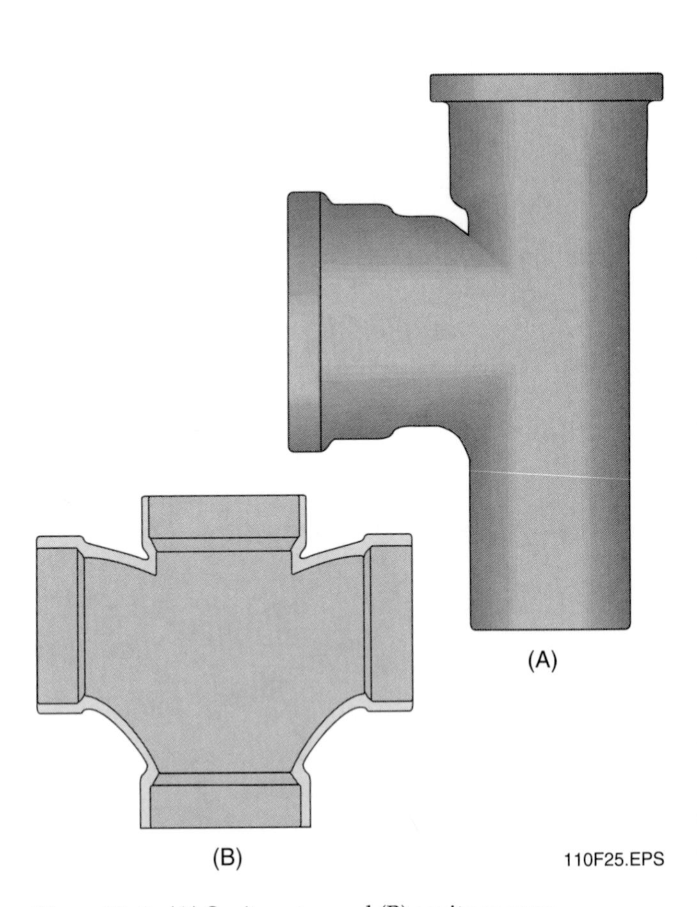

Figure 25 ◆ (A) Sanitary tee and (B) sanitary cross.

Model codes restrict their use to sanitary drainage systems where the flow of material is from the horizontal to the vertical.

Sanitary tees are available with side inlets. The side inlet allows drains of smaller sizes to be connected from the right or left. To determine whether a tee has a left or right inlet, place the spigot of the fitting down and look through the hub (or bell) end, as shown in *Figure 26*.

Vent tees (see *Figure 27*) are restricted by various codes for use only in vent lines and as cleanout fittings. They are prohibited for use in drainage systems because their design restricts the flow of material and may allow wastes and **scale** to collect within the fitting.

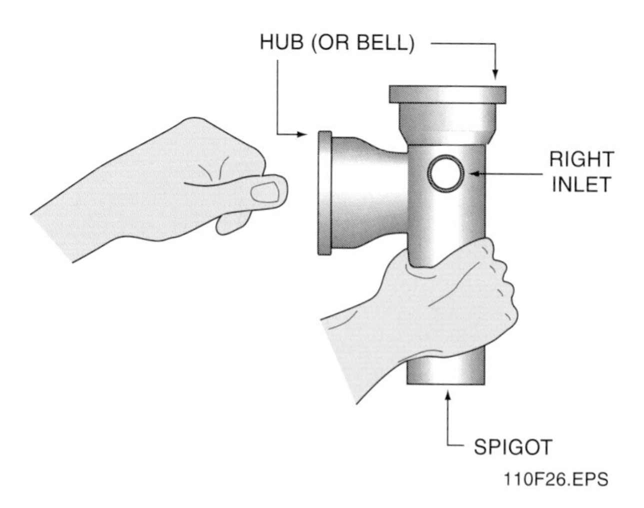

Figure 26 ◆ Determining hand of inlet.

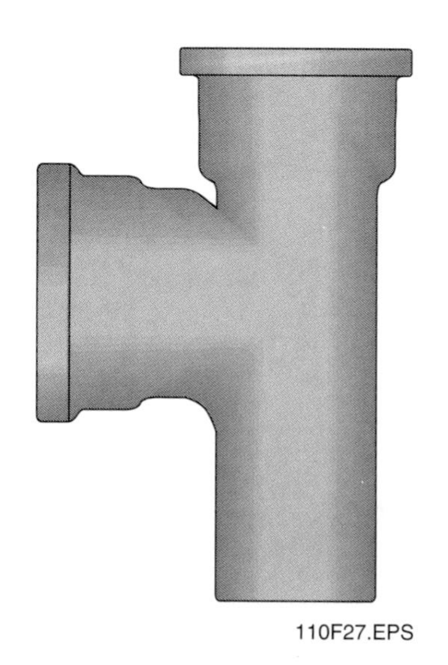

Figure 27 ◆ Tee.

Combination cleanout and test tees (see *Figure 28*) are installed in the DWV plumbing system as required by various codes. Placement of these and other fittings for cleanout purposes will be discussed later in your training. These tees serve as a test location from which to pressurize the system to test for leaks. After testing is completed, a plug is inserted in the fitting for future use as a cleanout.

7.3.5 Wyes

Sanitary wyes (see *Figure 29*) are used to provide a smooth-flowing DWV system in keeping with code requirements. Codes require vertical branch lines that intersect with horizontal branch lines to connect with long turn fittings such as a wye with a 45-degree angle or sweeps. If the correct fittings are not used, soil and wastes may collect on the pipe wall opposite the branch.

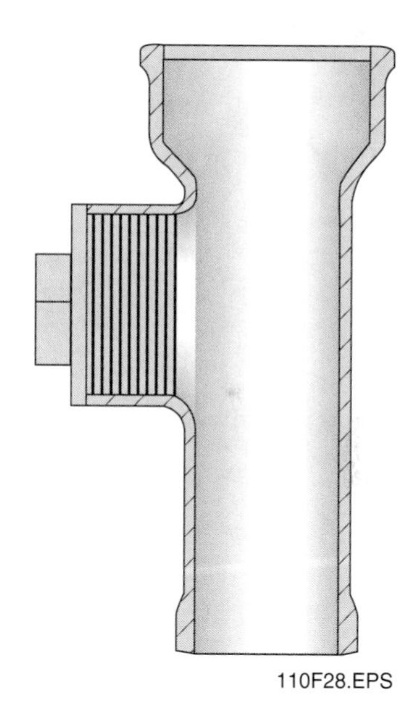

Figure 28 ◆ Combination cleanout and test tee.

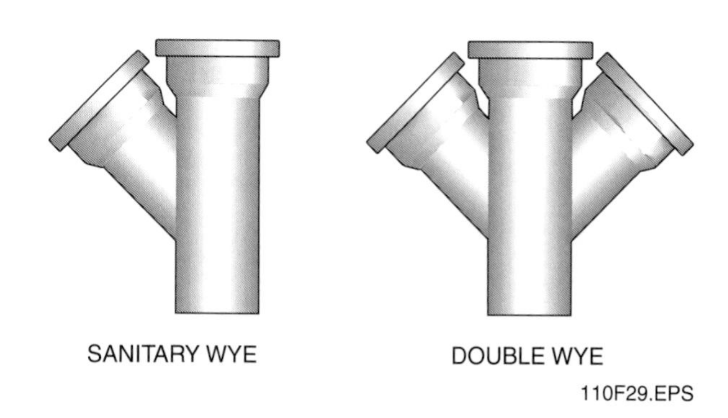

Figure 29 ◆ Sanitary wyes.

Sanitary upright wyes (see *Figure 30*) are used to connect the vent stack to the lower end of the soil and waste stacks.

Vent branches (see *Figure 31*) are used to join the upper end of the vent to the top of the soil and waste stacks.

Inverted wyes (see *Figure 32*) may be used in place of vent branches.

A **sanitary combination** is a fitting that combines a wye and a ⅛ bend. Also called a tee-wye, this fitting is available with single or double inlets (see *Figure 33*). Sanitary combinations are used to connect horizontal branch lines that intersect at 90 degrees with other horizontal branch lines. They also are used to connect a vertical stack with a horizontal drain. They are used in place of sanitary tees whenever space permits because they offer less resistance to the flow of materials than sanitary tees do. Combination fittings also reduce the number of fittings you need. As shown in *Figure 34*, a sanitary wye and a ⅛ bend would be needed to do the same job as a combination fitting. As you reduce the number of fittings, you also reduce the number of joints to be made and the amount of time needed to make them.

7.3.6 Miscellaneous Fittings

Sanitary increasers (see *Figure 35*) are used to enlarge the diameter of vent stacks and are usually placed at least 1 foot below the stack's intersection

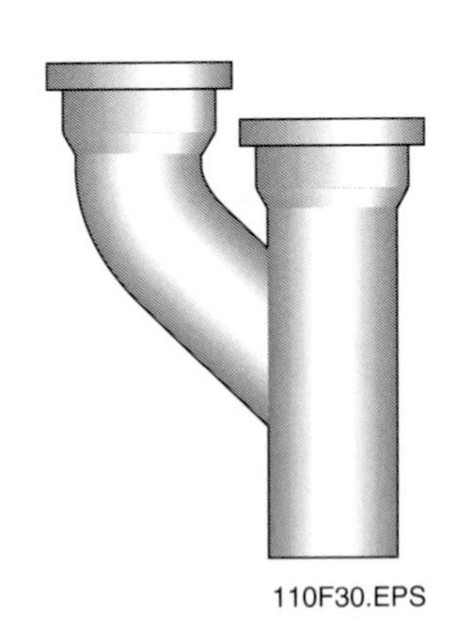

Figure 30 ◆ Sanitary upright wye.

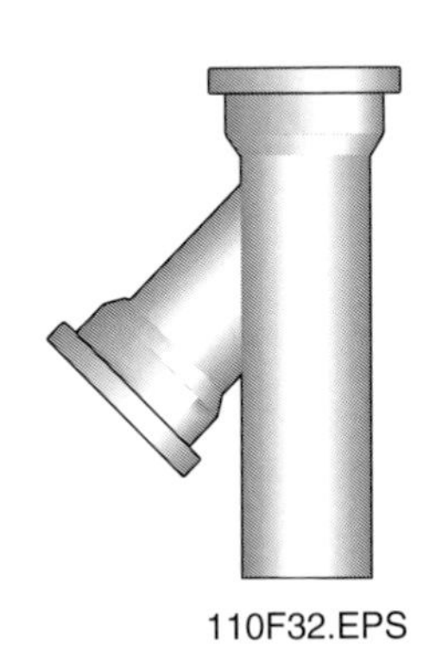

Figure 32 ◆ Inverted wye.

Figure 31 ◆ Vent branches.

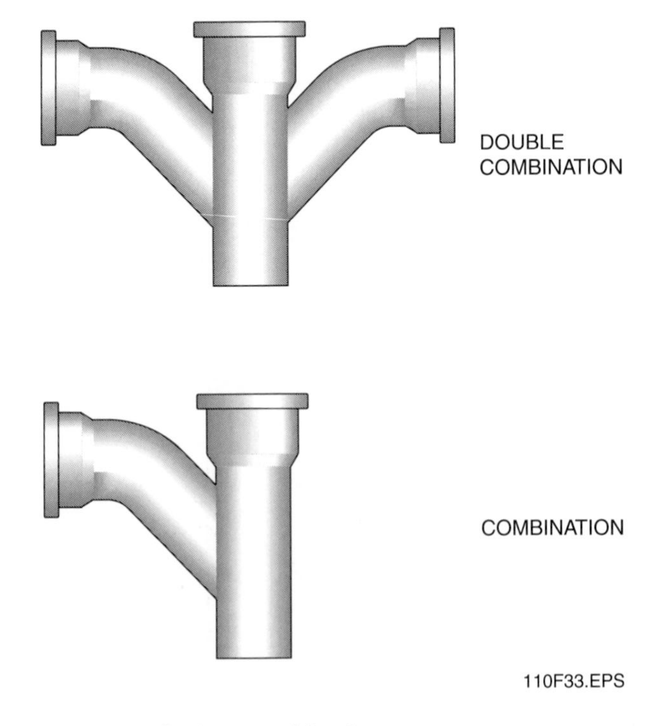

Figure 33 ◆ Sanitary combinations.

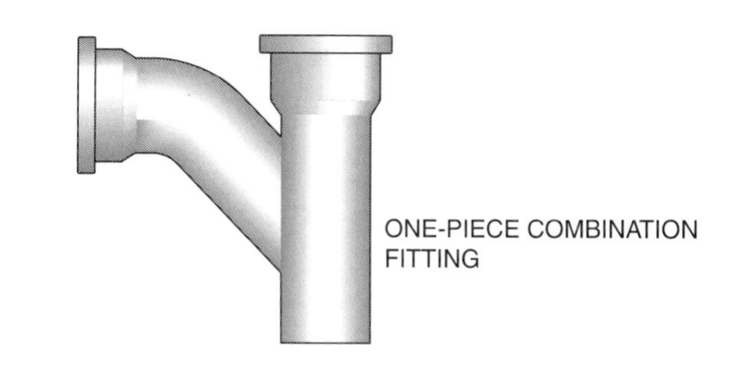

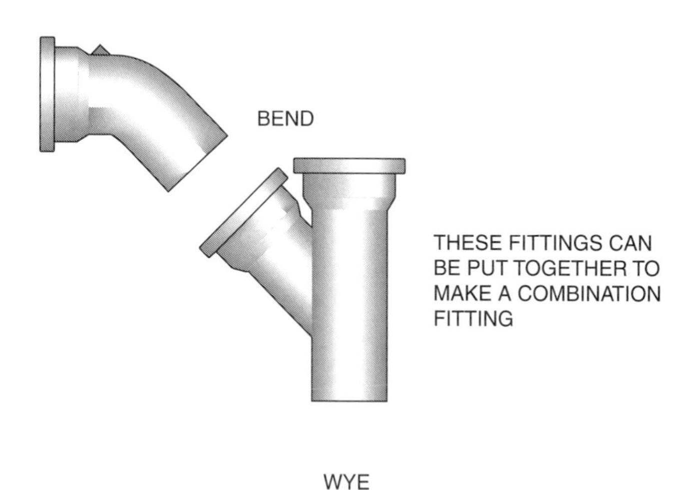

Figure 34 ◆ Combination fittings.

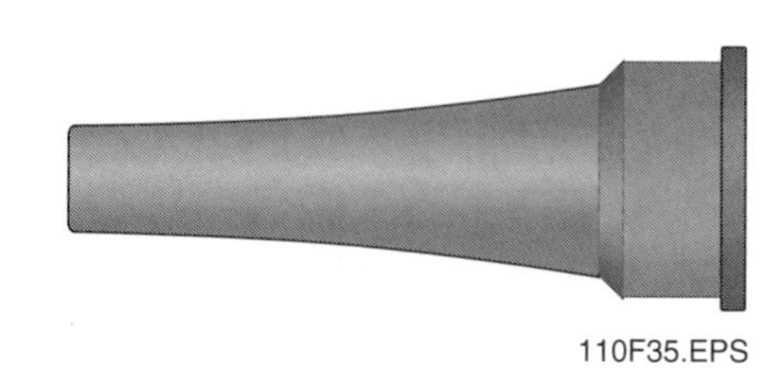

Figure 35 ◆ Sanitary increaser.

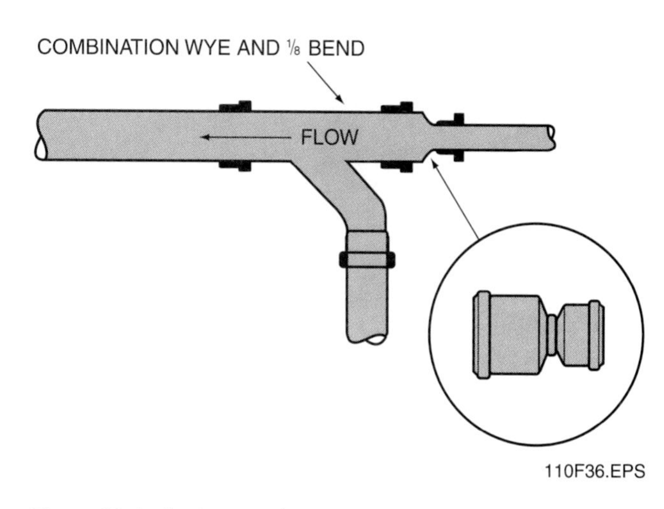

Figure 36 ◆ Sanitary reducer.

Figure 37 ◆ Offset.

with the roof, although this distance may vary by code. Sanitary increasers are necessary in cold climates to keep condensing water vapor from freezing and gradually closing the vent opening. The loss of the vent could cause the loss of the trap seals, allowing sewer gas to enter the building.

Sanitary reducers (see *Figure 36*) are used to reduce a larger pipe's diameter to that of a smaller pipe. They are mainly used to join a smaller branch line to the main drainpipe. You should never reduce the main drainpipe because this would restrict the flow of the system.

Offsets (see *Figure 37*) are used to change the path of the pipe to avoid obstruction. They can offset the run of the pipe from 2 inches to 12 inches.

Review Questions

Sections 7.0.0–7.3.0

1. Use a(n) _____ to join pipes of different materials in the DWV piping system.
   a. gasket
   b. increaser
   c. adapter
   d. tee

2. A _____ fitting has internal threads on the branch fitting to accept a threaded plug.
   a. cleanout
   b. sanitary
   c. wye
   d. reducer
3. For branches that run from horizontal to vertical, use a sanitary _____.
   a. fitting
   b. wye
   c. ell
   d. tee
4. To connect a vertical stack with a horizontal drain, use a(n) _____ fitting.
   a. elbow
   b. tee
   c. bend
   d. combination
5. _____ of a complete circle describes the degree of turn made by a ¼ bend.
   a. 45 degrees
   b. 90 degrees
   c. 30 degrees
   d. 22½ degrees

8.0.0 ◆ GRADE

Drainage and waste systems (see *Figure 38*) rely on gravity to move solid and liquid wastes, so these piping systems must be installed at a slope toward the point of disposal. In the plumbing industry, this slope is called **grade.** Drainage and waste piping systems are designed with the grade engineered into the system. The architects and engineers who design the piping system determine the grade. However, in some cases, such as in residential plumbing, the plumber specifies the grade.

8.1.0 The Importance of Grade

In a system with the proper grade, the liquid wastes will flow at the proper **velocity,** or speed, to scour the insides of the pipe, and the solids will be carried away. Proper grade is essential. If too much grade is used, the liquid wastes may flow too fast, leaving the solids behind. If too little grade is used, the liquid wastes will not flow fast enough to scour the pipe and remove the solid wastes. If the grade of a pipe does not remain constant, the velocity of the liquid wastes will change at the point where the grade increases or decreases. In any of these cases, the pipe will soon become blocked with solid wastes.

Figure 38 ◆ Grade.

Plumbers must determine the grade before they begin their work. This information may be in the local plumbing codes, in the **specifications (specs)** for the structure, or in the blueprints. If it is not in any of these places, contact the local plumbing inspector for a decision.

8.2.0 Torpedo Level

The torpedo level (see *Figure 39*) is similar to a general purpose level, but the torpedo level is much smaller—usually less than one foot long—and more streamlined. Its major advantage is in measuring grade for small runs of pipe—for example, fixture branch lines that run a short distance to the stack. The torpedo level is also light and easy to manipulate in tight places.

9.0.0 ◆ BUILDING DRAIN

The building drain is the main horizontal pipe inside a building. It carries all sewage and other liquid wastes to the building sewer. Codes usually define the building sewer as standing 2 to 3 feet outside the building foundation. The building drain also acts as the principal artery to which other drainage branches of the sanitary system may be connected.

Any vertical pipe, including the waste and vent piping of a plumbing system, is considered a stack. A soil stack is a vertical section of pipe that receives the discharge of water closets, with or without the discharge from other fixtures. The soil stack is connected to the building drain.

A **branch interval** is a section of a stack. The branch interval usually corresponds to a story height (the height of one floor in a building), but it can never be less than 8 feet in length.

A horizontal branch is the part of a drain pipe that extends laterally (sideways) from a soil or waste stack and receives the discharge from one or more fixture drains.

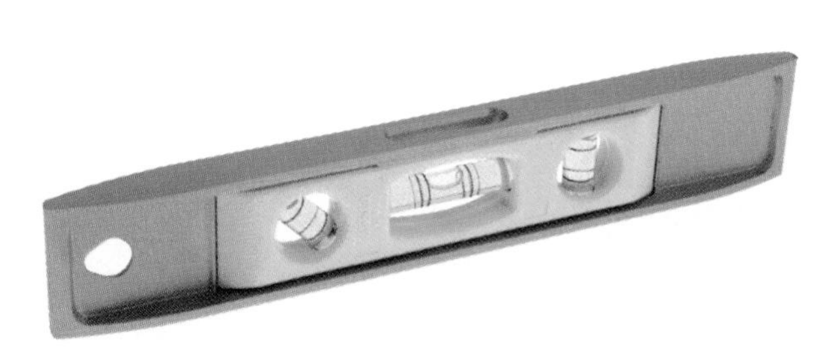

110F39.EPS

Figure 39 ◆ Torpedo level.

9.1.0 Cleanouts

Cleanouts, as shown in *Figure 40*, are fittings with removable plugs. The plugs provide access to the inside of drainage and waste piping systems so that blockages can be removed. As mentioned earlier, cleanout adapters may be used to convert other fittings so they can be used as cleanouts.

10.0.0 ◆ BUILDING SEWER

The building sewer or house sewer is the drainage piping that runs from the building's foundation to the sewer main or septic tank (private waste disposal system). It normally starts approximately 2 feet to 3 feet outside the building.

Building sewers are commonly made using **ABS** or **PVC** plastic, cast iron, or vitrified clay pipe (a hard, nonporous clay). When sewers are laid, the ground must be tamped to keep the pipes from settling and losing their grade. Sometimes the pipes are installed over a bed of gravel as a support for the pipe. In some parts of the country, plumbers lay the entire sewer from the foundation wall to the sewer main. In other locations, the

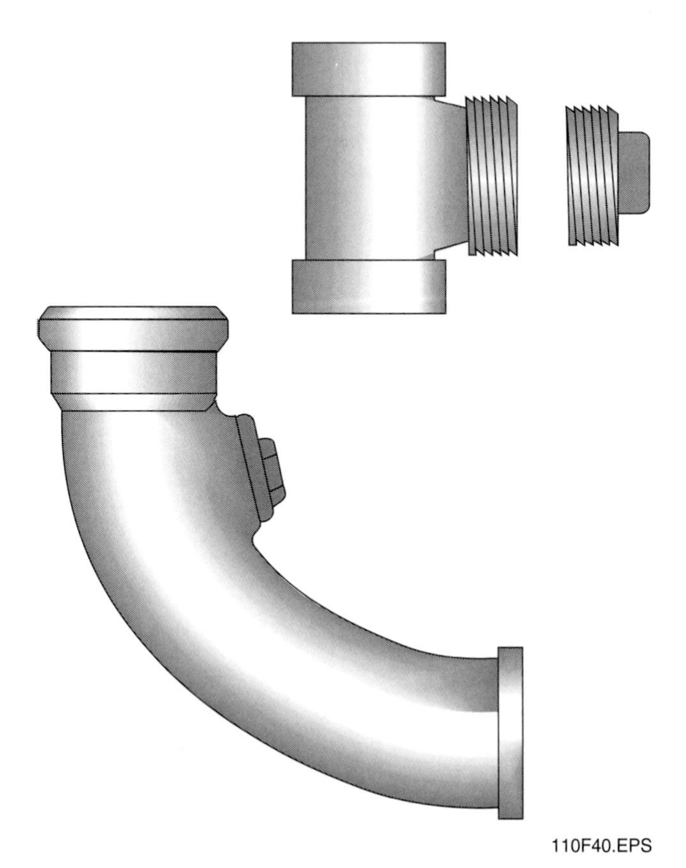

110F40.EPS

Figure 40 ◆ Cleanouts.

municipal sewer crew may be in charge of the installation from the property line to the sewer main. All operations associated with building sewers are regulated by code. Always check the local codes when you are working on this system.

Manholes must be provided for underground piping that is 8 inches or larger in diameter. They should be located at intervals not more than 400 feet apart and at every major change in direction, grade, **elevation**, or pipe size. To meet local codes for traffic and loading conditions, the manholes must have metal covers of sufficient weight.

11.0.0 ◆ SEWER MAIN

A public or municipal sewer is installed, maintained, and controlled by the local municipality or town. The sewer main is usually located in a street or alleyway or within an easement on privately owned land. Sewer mains carry waste to the treatment plant.

Municipalities or towns usually install a 6-inch sewer laterally from the sewer main to the edge of each building lot. This lateral pipe makes it possible to connect a building sewer to the public sewage system.

12.0.0 ◆ WASTE TREATMENT

Many municipalities have sewer systems in which wastes are collected and treated at a sewage plant, then discharged back into the ecosystem. Other municipalities require households to treat their waste individually in private waste disposal systems.

12.1.0 Municipal Waste Treatment Systems

Municipal waste treatment plants (see *Figure 41*) are highly sophisticated facilities, as public health and safety depend on their operation.

These systems are designed to handle thousands of gallons of sewage each day. The treatment facilities receive sewage into huge holding tanks, where heavier substances settle to the bot-

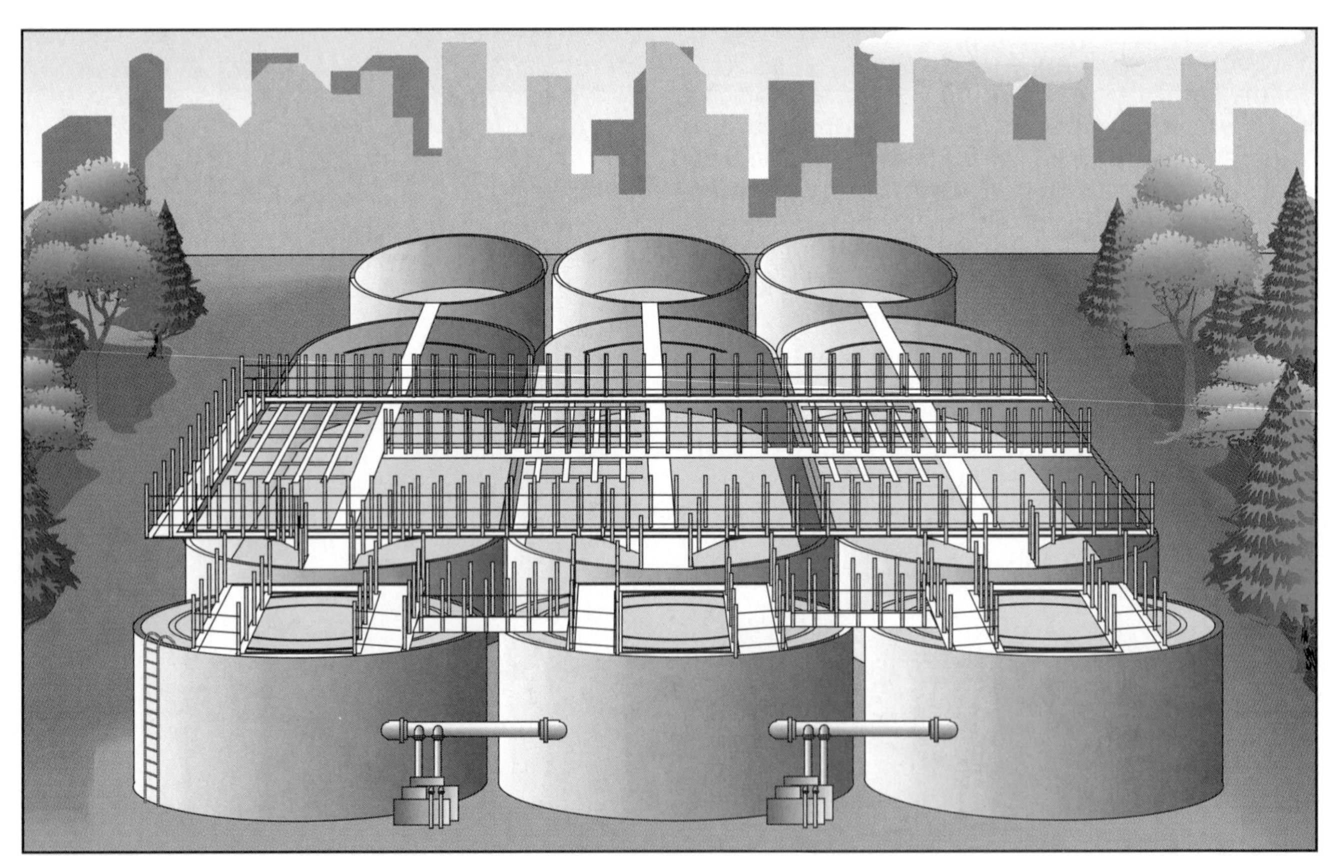

110F41.EPS

Figure 41 ◆ Municipal sewage treatment plant.

tom and lighter substances float to the top. The heavier substance is called **sludge.** Both the sludge and the wastewater are then treated.

12.2.0 Private Waste Disposal Systems

Private waste disposal systems are designed to meet the needs of individual households and the requirements of local codes and health departments. As in municipal waste treatment systems, but on a much more limited scale, the waste flows into a holding tank where the sludge settles out and is digested by bacteria. Liquid waste flows through a distribution box into a leachfield, where it then seeps down into the earth in a natural purification cycle.

Plumbers must know about different types of private disposal systems and the advantages and disadvantages of each. In addition, a plumber must be able to install private waste disposal systems correctly and according to code requirements. You will learn more about these systems as you advance through the plumbing curriculum. The following is a description of one basic component of private waste disposal systems.

12.2.1 Conventional Septic Tank System

A septic system consists of a septic tank, a distribution box, a leachfield, and piping between those parts (see *Figure 42*). A septic tank system provides partial treatment of raw wastewater. It protects the soil absorption system from becoming clogged by solids that are suspended in the raw wastewater. Local code and the local health department strictly regulate use of these systems.

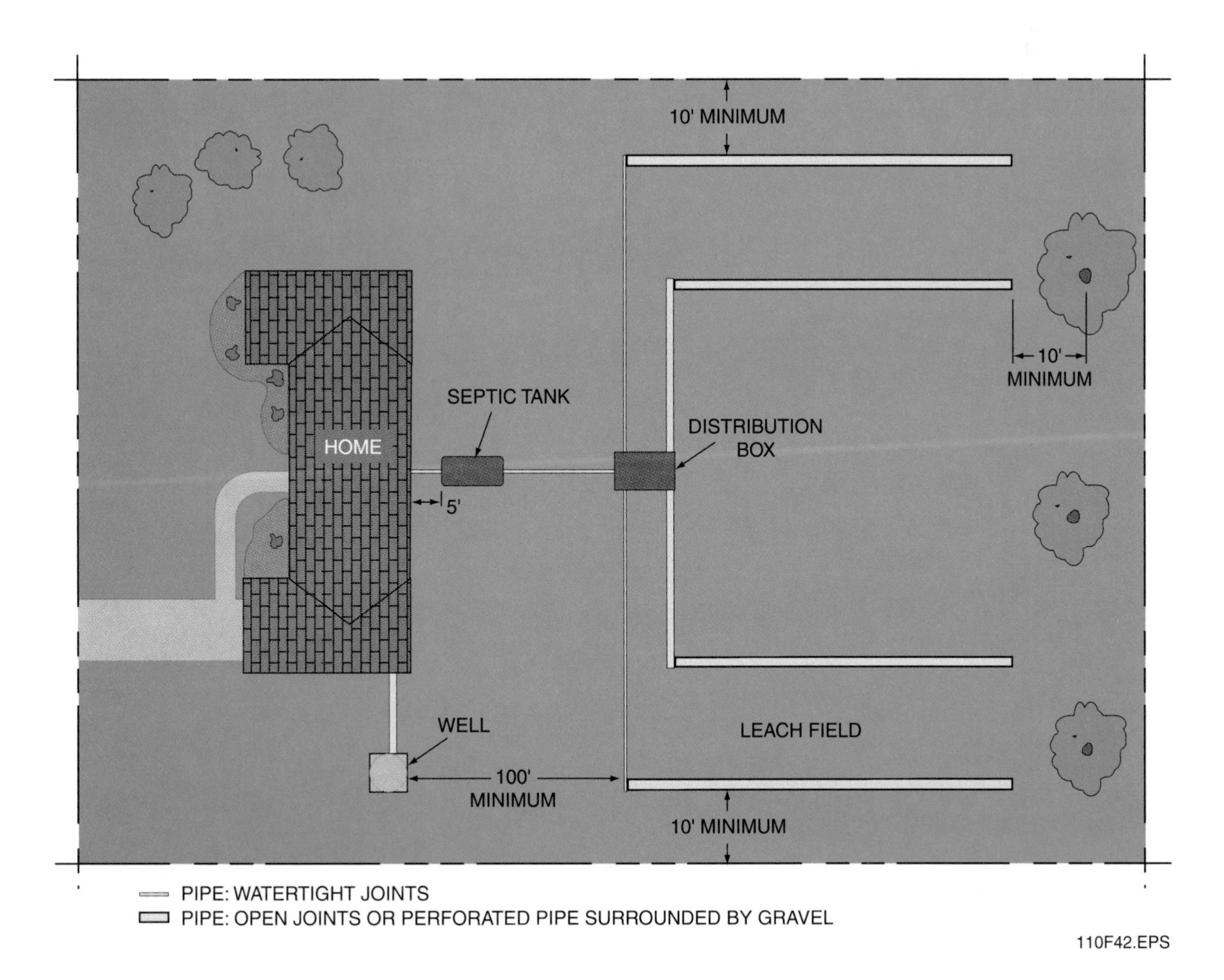

Figure 42 ◆ Typical layout for a septic tank system.

Did You Know?

The Babylonians had sewer systems more than 5,000 years ago. The Romans later built sewer systems for both storm water and wastewater. However, water treatment was unknown then, so the wastewater often was returned to the river just below the city or town.

Review Questions

Sections 8.0.0–12.2.1

1. Drainage and waste systems depend on _____ to move solid and liquid wastes through the pipes.
   a. velocity
   b. gravity
   c. specifications
   d. adapters

2. The main horizontal drain pipe located inside a building is called a _____.
   a. building drain
   b. sewer drain
   c. building sewer
   d. DWV stack

3. Fittings that work as removable plugs used to provide access to the inside of DWV piping are called _____.
   a. adapters
   b. sewers
   c. cleanouts
   d. manholes

4. In a septic tank _____ settles out of the raw wastewater.
   a. seepage
   b. bacteria
   c. sewage
   d. sludge

5. The _____ is the drainage piping that goes from the building's foundation to the sewer main or septic tank (private waste disposal system).
   a. horizontal branch
   b. building sewer
   c. building drain
   d. vent stack

Summary

The DWV system of a building is part of a plumbing system designed to protect the health and safety of the people who use its facilities. The system carries wastewater out of our buildings, treats it, and returns it to the ecosystem.

Drain (D) and waste (W) piping removes wastewater from a building. The type and size of the piping system selected depends on critical factors such as the amount of fluids expected to flow through the piping, the type of fluids that will be carried, and the grade. Drainage and waste systems rely on gravity to move solid and liquid wastes, so the plumber must install piping at the correct grade toward the building sewer where the wastes leave the building and enter the public or private waste disposal system. Grade determines the velocity of the liquid waste flowing through the piping. If too little grade is used, the liquid waste moves too slowly and it will not scour the pipe and remove solid wastes. If the grade is too steep, the liquid waste will flow too fast and leave solids behind in the piping.

The vent (V) system is an important part of the overall DWV system. Vent piping provides for the free flow of air in the drainage system. This air equalizes the atmospheric pressure inside the pipes and prevents back pressure or siphoning from destroying the water trap seals in the fixtures. These water trap seals keep sewer gas and odors out of the building. Traps provide a way for the wastewater or sewage to flow through the fixture and into the piping system while protecting the occupants of the building from bacteria and potentially explosive gases.

The DWV piping system relies on different kinds of fittings to join the lengths of pipe. These fittings may be made of copper, brass, lead, steel, cast iron, clay, glass, or plastic. The installation determines the best piping and fitting material to use. The fittings are designed so they do not block or slow the flow of materials in the pipe. The sweeping design of DWV fittings allows for the smooth flow of material within the system. Different fitting shapes serve specific purposes, depending on whether a pipe runs horizontally or vertically. Correct selection and installation of fittings is vital for the system to function properly.

PROFILE IN SUCCESS

William (Bill) Yeager

Owner
Yeager, Inc.
Newton, IA

Bill is a native of Allerton, Iowa. After high school, he moved to Newton, where he owns and operates Yeager, Inc. With 45 years in the plumbing and HVAC trades, Bill is committed to giving back to the industry by supporting education and training. Presently, he has seven apprentices in the NCCER program and another 13 employees who have graduated from the program. He loves his work and inspires others to pursue successful careers in construction.

How did you become interested in the construction industry?

When I finished high school, I went to college. The draft board came looking for me, and I joined the Navy where I spent the next four years. One day someone came through the barracks and asked who wanted to go to school. I jumped at the chance and didn't even ask what I was going to learn. The Navy sent me to boiler school in San Diego where I learned boiler tending and piping. I have always been mechanically inclined, and I fell in love with the trade.

Meanwhile, I got married and had two kids. When I got out of the Navy, I started back to college but soon found myself with three kids—I needed to get a job. Back in my hometown, I walked into a plumbing store and told them what I could do. They hired me on the spot. I have been there ever since—almost 45 years.

In 1980, I served on the National Education Committee at ABC when the Wheels of Learning got started. I firmly believe that apprenticeship training is essential to build a successful career in the trades. Currently, I serve as the chairman of the National Education Committee.

What do you think it takes to be a success in your trade?

It takes three things to be successful—good communication skills, a good work ethic, and the right attitude. If you have those, the Wheels of Learning program teaches you everything else you need to know. If you don't want to get up and work hard every day, then reconsider. You have to enjoy your profession.

I also tell people considering the trades to take as many math and computer classes as they can. This field is technology oriented. The more you learn, the better. I also tell apprentices that it is critical to write well. If you write up an order and no one can read your writing, it's a real problem.

I believe that you should think of this as a lifelong career, not just a job. I have three sons who are all licensed journeymen plumbers. Two came to work for me right out of high school. The other son went to college and got a government job. When budget cuts eliminated his job, he was left with a wife and kids and no income. He walked into a plumbing store and asked for a job. They asked what he knew. He told them about working with me every summer and they hired him on the spot. He is still working there. My daughter went into interior design. With all the emphasis on design/build, she is swamped with work.

What are some of the things you do in your role of company owner?

I am a liaison between the staff and the customers. We hold a meeting every morning. This is why communication skills are so important. During the meeting each employee gets to update the rest of us on the status of projects and to bring up any potential problems they see. This way, we work together to prevent problems and to find solutions as a team.

I also work with the apprentices. I try to help them out while they are learning the trade. For the last five years, I have also been teaching a class at the high school in exploring careers in construction.

What do you like most about your job?
I like the people. If I don't have good people working for me, the customers will not be satisfied. Dissatisfied customers don't call you back. This field is very customer-oriented. It really is a people business first of all.

I like the challenge. Every day brings something different. We work on a huge range of jobs. We do anything from a $45 service call to our award-winning $5 million project we just completed. The variety is a challenge. You have to solve new problems every day.

What would you say to someone entering the trades today?
Ask questions. Work safe. Find a mentor and learn from that person's experience. As an apprentice, you will have many qualified people working all around you. Don't hesitate to ask questions. Draw on the experience of others to learn more about your trade.

Think of this as a career. If you just want a job, you can go anywhere to find one. If you want to build a successful, meaningful career, then take your trades training seriously. In the construction industry, the possibilities are endless.

GLOSSARY

Trade Terms Introduced in This Module

ABS: The initials stand for acrylonitrile-butadiene-styrene, a tough, rigid plastic.

Adapter: A fitting that joins pipes of different sizes or materials, such as copper and galvanized pipe or cast iron and plastic pipe.

Back pressure: The compressing of trapped air, which holds back the flow of waste through the DWV piping in plumbing systems.

Bend: A fitting that changes the direction of piping.

Branch interval: The vertical distance between the connection of branch pipes to the main DWV stack.

Capillary attraction: The tendency for water to be drawn by a piece of paper or string that is stuck in a pipe. The water is drawn out of the trap and the seal is broken.

Cleanout: An access point to all parts of the drainage system for the removal of blockages.

Cleanout fitting: Removable drainage fitting, such as a plug, that allows access to the inside of draining pipes to remove blockages.

Combination cleanout and test tee: A tee installed as a test location for pressurizing the system to test for leaks, then plugged for later use as a cleanout.

Crown weir: The point in the curve of the trap that is directly below the crown. This is the point at which the water level will normally remain when the fixture is not discharging through the trap. Also called *trap weir.*

Double ¼ bend: A fitting used to collect and combine the flow from two opposite runs into a single run of pipe.

Drainage fitting: Any of a variety of fittings used in the DWV system to remove waste from a building.

Drum trap: An old kind of trap installed in the drainage piping. It is cylinder shaped, with an inlet near the bottom and an outlet near the top. Because the cylinder holds water, it prevents sewer gas from entering the building.

Elevation: The height above an established reference point, such as a grade reference point on a construction drawing.

Evaporation: Loss of water, especially in a drainage system, into the atmosphere.

Fall: The amount of slope given to horizontal runs of pipe.

Fixture drain: The drain extending from the trap of a plumbing fixture to a junction of that drain with any other pipe. Also called *trap arm.*

Grade: The slope of a horizontal run of pipe.

Heel inlet: A bend that has an inlet to connect a smaller pipe to the main line. The inlet is located at the base of the curve or heel of the bend.

Hydraulic gradient: The level line in which water tends to flow in a pipe.

Inverted wye: A fitting used to join the upper end of the vent to the top of the soil and waste stacks. It looks like an upside-down Y.

Long sweep ¼ bend: A bend used at the base of DWV stacks because the longer radius of its design greatly decreases flow resistance and back pressure. Use is regulated by code.

Nonsiphon trap: A trap with a deeper trap seal for installations that are subjected to abnormal pressure fluctuation or to conditions that promote evaporation.

Offset: A fitting that changes the path of a run of pipe to avoid obstruction.

P-trap: A P-shaped trap that provides a water seal in a waste or soil pipe, used mostly at sinks and lavatories.

PVC: The initials stand for polyvinyl chloride, a rigid plastic used in piping.

Run: One or more lengths of pipe that continue in a straight line.

S-trap: A trap with a long downstream leg, which tends to promote siphonage. These traps are not installed anymore but are still found in older buildings.

Sanitary combination: Also called a tee-wye, this fitting combines a wye and ⅛ bend. It is used to connect horizontal branch lines that intersect other horizontal branch lines. It offers less resistance to the flow of material than a sanitary tee.

Sanitary fitting: A fitting used to connect DWV branches to the main DWV system and to serve as a cleanout.

Sanitary increaser: A fitting used to enlarge the diameter of the vent stack; usually placed at least 1 foot below the intersection of the stack and the roof. Use is regulated by code.

Sanitary reducer: A fitting used to reduce a larger pipe's diameter to that of a smaller pipe. Used mainly to connect a smaller branch line to the main drain pipe.

Sanitary tee: A tee fitting for pipe that has a slight curve in the 90-degree transition to channel the flow of wastewater or sewage from a branch line to the main line.

Sanitary upright wye: A fitting used to connect the vent stack to the lower end of the soil and waste stacks.

Sanitary wye: A drainage fitting, shaped like the letter Y, that joins the main run of pipe at an angle.

Scale: Flaky material resulting from corrosion of metals, especially iron or steel. A heavy oxide coating on copper or copper alloys resulting from exposure to high temperatures and an oxidizer.

Short sweep ¼ bend: A bend fitting with a short radius used at the base of a DWV stack. Use is regulated by code.

Side inlet: An opening in an ell or tee fitting at right angles to the line of the run, used to connect smaller lines to the main line.

Siphonage: Loss of water in a trap seal caused by unequal pressure inside and outside DWV piping.

Sludge: Semi-liquid matter that settles out in a holding tank during the waste treatment process.

Specifications (specs): Written requirements that provide additional details or descriptions included with the drawings or blueprints of a construction project. Provide the technical standards to be met during construction.

Stack: The vertical pipe of soil, waste, or vent piping.

Trap: A device or fitting designed to provide a liquid seal that will allow the passage of sewage or wastewater through it but will prevent the back passage of air.

Trap weir: In a trap, the raised section in the fluid's flow path where the fluid is moved upward on its way to the drainage pipe.

Velocity: Speed of motion, such as the speed of sewage or wastewater through the drainage piping.

Vent branch (branch vent): A vent that connects a branch of the drainage piping to the main vent stack.

Vent ell: A plastic fitting with a sharp turn radius, used only in vent piping systems. Its use is regulated by code.

Vent tee: A fitting used in venting systems or as a cleanout. It may not be used in the drainage system because it restricts the flow of material. Use is regulated by code.

ANSWER KEY

Answers to Review Questions

Sections 2.0.0–4.4.0

1. c
2. d
3. b
4. a
5. b

Sections 5.0.0–6.0.0

1. b
2. a
3. d
4. b
5. c

Sections 7.0.0–7.3.0

1. c
2. a
3. d
4. d
5. b

Sections 8.0.0–12.2.1

1. b
2. a
3. c
4. d
5. b

NCCER CRAFT TRAINING USER UPDATES

The NCCER makes every effort to keep these textbooks up-to-date and free of technical errors. We appreciate your help in this process. If you have an idea for improving this textbook, or if you find an error, a typographical mistake, or an inaccuracy in the NCCER's Craft Training textbooks, please write us, using this form or a photocopy. Be sure to include the exact module number, page number, a detailed description, and the correction, if applicable. Your input will be brought to the attention of the Technical Review Committee. Thank you for your assistance.

Instructors – If you found that additional materials were necessary in order to teach this module effectively, please let us know so that we may include them in the Equipment/Materials list in the Instructor's Guide.

Write: Curriculum Revision and Development Department
National Center for Construction Education and Research
P.O. Box 141104, Gainesville, FL 32614-1104

Fax: 352-334-0932

E-mail: curriculum@nccer.org

Craft ______________________ Module Name ______________________

Copyright Date ______________ Module Number ______________ Page Number(s) ______________

Description

____

____

____

____

(Optional) Correction

____

____

____

(Optional) Your Name and Address

____

____

____

Module 02111-00

Introduction to Water Distribution Systems

Course Map

This course map shows all of the modules in the first level of the Plumbing Curriculum. The suggested training order begins at the bottom and proceeds up. Skill levels increase as you advance on the course map. The local Training Program Sponsor may adjust the training order.

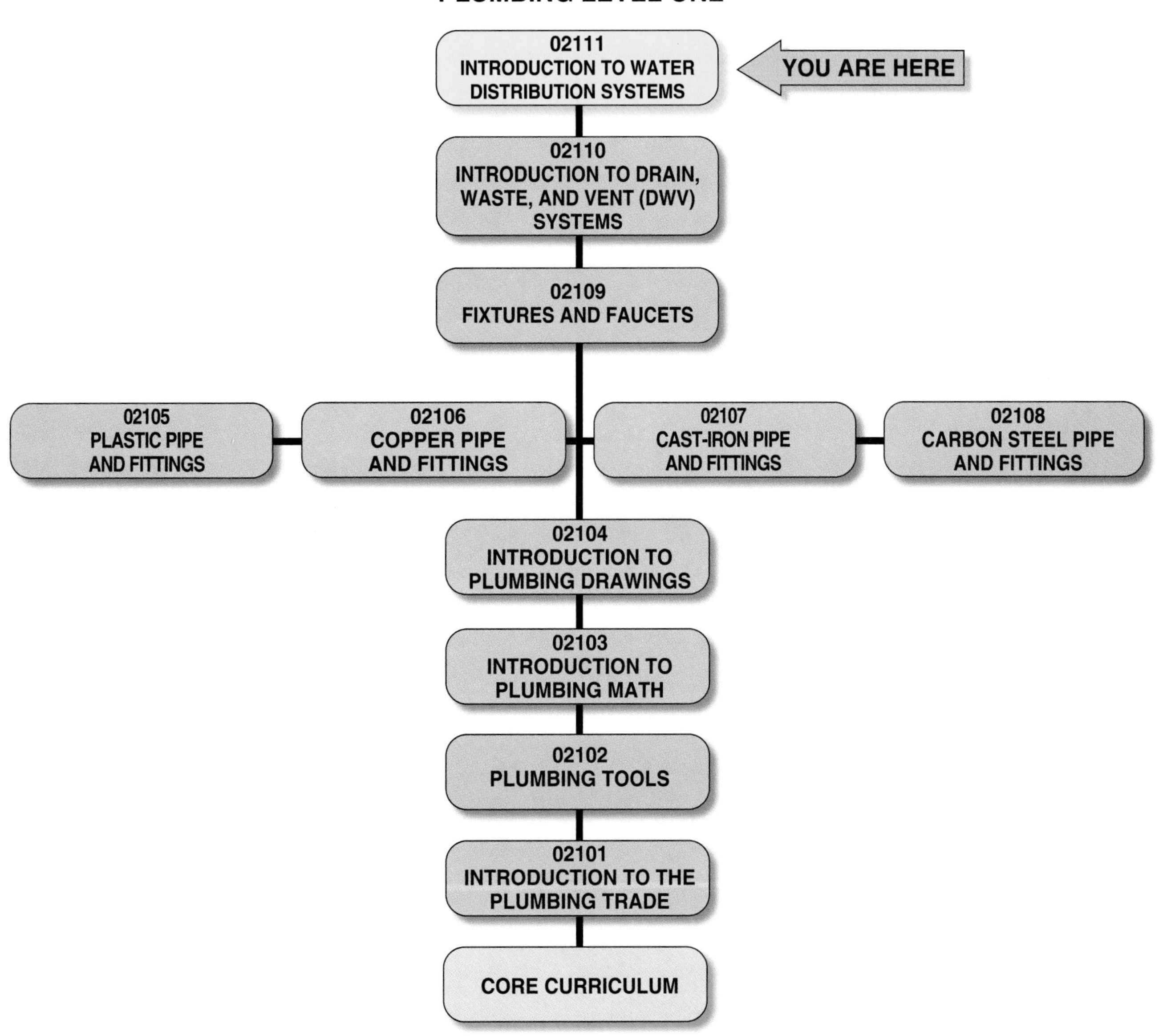

111CMAP.EPS

Copyright © 2000 National Center for Construction Education and Research, Gainesville, FL 32614-1104. All rights reserved. No part of this work may be reproduced in any form or by any means, including photocopying, without written permission of the publisher.

MODULE 02111 CONTENTS

Figures

MODULE 02111

Introduction to Water Distribution Systems

Objectives

When you finish this module, you will be able to:

1. Discuss how water moves from the source through the water distribution system to the fixture.
2. Identify the major components of a water distribution system and describe the function of each component.
3. Explain the relationships between components of a water distribution system.

Prerequisites

Before you begin this module, it is recommended that you successfully complete task modules: Core Curriculum; Plumbing Level One, Modules 02101 through 02110.

Materials Required for Trainees

1. This module
2. Appropriate personal protective equipment
3. Sharpened pencil and paper

1.0.0 ◆ INTRODUCTION

In this module you will learn about water supply and distribution. Water supply is either private (such as a well) or municipal (supplied through a public water distribution system). Components of the water distribution system include the pipes and fittings that carry hot and cold water into the building, the various valves used to regulate the flow of water to the fixtures and other outlets, and water heating and treatment equipment.

Any water distribution cycle begins with a water source. Water for private systems is located with a well sunk into an underground water supply and is usually pure enough to drink. Water for municipal systems is drawn from **reservoirs**, treated, and then distributed to individual homes and businesses.

City water (municipal systems) undergoes a purification process before it reaches the faucets in your home. Water is pumped to a treatment plant, where it is processed to remove harmful impurities. Chlorine, alum, and activated charcoal are added to the water during this cycle. The water, with its added chemicals, flows into a mixing basin where paddles mix the chemicals into the water. From here, the water moves to a settling basin where impurities are separated out. Moving from the basin, the water is filtered through sand and gravel to screen out most of the remaining impurities. Chlorine may be added again to make sure that the water is free of harmful bacteria. Some cities also add fluoride, to help prevent tooth decay in the general population. The filtered water is held in a reservoir until it is pumped into the main supply pipes that lead to building service lines. It is eventually delivered to sinks, bathtubs, showers, dishwashers, icemakers, hoses, and any other water outlet (see *Figure 1*).

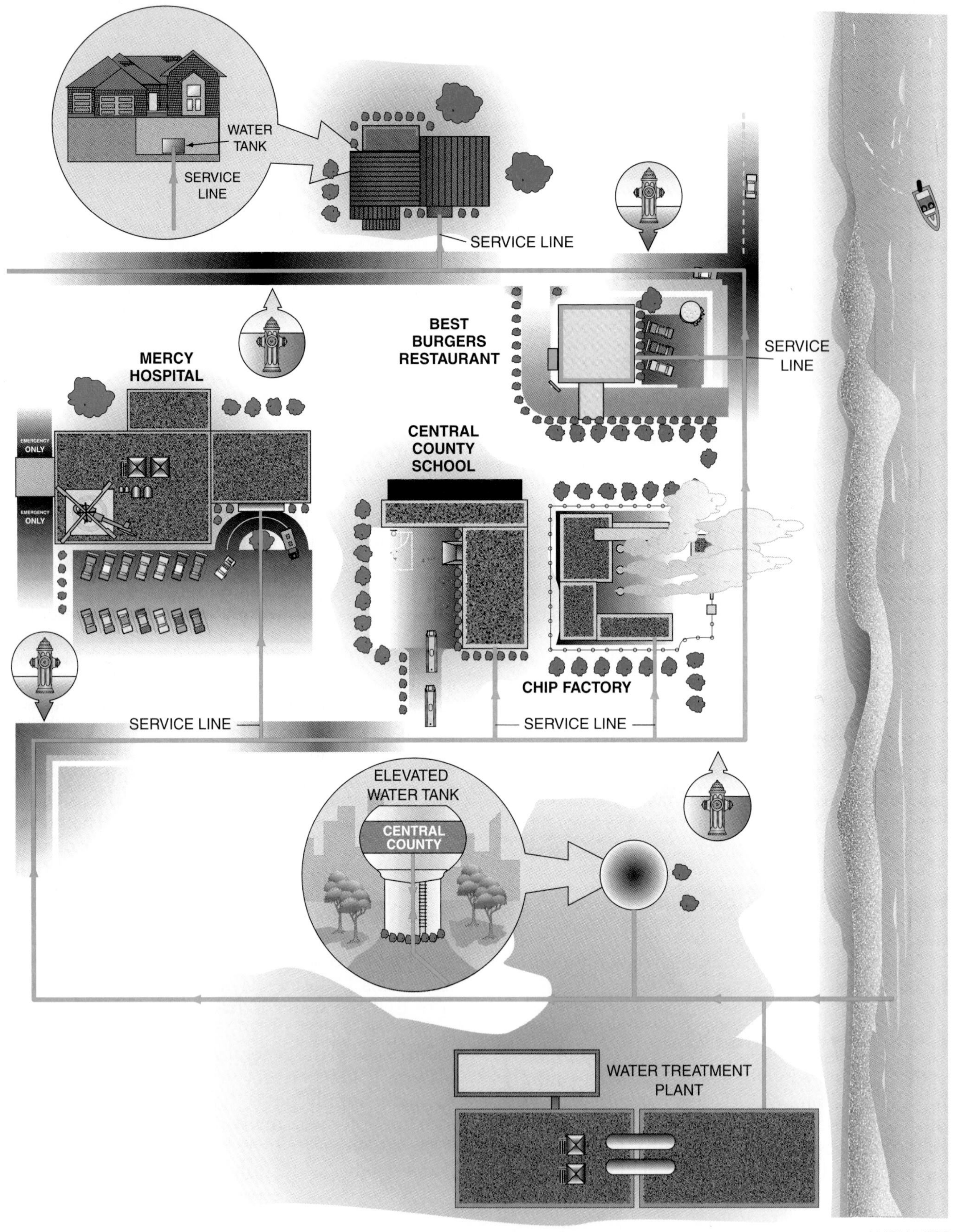

Figure 1 (1 of 2) ◆ (A) Water distribution system.

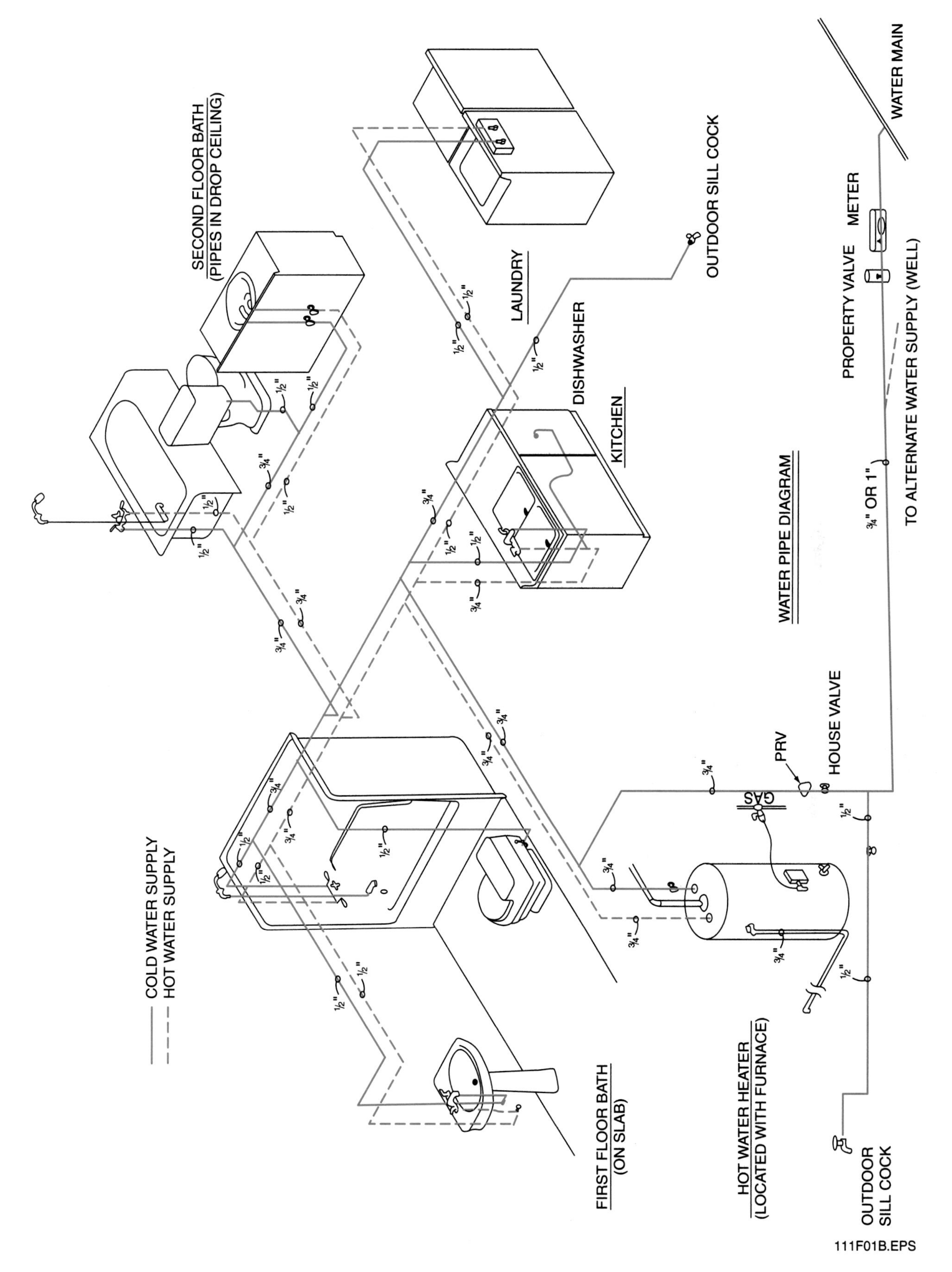

Figure 1 (2 of 2) ◆ (B) Water distribution inside a house.

Did You Know?

The water cycle begins in the oceans and seas (see *Figure 2*). Evaporation forms clouds filled with moisture. With the help of driving winds, clouds are spread across land, where they lose moisture in the form of rain or snow. Much of this **precipitation** collects and becomes streams, rivers, or lakes that return water to the sea to start the process over again. Some of this water is collected in reservoirs. The remaining water filters into the ground, where it collects under the surface in **aquifers** or other natural formations. These natural water reserves are then tapped by wells to provide water.

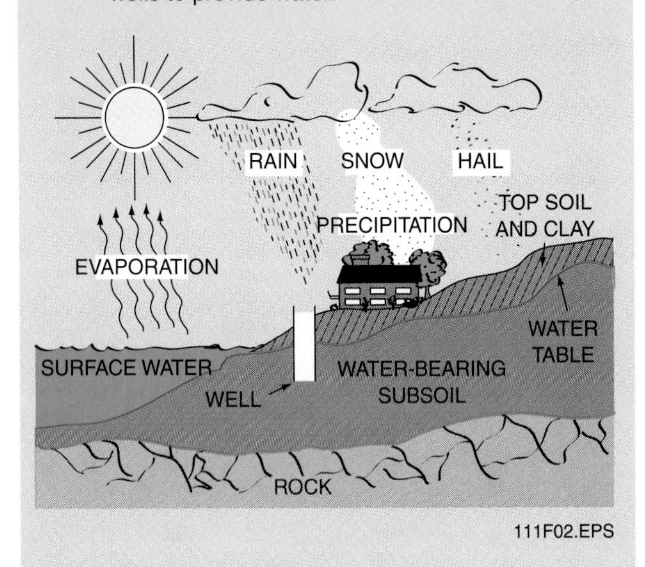

111F02.EPS

Figure 2 ◆ The water cycle.

Much of the water taken from streams and lakes is not suitable for use until it has been treated. Well water that is filtered through sand and rock is usually low in harmful bacteria. There are two types of water: potable and nonpotable. **Potable water** is safe for drinking. Nonpotable water is not safe for drinking but is used in some plumbing systems, such as irrigation systems.

2.0.0 ◆ SOURCES OF WATER

Nearly 18 percent of the U.S. population gets its water from private sources. Most of this water comes from wells. Wells are sunk (dug, driven, drilled, or bored) into the earth to extract the water (see *Figure 3*). Wells that are dug or driven are considered shallow wells. This type of well is generally used where the **water table** is within 20 to 50 feet of the earth's surface. Dug wells are easily contaminated by surface water. Driven wells are made by forcing a well point (see *Figure 4*) into the earth. These wells are only practical when the soil is fairly free of rocks.

DUG WELL / DRIVEN WELL / DRILLED WELL / BORED WELL / TOP SOIL / CLAY / PERMEABLE MATERIAL / GRAVEL AND WATER-BEARING SAND / SCREEN / WATER TABLE / ROCK

111F03.EPS

Figure 3 ◆ Types of wells.

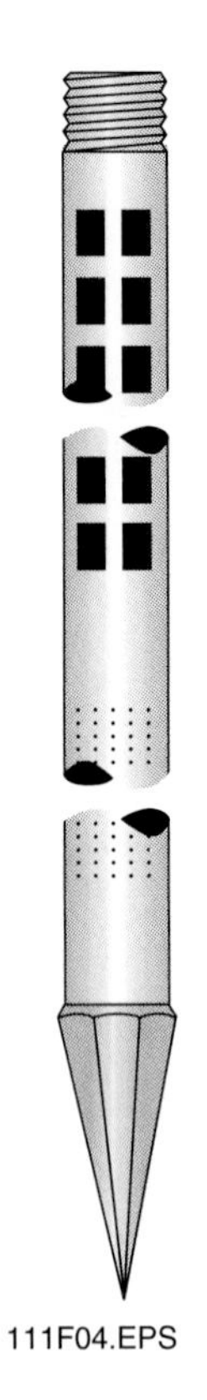

111F04.EPS

Figure 4 ◆ Well point.

Locating and Drilling a Well

Although plumbers seldom are directly involved in the well-drilling process, it is helpful to know and recommend a dependable well driller. This makes the plumber's job easier, because a good water supply and a trouble-free water supply system begin with a good well.

The first consideration before drilling or driving a well is state and local regulations. Wells may be approved for household use, domestic use, or commercial use. A local well driller is the best source of answers to the critical questions of what type of well is needed, what permits are required, and the history of water wells in the area. A variety of different aquifers (water channels) and water tables are available in various formations below the earth's surface. A history of wells drilled in a particular area provides important information necessary for the well driller to locate the best spot to drill.

Other considerations are the distance from property lines and the position of septic tanks and leachfields in relation to the well site. Most codes require at least 100 feet between the well and the leachfield. Because the plumber will make connections to both the well and the wastewater disposal, he or she must work closely with the owner, the well driller, and code officials.

Annual Testing

Wells must be tested annually to check for levels of bacteria, viruses, and microbes. These tests ensure the purity of the water. Usually, the local health department can do this testing.

Wells that are drilled are considered deep wells. They may extend hundreds of feet down through the earth. They are established using a drilling rig that penetrates subsurface materials. Bored wells are made using an earth auger (drill). Once the water table is reached and an adequate amount of water volume is determined, a **well casing** is inserted to protect the well from contamination. Casings are made of various materials and in different diameters (1½ to 2 inches) and depths, depending on where the water is beneath the surface.

After the well is established, water is pumped to the surface to a storage tank inside or near the building, where it can be treated and used by the owners. Generally, the deeper the well, the more gallons per minute can be pumped. The pump pressurizes the water in the storage tank so it provides enough water for each of the fixtures to work properly.

Another source of water is reservoirs. Most reservoirs are made by constructing dams across rivers or streams. Municipalities can then collect and store water for future use. This is particularly important in areas that receive little rainfall for part of the year. A pump is used to get the water from the reservoir to the water treatment plant.

3.0.0 ◆ WATER TREATMENT

Much of the water coming from open reservoirs, lakes, streams, and some wells is not ready for human use until it has been treated. Water must be tested to determine which problems (chemicals, **turbidity**, organic materials, contamination) are present in the water supply. Treatment removes impurities, odors, and unpleasant taste from the water. Following are some types of water treatment.

Private water treatment often consists of a water softener (see *Figure 5*). Depending on the hardness and the quality of the water, some homes may have a sediment filter or carbon filter.

Municipal water treatment is more complex because of the millions of gallons a day that are needed and the level of public safety that is required. A general process for treating water in a municipal water treatment plant includes the following steps (see *Figure 6*). The water treatment process begins when the water is pumped from a river or lake. The water is sent to the *aerators,* where dissolved carbon dioxide is removed. The aerator also removes iron and manganese by oxidizing the minerals and filtering them out of the water supply.

The aerated water is then pumped to the *clarifier,* where lime and soda ash are mixed in to create **coagulation,** which removes **precipitates** of calcium and magnesium. This process softens the water.

Carbon dioxide is injected to recarbonate the water and to stabilize and lower the acidity **(pH)** level. This water is passed through rapid sand *filters* to remove any remaining particles.

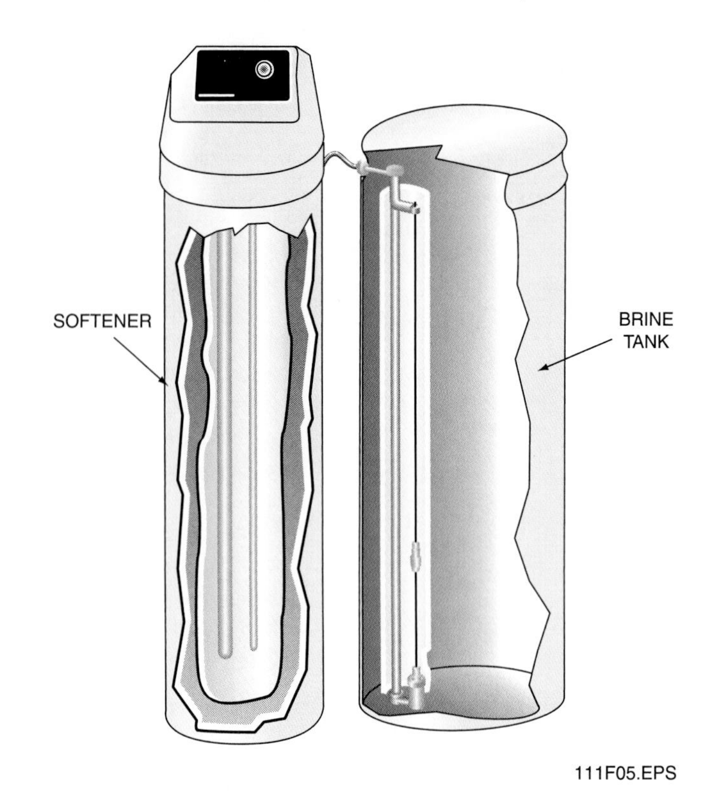

Figure 5 ◆ Water softener.

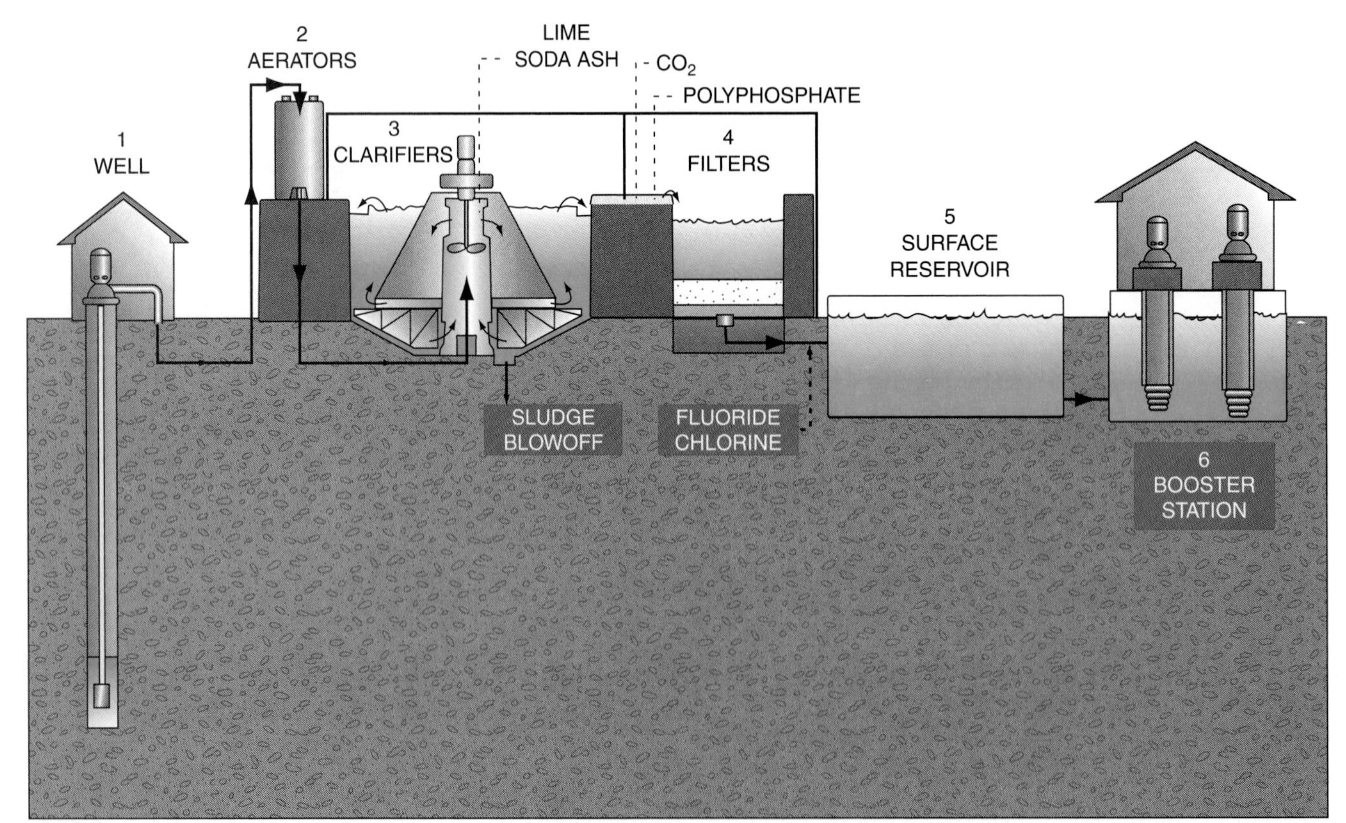

Figure 6 ◆ Municipal water treatment plant.

ON THE LEVEL

Water Pressure

The water pressure in some municipal systems may be either too low or too high for the plumbing system in a building. The plumber has ways to compensate. If the water pressure is too low, a booster system can be installed. One kind of booster has variable capacities to meet increased and decreased demand for water (see *Figure 7*). For example, it might have two pumps—one that functions at 33 percent of system capacity, and one that functions at 66 percent. When demand is low, the smaller pump operates. As demand increases, it shuts down and the larger pump starts up. At peak demand periods, both pumps operate.

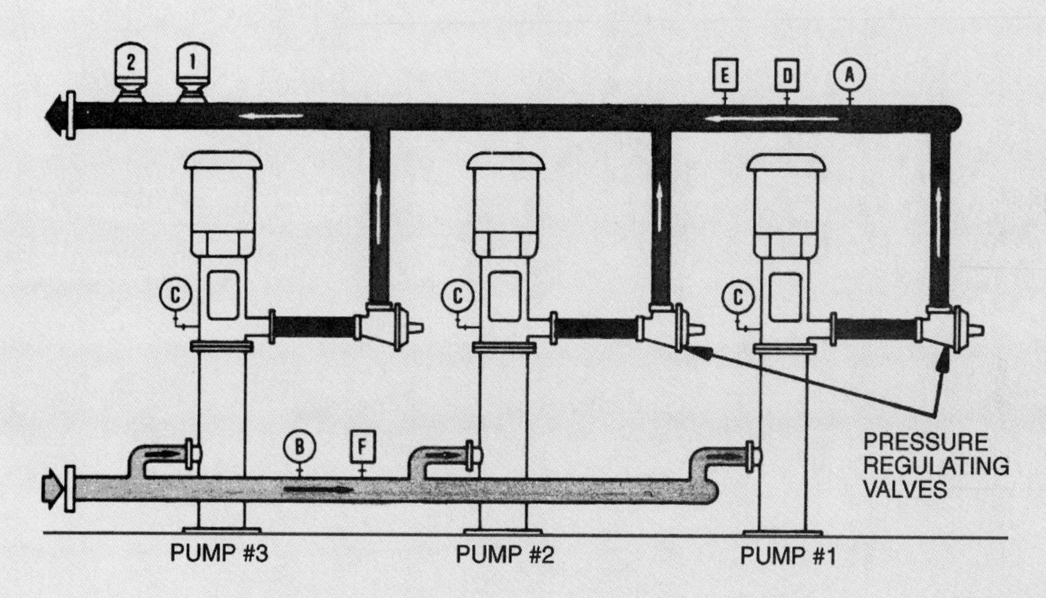

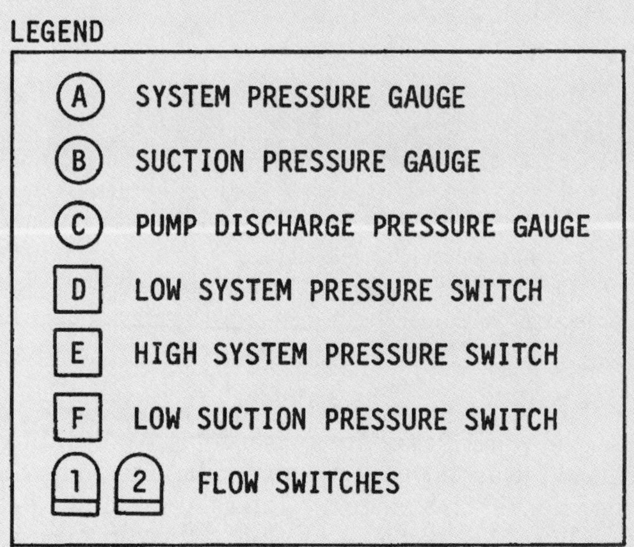

Figure 7 ◆ Variable capacity system.

On the other hand, if the water pressure in the system is too high, the plumber can use valves to lower it and to eliminate violent pressure changes (see *Figure 8*).

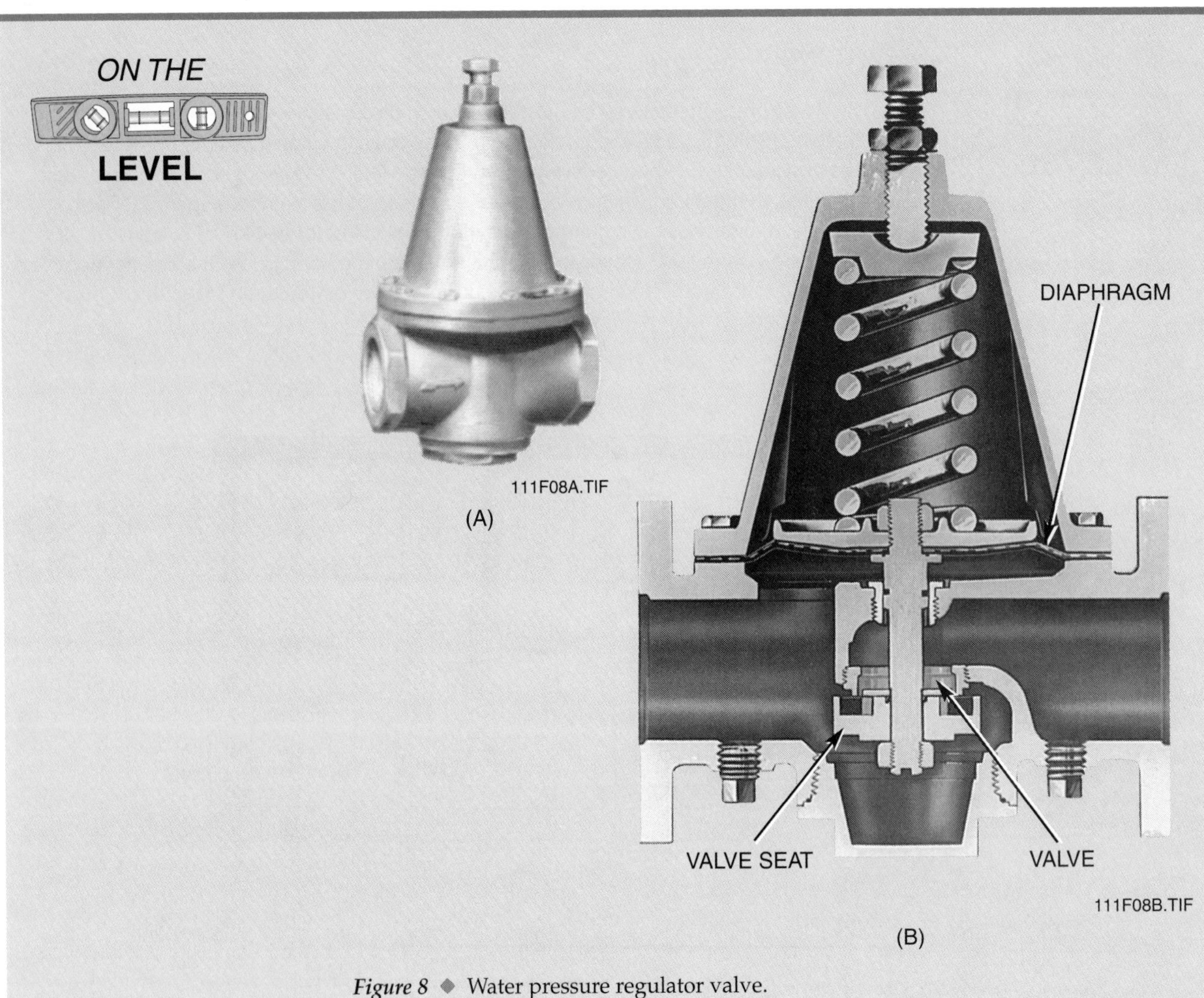

Figure 8 ◆ Water pressure regulator valve.

Valves are located according to the needs of the system. Some codes may require a pressure-reducing valve where pressure buildup at any time of the day exceeds 80 pounds per square inch (psi). A certain appliance (such as a dishwasher) may require a pressure-reducing valve because the operating pressure of the appliance is very different from the system's line pressure. Wherever they are located, the valves must be accessible for maintenance and must be protected from abuse or tampering.

After the filtering process, the water passes to the tanks, where it is disinfected using a *chlorine* solution (sodium hydrochloride). At this stage, *fluoride* is usually added to reduce the chance of tooth decay. This treated water is then stored in reservoirs until needed.

After water leaves the treatment plant, it is moved through pipes called water mains. The water mains usually run under the streets and serve many buildings. Permits are required to make connections to the water mains. In some areas, only municipal workers are authorized to make these connections. In other areas, licensed contractors or plumbers install the connections. The connections bring water from the main to the individual buildings through the building water service line.

Review Questions

Section 3.0.0

1. Most private water supply treatment consists of a(n)_____.
   a. fluoride additive
   b. water softener
   c. aerator
   d. clarifier

2. A water pressure _____ system is required if the water pressure available from the municipal main is inadequate.
   a. pump
   b. check valve
   c. booster
   d. regulator

3. One reason that municipal water treatment is more complex than private treatment is that _____.
   a. millions of gallons of water per day are needed
   b. it requires more water softener
   c. permits are required to make the connections
   d. fluoride must be added to reduce tooth decay

4. The pipes that carry municipal water from the treatment plant to buildings are called _____.
   a. sewers
   b. aerators
   c. connections
   d. water mains

5. After municipal water is aerated, it is pumped into a clarifier to be _____.
   a. recarbonated
   b. softened
   c. oxidized
   d. filtered

4.0.0 ◆ SUPPLY AND DISTRIBUTION

The water distribution system moves the water from its source to the building or structure where it is needed. This section traces that movement in both private wells and municipal water systems and provides information about the pipe materials, pumps, and valves that make this system work smoothly.

4.1.0 Materials

A pipe (the water service line for homes or the municipal water main for other buildings) is made of one of three popular materials: copper, plastic, or steel. Cost and life expectancy determine the choice of pipe materials. Other considerations include corrosion resistance, chance of freezing weather, working pressure, local plumbing codes, and ease of installation.

Galvanized steel pipe generally has a life expectancy of 30 years. Copper can last more than twice that long. Plastic pipe is considered the best for longevity. But plastics can absorb chemicals, such as insecticides, from the ground. In many cases, the life expectancy of galvanized and copper pipe depends on soil and water conditions. Acidic soils tend to act on galvanized pipe. Soil with carbon dioxide tends to limit the life of copper pipe. Hydrogen sulfide also has harmful effects on copper pipe. The corrosion caused by this gas breaks down the walls of metal pipe and causes the metal to thin or pit.

Another consideration in selecting the water supply pipe is the possibility of **galvanic corrosion,** which can develop when you join pipes of different materials, such as copper and galvanized steel pipe. Plumbers must take steps to reduce the likelihood of this happening. You will learn more about corrosion prevention as you advance in your training.

4.2.0 Service Line from a Private Water Supply

In a private water supply, the service line runs from the well to the house, and the pump can be located at the well or at the house (see *Figure 9*).

ON THE

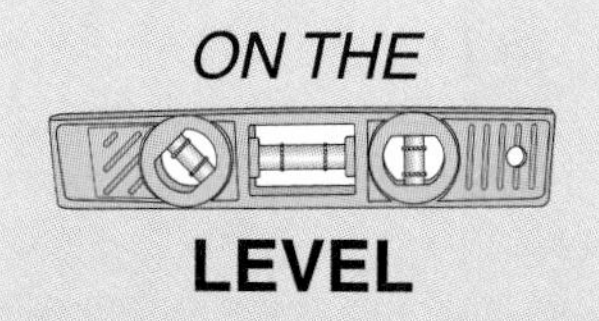

LEVEL

Copper pipe can be type K, L, or M, although type M is allowed in the three major codes (Standard Plumbing Code, Uniform Plumbing Code, and National Standard Plumbing Code). A minimum of type L usually is recommended, but plumbers must always check local codes.

Plastic water service lines must be rated a minimum of 160 psi at 73°F. Two common types of plastic used for water service lines are PVC (polyvinyl chloride) Schedule 40 and PE (polyethylene). Another type of plastic, PB (polybutylene) for cold water underground piping was affectionately known as Big Blue. In the 1980s and early 1990s, Big Blue was used for many service lines. Because of its failure rate, which led to many lawsuits, this type of pipe is no longer being produced.

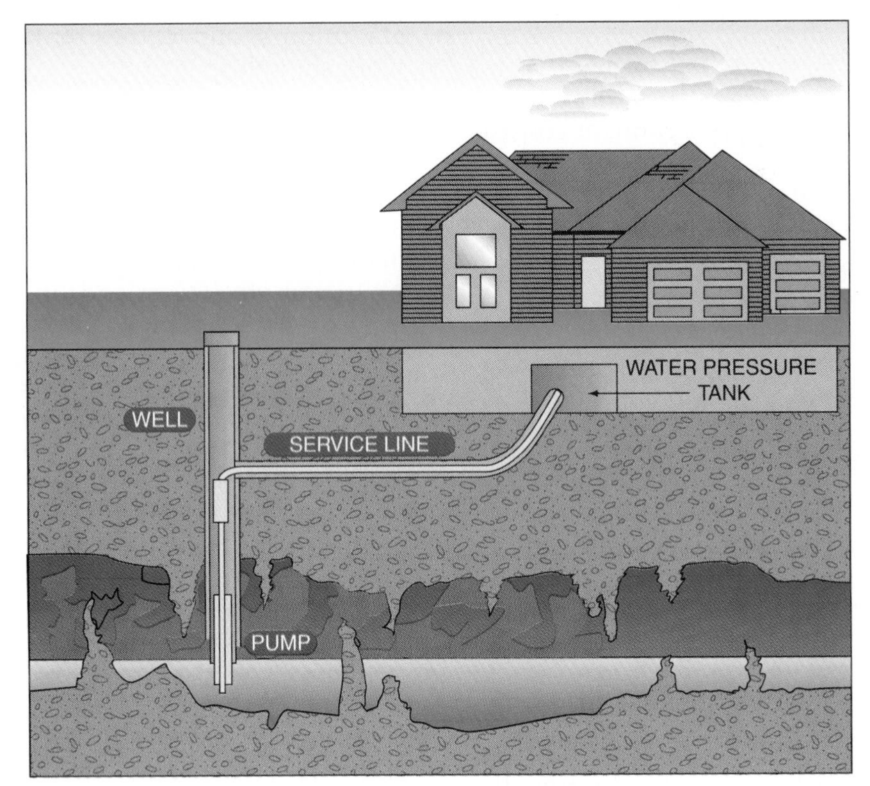

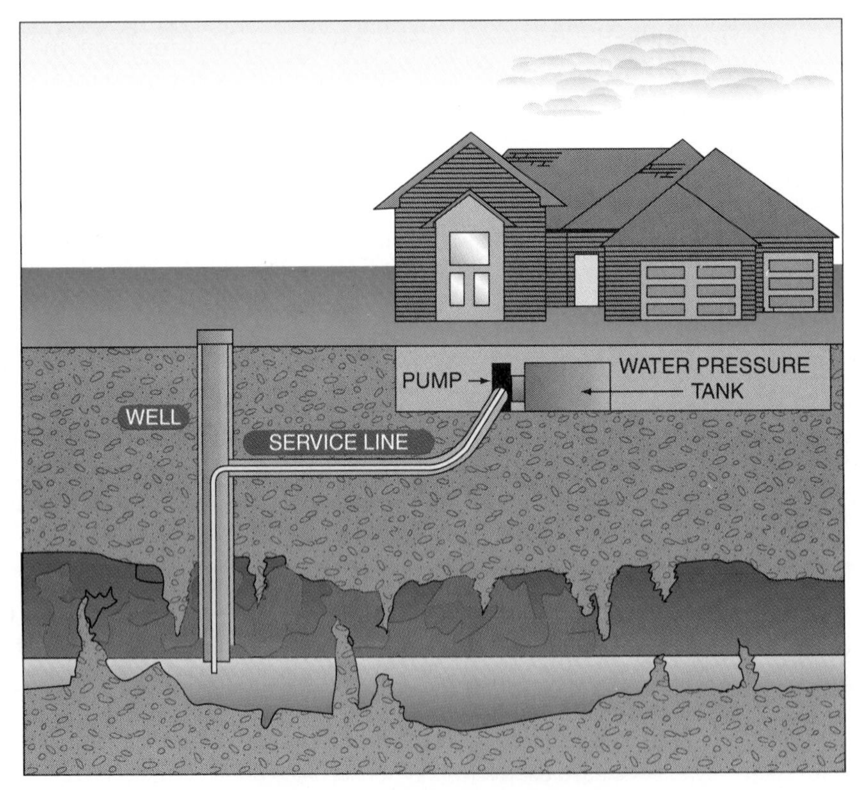

Figure 9 ◆ Private service line.

The installation of a private water supply system usually requires a well, although it may mean getting water from a spring or a cistern. A private water source is required when access to municipal water supplies is not available, such as in rural settings.

It may be the plumber's responsibility to set (install) the pump and complete the water service to the structure, or a certified well driller may do it. In some cases, a pump installer is required to set the pump, provide electrical service, and complete the water service to the pressure tank. There is a great deal of overlap in these responsibilities. Each state and plumbing code will have specific requirements for water service from a well. In most cases, a water sample must be drawn and tested by appropriate authorities after the well is **purged** and before the final tie-in.

Two kinds of pumps are used to draw water from a well: a jet pump for either a shallow or deep well and a submersible pump for deep wells. Shallow-well jet pumps are **self-priming** pumps. They produce a vacuum that allows normal air pressure to push water up the **suction pipe** and into a structure (see *Figure 10*). A deep-well jet pump uses higher than normal atmospheric pressure to force the water into the suction chamber. It must be submerged at least 5 feet below the water line (see *Figure 11*).

A submersible pump (see *Figure 12*) operates from inside the well and is installed below the water line. A submersible pump uses a **centrifugal pump** and an electric motor to drive the water out of the well. It pumps the water up a single pipe to the water supply storage. This kind of pump is popular for use in deep wells. It is more efficient and easier to install than a jet pump and is usually the first choice for well installation where physical conditions permit its use.

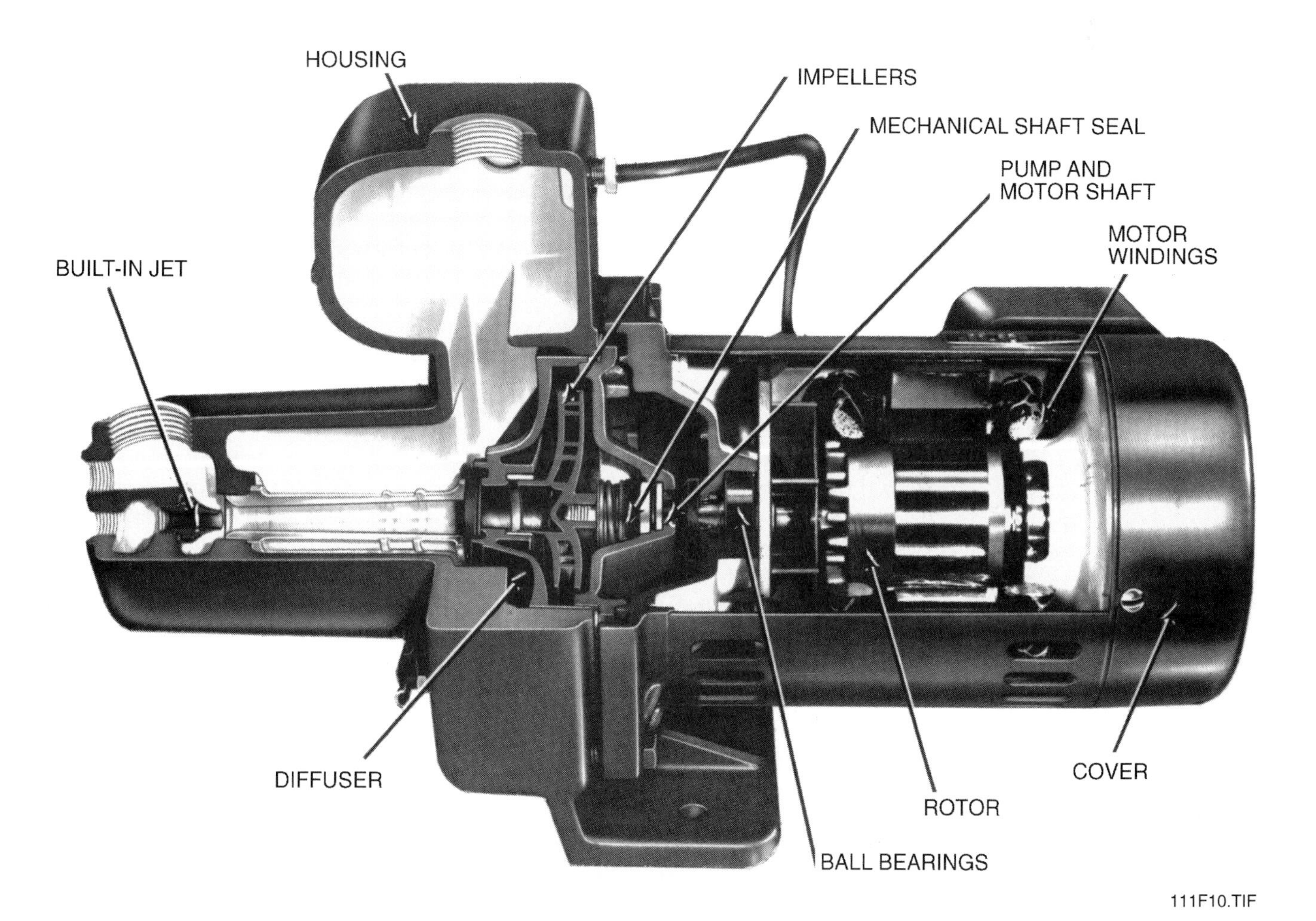

Figure 10 ◆ Shallow-well jet pump.

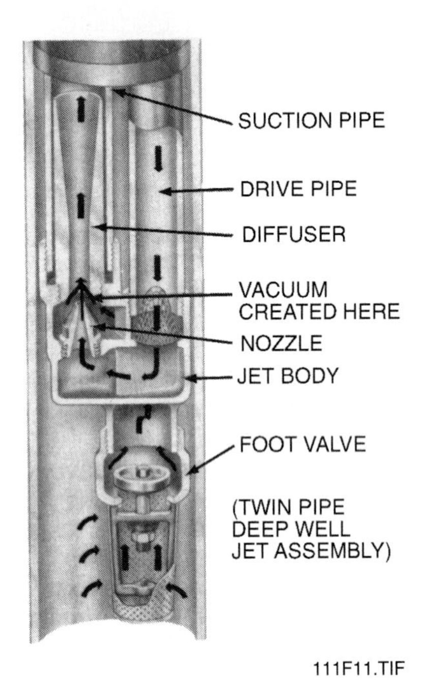

111F11.TIF

Figure 11 ◆ Deep-well jet pump.

111F12.TIF

Figure 12 ◆ Submersible pump.

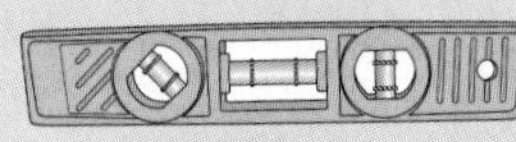

ON THE LEVEL

Selecting the Pump

The choice of pump is critical to the success of the system. Decisions are based on information from the well driller's log regarding the depth of the well, the level where water was first found, and the strata (layers) of earth that were penetrated. Knowing the well depth is the most important requirement for selecting a pump. The type and size of the pump are based on the following considerations:

- Total vertical height that the water must be pumped (see *Figure 13*)
- Rate at which the water is pumped
- Total distance (horizontal, vertical, angular) that the water must travel

The vertical distance that water must travel matters because as the water travels upward, it gets heavier. The higher it climbs, the heavier it gets, and its weight slows it down. The correct size pump must be selected to overcome this and bring the water to its usable level with enough service pressure.

The rate at which water is transported also affects the pump selection. As water speed increases, so does friction (resistance), which causes the water to lose pressure. In addition to flow rate, the number of fittings in the line increases the **turbulence,** which also causes friction-related losses. The pump size must compensate for pressure loss due to friction.

The total distance (horizontal, vertical, angular) that water is pumped is also an important consideration. The greater this total distance, the greater the friction-related losses, and the larger the pump must be.

Once these factors are considered, the plumber must determine whether the well requires a shallow- or a deep-well pump. A shallow-well pump uses atmospheric pressure to force the water up the suction pipe. These pumps usually are limited to a depth of 25 feet. When shallow wells are dug at high altitudes, the air pressure is lower, making it less likely that a shallow-well pump will do the job.

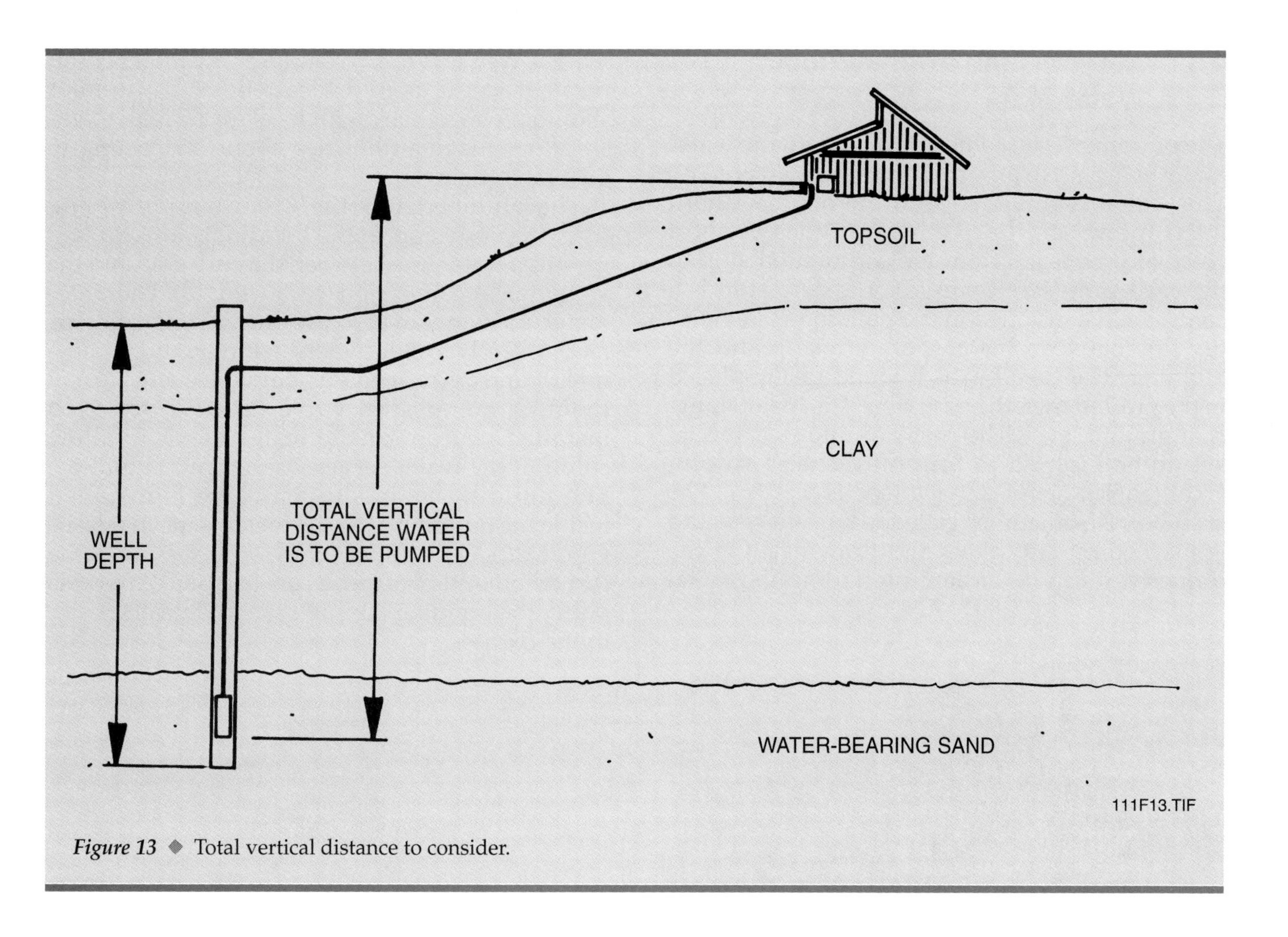

Figure 13 ◆ Total vertical distance to consider.

Review Questions

Section 4.2.0

1. A pump that uses an electric motor to drive water up a well's suction pipe is a _____ pump.
   a. shallow-well
   b. centrifugal
   c. jet
   d. deep-well
2. The most important requirement for selecting a pump is _____.
   a. knowing the depth of the well
   b. measuring the distance the water must travel
   c. estimating the weight of the water
   d. gauging centrifugal force
3. The greater the water speed, the greater the loss of pressure due to _____.
   a. pipe corrosion
   b. self-priming
   c. friction
   d. atmospheric conditions
4. The correct size and type of pump is selected based on the angular, horizontal, and vertical _____ the water must travel.
   a. speed
   b. slope
   c. height
   d. distance
5. The _____ produces a vacuum that allows normal air pressure to push the water up the suction pipe and into the building.
   a. submersible pump
   b. deep-well jet pump
   c. shallow-well jet pump
   d. centrifugal pump

4.3.0 Service Line from a Public Water Main

The building service line is tapped into the water main using a **corporation stop** (see *Figure 14*). This valve is threaded into the main without interrupting service to other locations. A short piece of pipe leads from the corporation stop to the **curb stop** (see *Figure 15*). The curb stop is a control valve installed in the building water supply line between the corporation stop and the building. The **curb box** is a round casing placed in the ground over the curb stop. The top opening is at ground level. The curb box allows a special key to be inserted to turn off the curb stop in emergencies or for service (see *Figure 16*). The curb box is sometimes called a **buffalo box** and is marked on city plans and drawings. On the property, it can be located by finding the metal marker (see *Figure 17*). Because of differences in local procedures and local codes, a plumber should always check with local municipal, township, or village authorities about connecting to the water main.

The plumber runs the water supply piping from the curb stop to the building's interior and installs a water line between the curb stop and the **water meter** connection (see *Figure 18*). The water meter measures water use (in gallons or cubic feet). Usually, water meters belong to the municipality, and a city or county employee must install them. A meter stop valve (see *Figure 19*) is installed in front of the water meter to allow cutoff of the water service to the entire building. The building main shutoff valve is installed inside the building after the meter is installed (see *Figure 20*). This valve allows the service to be turned off and on during construction. After this point, the building supply system delivers water to the fixtures.

111F14.TIF

Figure 14 ◆ Corporation stop.

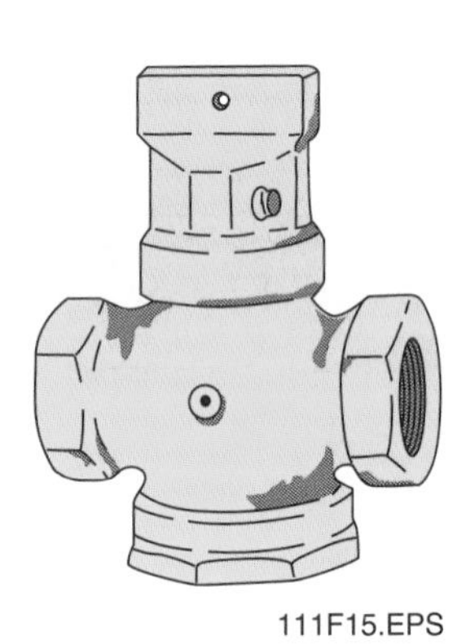

111F15.EPS

Figure 15 ◆ Curb stop.

Frost Protection

Frost protection is an important consideration in many parts of the country. Frost depths called **frost lines** are fairly standard for different areas. The plumber must install the water service line below the frost line to make sure that the piping is not affected by freezing and thawing temperatures in the earth. If the water supply line enters the building above grade, it must be insulated to prevent freezing.

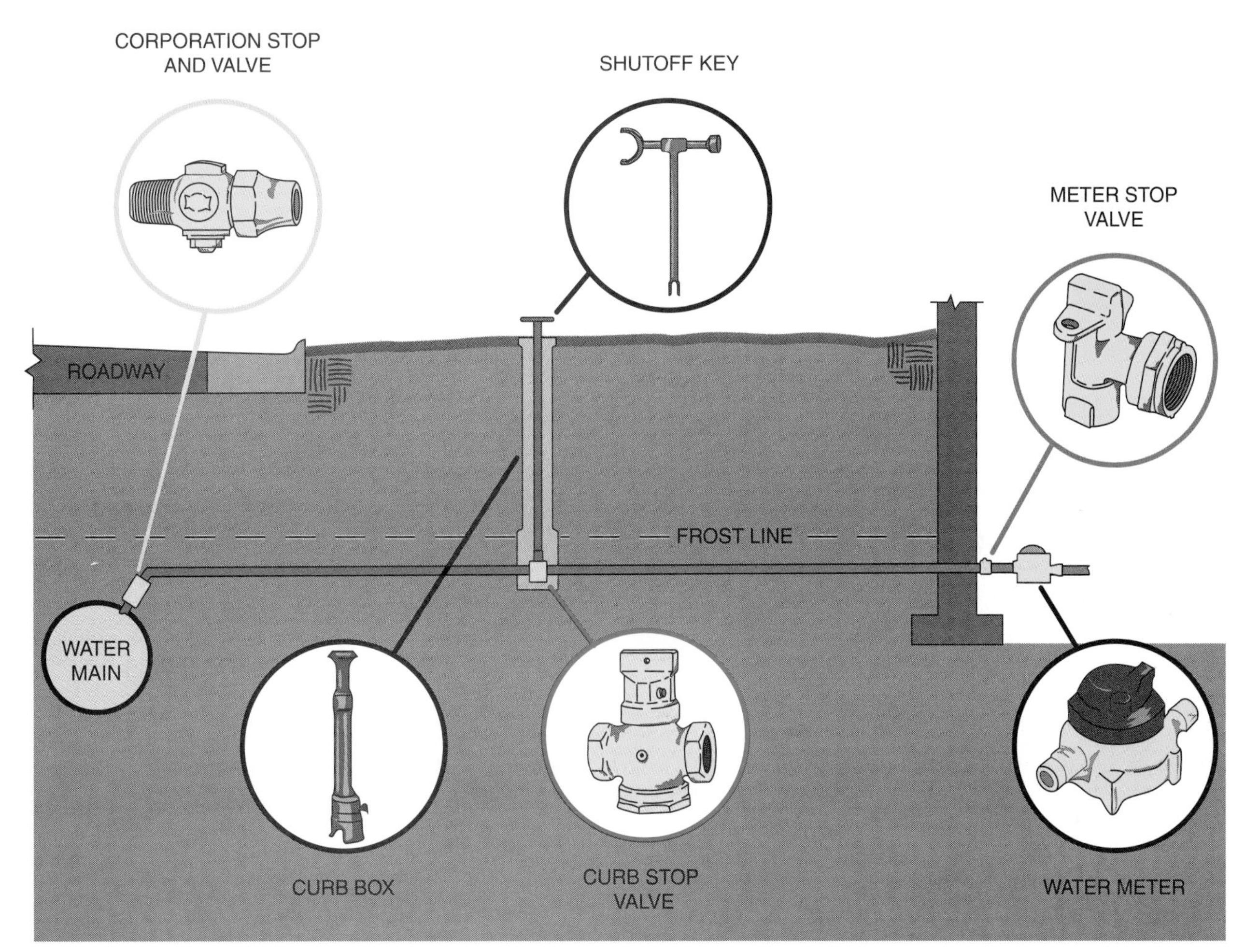

Figure 16 ◆ Municipal water supply connection.

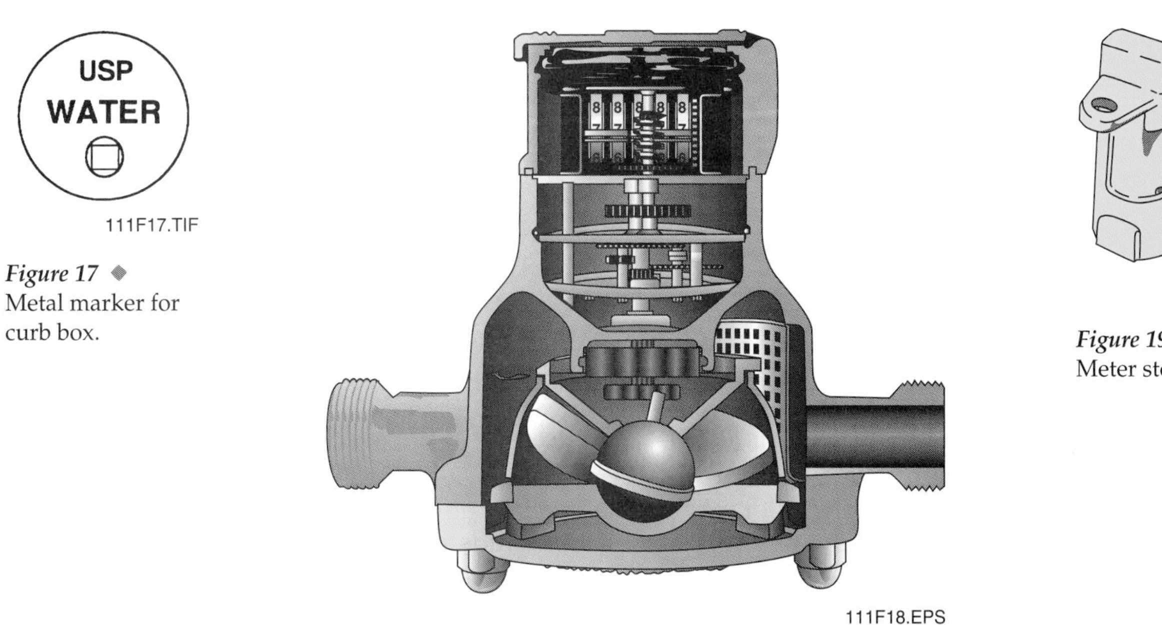

Figure 17 ◆ Metal marker for curb box.

Figure 18 ◆ Water meter.

111F19.EPS

Figure 19 ◆ Meter stop valve.

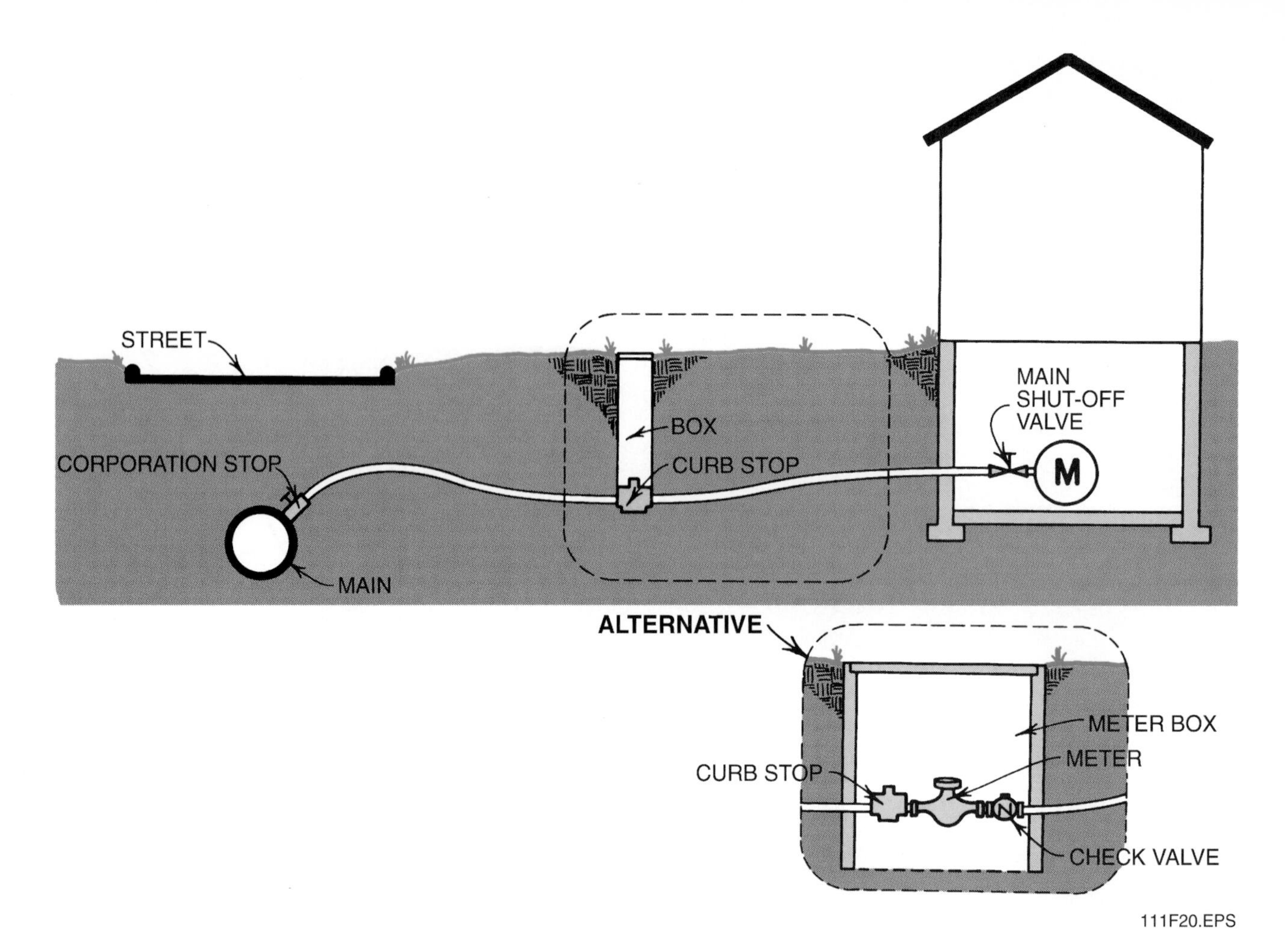

Figure 20 ◆ Main shutoff valve.

Review Questions

Section 4.3.0

1. The building service line is tapped into the main using a _____.
   a. curb stop
   b. curb box
   c. corporation stop
   d. check valve

2. The _____ is a round casing placed in the ground over the curb stop.
   a. buffalo box
   b. water main
   c. water meter
   d. meter box

3. The _____ measures water use in gallons or cubic feet.
   a. check valve
   b. pressure valve
   c. water meter
   d. corporation stop

4. Usually the water meter is installed by _____.
   a. the person who installs the curb box
   b. a plumbing contractor
   c. a meter reader
   d. a municipal employee

5. In colder climates pipes above the _____ must be insulated.
   a. frost lines
   b. thaw lines
   c. tree lines
   d. meter lines

5.0.0 ◆ BUILDING SUPPLY

Backflow prevention is required in many plumbing installations to keep contaminated water or other liquids from flowing back into the potable water system. This reverse flow occurs as a result of **cross-connection,** which is a direct link between a contaminated liquid and a potable water supply. This condition may occur as a result of an improper or altered plumbing hookup or when a garden hose is left in a pool of contaminated water (see *Figure 21*).

These conditions do not in themselves cause a significant hazard, but they do create the potential for a serious problem. If a break in the water main or other vacuum potential exists somewhere in a water line, the water could be drawn backward through the water supply line. If a hose were left in a pool of wastewater, that waste could be siphoned back into the system. Codes require installation of backflow preventers in every new structure.

Basic backflow prevention devices are designed to safeguard against dangerous cross-connections. In a bathroom sink, for example, the faucet line is set above the sink's flood level rim (the point at which water begins to overflow the top of the sink). In addition, the top drain returns water to the bottom drain before the water can rise to the faucet. The **vacuum breaker** in a **hose bibb** connection also acts as a backflow preventer by interrupting the vacuum that could cause a backflow (see *Figure 22*). Each type of backflow preventer has specific applications and limitations, which you'll learn more about as you continue in your training. These factors must be considered in any type of system design using backflow preventers.

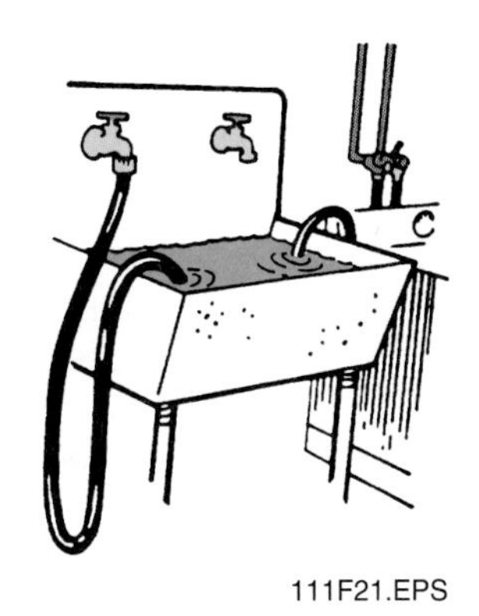

Figure 21 ◆ Cross-connection.

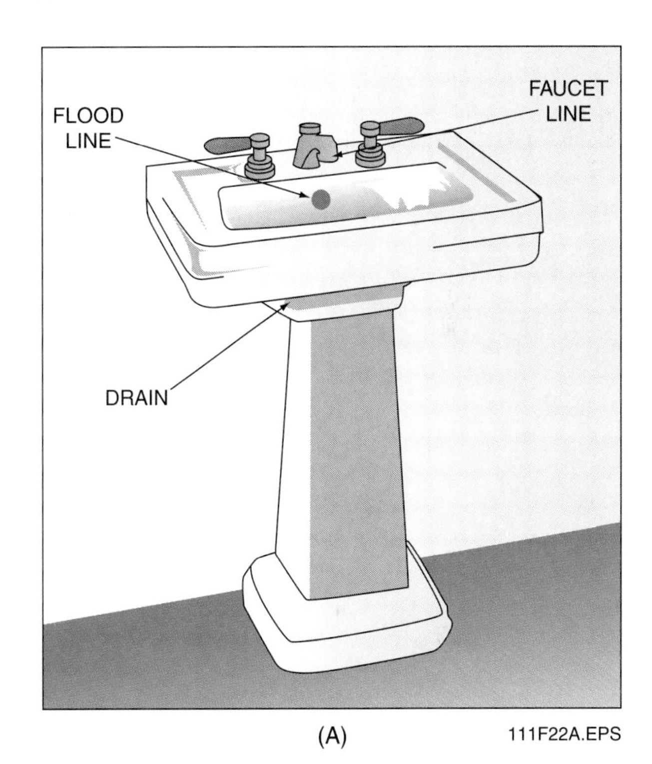

(A) 111F22A.EPS

(B)

Figure 22 ◆ Backflow prevention devices. (A) Faucet above the flood line. (B) Hose bibb connection vacuum breaker.

5.1.0 Valves

A valve or faucet regulates the flow of water in a water distribution system. It may be used to turn water service on and off, act as a throttling device that controls the rate of water flow, regulate the pressure, or prevent a reversal of flow through a line.

5.1.1 Valve Terminology

Plumbers must understand the following terms to discuss valves.

- **Trim** refers to the parts of a valve that receive the most wear and are, consequently, replaceable. Trim includes the stem, disc, seat ring, disc holder or guide, wedge, and bushings.
- **Straight-through flow** describes a flow that is not restricted as it passes through a valve. The element that closes the valve is pulled back entirely clear of the passage.
- **Full flow** describes the relative flow capacities of various valves.

ON THE LEVEL

Cross-Connection

The deadly outbreak of Legionnaires' disease at a hotel in Philadelphia in 1976 was linked to a cross-connection problem. The bacterial infection may have entered the ventilation system as contaminated moisture. As a result of breathing the contaminated hotel air, 34 people staying at the hotel died and 221 more became ill. In 1982, a microbiologist identified the source of the bacterium, which was contaminated water in the air-conditioning system. Water with this bacterium may come from lakes and rivers or from commercial water systems such as air-conditioning cooling towers. The bacterium is most often found in the water systems of large buildings.

In the fall of 1996, the Centers for Disease Control and Prevention identified another outbreak of the disease that was caused by contaminated water sources at a home improvement center in Virginia. Samples were collected from a whirlpool spa basin, spa filters, a greenhouse sprinkler system, a decorative fish pond, potable water fountains, urinals, and hot and cold water taps in the store's restrooms. The bacterium showed up in one of the spa filters. Several people were hospitalized, but no one died.

- **Throttled flow** describes a flow rate that can be controlled through the valve. Not all types of valves are suitable for throttled flow.

5.1.2 Types of Valves

The following are common types of valves:

- **Gate valves**
- **Globe valves**
- **Angle valves**
- **Ball valves**
- **Check valves**
- **Pressure regulator valves**
- **Supply stop valves**
- **Temperature and pressure (T and P) relief valves**

These valves are used in different parts of the water distribution system, as shown in *Figure 23*.

A gate valve is a valve in which the flow is controlled by moving a *gate* or disc that slides in machined grooves at right angles to the flow. The gate is moved by the action of the threaded stem on the control handle (see *Figure 24*).

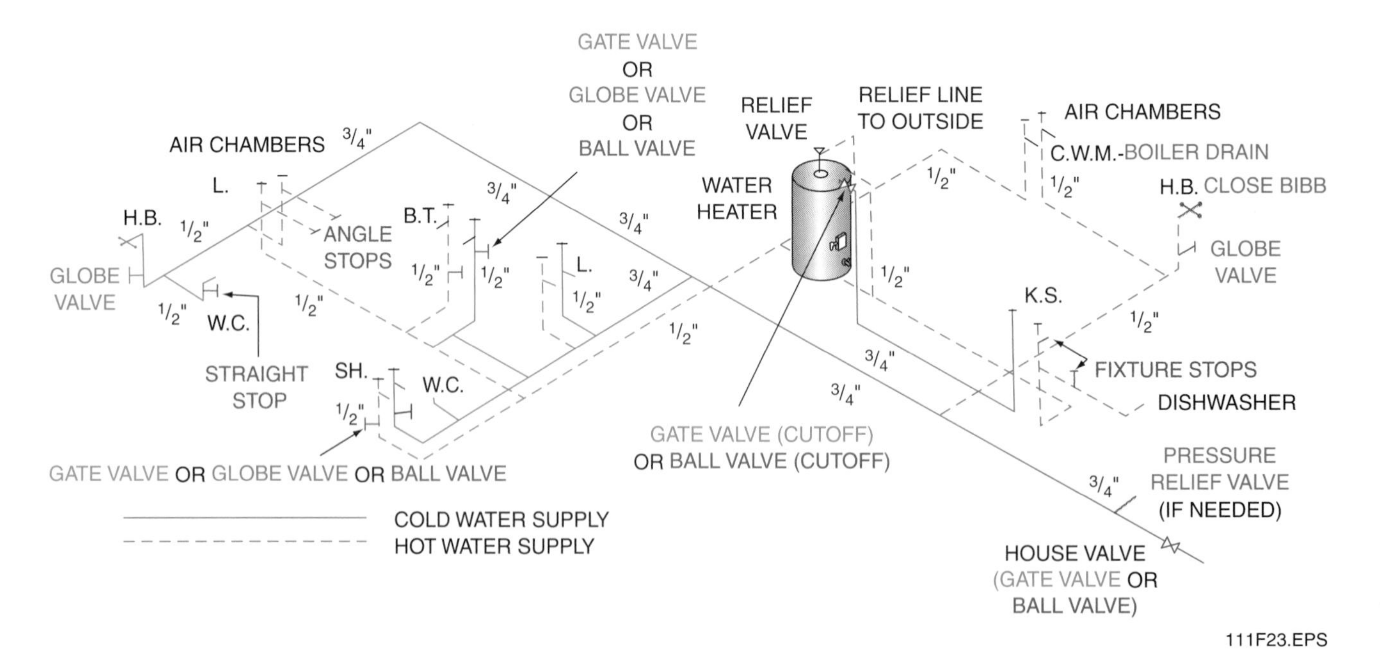

Figure 23 ◆ Valves in the distribution system.

Gate valves are best suited for main supply lines and pump lines. They provide an unobstructed passageway when the gate is fully opened. They can be used on lines containing steam, water, gas, oil, and air.

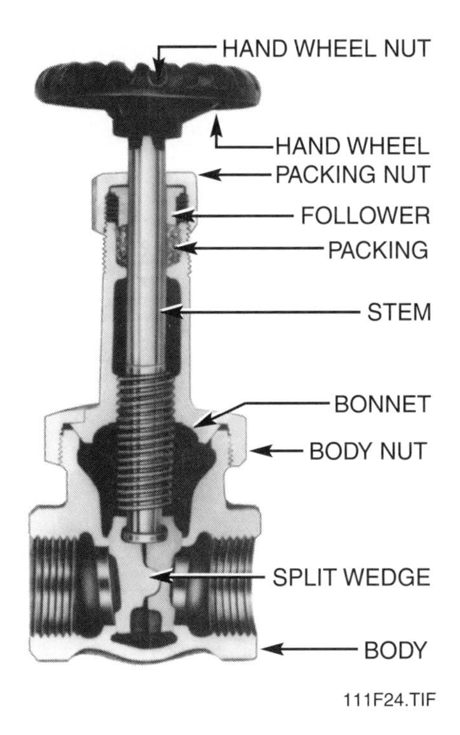

Figure 24 ◆ Gate valve.

A globe valve (see *Figure 25*) is a valve in which the flow is controlled by moving a circular disc against a metal seat that surrounds the flow opening. The disc is forced onto the seat or withdrawn from it by screw action as the handle is turned.

Globe valves are recommended for general service on lines containing steam, water, gas, and oil, where frequent operation and close control of flow is required. Inside the globe-shaped body of the valve is a partition. This partition closes off the inlet side of the valve from the outlet side, except for a circular opening called the *valve seat*. The upper side of the valve seat is ground smooth so that a proper and complete seal can be made when the valve is in the closed position. Turning the handle clockwise until the valve stem firmly seats the washer or disc between the valve stem and the valve seat closes the valve. This stops the flow of gas or liquid.

The angle valve (see *Figure 26*) is similar to the globe valve, but it can serve as both a valve and a 90-degree elbow. Because flow changes direction only twice through an angle valve, the angle valve is less resistant to flow than the globe valve, in which flow must change direction three times. Angle valves are available with conventional, plug-type, or composition discs.

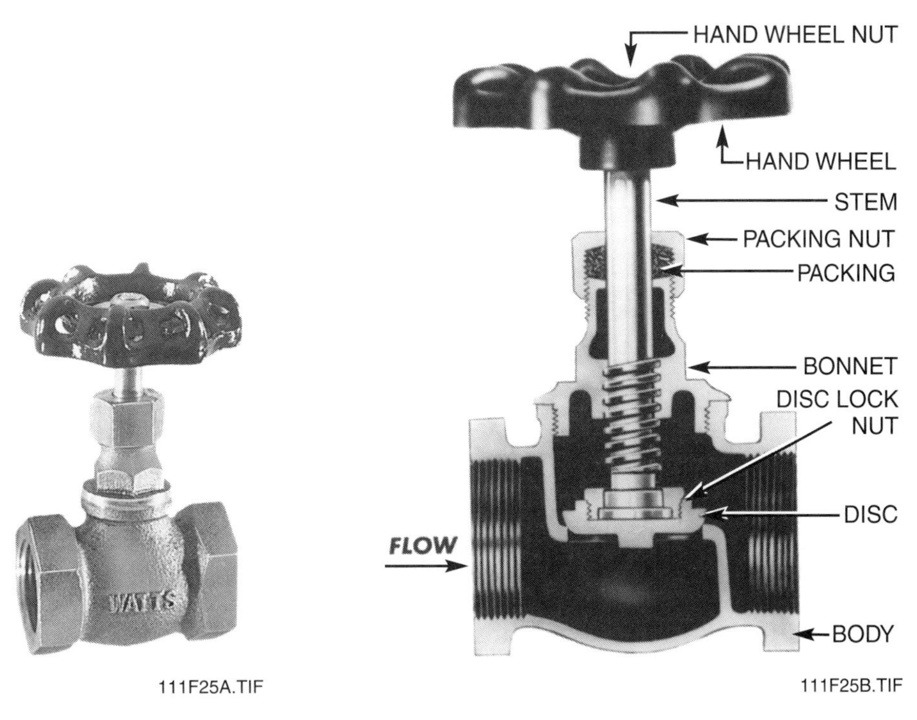

Figure 25 ◆ Globe valve.

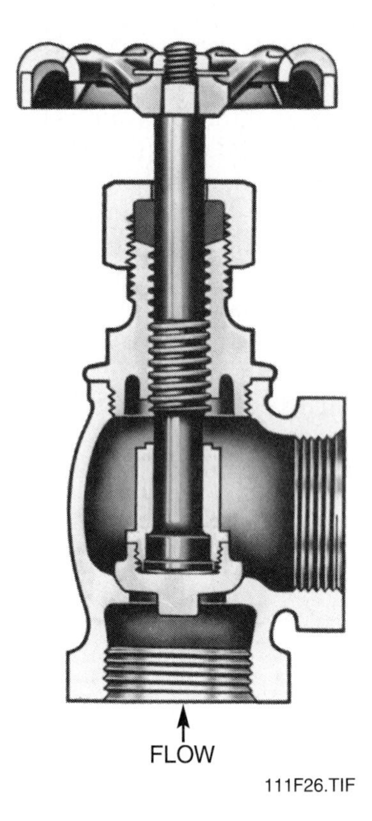

Figure 26 ◆ Angle valve.

The ball valve (see *Figure 27*) is used to control the flow of gases and liquids. It is installed in piping systems where quick shutoffs for in-line maintenance may be necessary, or in lines used for mixing various liquids and gases. The ball part of the valve is rotated into the open or closed position by a handle located on the outside of the valve body. These valves allow quick action in controlling the flow in piping systems.

A check valve is used to prevent reversal of flow in a piping system. Pressure in the line keeps the valve open. The valve is automatically closed by the reversal of flow or by the weight of the disc mechanism.

Three types of check valves are available: the ball-check valve, the swing-check valve, and the lift-check valve.

The ball-check valve (see *Figure 28*) allows one-way flow in water supply or drainage lines and can be used with extremely low back pressure.

The swing-check valve (see *Figure 29*) has a low flow resistance that makes it well suited for lines containing liquids or gases with low to moderate pressure. The swing-check valve is available in up to four different types, depending on the manufacturer.

The lift-check valve (see *Figure 30*) can be used for gas, water, steam, or air. It is recommended for

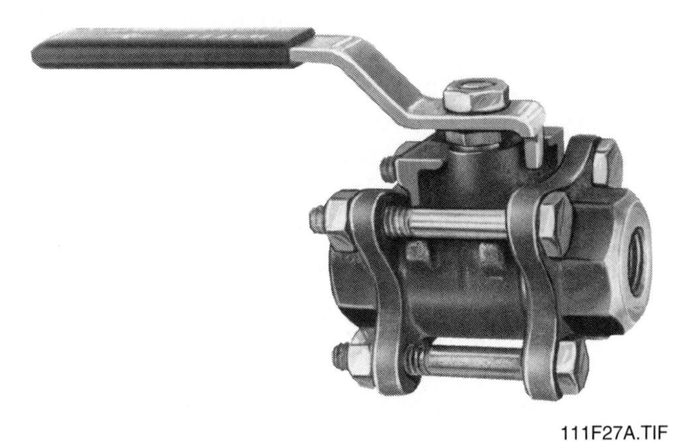

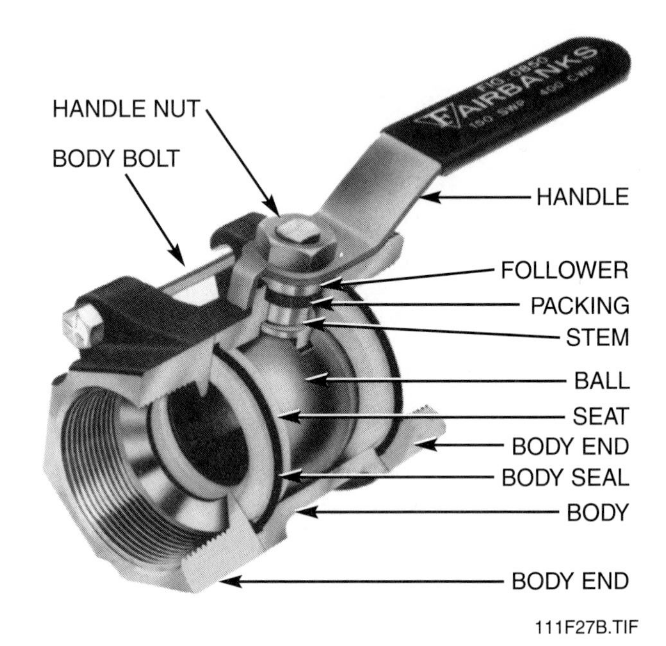

Figure 27 ◆ Ball valve.

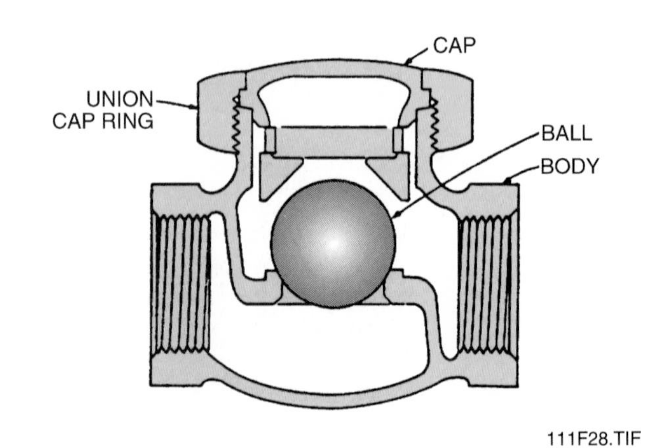

Figure 28 ◆ Ball-check valve.

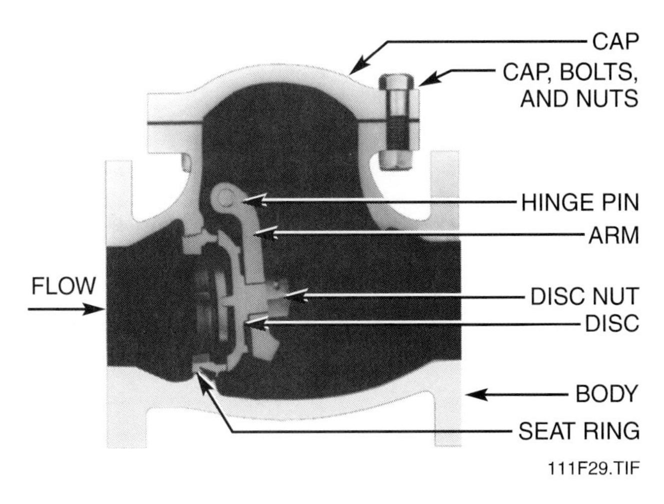

Figure 29 ◆ Swing-check valve.

lines that have frequent fluctuations in flow. These valves are available in either horizontal or vertical styles. The horizontal type has an integral construction similar to the globe valve, while the vertical lift-check valve has a straight-through flow.

The pressure regulator valve (see *Figure 31*) is used to reduce water pressure in a building. The valve is activated by changes in pressure within the system. As the pressure changes, a spring (located in the dome of the valve) acts on the diaphragm to move the valve up or down. The valve is in the open position when the valve is pressed down away from the valve seat, and it remains open until the pressure in the building reaches a set level. The valve then closes and remains closed until the pressure in the building begins to drop.

Steam safety valves (see *Figure 32*) are used with gases, including air and steam. Their design always includes a chamber that uses the force the gas creates when it expands to quickly open or close the valve. The difference between the opening and closing pressures is called *blowdown.*

Temperature and pressure (T and P) relief valves (see *Figure 33*) are normally used for liquid service, although safety valves also may be used. Ordinarily, **pressure relief valves** do not have a chamber or a regulator ring for varying or adjusting blowdown, so they operate with a relatively lazy motion. As pressure increases, they slowly open; as pressure decreases, they slowly close.

Supply stop valves, or supply valves, are commonly used to disconnect the hot or cold water supply to water closets and sinks. These valves make it easy to control the water connection at an individual device for repair work. They are available in either right angle (see *Figure 34*) or straight design and are usually chrome plated to look attractive.

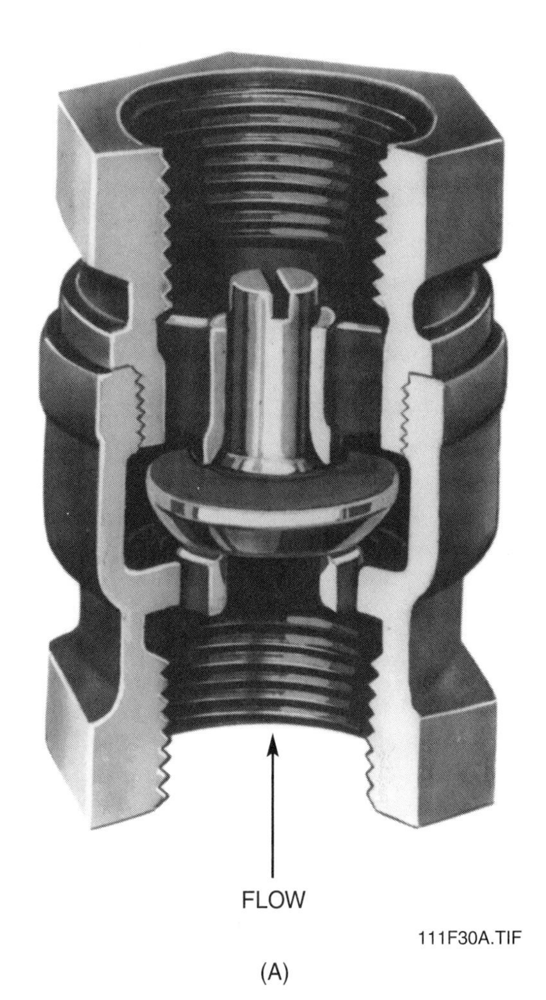

(A)

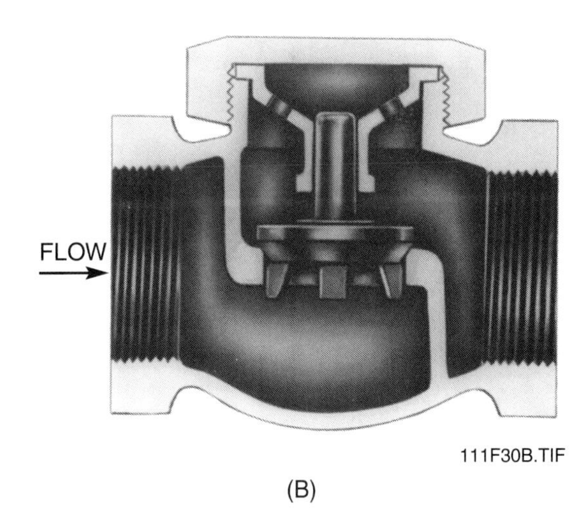

(B)

Figure 30 ◆ Lift-check valve.

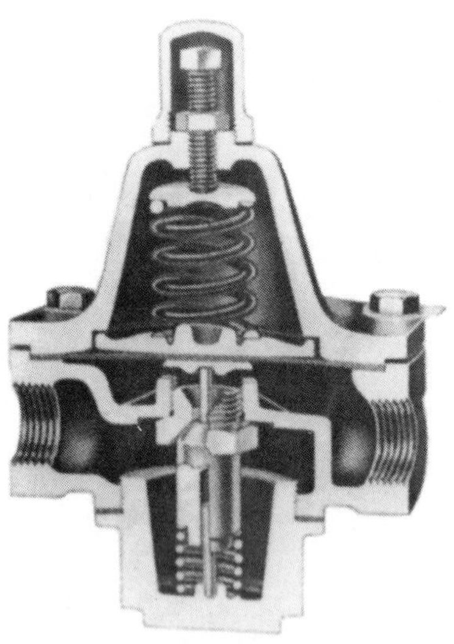

Figure 31 ◆ Pressure regulator valve.

Figure 32 ◆ Steam safety valve.

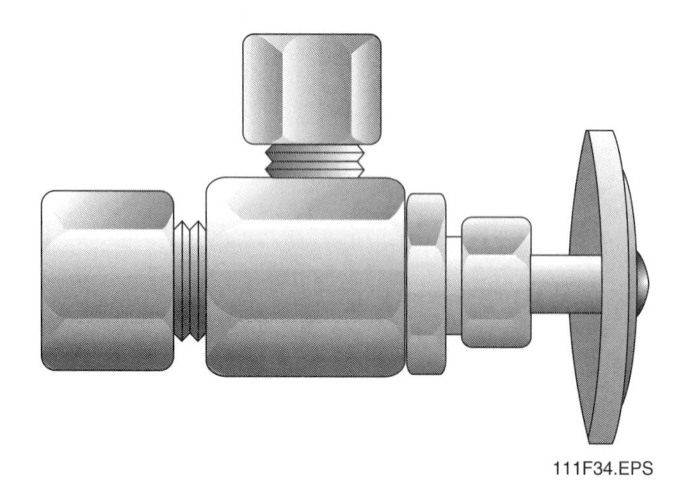

Figure 34 ◆ Supply stop valve.

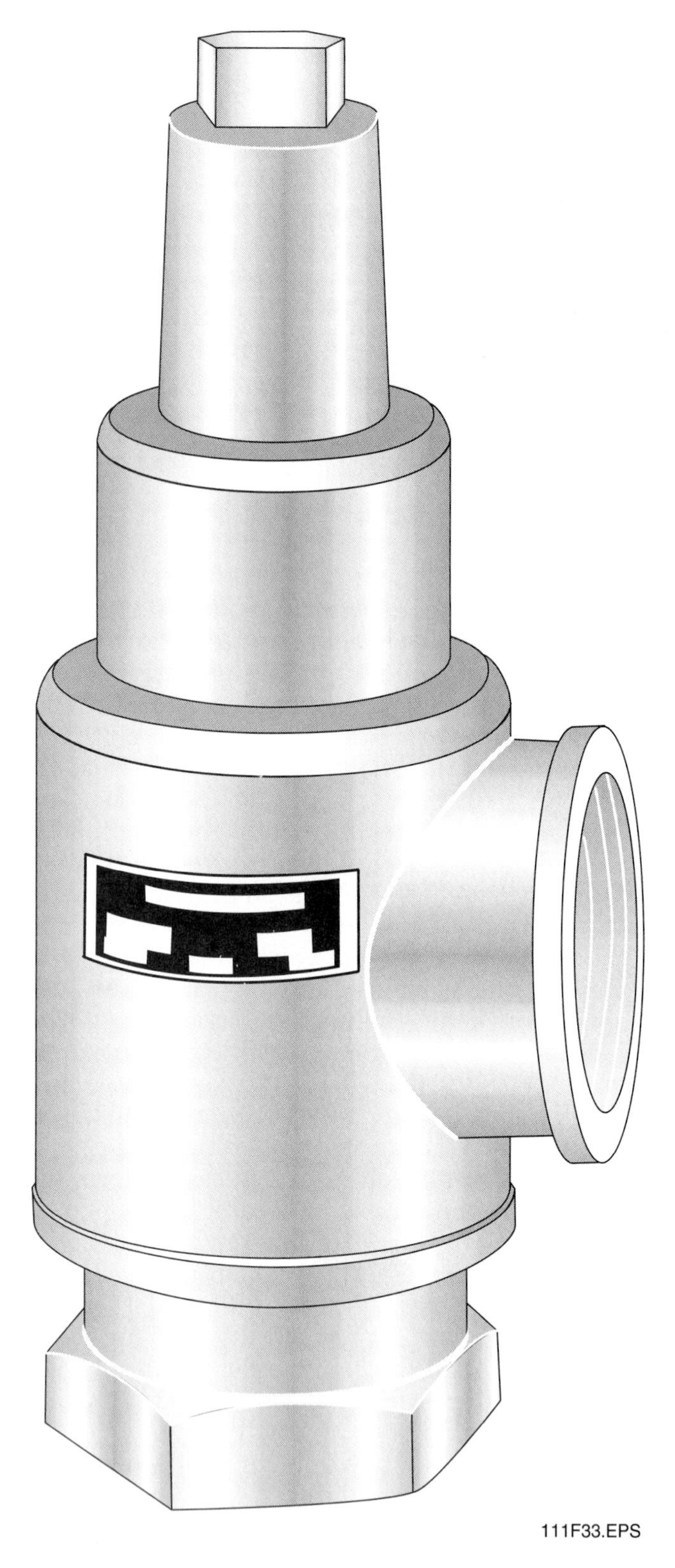

Figure 33 ◆ Pressure relief valve.

Review Questions

Section 5.0.0

1. The reverse flow of contaminated water or other liquids into the potable water system is caused by _____.
   a. faulty throttling devices
   b. connection pressure
   c. cross-connection
   d. backflow connection

2. Codes require installation of _____ in every new structure.
   a. vacuum breakers
   b. backflow preventers
   c. pressure reducers
   d. water mains

3. If a break in the water main or other potential for a vacuum exists in a water line, _____ could be drawn backward into the water supply system.
   a. wastewater
   b. dust particles
   c. pressure
   d. excess air

4. The stem, disc, seat ring, disc holder or guide, wedge, and bushings of a valve make up the _____.
   a. flow
   b. gate
   c. trim
   d. throttle

5. _____ are best suited for main supply lines and pump lines because they provide an unobstructed passageway for the water.
   a. Globe valves
   b. Angle valves
   c. Check valves
   d. Gate valves

6. A _____ valve is used to prevent the reverse flow of liquid in a piping system.
   a. ball
   b. check
   c. gate
   d. globe

7. As pressure increases, a _____ valve opens slowly; as pressure decreases, it closes slowly.
   a. swing-check
   b. pressure regulator
   c. blowdown
   d. temperature and pressure relief

8. For lines that have frequent changes in flow, a _____ valve can be used in either a vertical or horizontal style.
   a. lift-check
   b. swing-check
   c. gate-check
   d. ball-check

9. A _____ valve is commonly used as a cutoff valve for the hot or cold water supply to water closets and sinks.
   a. gate
   b. globe
   c. supply stop
   d. swing-check

10. For quick shutoffs for in-line maintenance or for lines mixing various liquids and gases, use a(n) _____ valve.
    a. ball
    b. check
    c. gate
    d. angle

6.0.0 ◆ BUILDING DISTRIBUTION

Once the water supply piping is installed to bring the water from the main into the building, the next step is to install the water heater, hose bibbs, water softener (if needed), and fixtures. These components complete the water distribution system (see *Figure 35*).

6.1.0 Locating the Water Heater

Place the water heater in the most efficient location, which is usually as close as possible to the greatest number of hot water outlets. This minimizes the length of hot water piping runs between the heater and the fixtures. You should also consider the location of the gas supply if installing a gas water heater or the power supply if installing an electric water heater.

6.2.0 Locating the Water Softener

If the plumbing system includes a water softener, the plumber must first determine the expected use of various fixtures. Not all fixtures or outlets require access to softened water. For example, the hot water supply will always be softened, but water flowing to the hose bibbs may not need to be softened. This decision affects the placement of pipe runs.

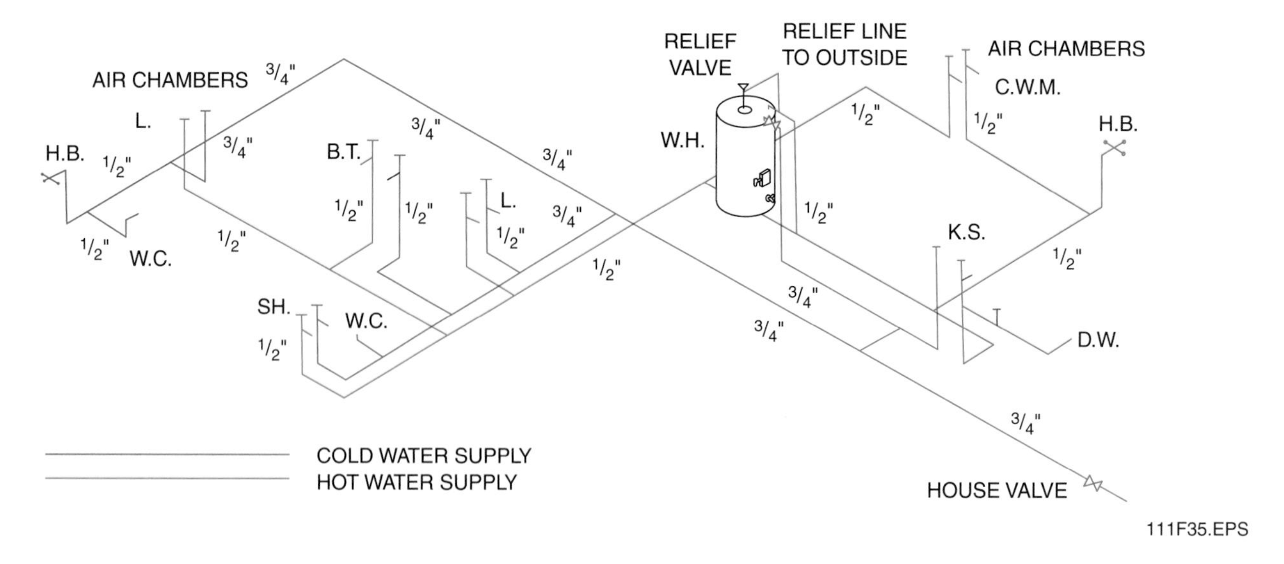

Figure 35 ◆ Sample water distribution piping diagram.

6.3.0 Locating Hose Bibbs

At least one exterior water outlet or hose bibb (see *Figure 36*) is required in residential structures. More outlets are beneficial. The piping runs serving hose bibbs may be long, and they will probably bypass the water softener.

6.4.0 Locating Fixtures

After locating the water heater, hose bibbs, and water softener, the plumber locates the water supply piping that serves individual fixtures and appliances in the building. The plumber then locates the **fixture risers** and **stub-outs.** When you are installing the plumbing for a structure, you must first install the drain, waste, and vent (DWV) system. Because of the large size of the pipe and limited flexibility in installation, taking care of the DWV first can eliminate most problems. Water supply piping is located in relation to the already installed DWV piping, not to the walls or corners of the building. Water supply piping is also located in relation to drains and fixture controls for the fixtures. When you are installing water supply piping, recheck the location of the drains to make sure you are positioning the piping correctly.

The plumber constructs a vertical section of pipe (also known as a *fixture riser*) that goes inside the wall as the connection from the fixture to the supply pipe beneath the flooring. The plumber assembles fixture risers and stub-outs and mounts them on a **backing board** inside the wall behind the fixture (see *Figure 37*). When the water supply lines have been located at the fixtures and the stub-outs assembled, the assembly is placed through the access hole in the floor and connected to the **feeder lines** or **branches.**

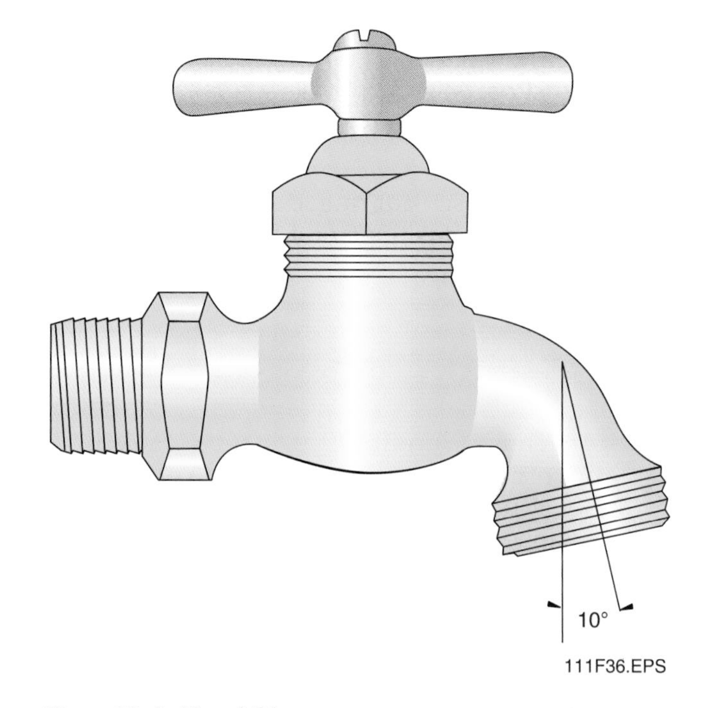

Figure 36 ◆ Hose bibb.

In large water lines or where fixture controls have quick-closing valves, **hammer arresters** or shock arresters must be installed to absorb the

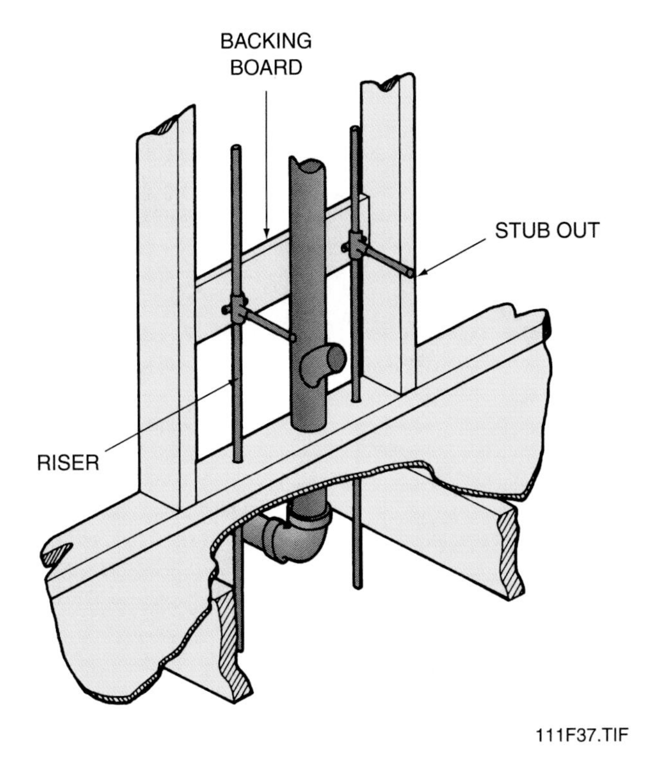

Figure 37 ◆ Installation of fixture riser and stub-outs.

energy when the flow of water suddenly stops. These are usually placed near the fixture as an extension of the water pipe riser (see *Figure 38*). If hammer arresters are not installed, installed incorrectly, or fail, the pipe may hammer (vibrate and rattle) when the quick-closing fixture control is used. This hammering could eventually lead to pipe, joint, or valve failure in the water distribution system.

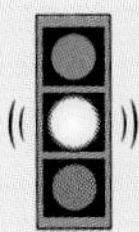

CAUTION

Make every effort to keep the water supply piping as clean as possible before it is used. Clean and dry storage of pipes helps prevent contamination. Cap all ends of pipe at the end of each workday.

6.5.0 Main Supply Lines

Plumbers must size the supply pipes to the fixtures and appliances correctly and use sizing tables supplied by the state or local plumbing codes. These tables estimate the anticipated demand for water as measured by **water supply fixture units (WSFUs)**. Sizing pipes must take into account:

Figure 38 ◆ Shock arresters.

- Type of flush devices used on different fixtures
- Water pressure in psi at the source
- Length of pipe in the building
- Types and number of different fixtures installed
- Total number of fixtures in use at any one time

For residential installations, the main feeder line beyond the water heater must be sized to supply the required flow and pressure. Smaller pipe sizes can save both energy and money. Larger pipe sizes carry more water, but water left in the piping system when the fixture is turned off cools down. When the user turns on the faucet again, more cool water must clear the pipe before hot water reaches the fixture. This wastes water and energy for the user. Always consult local codes to determine the water- and energy-saving measures possible for each job.

At the end of the hot water line, a tee and a pump can be installed to return water to the heater into the cold side or into the drain valve opening. A check valve needs to be installed on the cold incoming water so the return water will flow in only one direction. This is called a hot water recirculating system and is needed when the fixtures are a long distance from the hot water source. The complete loop of a recirculating system should be installed.

When you are installing cold water lines, you must consider possible branches to hose bibbs and plan efficiently to save pipe and time. It is not economically feasible or necessary to soften the cold water that is used at the hose bibbs for lawn sprinklers or for other outdoor applications.

7.0.0 ◆ FIXTURES AND FAUCETS

Various types of fixtures are used in all residential, commercial, and industrial buildings. For this reason, plumbers need to know more than just installation techniques. They should be aware of the construction and working principles of each type of fixture to make the proper water supply connections. The same is true for different types of faucets. Fixtures and faucets are covered in depth in the *Fixtures and Faucets* module. This section contains information about regulating the temperature of water coming out of a faucet.

Hot and cold water can be mixed to prevent scalding at faucets such as shower valves or sweating at water closet tanks. The best way to do this is to use a tempering valve. Tempering valves are available for both residential and commercial/industrial applications (see *Figure 39*). These valves mix the water coming from the hot water heater to a predetermined temperature. The temperature is set using an adjustable thermostat incorporated into the valve.

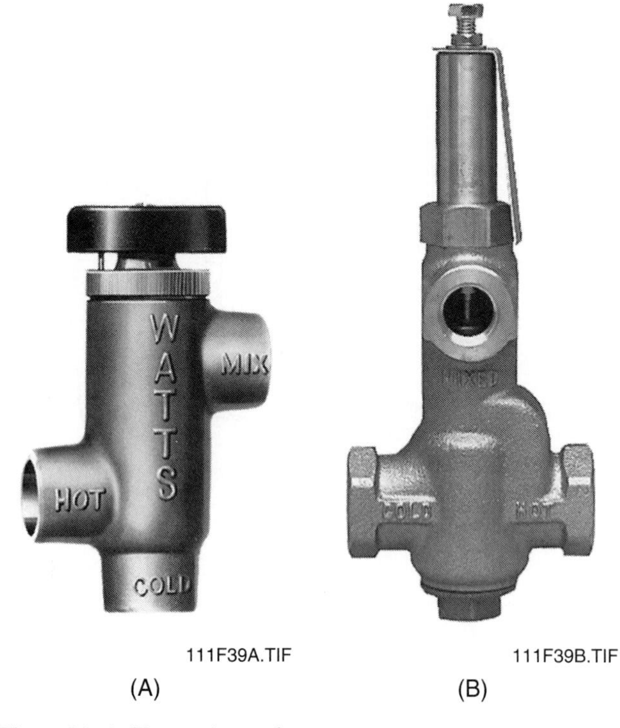

Figure 39 ◆ Tempering valves.

Review Questions

Sections 6.0.0–7.0.0

1. Water supply piping is located in relation to the already installed _____ piping.
   a. cold water
   b. PVC
   c. DWV
   d. hot water

2. The plumber assembles the _____ and stub-outs and mounts them to a backing board.
   a. branches
   b. hose bibbs
   c. feeder lines
   d. fixture risers

3. Sizing tables estimate the anticipated demand for water as measured by _____.
   a. water supply fixture units
   b. water pressure fixture counts
   c. water supply fixture risers
   d. water pressure fixture valves

4. To make sure the water flows in only one direction, install _____ in a hot water recirculating system.
   a. a water softener
   b. a check valve
   c. branches
   d. hose bibbs

5. A _____ valve is used to mix hot and cold water to prevent scalding.
   a. regulator
   b. tempering
   c. riser
   d. fixture

Summary

Water comes from a private source (such as a well) or a municipal source (a public water distribution system) which draws water from a reservoir. While well water is usually drinkable, water supplied by municipalities must be treated to remove harmful impurities. Water service lines (both private and public) move water from its source to homes and other buildings. These lines, or pipes, may be made of copper, plastic, or steel. Factors such as weather, resistance to corrosion, working pressure, local codes, ease of installation, cost, and life expectancy determine the choice of pipe materials. Pumps move water from wells into the water distribution system, and plumbers choose the correct pump based on the depth of the well, the rate at which the water is pumped, and the total distance that the water must travel. The flow of water in a water distribution system is regulated by valves or faucets, which are used to turn water on and off, to control the rate of water flow, to regulate the pressure, or to prevent a reversal of flow through a line. After a water supply to a home or other building is set up, plumbers install the water heater, hose bibbs, water softener (if needed), and fixtures. Plumbers must locate and connect these items efficiently to avoid softening water that does not need to be softened and to save on materials and time.

PROFILE IN SUCCESS

Dan Warnick

Master Plumber
Yeager, Inc.
Newton, IA

Dan was born and raised in Newton, Iowa. He started working part-time at Yeager, Inc., while still in high school. Although jobs at the Maytag washer and dryer plant were available, Dan decided to pursue a career in construction because he didn't want to be cooped up inside all day like plant workers. His hard work and dedication paid off in 1984 when he became a Master Plumber. He has been an NCCER Standardized Craft Training Curricula instructor for more than five years.

How did you become interested in the construction industry?
I really started in 1973 while I was still in high school. I worked part-time at Yeager and then continued on with them after I graduated. I went through an apprenticeship program and in four years had my journeyman's license. During those years of training I worked in welding and heating and cooling before finally switching back to plumbing.

There was a natural choice in Newton to go work at the Maytag plant, but I decided that I wanted to see the daylight every day and not be stuck inside a factory. Factory work can be very repetitive. With plumbing you handle new tasks every day. I always enjoyed working with my hands, so craftwork was a logical career.

What do you think it takes to be a success in your trade?
Hard work. You have to be willing to get up in the morning and go to work. It is critical to want to do your best every day. Only your best effort will enable you to succeed. Some of the basic skills you need are math, critical thinking, and decision making. You have to be able to think on your feet, too.

You also need good people skills. A hot head will not last long in a job. You have to know how to work with others and how to deal effectively with your customers. Pleasing people is the biggest job you have in craftwork.

What are some of the things you do in your role of Master Plumber?
The Master Plumber designation represents a skill and experience level above the average plumber. It is a difficult task to get the Master Plumber license. It requires learning advanced information and taking a tough test. Then you have to keep up your license through continued study.

At Yeager, I work with plumbing system installations from underground and go up. There are so many different types of jobs—commercial, industrial, residential—that it is hard to generalize. I work with other plumbers, journeymen, and apprentices. One of the most challenging and interesting jobs we did was piping a boiler room. The boiler room is the headquarters for all distribution throughout the building. By the time you finish, the room is a mass of pipes.

I also teach. I have worked with apprenticeship training and the NCCER Standardized Craft Training program for more than five years. It is a way to give back to the industry.

What do you like most about your job?
The challenges I meet make this job what it is. I fit new systems into old systems and make old systems work with the new. There is constant variety in types of jobs and work to be done, and each situation has its unique characteristics. Machines can never take my place. Someone will always need a sink or a faucet installed or a system repaired. These are things that I can do with my own hands. It is very satisfying. You can travel, too. Once you develop your skills, you can go anywhere in the United States. There is no limit to where you can work or how much you earn. If you put in the time to learn the trade to the best of your ability, there is money to be made.

Opportunity is also a key factor in my job. In 1998 I began to buy the Yeager business. One day soon, I will become the company president. Not bad—

starting part-time in high school and ending up owning the company.

What would you say to someone entering the trades today?
Give it a try. You won't know if you like the trades until you get some experience working with your hands and using your brain every day. I personally like the apprenticeship training approach for getting a great start in a construction career. The biggest advantage to a formal training program is that you are taught by craftspeople while you are learning the trade. And remember, the schooling never ends. New technologies and materials develop yearly. It is a great challenge to keep learning new skills. Once you have your license, you must keep up with all the developments as well. Trades allow you to excel; the only limits are those you set yourself.

GLOSSARY

Trade Terms Introduced in This Module

Angle valve: A valve that is similar to the globe valve but can serve as both a valve and a 90-degree elbow.

Aquifer: A rock formation (or group of formations) that holds water.

Backing board: The unfinished side of the inside wall, used to mount fixture risers and stub-outs behind the fixture.

Ball valve: A valve used to control the flow of gases and liquids. It is installed in piping systems where quick shutoffs for in-line maintenance may be necessary.

Branch: A pipe stemming off from the main pipe run to bring hot or cold water to a fixture.

Buffalo box: See *curb box.*

Centrifugal pump: A device that moves fluid by pressure in an outward direction away from the center of the pump.

Check valve: A valve that allows flow of liquid in only one direction. Pressure within the line keeps the valve open. Closure of the valve is automatically activated by the reversal of flow or by the weight of the disc mechanism.

Coagulation: Thickening of a liquid into a soft, semi-solid, or solid mass.

Corporation stop: A valve installed in the building water service line to connect to the water main.

Cross-connection: A connection between two separate piping systems that can create contamination of potable drinking water by waste or other harmful liquids within the water supply system.

Curb box: A cylindrical casing placed in the ground over the curb stop that allows a special key to be inserted to turn off the curb stop; also called *buffalo box.*

Curb stop: A control valve installed in building water supply lines between the corporation stop and the building.

Feeder line: A pipe stemming off from the main pipe run to bring hot or cold water to a fixture.

Fixture risers: Vertical section of pipe located inside the wall to connect the fixture to the supply pipe beneath the flooring.

Frost line: Depth from the surface where frost can penetrate the earth and cause freezing and thawing of the dirt. This freezing and thawing cycle, called *heave,* can disrupt pipes and cause leaks and breaks in the piping.

Full flow: Describes the relative flow capacities of various valves.

Galvanic corrosion: Corrosion of metal as a result of electrical conductivity differences between different types of joined metal pipe, such as copper and galvanized steel.

Gate valve: A valve in which the flow is controlled by moving a *gate,* or disc, that slides in machined grooves at right angles to the flow.

Globe valve: A valve in which the flow is controlled by moving a circular disc against a metal seat that surrounds the flow opening.

Hammer arrester: A device installed in a piping system to absorb hydraulic shock waves and eliminate water hammer, a loud thumping that results when the flow of water is suddenly stopped.

Hose bibb (or bib): A faucet with a threaded outlet used to connect a hose. Usually located on the outside of a building.

pH: Potential of hydrogen; a measure of the acidity or alkalinity of a solution. A pH number of 7 is considered neutral; higher numbers are more alkaline, lower numbers are more acidic.

Potable water: Water that is safe to drink.

Precipitate: (noun) A solid that is chemically separated from a solution.

Precipitation: Water droplets or ice particles condensed from atmospheric water vapor that fall to the earth's surface. The production of a precipitate.

Pressure regulator valve: A valve used to reduce water pressure in a building. The valve is activated by changes in pressure within the system.

Pressure relief valve: A valve normally used for liquid service that operates with a relatively lazy motion. As pressure increases it slowly opens, and as pressure decreases it slowly closes.

Purge: To remove impurities or cleanse.

Reservoir: A body of water collected and stored in a natural or artificial (man-made) lake.

Self-priming: A mechanism that injects a small amount of water into a pump to get it started.

Straight-through flow: A flow that is not restricted as it passes through a valve. The element that closes the valve is retracted entirely clear of the passage.

Stub-out: A short pipe extension running from the supply piping through the wall in anticipation of connecting a fixture. The pipe is capped until it is connected to the fixture valve.

Suction pipe: Pipe where the water flows up from the well.

Supply stop valve: A valve that is commonly used to disconnect the hot or cold water supply to water closets and sinks. They are available in either right angle or straight design.

Temperature and pressure (T and P) relief valve: Combines the pressure control functions of a pressure relief valve and temperature control through a safety valve that shuts off the water flow when the water temperature gets higher than a preset limit.

Throttled flow: Describes a valve that provides a control of the rate of flow through the valve.

Trim: The parts of a valve that receive the most wear, including the stem, disc, seat ring, disc holder or guide, wedge, and bushings.

Turbidity: Presence of particles (sand, mud, silt) suspended in water that gives the water a cloudy appearance.

Turbulence: Force or motion of a fluid that has changing speeds and pressures.

Vacuum breaker: A backflow preventer, which prevents a vacuum in a water-supply system from causing backflow.

Water meter: A device to measure water flow (in gallons or cubic feet) into an individual building.

Water supply fixture unit (WSFU): A design factor to determine the load that different plumbing fixtures produce on the overall plumbing system.

Water table: The distance down from the soil surface to the soil that contains moisture.

Well casing: Outer tube or pipe sunk into the ground when drilling or driving a well.

ANSWER KEY

Answers to Review Questions

Section 3.0.0
1. b
2. c
3. a
4. d
5. b

Section 4.2.0
1. b
2. a
3. c
4. d
5. c

Section 4.3.0
1. c
2. a
3. c
4. d
5. a

Section 5.0.0
1. c
2. b
3. a
4. c
5. d
6. b
7. d
8. a
9. c
10. a

Sections 6.0.0–7.0.0
1. c
2. d
3. a
4. b
5. b

Additional Resources

Advanced Home Plumbing, Black & Decker Home Improvement Library. Minnetonka, MN: Cowles Creative Publishing, Inc., 1997.

Basic Plumbing with Illustrations, Howard C. Massey. Carlsbad, CA: Craftsman Book Company, 1980.

Modern Plumbing, E. Keith Blankenbaker. Tinley Park, IL: The Goodheart-Willcox Company, Inc., 1997.

NCCER CRAFT TRAINING USER UPDATES

The NCCER makes every effort to keep these textbooks up-to-date and free of technical errors. We appreciate your help in this process. If you have an idea for improving this textbook, or if you find an error, a typographical mistake, or an inaccuracy in the NCCER's Craft Training textbooks, please write us, using this form or a photocopy. Be sure to include the exact module number, page number, a detailed description, and the correction, if applicable. Your input will be brought to the attention of the Technical Review Committee. Thank you for your assistance.

Instructors – If you found that additional materials were necessary in order to teach this module effectively, please let us know so that we may include them in the Equipment/Materials list in the Instructor's Guide.

Write: Curriculum Revision and Development Department
National Center for Construction Education and Research
P.O. Box 141104, Gainesville, FL 32614-1104

Fax: 352-334-0932

E-mail: curriculum@nccer.org

Craft ______ Module Name ______

Copyright Date ______ Module Number ______ Page Number(s) ______

Description ______

______

______

______

(Optional) Correction ______

______

______

(Optional) Your Name and Address ______

______

______

Photo Credits

Module 02102

Bosch Power Tools, 102F36A, 102F36B
Cooper Tools, 102F01O, 102F02, 102F03A, 102F62
DeWALT Industrial Tool Company, 102F21, 102F22, 102F34, 102F35A, 102F35B, 102F38
ESAB Group, 102F01H
Klein Tools, 102F56
Laser Reference, Inc., 102F14
L.S. Starrett Co., 102F13A, 102F13B
Milwaukee Electric Tool Corporation, 102F01P, 102F23, 102F39A, 102F39B, 102F39C, 102F39D, 102F39E, 102F39F
North Safety Products, 102F01S
Reed Manufacturing Company, 102F01C, 102F01D, 102F01E, 102F49A, 102F49B, 102F49D, 102F53, 102F54, 102F57, 102F58, 102F61, 102F69
Ridge Tool Company, 102F01K, 102F01L, 102F01M, 102F01N, 102F01U, 102F10, 102F18A, 102F18B, 102F24, 102F28, 102F31A, 102F31B, 102F31C, 102F32A, 102F32B, 102F33A, 102F33B, 102F40, 102F43A, 102F44, 102F46, 102F49C, 102F52, 102F55, 102F59, 102F60, 102F66B, 102F67A, 102F67B, 102F67C, 102F70, 102F71, 102F72
The Stanley Works, 102F01A, 102F01B, 102F01F, 102F01G, 102F01I, 102F01J, 102F01Q, 102F04A, 102F04B, 102F05, 102F06, 102F07, 102F08, 102F09, 102F11, 102F12, 102F15, 102F17A, 102F17B, 102F18C, 102F18D, 102F26, 102F63, 102F65, 102F66A, 102F68
Texas Instruments, 102F01T

Module 02105

Charlotte Pipe and Foundry Company, 105F01, 105F02, 105F03, 105F17A, 105F17B, 105F17C, 105F17D, 105F17E, 105F17F, 105F17G, 105F17H
Sioux Chief Manufacturing Company, Inc., 105F12A, 105F12B, 105F12C, 105F12D, 105F13A, 105F13B, 105F14A, 105F14B, 105F14C, 105F14D, 105F14E

Module 02106

Black & Decker Corporation, 106F16
ESAB Group, 106F19A
Ridge Tool Company, 106F14, 106F15, 106F18A, 106F18B, 106F18C, 106F19B, 106F19C, 106F24
Sioux Chief Manufacturing Company, Inc., 106F06A, 106F06B, 106F06C, 106F06D
Becki Swinehart, 106F02A–P, 106F03D, 106F03F
Victaulic Company of America, 106F02Q
Watts Regulator, 106F03A, 106F03B, 106F03C, 106F03E
Wheeler-Rex Pipe Tools, 106F27A, 106F27B

Module 02107

Charlotte Pipe and Foundry Company, 107F02, 107F55B
Milwaukee Electric Tool Corporation, 107F24
Ridge Tool Company, 107F35

Module 02108

Reed Manufacturing Company, 108F34, 108F35, 108F37, 108F39
Ridge Tool Company, 108F27, 108F28A, 108F28B, 108F30, 108F31, 108F32, 108F33, 108F36, 108F38, 108F41

Module 02109

Eljer Plumbingware, Inc., 109F03A, 109F03C, 109F07, 109F11B, 109F19
Elkay Manufacturing Co., 109F30, 109F31
In-Sink-Erator, 109F22
Kohler Co., 109F01, 109F02A, 109F02B, 109F02C, 109F02D, 109F03B, 109F03D, 109F04, 109F05, 109F06, 109F08A, 109F09A, 109F09B, 109F10, 109F11A, 109F18, 109F21, 109F28, 109F29, 109F37B, 109F41B, 109F46B
Maytag Corporation, 109F24

Module 02110

The Stanley Works, 110F39

Module 02111

Cary Mandeville, 111F01B
Culligan International Company, 111F05
Sioux Chief Manufacturing Company, Inc., 111F38
Watts Industries Inc., 111F08A, 111F08B, 111F25A, 111F27A, 111F32, 111F39A, 111F39B